Springer Tracts in Advanced Robotics

Volume 30

Davide Brugali (Ed.)

Software Engineering for Experimental Robotics

With 204 Figures

Springer

Professor Bruno Siciliano, Dipartimento di Informatica e Sistemistica, Università degli Studi di Napoli Federico II, Via Claudio 21, 80125 Napoli, Italy, email: siciliano@unina.it

Professor Oussama Khatib, Robotics Laboratory, Department of Computer Science, Stanford University, Stanford, CA 94305-9010, USA, email: khatib@cs.stanford.edu

Professor Frans Groen, Department of Computer Science, Universiteit van Amsterdam, Kruislaan 403, 1098 SJ Amsterdam, The Netherlands, email: groen@science.uva.nl

Editor

Prof. Davide Brugali

Università degli Studi di Bergamo
Dipartimento di Ingegneria Informatica e Metodi Matematici
viale Marconi 5
24044 Dalmine BG
Italy

brugali@unibg.it

ISSN print edition: 1610-7438
ISSN electronic edition: 1610-742X

ISBN-10 3-540-68949-4 **Springer Berlin Heidelberg New York**
ISBN-13 978-3-540-68949-2 **Springer Berlin Heidelberg New York**

Library of Congress Control Number: 2006938802

Springer is a part of Springer Science+Business Media
springer.com

Printed in Germany

Typesetting: Digital data supplied by editor.
Data-conversion and production: PTP-Berlin Protago-TEX-Production GmbH, Germany (www.ptp-berlin.com)
Cover-Design: WMXDesign GmbH, Heidelberg
Printed on acid-free paper 89/3141/Yu - 5 4 3 2 1 0

STAR (Springer Tracts in Advanced Robotics) has been promoted under the auspices of EURON (European Robotics Research Network)

To Giovanna, Giacomo, and Sofia

Foreword

At the dawn of the new millennium, robotics is undergoing a major transformation in scope and dimension. From a largely dominant industrial focus, robotics is rapidly expanding into the challenges of unstructured environments. Interacting with, assisting, serving, and exploring with humans, the emerging robots will increasingly touch people and their lives.

The goal of the new series of Springer Tracts in Advanced Robotics (STAR) is to bring, in a timely fashion, the latest advances and developments in robotics on the basis of their significance and quality. It is our hope that the wider dissemination of research developments will stimulate more exchanges and collaborations among the research community and contribute to further advancement of this rapidly growing field.

The edited volume by Davide Brugali is the outcome of a well-attended workshop which took place during the 2005 IEEE International Conference on Robotics and Automation.

The authors are recognised as leading scholars internationally, and have contributed twenty-seven chapters, These are keenly organized in four parts. Noticeably, the first chapter of each part is co-authored by the editor with some of the authors of the subsequent chapters in the part, allowing an effective fruition of the contents by the average readership. The result is an effective blend of fundamental and innovative results on research and development in software for robotic systems, where one common factor is the integration of reusable building blocks.

Besides the advancement in the field, most contributions survey the state-of-the-art, report a number of practical applications to real systems, and discuss possible future developments.

As the first focused STAR volume in the area of robot software engineering, this title constitutes a fine addition to the series!

Naples, Italy — *Bruno Siciliano*
October 2006 — *STAR Editor*

Preface

The development of robotic systems has always been a severe challenge, due to the heterogeneous technological issues involved. The process of bringing intelligence to a robot, requires tightly coupled capabilities in sensing, processing, and acting. In this scenario software plays a key role, as it is the medium to embody intelligence in the machine.

Today, most research and development in software for robotic systems is based on proprietarily designed architectures invented from scratch each time. Many valuable robotic applications are monolithic systems that have been developed to solve a specific class of problem, and are often not reusable, even in only slightly differing scenarios.

There is therefore a need to identify techniques and methodologies that will allow the reuse in future applications of the large corpus of existing software implementations. This will promote a new style for developing robotic software systems, based on the systematic reuse and integration of their constituent parts.

In this context, synergy between Robotics and Software Engineering is strategic. Their mutual benefit is not merely to develop bigger, faster, and cheaper software systems, but rather to make it possible to build and evolve new software systems from the integration of reusable building blocks. The key word is "system"; and the robotics field contains good and challenging examples of such systems: heterogeneous structures where properties emerge from the interaction of constituent parts.

This book collects contributions that describe the state of the art in software development for the Robotics domain. It reports on innovative ideas that are progressively introduced in the software development process, in order to promote the reuse of robotic software artifacts: domain engineering, components, frameworks and architectural styles. It illustrates the results of the most successful and well-known research projects which aim to develop reusable robotic software systems. Most of the chapters report on concepts and ideas discussed at the ICRA2005 Workshop on "Principles and Practice of Software Development in Robotics", Barcelona, Spain, April 18 2005.

Who Should Read This Book

In the Software Engineering literature there is no text book that refers to the robotic domain to exemplify software development methodologies and techniques. In the Robotics literature there is no professional guide that addresses software development issues in the construction of robotic systems.

The aim of this book is to point to the strategic role of advanced software development techniques as the basis of new robotic systems, and to identify strategic directions for Software Engineering research, in order to meet the requirements of real world robotic systems.

Thus, this book is intended for researchers in Robotics who want to learn more about advanced techniques for developing robotic software systems, and for professionals developing advanced software systems for robotic applications. This book is appropriate for senior level courses on Advanced Robotics in departments of Computer Science.

Organization of This Book

The book is structured in four parts. Each part collects contributions from different research groups that have exploited similar approaches in robotic software development. The first chapter of each part discusses trends in the development of software for robotic systems. It is written in collaboration with the authors of the subsequent chapters and serves as a guide to relate their innovative contributions to common research issues. A sidebar that gives a working knowledge of an advanced topic completes each book part.

Part I, *Robot Software Development*, discusses issues and challenges in software development for the Robotics domain, analyzes techniques to identify, document, and cope with them. The sidebar reports on several software development environments for commercial robots.

Part II, *Component-based Robotics*, introduces and tackles the problem of enriching the general component based software engineering (CBSE) approach, by taking into consideration the specific requirements of robotic application domains. The sidebar discusses the concept of the software architecture, and its role in the software development process.

Part III, *Robotic Software Frameworks*, introduces and includes innovative contributions on how to apply framework development concepts in the robotics domain. The sidebar discusses and compares commercial middleware frameworks for distributed computing.

Part IV, *Software Environments for Networked Robotics*, introduces and exemplifies the adoption of advanced software development techniques in an emerging robotic research field. The sidebar illustrates some examples showing how Sun Microsystems Java3D library can be exploited for robot teleoperation.

Acknowledgement

This editorial project has been a really exciting adventure. The result has been achieved thanks to the commitment of all the authors, the valuable comments of the reviewers, and the encouragement of Bruno Siciliano.

A special thank to Monica Reggiani, who contributed to the organization of the workshop on "Principles and Practice of Software Development in Robotics" at ICRA 2005 where this book has been conceived.

Dalmine, Italy *Davide Brugali*
October 2006 *Editor*

Contents

List of Contributors

Arvin Agah
University of Kansas, USA
agah@ku.edu

Eric L. Akers
University of Kansas, USA
eakers@ku.edu

Moisés Alencastre-Miranda
Tecnológico de
Monterrey,
México malencastre@itesm.mx

Vincenzo Ambriola
Universitá degli Studi di
Pisa, Italy
ambriola@di.unipi.it

Igor Belousov
Hewlett-Packard Moscow, Russia
igor.belousov@hp.com

Geoffrey Biggs
University of Auckland, New Zealand
g.biggs@auckland.ac.nz

Andrea Bonarini
Politecnico di Milano, Italy
bonarini@elet.polimi.it

Debora Botturi
University of Verona, Italy
debora@metropolis.sci.univr.it

Alex Brooks
University of Sydney, AUSTRALIA
a.brooks@cas.edu.au

Yannick Brosseau
Université de Sherbrooke,
CANADA
Yannick.Brosseau@USherbrooke.ca

Greg Broten
Defence Research and Development,
Canada
greg.broten@drdc-rddc.gc.ca

Davide Brugali
Universitá degli Studi di
Bergamo, Italy brugali@unibg.it

Jorge Cabrera-Gámez
Universidad de Las Palmas
de Gran Canaria, Spain
jcabrera@dis.ulpgc.es

José María Cañas
Universidad Rey Juan
Carlos, Spain
jmplaza@gsyc.escet.urjc.es

Andrea Castellani
University of Verona, Italy
castellani@metropolis.sci.univr.it

Luiz Chaimowicz
Federal University of Minas Gerais, Brazil chaimo@dcc.ufmg.br

Antonio Cisternino
Universitá degli Studi di Pisa, Italy cisterni@di.unipi.it

Toby H.J. Collett
University of Auckland, New Zealand
t.collett@auckland.ac.nz

Diego Colombo
IMT Lucca, Italy
diego.colombo@imtlucca.it

Marco Combetto
Microsoft Research ltd., UK,
marcomb@microsoft.com

Carle Côté
Université de Sherbrooke, CANADA
Carle.Cote@USherbrooke.ca

Anthony Cowley
University of Pennsylvania, USA
acowley@grasp.cis.upenn.edu

Liam Cragg
University of Essex
lmcrag@essex.ac.uk

Antonio C. Domnguez-Brito
Universidad de Las Palmas de Gran Canaria, Spain
adominguez@iusiani.ulpgc.es

Giorgio Ennas
Universitá degli Studi di Pisa, Italy ennas@di.unipi.it

Paolo Fiorini
University of Verona, Italy
paolo.fiorini@univr.it

Jannik Fritsch
Bielefeld University, Germany
jannik@techfak.uni-bielefeld.de

Stefano Galvan
University of Verona, Italy
galvan@metropolis.sci.univr.it

Brian P. Gerkey
SRI International, USA
gerkey@ai.sri.com

Hans P. Harmon
Pinnacle Technology, Inc.
hharmon@pinnaclet.com

Daniel Hernndez-Sosa
Universidad de Las Palmas de Gran Canaria, Spain
dhernandez@iusiani.ulpgc.es

Huosheng Hu
University of Essex
hhu@essex.ac.uk

Ulrich Kaufmann
University of Ulm, Germany
ulrich.kaufmann@uni-ulm.de

Tobias Kaupp
University of Sydney, AUSTRALIA
t.kaupp@cas.edu.au

Gerhard Kraetzschmar
Fraunhofer Institute for Autonomous Intelligent Systems Behavior Engineering, Germany
gerhard.kraetzschmar@ais.fraunhofer.de

Oscar Kuo
University of Auckland, New Zealand
o.kuo@auckland.ac.nz

Josep Isern-Gonzlez
Universidad de Las Palmas de Gran Canaria, Spain
jisern@dfm.ulpgc.es

Patricia Lago
Vrije Universiteit, The Netherlands
{patricia, hans}@cs.vu.nl

Dominic Létourneau
Université de
Sherbrooke, CANADA
Dominic.Letourneau@USherbrooke.ca

Bruce A. MacDonald
University of Auckland, New
Zealand
b.macdonald@auckland.ac.nz

Alexei Makarenko
University of Sydney, AUSTRALIA
a.makarenko@cas.edu.au

Vicente Matellán
Universidad Rey Juan Carlos,
Spain vmo@gsyc.escet.urjc.es

Matteo Matteucci
Politecnico di Milano, Italy
matteucci@elet.polimi.it

Gerd Mayer
University of Ulm, Germany
gerd.mayer@uni-ulm.de

François Michaud
Université de Sherbrooke,
Québec
Francois.Michaud@USherbrooke.ca

Davide Moschini
University of Verona, Italy
dado@metropolis.sci.univr.it

Lourdes Muñoz-Gómez
Tecnológico de
Monterrey, México
lmunoz@itesm.mx

Issa A.D. Nesnas
Jet Propulsion Laboratory, USA
nesnas@jpl.nasa.gov

Carlos Nieto-Granda
Tecnológico de Monterrey,
México A00461680@itesm.mx

Anders Orebäck
Royal Institute of Technology,
SWEDEN oreback@nada.kth.se

Clément Raïevsky
Université de
Sherbrooke, CANADA
Clement.Raievsky@USherbrooke.ca

Marcello Restelli
Politecnico di Milano, Italy
restelli@elet.polimi.it

Isaac Rudomin
Tecnológico de Monterrey, México
rudomin@itesm.mx

Christian Schlegel
University of Applied Sciences,
Ulm, Germany schlegel@fh-ulm.de

William D. Smart
Washington University in
St. Louis, USA wds@cse.wustl.edu

Richard S. Stansbury
University of Kansas, USA
richss@ku.edu

Ricardo Swain-Oropeza
Tecnológico de Monterrey,
México rswain@itesm.mx

Camillo J. Taylor
University of Pennsylvania, USA
cjtaylor@grasp.cis.upenn.edu

Hans Utz
Ames Research Center, USA
hutz@mail.arc.nasa.gov

Richard T. Vaughan
Simon Fraser University, Canada
vaughan@cs.sfu.ca

Hans van Vliet
Vrije Universiteit, The Netherlands
hans@cs.vu.nl

Norbert Voelker
University of Essex, UK
norbert@essex.ac.uk

Stefan Williams
University of Sydney, AUSTRALIA
s.williams@cas.edu.au

S. Wrede
Bielefeld University, Germany
swrede@techfak.uni-bielefeld.de

Part I

Robot Software: Principles and Challenges

Trends in Robot Software Domain Engineering

Davide Brugali[1], Arvin Agah[2], Bruce MacDonald[3], Issa A.D. Nesnas[4], and William D. Smart[5]

[1] Universitá degli Studi di Bergamo, Italy brugali@unibg.it
[2] University of Kansas, USA agah@ku.edu
[3] University of Auckland, New Zealand b.macdonald@auckland.ac.nz
[4] Jet Propulsion Laboratory, USA nesnas@jpl.nasa.gov
[5] Washington University in St. Louis, USA

1 Introduction

Domain Engineering is a set of activities aiming at developing reusable artifacts within a domain. The term domain is used to denote or group a set of systems or functional areas within systems, that exhibit similar functionality.

The fundamental tenet of Domain Engineering is that substantial reuse of knowledge, experience, and software artifacts can be achieved by performing some sort of commonality/variability analysis to identify those aspects that are common to all applications within the domain, and those aspects that distinguish one application from another [CHW98].

In some application domains, such as telecommunications, factory automation, and enterprise information systems, large companies or international committees have defined standards for reference architectures (e.g. the ISO-OSI model or the USA-NBS Reference Model for Computer Integrated Manufacturing [MMB83]) and software frameworks (e.g. TINA [TIN], CIMOSA [ZK97]). We argue that this is not a viable approach for solving the problem of developing reusable and interoperable robotic system.

The reason can be found in the peculiarity of the robotic domain: a robot is a multi-purpose (what it is for), multi-form (how it is structured), and multi-function (what it is able to do) machine. It is this diversity in form and function that requires a model for flexibility and efficiency beyond those developed in other domain application. The goal of defining a one-size-fits-all architecture or framework for every robotic application is elusive.

In contrast, we believe that robotic software experts should pursue the goal of identifying stable requirements, common design issues, and similar approaches to recurrent software development problems while developing every new robotic application. If the robotic community shares this common vision, then reusable software solutions can naturally emerge from the profitable exchange of knowledge and experience and from common practice.

D. Brugali (Ed.): Software Eng. for Experimental Robotics. STAR 30, pp. 3–8, 2007.

2 Opportunities to Characterize the Robotics Domain

We propose five opportunities to characterize the Robotics domain, namely the need of conceptual tools to identify the stable requirements of robotic systems, the need of abstract models to cope with hardware and software heterogeneity, the need of development techniques to enable a seamless transition from prototype testing and debugging to real system implementation and exploitation, the need of a methodology to manage the complexity of real world deployment of robotic applications, and the need of tools to assist the software engineer in the development of robotic applications.

These opportunities have led to the definition of innovative software development approaches and solutions that are thoroughly described in the five chapters of the first part of this book.

Thereafter, the Sidebar *Programming Commercial Robots* by José María Cañas et al. reports on several software development environments for commercial robots.

2.1 Assessing and Documenting a Robotic Domain Model

In order to make the large software corpus available today within the Robotics community reusable, there is a need to make the domain knowledge and design experience behind it transparent to the developers of new robotic systems.

We argue that this can be achieved if the robotics community pursues two main objectives:

1. To identify recurrent and stable aspects in the robotic domain, which emerge from the analysis of the current practice of software development in robotics. A domain reference model, and a development methodology which incorporates reuse both of design and code, are essential vehicles for the successful development of large and complex robotic control systems.
2. To formulate these stable and reusable aspects in a common language that can allow the understanding of the proposed solutions and make them reusable for the resolution of new problems.

The Chapter "Stable Analysis Patterns for Robot Mobility Software" by Davide Brugali formulates the problem of developing stable software systems in the robotics domain and in particular of stable robot mobility software. It presents a set of conceptual tools (Stable Analysis Patterns) that help the robot software developers identify the stable aspect of a robotic application and guide them in the process of building stable and reusable software components (Stable Design and Architectural Patterns) and systems.

2.2 Defining Abstract Models to Manage Heterogeneity

Developing reusable robotic software is hard because of the inherent complexity and multi-disciplinary nature of the robotics domain. Developing robotic capabilities with an overarching objective of supporting new platforms and algorithms that are not known a priori is a real challenge.

One of the primary difficulties stems from the heterogeneous nature of robotic hardware. Today's robots have different physical capabilities, sensor configuration, and hardware control architectures. On the physical differences, robots can be aerial, surface, or underwater vehicles. Many are mobile but some can be stationary such as manipulators. Some mobile platforms can be legged, wheeled, or hybrid of the two. Sensor suites can also significantly differ on various platforms. Some have stereovision cameras mounted on a pan tilt unit for 360 degrees coverage, while others have stationary stereo cameras that are only pointed by moving their mobile base. It is this heterogeneity of the physical capabilities that often makes robotic software unique to each platform.

Another set of challenges stems from the heterogeneity of robotic software. First, new algorithms developed in isolation (without a reference model) are difficult to integrate because they would often use different representations of information. Second, it is both desirable and necessary to develop robotic software in a modular fashion without sacrificing performance. Modularity is necessary for both reuse and for testing the individual components. Third, robotic software needs to be adapted to hardware or simulators at different levels of granularity. This is needed because functionality, in some cases, can be migrated to hardware.

Chapter *The CLARATY Project: Coping with Hardware and Software Heterogeneity* by Issa A.D. Nesnas, describes a methodology that uses a commonality / variability analysis coupled with iterative approach to define, design, and prototype software components with reuse as one of the primary objectives. By adapting these generic components to various heterogeneous platforms over several years, one hopes that some of the more stable components of a robotics framework mature to the point of becoming reusable robotic entites.

2.3 Managing the Evolution from Simulation to Reality

Development of software for autonomous field mobile robots can be a challenging task due to the difficulties associated with testing of the robot. Because testing is an integral part of all phases of software development, the robot and its software must undergo extensive testing during its software life-cycle. However, testing a field mobile robot requires access to the field, logistics, maintenance of the robot, and adhering to safety standards. .

Simulating some parts of the system can make testing and debugging much easier; realistic simulation can help in determining numerous hardware design parameters, such as physical dimensions, propulsion, payload distribution, etc.

When parts of the system are simulated, a seamless evolution to the real system is fundamental; also, the changes involved must be of a local nature.

If we make changes to the non-simulated components, we can no longer trust the results obtained by the simulation, and its whole value would be lost.

A twofold approach to address these issues is described in Chapter *Simulation and Testbeds for Autonomous Mobile Robots in Harsh Environments*, by Arvin Agah *et al.* It consists of (1) developing a software architecture that is reusable and platform independent, and (2) building a virtual prototype and a small mobile robot to serve as testbeds for the software development. The small robot is a scaled down version of the field robot with as many identical sensors and actuators to the field robot, as possible.

2.4 Coping with Software Deployment and Maintenance

When writing robot control software, we are not simply trying to control the robot itself. Rather, we are controlling the interaction between the robot and its environment. This makes development and testing in a laboratory very different from running the system in the real world. In particular, if the environment in which the system is to operate is different from that in which it was developed, we cannot guarantee that it will function as expected. To accommodate these changes, we write our software to be as general as possible. In practice, this often means a large number of environmentally-dependent parameters have to be set in order for the system to work properly. However, this leads to a problem: How do we set the parameters correctly, while meeting the deadlines often imposed by robot deployments?

Even well-parameterized software will not be able to deal with all contingencies, and it is almost certain that new software will have to be written on-site to make the deployment succeed. Often this software is written under extreme time pressure, against unmovable deadlines (the museum opens, the event to be monitored is about to happen). Software written under these conditions cannot be expected to be as robust as that developed and tested over time.

Chapter *Writing Code in the Field: Implications to Robot Software Development*, by William D. Smart examines these issues in more depth. How can we design robot control software to make the process of deploying robots in the real world more efficient and effective? How can we mitigate the effects of hastily-written software during the deployment? How can we streamline the process of tuning and adapting the software to the deployment environment?

2.5 Software Environments for Robot Programming

The goal of building effective robotic assistants for humans has never been more relevant, yet robot system developers face the complexity of mobile robotic systems, the difficulties of variable and unpredictable environments, and the challenge of making human–robot interaction effective and safe. Unfortunately, robotic programming tools have not advanced as rapidly as the robots themselves, nor as rapidly as the demands for complex robotic tasks.

Existing tools are often *ad hoc*; they may be specific to the robot hardware, lack open standards, and ignore the human involvement.

We believe robot programming tools must be targeted more closely to robotics, paying attention to the needs of the robot developer and therefore the nature of typical robot programs, the typical skills of robot programmers, the interactions between humans and robots, and the programming constructs that prevail in robotic applications. In designing the developer's toolchain and choosing the underlying technologies for robotic programming, we must account for the presence of human developers in the immersive robot environment, and recognise that it is the *robot's interaction with the environment* that makes robot programming different and challenging. It is *programmer's lack of understanding of the robot's world view* that makes robot programs difficult to code and debug. It is the *expressiveness of the language* that the human programmer uses to describe the robot's behaviour.

Chapter *Software Environments for Robot Programming* by Bruce MacDonald *et al*, focuses on case studies of improvements in three areas related to the human programmer; the expressiveness of the robot programming language, the expressiveness of the robot software framework, and the need for human–oriented robot debugging tools.

3 Conclusion

Software development for robotics is somewhat different from (and more difficult than) software development for other computer systems.

1. Robotics is more a research field than an application domain and robot researchers seem to shun standardization, since everyone wants to write their own architectures and code. Robotic software is quite complex and custom crafted solutions have been more dominant because they initially appear less complex. However, custom crafted solution do not scale well and can lead to more complex systems when integrating multiple capabilities. The development approach most likely to succeed is a very generic one that leverages common functionality and where everyone can integrate their component technology. Domain analysis is a promising approach to identifying commonalties.
2. The multi-disciplinary nature of the robotics domain and the heterogeneity of robotic technologies (hardware platforms, control paradigms, capabilities) make the reuse of existing software solutions very hard. Nevertheless, by looking at robots as system with multiple levels of abstraction, we can develop reusable components at every level of the hierarchy and reuse appropriate components in different platforms. Abstractions can exist at the hardware interfaces, the low-level capabilities as well as at the higher-level capabilities.

3. In a robot system, we are designing the robot-environment interaction, and we often don't know the environment side of the system in detail as we are developing the robot control application. If we have a large, parameterizable library of robot code that implement consolidated models (both for simulated and real robots), this will make deployments easier.
4. Designers of robotic systems ought to include the human developer as a significant role. Robot development tools should provide expressive programming languages and frameworks that enhance the opportunity for human developers to describe robot behaviour. Debugging tools should be human–oriented and improve the immediate visualisation of robot data for the human developers.

These characteristics are what makes software development for robotics more an art than an engineering discipline. The goal of this first part of this book is to present four approaches that exemplifies the art of crafting software for robotic systems in an engineering way.

References

[CHW98] J Coplien, D Hoffman, and D Weiss, *Commonality and variability in software engineering*, IEEE Software **15** (1998), no. 6.
[MMB83] C. McLean, M. Mitchell, and E. Barkmeyer, *A computer architecture for small batch manufacturing*, IEEE Spectrum (1983).
[TIN] TINAC, Tech. report, http://www.tinac.com/.
[ZK97] M. Zelm and K. Kosanke, *Cimosa and its application in an iso 9000 process model*, Proceedings of the IFAC Workshop-MIM'97 (1997).

Stable Analysis Patterns for Robot Mobility

Davide Brugali

Universitá degli Studi di Bergamo, Italy brugali@unibg.it

1 Introduction

During the last few years, many ideas for software engineering (modularity, information hiding, Component-based development, and architectural styles) have progressively been introduced in the construction of robotic software systems, to simplify their development, and to improve their reusability. All of these techniques offer partial (sometimes overlapping) views and solutions to the problem of developing reusable software.

While a universal reuse solution remains elusive, great improvements can be made by focusing on well-defined areas of knowledge (domains). The term domain is used to denote or group a set of systems (e.g. mobile robots, humanoid robots) or functional areas (motion planning, deliberative control), within systems, that exhibit similar functionality.

Domain Engineering is a set of activities aiming at developing reusable artifacts within a domain. The fundamental tenet of Domain Engineering is that substantial reuse of knowledge, experience, and software artifacts can be achieved by performing some sort of commonality/variability analysis to identify those aspects that are common to all applications within a domain, and those aspects that distinguish one application from another [CHW98].

Recently, the focus of domain engineering has moved from commonality/variability analysis, toward the new concept of stability analysis. Software stability can be defined as a software system's resilience to changes in the original requirements specification. Stability analysis enhances reuse as it focuses on those aspects of a domain that will remain stable over time. Such an approach ensures a stable core design, and thus stable software artifacts [FA01].

In order to support stability analysis, the concept of an Enduring Business Theme (EBT) was first introduced in [CG00], and further investigated in [Fay02] and other papers by the same authors. An EBT is an aspect of a domain that is unlikely to change, since it is part of the essence of the domain. For example, the robot property of having a physical body is a an essential

D. Brugali (Ed.): Software Eng. for Experimental Robotics. STAR 30, pp. 9–30, 2007.

aspect of every robotics application. An EBT is modeled as software entity that represents the stable property at a high level of abstraction. At design level, EBTs are refined by more concrete software entities called Business Objects (BO), which offer application-specific functionality. BO have stable interfaces and relationships between them. At implementation level, BOs are internally implemented on top of more transient software entities called Industrial Objects (IO). The identification of EBTs, BOs, and IOs, requires a deep knowledge and understanding of the domain.

In a previous paper [BR05], we have formulated the problem of developing stable software systems in the mobile robotics domain, and analyzed the issues that make the problem hard. In [BS06] we have identified three Enduring Business Themes related to Embodiment, Situatedness, and Intelligence, and we have modeled them as stable analysis patterns. In this chapter, we present three Analysis Patterns that document these three EBTs and a framework of BOs that refine at design level the Embodiment EBT.

The three Analysis Patterns are documented using the the template defined in [CHF05], which is structured in four sections. The *Context* describes the possible scenarios for situations in which a pattern may recur. The *Problem* presents the problem a pattern concentrates on. The *Solution* illustrates a stable object model in terms of participants and relationships. There are two main participants in a solution model: classes and patterns. Classes are the same as those which are used in traditional Object-Oriented class diagrams. Patterns are themselves models that contain classes, and in some cases other patterns. In the model, BOs names start with "Any" indicating that they are standalone stable patterns, and satisfy any application. The fourth section, *Consequences*, analyzes the trade-off and results of using a pattern.

The remainder of this chapter is organized as follows: Section 1.1 introduces three stable aspects related to robot mobility. Sections 2, 3, and 4, present the Intelligence EBT, the Situatedness EBT, and the Embodiment EBT respectively. Section 5 documents the design of the BOs that refine the Embodiment EBT. Finally, Section 6 draws relevant conclusions, and highlights future steps for this research.

1.1 Robot Mobility

A milestone paper by Rodney Brooks [Bro91], identifies a set of properties for every mobile robotics system, among which three are of interest in our discussion:

Intelligence: Robot intelligence refers to the ability to express adequate and useful behaviors while interacting with a dynamic environment. Intelligence is perceived as "what humans do, pretty much all the time" [Bro91]; and mobility is considered at the basis of every ordinary human behavior.

Robot mobility is related to Intelligence as explained in terms of robot behaviors and tasks. These concepts help to answer two questions: How does a robot move? Why does a robot move?

Situatedness: Robot situatedness refers to robots existing in a complex, dynamic and unstructured environment, which strongly affects the robots behavior. For example, the environment could be a museum full of people where a mobile robot guides tourists and illustrates masterworks, or a game field where two robot teams play soccer, or a manufacturing workcell where an industrial manipulator handles workpieces. Situatedness implies that a robot is aware of its own posture in one place at a given time. It does not deal with abstract descriptions, but with the here-and-now of the environment that directly influences robot behavior.

Robot mobility is related to Situatedness, as it is a form of robot-environment interaction. It refers to the ability of a robot to change its own position with regard to the environment, and thus of being in different places at different times. Situatedness helps to answer two questions: With respect to what does a robot move? When and where does it move?

Embodiment: Robot embodiment refers to the consciousness of having a body (a mechanical structure with sensors and actuators) that allows a robot to experience the world directly. A robot receives stimuli from the external world, and executes actions that cause changes in world state. Simulated robots may be 'situated' in a virtual environment, but they are certainly not embodied.

Robot mobility is related to Embodiment, as it applies to both a robot and its constituent components, which move with respect to each other. For example, a humanoid robot is a complex structure with many limbs that change their relative position while it is walking. This also applies to parts or devices of a robot that move with respect to the environment: For example, when a robot stands but changes its head orientation to track a moving object (e.g. a human being).Embodiment helps to answer two questions: Which part of a robot does move? How much does it move?

Simon's "ant on the beach" [Sim69], is a classical example which demonstrates the intrinsic relation of mobility with Intelligence, Situatedness, and Embodiment. An ant's behavior control mechanism (its intelligence), is very simple: obstacle right, turn left; obstacle left, turn right. On a beach with rocks and pebbles (situatedness), an ant's trajectory will be a zigzag line (mobility). But if the size of an ant (embodiment) were to be increased by a factor of 1000, then its trajectory would be much straighter.

2 Intelligence Analysis Pattern

Robotics is an experimental science that can be analyzed from two perspectives. From one perspective it is a discipline that has its roots in mechanics, electronics, computer science and cognitive science. In that regard, robotics plays the role of an integrator of the most advanced results in order to build complex systems. From another perspective, robotics is a research field which pursues ambitious goals, such as the study of intelligent behavior in artificial systems. Many achievements have found application in industrial settings and everyday life.

Context

The concept of intelligence (the ability of expressing useful behavior) is quite elusive. It is usually associated with other concepts, such as autonomy, i.e. a robot's ability to control its own activities, and to carry out tasks without the intervention of a human operator [Ark98] (e.g. navigating towards a target position without colliding with obstacles); deliberativeness, i.e. the ability to plan and revise future actions in order to achieve a given goal while taking into account the changeable conditions of an external environment [KDR04] (e.g. planning the shortest path between two locations in an indoor environment); and adaptability, i.e. the ability to change behavior in response to external stimuli, according to past interactions with the real world (e.g. recognizing already visited places).

Problem

The key ingredients in describing robot autonomy (deliberativeness, and adaptability, and thus the ability to move), are the actions that a robot executes, the behaviors that it exhibits, and the control schema that it implements.

The first goal of this pattern is to define these three concepts and to identify their relationships.

Robot intelligence is enacted by its control architecture. In [Mat02] Maja Mataric summarized the concept of robot control as "the process of taking information about the environment, through the robot's sensors, processing it as necessary in order to make decisions about how to act, and then executing those actions in the environment". This definition implicitly refers to the concepts of situatedness and embodiment.

The second goal of this pattern is to insulate software abstractions that refer to robot control from those that refer to robot-environment interaction and robot structure. This will allow robot control architectures to be developed, which are not tied to any specific robot platform, so that different control architectures can be experimented with on the same robot platform.

Solution and participants

A solution consists of the definition of the Intelligence EBT, and a set of related analysis patterns (BOs). A stable object model is depicted in Figure 1. Three BOs represent stable abstract concepts that characterize robot intelligence.

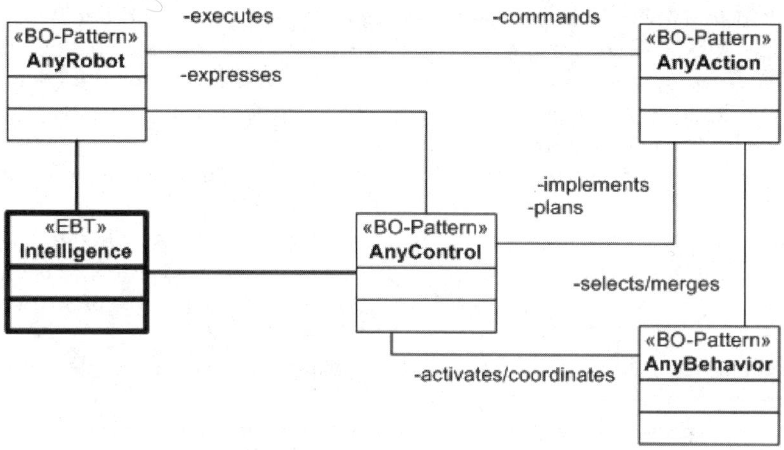

Fig. 1. The Intelligence Analysis Pattern

BO: AnyAction - This represents a generic action that a robot can execute. Robot actions can be defined as those activities which produce modifications in a robot's hardware configuration and cause robot-environment interaction. For example, actions are commands issued to a robot's actuators in order to change the relative position of rigid components, and therefore their position with regard to the external environment. Actions can also include the activation/deactivation of a robot's sensors by which physical interactions with an environment can be determined (e.g. the emission of a sonar pulse that detects the proximity of an obstacle).

The AnyAction pattern is intended to define an abstract representation of a robot's actions, independent of a specific robot structure and its equipment. Key features of this representation are for example timing requirements, expected results, a probability of success, possible types of fault, and recovery procedures.

BO: AnyBehavior - This represents the generic behavior that a robot can exhibit. A robot behavior is a schema of actions that interpret the current status of the robot, environment, and external stimuli, in order to select the next command to issue to a robot.

The AnyBehavior pattern is intended to define an abstract representation of a robot's behaviors which is independent of the specific sensor and actuator system of the robot. Behaviors are constructed by composing available robot actions, according to possible robot-environment interactions.

BO: AnyControl - This represents generic control schema that a robot can perform. Control of robots is one of the traditional testbeds for several areas of artificial intelligence, such as problem solving, planning, learning and coordination, etc. Four main classes of robot control methodology have been identified for mobile robots: reactive, deliberative, hybrid, and behavior-based [Mat02]. All of these can be described in terms of actions and behaviors.

Reactive control is a technique for tightly coupling sensor data and command actions. It builds on the concept of closed-loop feedback, derived from control theory. For example, the ant's control law [Sim69] is described by two simple perception/action patterns: obstacle at right, turn left; obstacle at left, turn right.

Deliberative control refers to the use of symbolic reasoning in classical Artificial Intelligence. A robot constructs and evaluates alternative sequences of actions (plans), and executes the one that best satisfies its constraints and goals.

Hybrid control merges the previous approaches by coordinating immediate responses to external stimuli, whilst also executing goal-directed planning.

Behavior-based control deals with the activation and coordination of robot behaviors that simultaneously govern robot motion in an environment. For example, continuous robot motion can be determined by a simultaneous fusion of command actions that guide a robot towards a goal (e.g. a moving ball) and away from the obstacles (e.g. other robots).

Consequences

The Intelligence Pattern has the following pros and cons:

Mobility-oriented: it focuses only on those aspects of robot intelligence that are more related to mobility, i.e. on how to control a robot in order to exhibit a given behavior.

Completeness: it analyzes the conceptual building blocks of four classical robot control paradigms, i.e. reactive, deliberative, hybrid, and behavior-based.

Flexibility: it provides a clear separation of aspects related to actions, behaviors, and control, which allows the development of flexible robot architectures that encompass the entire spectrum of robot control architectures.

3 Situatedness Analysis Pattern

Robot situatedness means that a robot is located in an environment in which it operates, both from spatial and temporal points of view. The aim of a Situatedness Pattern, is to capture knowledge related to the here and now aspects of the interaction between a robot and its environment.

Context

Robotic devices have been conceived and developed to operate in a variety of environments. Service and humanoid robots operate in indoor environments designed for and occupied by humans. Industrial robots perform repetitive tasks in structured manufacturing plants. Space robots explore the surface of planets and satellites, or inspect and repair space shuttles. These are just a few examples.

Robots interact with their operating environment according to a variety of spatial-temporal patterns, which depend on the type of task the robot has to perform. Both time and space can be represented as continuous or discrete dimensions. For example, the trajectory traveled by a mobile robot is a continuous spatial-temporal interval; past, current, and future robot locations are discrete spatial-temporal positions.

Problem

Robot systems are usually developed both from a hardware and software point of view for a given target environment. Nevertheless, many design solutions conceived for highly different environments share relevant commonalities, at least from a conceptual point of view. For example, the concept of obstacle avoidance is the same, whether a robot has to avoid rocks on the surface of Mars, or wastebaskets in an office. Therefore patterns are used to concentrate on the problem of capturing the basic elements, and relationships, among these elements of robot situatedness; in order to develop a conceptual foundation for the description of every robot environment.

Solution and Participants

A solution consists in the definition of a Situatedness EBT, and of a set of related analysis patterns (BOs). A stable object model related to a Situatedness EBT is depicted in Figure 2; it shows participants (BOs), and the relationships between them. The BOs represent stable abstract concepts that characterize robot situatedness. In order to analyze and design concrete robot systems, every BO has to be further detailed in order to capture the requirements of dynamic robot-environment interactions (e.g. real-time constraints, historic data log, etc.); and can be documented as a pattern by themselves.

EBT: Situatedness - This represents the concept of the situatedness of any robot operating in an environment. This class consists of behaviors and attributes that allow the customization of a system with regard to global properties that cut across individual BOs, such as the role and characteristics of an external observer.

BO: AnyRobot - This is the actor that expresses robot behaviors while it interacts with an environment. Robot-environment interaction is governed by three major components: a) a robot itself, its sensors and actuators, b) an environment's perceptual properties, and c) a task, usually a control program being executed by a robot. A robot's behavior emerges through the interaction of these three aspects. We intentionally concentrate our attention only on the role of a robot as an active player in environmental interactions. Other

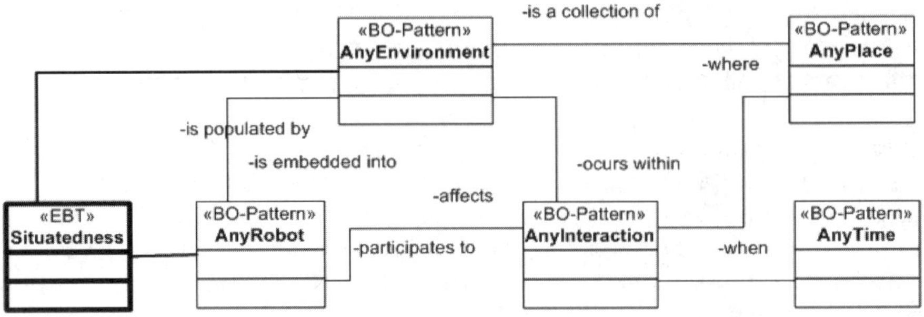

Fig. 2. The Situatedness Analysis Pattern

players, such as a human operator, or other dynamic systems populating an environment, have to be individually described by other BOs, and patterns that are beyond the scope of this chapter. The relationship between a model's participants, state that AnyRobot is embedded in AnyEnvironment, and participates in several AnyInteractions.

BO: AnyInteraction - This represents the type of interaction that occurs between a robot and environment. Basically, we can classify interactions into two main categories: physical and behavioral interactions. Moving within a room, grasping an object, and measuring physical quantities such as distances, forces, and energy are examples of physical interactions. Physical interactions are non-deterministic, at least from a robot's point of view, because sensors and actuators are imperfect, therefore measurements and actions are affected by uncertainty. Complex combinations of physical interactions can be abstracted into behavioral interactions. For example, supervision is a kind of human-robot interaction that requires a robot to interact with a human operator via a variety of media, such as speech, vision, force, etc. Collaboration and competition are typical robot-to-robot interactions, while learning is a high level robot-environment interaction. The relationships between a model's participants, state that AnyInteractions affect the behavior of AnyRobot, can occur in AnyEnvironment, and take place in AnyPlace and at AnyTime.

BO: AnyPlace - This represents where an interaction takes place. It can be meant as the portion of an environment affected by an interaction. For example, a humanoid robot walks on a floor, or a manipulator robot operates in a workcell. The computational representation of places, depends on the tasks that robots have to perform in the environment, and on the kind of interactions that have to be modeled. For example, a robot manipulator needs a detailed 3D geometric representation of its operational environment in order to accurately grasp an object, while an autonomous mobile robot is able to navigate in an indoor environment by localizing with relation to known landmarks. The relationships between a model's participants, state that AnyEnvironment is a collection of AnyPlaces which always belong to AnyEnvironment. AnyPlace is where AnyInteraction can occur, which is always spatially situated in AnyPlace.

BO: AnyTime - This represents the time at which an interaction takes place. The timing characteristics of a robot-environment interaction can be described basically in terms of discrete events, or continuous intervals that determine its occurrence (e.g. a clock tick fires a sonar reading, operator input stops a movement, a grasping tool holds a workpiece for 10 seconds, etc.). The relationship between a model's participants, state that AnyTime is when AnyInteraction occurs.

BO: AnyEnvironment - This represents the totality of the surrounding physical conditions that affect, enable and limit a robot, while it is performing a task. The environment is a continuum of physical configurations, but from a computational point of view it can be represented as a discrete spatial-temporal milieu, that is, made up of robots and any other dynamic or static system such as people, equipment, buildings, all of which have mutual interactions governed by the laws of Physic's. An environment is partially observable due to the limited capabilities of a robot's sensors (vision systems do not work in a dark environment, laser range finders do not detect transparent surfaces, etc.). An environment is unstructured, as it has usually not been specifically designed to host a robot in order to simplify task execution. An environment is unpredictable since it is populated by systems whose dynamic behavior is not under a robot's control. The relationships between a model's participants, state that AnyEnvironment is populated by AnyRobot, and is a collection of AnyPlace and hosts AnyInteraction.

Consequences

The Situatedness Pattern has the following pros and cons:

Abstraction: it captures the fundamental elements and relationships that are needed to describe the situatedness of every robot in every environment, but has to be refined to capture the qualitative and quantitative factors that characterize any specific situated robot.

Extensibility: a pattern is modular and allows the seamless introduction of new concepts, such as those of human operator and generic dynamic systems.

Common language: a pattern facilitates the easy description of a variety of systems, developed by different researchers, using a common language.

4 Embodiment Analysis Pattern

The concept of embodiment was thoroughly investigated during the second half of the last century by many researchers in Cognitive Science, Artificial Intelligence, and Robotics (see [CZ00] for a survey).

In the context of our discussion, a robot's physical body, i.e. its electro-mechanical structure, is a medium employed to experience a sense of time and space, i.e. by being situated in a physical environment.

Context

Physical interactions with the surrounding world occur via sensors and actuators. Sensors allow a robot to perceive environmental changes and to react by changing its own behavior (e.g. the detection of a very close obstacle causing a robot to switch from an operating to an emergency mode). By means of its sensors, a robot experiences the effect of its actions on an environment via the effects that such actions produce. Actuators are the tools with which a robot changes its environment. The spatial-temporal interaction of sensors and actuators with an environment is a consequence of their physical situatedness, which is dependent on a robot's structure.

Problem

Robots substantially differ from one other in their mechanical structure. Nevertheless robots are heterogeneous compositions of a limited number of basic building blocks (sensors, motors and actuators, control and processing units, and communication interfaces). Robot control applications strongly depend on the type of robot used to carry out a task, i.e. a robot's mechanical structure greatly influences the dynamics of its body structure and the requirements of software applications that control it. For example, a video camera mounted on a mobile robot can be used for site surveillance, while the same video camera mounted on a room ceiling can help a robot to self-localize.

This pattern concentrates on the problem of capturing the basic elements and relationships between the elements of robot embodiment, in order to develop a conceptual framework for a description of all robot mechanisms. Therefore, the goal of this pattern is to define mechanical abstractions, which enable the development of stable software that can be used to control and expose robot hardware.

Solution and participants

A solution consists of the definition an Embodiment EBT, and a set of related analysis patterns (BOs). A stable object model is depicted in Figure 3. Three BOs represent stable abstract concepts that characterize robot embodiment.

BO: *AnyBody* - This represents the electro-mechanical body of any robotic mechanism, i.e. the ability to perceive its own state in terms of relative position, orientation, and movement of a robot's body and its parts, level of battery charge, fault conditions of its devices (proprioception), and the ability to change its internal state and interaction with the external world

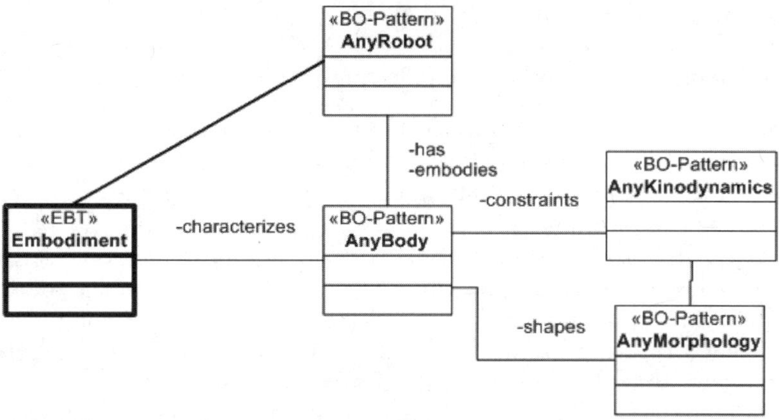

Fig. 3. The Embodiment Analysis Pattern

(actuation). The relationships between the model's participants, state that *AnyBody* embodies AnyRobot.

BO: AnyMorphology - This describes the shape of a robot's body, and its components and their structural relationships. Typically, robot mechanical design is based on serial link chains with actuated rotary or prismatic joints (industrial manipulators). More advanced bio-inspired robotic structures include both active and passive joints, compliant structures, and rigid components connected by a network of tensile elements. Reconfigurable or evolvable robots have the ability to adapt both body and control mechanisms to their environment. Several types of robot structures aim to replicate the morphology of animals (ant, dog, snake), and human beings (humanoid robots). The majority of robots resemble human vehicles: cars, tanks, blimps, aircraft. The relationship between the model's participants, state that AnyMorphology shapes *AnyBody* .

BO: AnyKinoDynamics - This describes the kinematics (position and velocity) and dynamic (acceleration, force) constraints that limit the relative movement of a robot's mechanical components and body in an environment. These constraints are determined by a robot's morphology (e.g. links and joints), by the mechanical power of actuators (e.g. torque force), and by the laws of Physics (e.g. gravity). For example, holonomic constraints reduce the number of degrees of freedom for a set of linked rigid bodies from a number of original coordinates. A holonomic robot is capable of rotating (turning) while moving forward or backward and left or right. In other words, it has a 3 degree-of-freedom base, which is the maximum possible degrees-of-freedom for a mobile robot. The relationships between the model's participants, state that AnyMorphology shapes *AnyBody* .

Consequences

The Embodiment Pattern has the following pros and cons.

Specificity: it focuses only on a specific but fundamental aspect of embodiment, i.e. the physical structure of a robot body. It does not address issues related to the role of the body in cognition, or the structure of an animal brain as a connection between mind and body.

Modularity: it explicitly separates the structural and dynamic features of a robot's body, although they are interdependent. This choice allows the expression of kinodynamics properties of the same robot components in different forms (e.g. geometric or differential equations, logical constraints, etc.) depending on the control application that uses them.

Abstraction: Modularity also facilitates the representation of properties that do not refer to any specific part or component of a robot, but to its body as a whole. For example, despite vastly different morphologies, it is possible to abstract based on functional characteristics the motion of a robot body (e.g. a translation through space).

5 From Analysis to Design: BOs for Embodiment

Robot control applications strongly depend on the type of robot used to carry out a task, i.e. the robot mechanical structure greatly influences the requirements of the software applications that control it. Typically, the mechanical model information are scattered among different functional modules: the locomotor uses the kinematics models of wheels or legs; the obstacle avoider and the path planner need information about the robot shape; the visual odometer stores the position of the stereo vision system with respect to the robot reference frame. As a consequence, most implementations of robot functionality are tied to specific robot hardware, which might substantially differ from one other in their mechanical structure.

In order to make control applications reusable across different robot platforms, there is the need to make software dependencies to the mechanical structure explicit. This means to define a common mechanism model that is used by every functional module to refer to the specific characteristics of the robot hardware. A few research projects have dealt with this issue, as documented in [OSCAR, OROCOS, Nesnas06].

Such a model should enable the description of the structural and behavioral properties of each mechanism part or subsystem, the relationships and constraints among the parts, and the topology of the system. From a software development perspective, the mechanism model should be described in terms of software abstractions that have the following software quality factors:

- *Expressiveness.* The model enables the description of different mechanism types, such as open loop and closed loop chains.

- *Stability.* The model representation is resilient to changes in the application requirements specification.
- *Scalability.* The model allows the seamless composition of simple mechanisms into more complex mechanisms.
- *Efficiency.* The model supports the implementation of efficient query and update operations.

In this section we illustrate the ANYMORPHOLOGY Design Pattern which describes a set of software abstractions used to model robotic mechanisms. The model has been used to develop control applications of both industrial manipulators and mobile rovers.

5.1 Related Works

Many approaches to mechanisms modeling have been documented in the literature. They adopt different representation paradigms which are tailored to specific application domains, such as multibody systems simulation [Otter96], and 3D mechanical design [SPK00]. A few attempts at defining a unified representation for robot mechanisms have been documented in the literature, such as DARTS/DShell [DDS, Jain91], OSCAR [OSCAR, PTKT02], and RoboML [RML]. The OROCOS [BSK03] and CLARAty [NWBS03, CLR] projects have defined software abstractions to represent any kind of robotic mechanism.

The Open Robot Control Software (OROCOS) project represents an attempt to develop a general-purpose, open-source, modular framework for robot and machine control, whose ongoing development has started in year 2002. Currently, the project has released three software toolkits, which include the "Kinematics and Dynamics Library", and a set of requirement documents, which describe concepts related to the semantic decoupling between physical properties, mathematical representations, symbolic attributes, and interconnection information. In particular, the concept of interconnection decoupling has been described in the "Object-Port-Connector" (OPC) Software Pattern [OROCOS].

The Coupled Layered Architecture for Robotic Autonomy (CLARAty), which has started in 2000, is a framework for generic and reusable robotic components that can be adapted to a number of heterogeneous robot platforms at NASA. It defines and implements the Mechanism Model framework [DNNK06] for modeling mechanisms for robotic control.

5.2 The AnyMorphology Design Pattern

The ANYMORPHOLOGY Design Pattern pursues software stabilities by defining software abstractions that directly map to concepts found in problem domain and by modeling them in the same way as corresponding physical entities are described. Five software abstractions characterize the Pattern: *AnyElement* , *AnyPort* , *AnyPair* , *AnyMechanism* , and *AnyBody* . Their

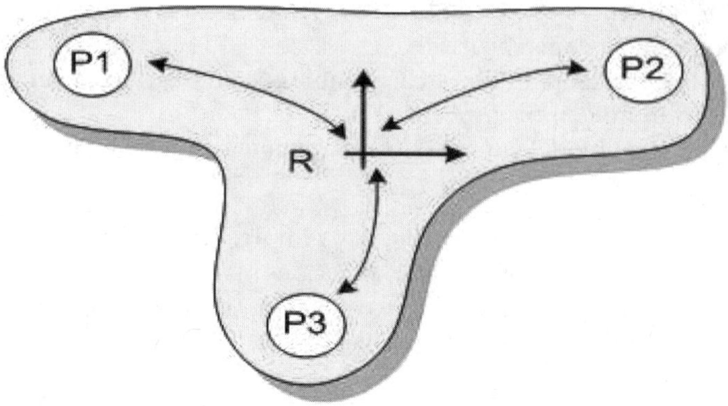

Fig. 4. An example of mechanical element with three ports

names starts with "Any" in order to emphasize their independence of any specific application.

AnyElement represents elemental mechanism parts, such as manipulator links, rover chassis, linear springs, etc. It has one reference frame relative to which all the kinematics properties of an element are expressed. It holds a symbolic attribute describing information related to its physical position on the element, e.g. it is coincident with the centre of mass or with one of its ports. *AnyElement* is intended to store information related to several types of physical properties, such as the mass distribution, the elasticity, and the 3D shape. Our current implementation defines the 3D shape only as a hierarchy of bounding boxes. *AnyElement* does not record any information about external relationships, such as the transformation between an element's reference frame and the ground inertial frame. This information is represented outside the mechanism model in the application that uses it.

AnyPort represents the place where the physical interaction between an element and other elements occurs, such as the mounting points of a link, the working point of an end-effector, the optical centre of a camera, or the contact point of a wheel. *AnyElement* may have any number of *AnyPort* instances, which can represent any kind of physical interaction. Our current implementation defines only the geometric position and orientation of an element's port with respect to the element's reference frame. This allows the computation of forward and inverse kinematics for chains of interconnected elements.

Figure 4 shows an example of *AnyElement* with three *AnyPort* objects P1, P2, and P3 depicted as circles. The arrows represent the rigid transformations between the ports and the element's reference frame R. In order to navigate the geometric relationships between ports efficiently, each port records both direct and inverse transformations. Thus the relative position of two ports, e.g. P1 and P2, is computed as the product of the transformation from P1 to R) and the transformation R to P2. If the element is rigid, these transformations are constant and need to be loaded or computed at initialization time only. Stiffness matrix can be associated to *AnyPort* objects matrices to describe how a vector of external forces applied to an *AnyElement* object cause deformation in *AnyPort* displacement with respect to the *AnyElement* reference frame.

AnyPair represents the interconnection of two elements through their ports. It stores the links to the interconnected ports, the type of the interconnection (e.g. revolute, prismatic, point-contact, etc.) and the value of the articulated position between the two interconnected elements. Typically, *AnyPair* corresponds to a joint between the surfaces of two *AnyElement* objects that keeps them in contact and constraints their relative motion.

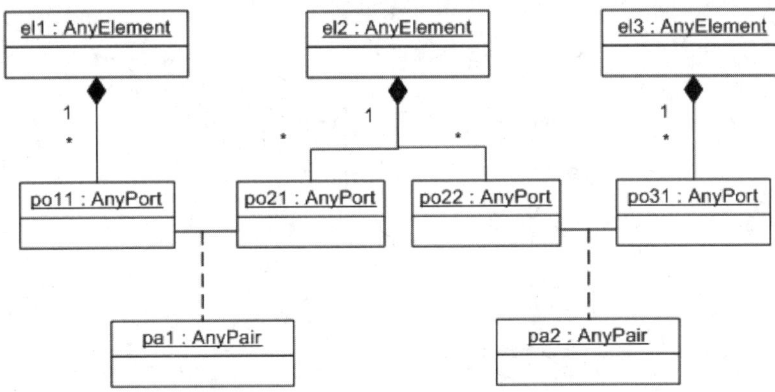

Fig. 5. UML diagram of three interconnected elements

Figure 5 shows the UML diagram that represents the interconnection of elements el1, el2, and el3 through their ports po11 and po21, po22, and po31 by means of pairs pa1 and pa2.

The diagram indicates that class *AnyElement* has a composition relationship with class *AnyPort* , that is, an instance of class *AnyPort* belongs to and is created / destroyed by only an instance of class *AnyElement* . The relationship is bidirectional, thus *AnyPort* objects have a reference to their container object. Class *AnyPair* explicitly represents the association between two *AnyPort* objects, both of which are aware of the relationship between them. As a consequence, when an association between two *AnyPort* objects is created or destroyed, both objects must be informed. If one of the *AnyPort* objects is deleted, the *AnyPair* instance that link them is dissolved as well.

AnyMechanism represents complex mechanisms made up of several *AnyElement* objects, which are concatenated by means of *AnyPort* and *AnyPair* objects. *AnyMechanism* is thus a container software module that encapsulates a set of *AnyElement* objects and maintains the graph of interconnections that make up the mechanism topology. It is in charge of creating and destroying

AnyPair instances and to link/unlink *AnyPort* objects. It embeds algorithms for graph iteration, update, and query.

AnyMechanism supports the hierarchical composition of mechanisms to build more complex mechanisms. For example, a mobile manipulator is the composition of a mobile rover and an articulated arm. In order to allow the definition of algorithms that uniformly iterate the mechanism hierarchical structure, we have applied the Composite Design Pattern [GHJV95]. The result is the data structure depicted as UML diagram in Figure 6, which represents the ANYMORPHOLOGY Design Pattern.

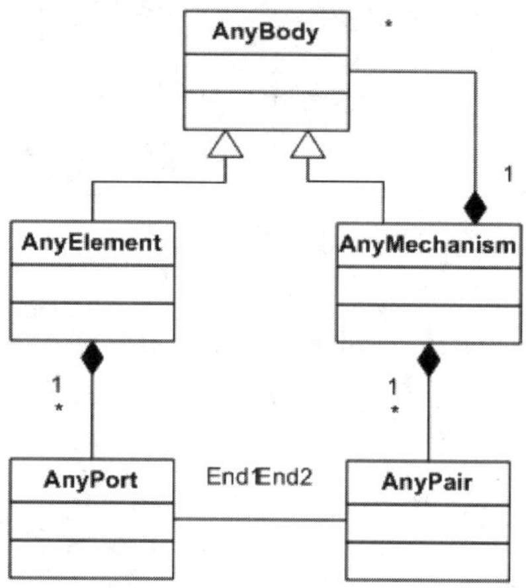

Fig. 6. UML diagram of the AnyMorphology Pattern

The *AnyBody* software abstraction represents both simple *AnyElement* objects and their *AnyMechanism* containers of arbitrary complexity. This is an abstract class that does not need to be instantiated. It implements methods that allow client applications to retrieve the physical properties of individual elements or mechanisms in a uniform way, such as their 3D shape. Thus, clients may even not know if a body is an element or a mechanism.

In turn, *AnyMechanism* objects are made up of an arbitrary number of *AnyBody* objects, which can be *AnyElement* or other *AnyMechanism* objects. *AnyBody* encapsulates the algorithms for iterating the mechanism hierarchical structure.

Figure 7 shows an example of a simplified mobile manipulator structure, where solid line rectangles represent *AnyElement* objects (e.g. chassis, link, ...), small circles represent *AnyPort* objects, small crosses represent *AnyPair* objects, and dashed line rectangles represent *AnyMechanism* objects. The Robot mechanism encapsulates the Manipulator mechanism and the Rover mechanism. The Manipulator encapsulates three Link elements, while the

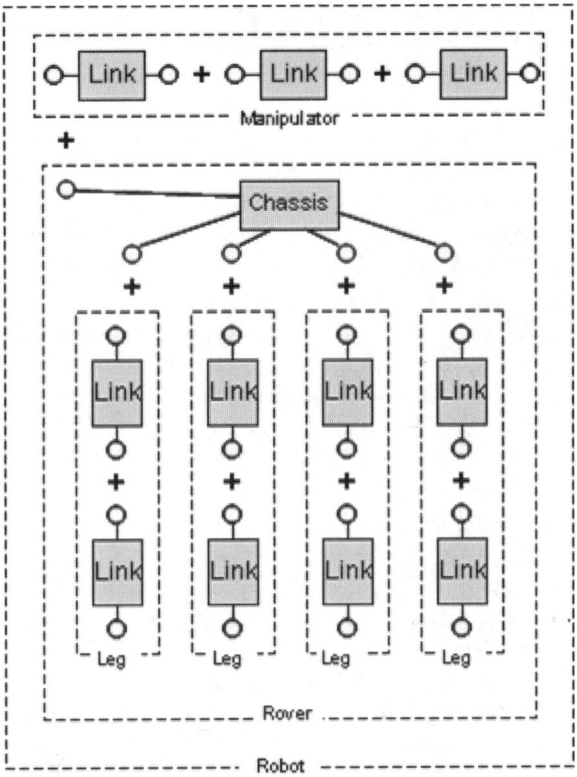

Fig. 7. The topology of a simplified mobile manipulator structure

Rover is made up of the Chassis element and four Leg mechanisms. Finally, the Leg mechanisms are made up of two Link elements each.

The resulting structure can be observed from two distinct perspectives.

First, it represents a flat interconnection of *AnyElement* objects. This structure directly maps the physical chains of mechanical components, which link for example the end-effector of a manipulator arm to the wheels of the mobile rover carrying it. This representation is useful to analyze physical properties of complex mechanisms: an algorithm can iterate through *AnyPort* and *AnyPair* objects in order to compute collision-free relative positions between *AnyElement* objects.

Second, it represents a hierarchical composition of *AnyBody* objects. This structure directly maps the logical encapsulation of partial views on the complete graph, e.g. the Manipulator, the Rover, and the Robot.

This representation is useful to allow a client control application to access mechanical information related to individual mechanisms, e.g. the path planner is concerned only in the rover kinematics.

Generic direct and inverse kinematics and collision detection algorithms are implemented in *AnyMechanism* . Current implementation of kinematics algorithms only work for serial chains. Additionally, *AnyMechanism* objects can be customized with specific algorithms to compute inverse kinematics for specific mechanical structures (e.g redundant or parallel robots [YCLY01]). Customization is done by designing *AnyMechanism* according to the Command Design Pattern [GHJV95]. This pattern defines a way of encapsulating an action and its parameters into an object. This allows adding new operations to existing object structures without modifying those structures.

According to the Command Pattern, the *AnyMechanism* class defines a set of standard interfaces for common algorithms. For examples, forward kinematics algorithms compute the transformation between any two *AnyElement* reference frames. Inverse kinematics algorithms compute the articulation values of a chain of *AnyPort* objects, knowing the transformation between the reference frames of the *AnyElement* endpoints in the chain. Collision detection algorithms compute the distance between the shapes of two *AnyBody* objects. *AnyMechanism* defines the addCommand method that accepts concrete Command objects implementing specific versions of those algorithms. If a new algorithm is defined later, a new Command object is implemented, which can work with *AnyMechanism* automatically.

The Command Pattern allows to encapsulate a single algorithm in one object and to customize a mechanism behavior without the need of implementing a specific adaptation of *AnyMechanism* . Thus, algorithms for complex mechanisms can be implemented taking advantage of the uniform composition of simple mechanisms.

5.3 Software Quality Factors

The ANYMORPHOLOGY Pattern has the following pros and cons. Expressiveness. The proposed model supports the representation of different mechanism topologies, such as the chain of links that make up a serial manipulator, or the tree of devices that make up a mobile rover (transforming chassis, wheel bogies, suspension systems, etc.) or even the graph of serial kinematics chains (legs) that make up a parallel robot.

An interesting consequence of representing the relative positions of *AnyPort* objects explicitly is the possibility to model elastic properties of mechanical elements seamlessly. For example, it is possible to attach a dynamic behaviour to *AnyPort* transformations in order to represent an element deformation due to external or internal forces (e.g thermal dilatation).

Stability. The proposed model clearly separates the invariant properties of a mechanical element from the variable attributes that define how the element is interconnected with other elements to form composite mechanisms.

OROCOS has a similar approach, but the ANYMORPHOLOGY model keeps the representation of the element interconnections as simple as possible. This makes the model resilient to changes in the application requirements specifications, as it supports the representation of new mechanism topologies seamlessly, such as in the case of reconfigurable robots.

Scalability. The model is highly scalable thanks to its modular structure. Every mechanism is modelled as a separate entity, whose properties are defined within an individual module. AnyMecahnism objects encapsulate the constituent elements, their interconnections, and the algorithms that manage them. Complex *AnyMechanism* objects are assembled from building-blocks mechanisms that completely hide their internal structure.

Efficiency. The proposed model supports efficient traversal of complex mechanical structures thanks to the hierarchical organization of views on the flat elements graph. Even if the graph size is very large, search and update algorithms can exploit the mechanism composition hierarchy to retrieve information on every element.

The design choice of representing geometric transformations as *AnyPort* objects limits the performance of kinematics algorithms since every transformation between two elements' reference frames is represented as the composition of two *AnyPort* transformations. Nevertheless, most of the typical mechanical elements used to build robots (such as manipulator links) are represented by two-ports *AnyElement* objects. The reference frame of these elements is usually set in correspondence of either of their ports, thus, they relative position is represented by a single transformation.

5.4 How to Use the ANYMORPHOLOGY Mechanism Model

As stated in Section 5, the goal of the ANYMORPHOLOGY Pattern is to support the development of software control applications that are reusable across different robotic platforms, by making software dependencies to the robot mechanical structure explicit.

This means, that typical functionality such as motion control, path planning, and navigation are implemented on top of the mechanism model by using its algorithms to retrieve and compute kinematics information. If the mechanical structure changes, the mechanism model and its algorithms need to be update, but the implementation of the robot functionality is not affected.

Four alternative usage scenarios demonstrate the flexibility of the proposed model. In the first scenario, a copy of the mechanism model is instantiated for each software functional component. The model records the current configuration state of the robotic mechanism (i.e. the values of the *AnyPair* articulations), while the functional component is in charge of keeping it up-to-date (e.g. by reading the robot sensor). This scenario allows the component to optimize the update rate and to have exclusive access to the model state. Components might even maintain only a partial view of the mechanism state (e.g. only the rover state). The main disadvantage is the waste of memory to load several copies of the same mechanism model and the difficulty of synchronizing different components.

In the second scenario, a single instance of the mechanism model acts as a central repository of the mechanism state for all functional modules. It is in charge of managing the synchronized access to the state by different components. The main disadvantage is the limited performance due to the concurrent access to the central repository.

In the third scenario (similar to the CLARAty approach), a central repository maintains only information related to the invariant properties of a robot mechanism, i.e. the values of *AnyElement* and *AnyPort* attributes. Every functional component that uses the model is in charge of providing a the mechanism state information, i.e. the values of *AnyPair* attributes. The current state is passed to the methods of the mechanism model every time an algorithm has to be executed. The main advantage is that multiple clients can share the same mechanism model efficiently. This is also powerful for reconfigurable robots where a single copy of the robot topologies is maintained. The main disadvantage is that every client (i.e. functional component) has to manage and provide mechanism state information. Different components may share the mechanism state if they agree on a common representation.

In the fourth scenario, not yet supported by the current implementation, the central repository instantiates as many copy of the mechanism state as needed. The idea is to extend the design of the ANYMORPHOLOGY Pattern according to the Iterator Design Pattern [GHJV95], which allows a way to traverse the elements of a composite object without exposing its internal structure. Every functional component holds an instance of an Iterator object, which stores the current state of the robotic mechanism. The component is in charge of updating the Iterator image of the mechanism state. Iterator objects are returned by the mechanism model on demand and can be easily cloned in order to share the same mechanism state with other functional components.

Another intriguing aspect related to how to use the ANYMORPHOLOGY Pattern is the relationship between the model of a mechanism and the external inertial reference frame. It is clear that the model records the state of a mechanism only in configuration space, i.e. the values of the *AnyPair* articulations. Once the transformation between the reference frame of at least one mechanism element and the inertial frame is known, the state of any mechanism element in Cartesian space can be derived from the mechanism state by computing the direct kinematics.

A question rises on how to represent the relationship with the inertial frame. The first solution requires the client functional components (e.g. the navigator) to record the global positions of any of the mechanism elements and to compute the positions of the other ones using the mechanism model internal configuration state and the kinematics algorithms.

The second solution, which actually does not adhere to the overall stability-oriented approach of the ANYMORPHOLOGY Pattern, consists in adding a "virtual" *AnyPort* objects to every *AnyElement* object to represent the transformation between its reference frame and the global inertial frame.

6 Conclusions

The three EBTs presented in this chapter and the ANYMORPHOLOGY Pattern have been applied to model different type of robotic systems. In particular, a Kuka KR16 industrial manipulator, a Pioneer 3DX rover, a Labmate mobile platform, and several simulated robots. Actually, the benefits of appling the proposed Software Patterns can be appreciated more from a software development perspective. It simplifies the development of new control applications and the reuse of an application on different robotics platforms.

References

[Ark98] R.C. Arkin, *Behavior-based robotics (intelligent robotics and autonomous agents)*, The MIT Press, 1998.

[Bro91] R.A. Brooks, *Intelligence without reason*, in Proc. of the 12th Int. Joint Conference on Artificial Intelligence (1991).

[CG00] M. Clien and M. Girou, *Enduring business themes*, Communications of the ACM **43** (2000), no. 5, 101–106.

[CHF05] Y. Chen, H.S. Hamza, and M.E. Fayad, *A framework for developing design models with analysis and design patterns*, The IEEE Int. Conference on Information Reuse and Integration IEEE IRI-2005, Las Vegas, 2005.

[CHW98] J. Coplien, D. Hoffman, and D. Weiss, *Commonality and variability in software engineering*, IEEE Software **15** (1998), no. 6.

[CZ00] R. Chrisley and T. Ziemke, *Embodiment*, Encyclopedia of Cognitive Sciences (2000).

[FA01] M.E. Fayad and A. Altman, *An introduction to software stability*, Communications of the ACM **44** (2001), no. 9.

[Fay02] M.E. Fayad, *Accomplishing software stability*, Communications of the ACM **45** (2002), no. 1.

[GHJV95] E. Gamma, R. Helm, R. Johnson, and J. Vlissides, *Design patterns: Elements of reusable object oriented software*, Addison-Wesley, NY, 1995.

[KDR04] K. Kawamura, W. Dodd, and P. Ratanaswasd, *Robotic body-mind integration: next grand challenge in robotics*, 13th IEEE Int. Workshop on Robot and Human Interactive Communication (ROMAN2004), September 2004.

[Mat02] M. Mataric, *Situated robotics*, Encyclopedia of Cognitive Sciences (2002).

[Sim69] H. A. Simon, *The architecture of complexity*, The Sciences of the Artificial (1969), 192–229.

[BR05] D. Brugali and M. Reggiani, *Software Stability in the Robotics Domain: Issues and Challenges.* In Proceedings of the IEEE International Conference on Information Integration and Reuse (IRI2005), Las Vegas, Nevada, August 2005

[BS06] D. Brugali and P. Salvaneschi, *Stable Aspects in Robot Software Development*, Journal on Advanced Robotic Systems, Vol.3, N.1, March 2006, pp. 17-22.

[OSCAR] *The OSCAR Project.* Online resources at http://www.robotics.utexas.edu/rrg/research/oscarv.2/

[OROCOS] *The OROCOS Project.* Online resources at http://www.orocos.org

[Nesnas06] I.A. Nesnas, R. Simmons, D. Gaines, C. Kunz, A. Diaz-Calderon, T. Estlin, R. Madison, J. Guineau, M. McHenry, I. Shu, and D. Apfelbaum, *CLARAty: Challenges and Steps Toward Reusable Robotic Software,* International Journal of Advanced Robotic Systems, Vol. 3, No. 1, pp. 023-030, 2006.

[Otter96] M. Otter, H. Elmqvist, F.E. Cellier, *Modeling of Multibody Systems with the Object-Oriented Modeling Language Dymola*, Nonlinear Dynamics, Vol. 9, pp. 91-112, 1996

[SPK00] R. Sinha, C.J.J. Paredis, and P.K. Khosla, "Integration of Mechanical CAD and Behavioral Modeling," IEEE/ACM Workshop on Behavioral Modeling and Simulation, Orlando, FL, USA, 2000.

[PTKT02] Mitchell W. Pryor, Ross C. Taylor, Chetan Kapoor, and Delbert Tesar, *Generalized Software Components for Reconfiguring Hyper-Redundant Manipulators*, IEEE/ASME Transactions on Mechatronics, Vol. 7, No. 4, Dec. 2002, pp. 475-478

[RML] *The RoboML Project.* Online resources available at: http://www.roboml.org/

[BSK03] H. Bruyninckx, P. Soetens, and B. Koninckx, *The Real-Time Motion Control Core of the Orocos Project*, In Proceedings of the 2003 IEEE International Conference on Robotics & Automation, pages 797-802, Taipei, Taiwan, 2003.

[NWBS03] A. Nesnas, A. Wright, M. Bajracharya, R. Simmons, T. Estlin, Won Soo Kim, *CLARAty: An Architecture for Reusable Robotic Software,* SPIE Aerosense Conference, Orlando, , April 2003.

[CLR] *The CLARAty project.* Online resources at http://claraty.jpl.nasa.gov

[DNNK06] A. Diaz-Calderon, I. A. D. Nesnas, H. Das Nayar, and W. S. Kim, *Towards a Unified Representation of Mechanisms for Robotic Control Software,* Journal on Advanced Robotic Systems, Vol.3, N.1, March 2006, pp. 61-66.

[DDS] *The DARTS/DShell Project.* Online resources at http://dartslab.jpl.nasa.gov/ROAMS/index.php

[Jain91] A. Jain. *Unified Formulation of Dynamics for Serial Rigid Multibody Systems*, Journal of Guidance, Control and Dynamics, vol. 14, 1991, pp. 531-542.

[YCLY01] Guilin Yang, I-Ming Chen, Wee Kiat Lim, Song Huat Yeo, *Self-calibration of three-legged modular reconfigurable parallel robots based on leg-end distance errors*, Robotica (2001) volume 19, pp. 187-198. Cambridge University Press

The CLARAty Project: Coping with Hardware and Software Heterogeneity

Issa A.D. Nesnas

Jet Propulsion Laboratory, California Institute of Technology
nesnas@jpl.nasa.gov

1 Introduction

Developing reusable robotic software is particularly difficult because there is no universally agreed upon definition of what a robot is. Over the past half-century, robots took many forms. Some were inspired by their biological counterparts such as humanoids, dogs, snakes, and spiders. Others were designed for particular domains such as manufacturing, medical, service, military, or space applications. They took the form of arms, wheeled robots, legged robots, hoppers, blimps, underwater vehicles, sub-surface diggers, and even reconfigurable robots.

While it is neither practical nor possible to address heterogeneity across all types of robots, there are some common themes that recur in robotic software. In this chapter, we will summarize the challenges of developing reusable software for heterogeneous robots and present some principles for coping with this variability.

We arrived at these principles by doing a variability and commonality analysis and building an application framework for a class of heterogeneous robots. Our goal is to improve the interoperability of advanced robotic algorithms through the reuse of the software that implements these algorithms. This framework is called *CLARAty* , which stands for *Coupled-Layer Architecture for Robotic Autonomy* [cla06]. It is a joint collaboration among the Jet Propulsion Laboratory, NASA Ames Research Center, Carnegie Mellon, and the University of Minnesota. CLARAty is the framework for integrating, maturing, and validating new technologies developed by institutions and universities under NASA's Mars Technology Program.

We have built this framework by generalizing legacy software of multiple robots developed over a decade. We considered different architectural styles to improve software reuse and to ensure scalability of the framework. Later, we extended our framework to a larger set of platforms expanding both the scope and capability, which required a more general domain analysis. This iterative process of design, implementation, deployment, testing, and capturing

D. Brugali (Ed.): Software Eng. for Experimental Robotics. STAR 30, pp. 31–70, 2007.

of lessons learned from real systems is critical to reaching reusable and stable robotic software.

Finding the right level of software generalization depends on the scope of their applicability and the life cycle of the robots. Writing application software against generalized and stable components allows upgrading robot hardware without having to rewrite the application software. This has been true over the life cycle of several of our robots.

Efforts to build reusable robotic software and frameworks date back several decades. These efforts have primarily been driven by a pragmatic need to structure the development of software to simplify the building of larger systems. These included developing reusable robotic software libraries [HP86] and developing application frameworks [AML87] [PCSCW98] [SVK97].

Despite earlier efforts, integrating robotic capabilities on different platforms remained quite difficult. The desire for interoperability of robotic software continued, which led to renewed efforts [jau06] [NWB03] [Hat03] [ACF98] [Alb00] [VG06] [KT98], to name a few. While many techniques have been proposed over the years, the primary challenge remains in the poor scalability and lack of flexibility to handle the heterogeneity of robotic software and hardware.

2 Challenges

Two paradigms have emerged for reusing software in robotics: (a) a component-based approach where the components are concrete reusable elements and (b) an object-oriented approach where the reusable components are generic abstractions and interfaces that get adapted to a particular context. In the first model, components are concrete building blocks that achieve specific functionalities. They are then connected to each other statically or dynamically using architecture description languages (ADLs). In the second model, components are separated into generic base classes, which define the generic interfaces and interactions with other classes, and specialized classes that adapt the generic functionality to a given platform. Classes are connected through interfaces that either statically or dynamically bind to other classes. In this Chapter, we will primarily focus on the second model. We will present the challenges in developing reusable robotic software and present techniques for coping with the heterogeneity of hardware and software.

The rest of the chapter is structured into two main parts. The first part addresses the challenges of software variability that stems from (a) software complexity, (b) algorithm integration, (c) architectural mismatches, (d) software efficiency, and (e) multiple operating systems and tools. The second part addresses the challenges of hardware variability that stems from (a) hardware architectures, (b) hardware components, (c) sensor configurations, and (d) different mechanisms.

Fig. 1. From top left and clockwise: *Rocky 8*, FIDO, *Rocky 7*, ROAMS simulation, Lemur II, and ATRV Jr. robots

Each part briefly describes the challenge and proposes one or more solutions to cope with the challenge. We illustrate these with examples from our systems. The list of challenges is not intended to be exhaustive, but rather characteristic of the key areas that are common in standardizing the development of robotic software. A more comprehensive discussion of these challenges can be found in [NSG06]

We will conclude with a brief description of the software capabilities that we integrated into the *CLARAty* framework using these techniques. We have successfully interoperated these capabilities on most of the systems shown in Fig. 1, which include the *Rocky 7*, *Rocky 8*, FIDO, K9, Dexter, and ATRV Jr. rovers as well as the ROAMS simulation.

In this chapter, we will use the term *component* to refer to the software that implements a concrete capability. We will use *generic abstraction* (or abstract class) to define a capability and an interface devoid of its implementation. We will use the term *module* to refer to a collection of abstractions (i.e. classes) that have high cohesion and *package* to define a collection of modules that represent a domain.

3 Software Variability

3.1 Software Complexity

Challenge: How to Decompose a Robotic System

Developing robotic software is already complex because of its multi-disciplinary nature, but doing so with an objective of supporting future platforms and algorithms is a real challenge. This process requires deep knowledge, broad experience and an anticipation of future capabilities and platforms.

Solution: Decompose to highlight stable behaviors and not run-time implementations

Some system decompositions highlight the runtime model of the system, while others highlight the abstract behavior of the components hiding the runtime implementation. It is very likely that future platforms will have different hardware architectures, which would require different runtime models. Therefore, it necessary to encapsulate runtime models and highlight abstract behaviors of components, which are more stable across applications.

Example: Image acquisition

To illustrate this point, consider the example of an imaging system. The primary function of such a system is to acquire images. How the imaging system acquires the images depends, largely, on the underlying hardware. In some systems, an analog camera is connected to a frame grabber that is mounted in a backplane bus (e.g. cPCI). In other systems, a digital camera is used to transmit images over a fast serial interface directly to the host's memory. In either case, the primary function of the imaging system remains the same, i.e. to acquire images or a video stream. We can represent such a system by a camera abstraction that publishes a uniform interface but hides the details of its implementation and the runtime models that accomplish the image acquisition.

Challenge: How to Organize the Software

Complex applications require a large amount of software to be managed, integrated and deployed. The primary challenge in decomposing the system is to define where to draw the lines. This largely depends on what elements of the software are targeted for reuse by future applications.

Solution 1: Decompose into small modules rather than large packages

It is natural to think of decomposing the system into packages that reflect the robotic domains. However, such decomposition would overlook common themes across domains. Working at the granularity of packages would also incur a larger overhead especially when few software capabilities are needed by an application. The guiding principle here is to ensure that simple functionalities are light-weight and easy to implement, while complicated functionalities can be more complex.

Therefore, it is preferable to decompose robotic software into smaller modules as opposed to larger packages. Each module provides primarily a single capability but contains a collection of software with high cohesion and fewer interactions with other modules. A system will then be composed of a number of inter-dependent modules. A module usually contains a set of abstractions that are closely related to one another and that are managed as a group. A module defines the smallest deployable software collection. It is managed as a unit for repository access, building, and testing. Packages are then a loose grouping of modules where multiple packages share common modules. These are primarily used to simplify the description of an overall system.

Example: Vision package

Consider a vision package that contains the following modules: camera models, stereo processor, visual odometer, structure-from-motion, visual tracker, object finder and template matcher. An application that only requires a stereo-vision algorithm would only need to use three modules: stereo processor, template matcher and camera models. There would be no need to checkout and build the remaining modules. Without the ability to work at the granularity of modules, one would have to carry the weight of irrelevant modules in a package. This problem gets compounded for multiple domains such as vision-guided manipulation.

Solution 2: Use explicit inter-module dependencies

Communicating using strongly typed messages ensures properly matched interfaces for information flow across modules. It also enables the tracking of interface changes at compile-time rather than at run-time. Modules that are not interdependent, on the other hand, would not share a common infrastructure and would end up with implicit dependencies on data format for information exchange. Both the format and context of these data packets would need to be tracked and verified to ensure compatibility between senders and receivers. While a loosely coupled system may be easier to initially build and interface with, its software would be harder to extend and maintain over its life cycle.

Using explicit inter-module dependencies also enables the automation of inter-module dependencies. This is important as the number of modules grow in the system. Without the ability to check out and build parts of a generic robotic repository, it becomes too complex and unwieldy to use.

Example: Checking out modules

Because robotics is multi-disciplinary, it is not unreasonable to expect hundreds of modules especially as each capability may have multiple implementations from different institutions. Today, the *CLARAty* software repository contains over 400 modules, which includes hardware adaptations to half a dozen platforms. Because any given application only exercises parts of this reusable software repository, we use an automated process for tracking module dependencies to hide the complexity of managing these dependencies. For example, checking out and building the robotic manipulation software requires modules for arm control, motor, trajectory generator, and mechanism model. It would not be necessary to include other modules.

In a second example, a rover navigation component can use one of three components for estimating the rover pose: wheel-odometry, visual-odometry, or an inertial-based pose estimator. Depending on the desired capability, the software checkout and build will be different for each configuration.

Solution 3: Define a common vocabulary

Decomposing the system into modules and further into components defines a common vocabulary that describes the entities and their functionality. There are two types of modules: generic and specific. Generic modules declare abstract interfaces that describe the language of inter-module interactions. The interactions of these abstractions define the generic framework. Specific modules adapt the abstractions and hence the framework to various algorithms and platforms.

Example: Locomotor and stereovision

A *locomotor* module describes the generic abstractions and interfaces for mobile robots. This module with its limited interface is general enough to support any wheeled, legged or hybrid robot. The more specific *wheel locomotor* module provides a richer interface for all wheeled vehicles (Fig. 15).

An example of generic and specific modules relates to the stereovision capability. A generic *stereovision* module contains the data structures and interfaces that define the stereovision capability. The specific stereovision modules contain different stereovision algorithms that implement that interface. Each stereovision implementation resides in its own module.

Solution 4: Avoid unnecessary code duplication and overgeneralization

To keep the complexity manageable, and to simplify the software, it becomes necessary to reduce code duplication as much as possible across domains. This raises the question of when it is appropriate to encapsulate a new algorithm vs. refactor it to leverage a common infrastructure. The decision is often influenced by non-technical factors involving the nature of the technology, the expertise necessary to re-implement the algorithm, the return on investment, and the long-term plan to support the algorithm as part of a common framework. Because any reusable robotic system is doomed to become enormous, it is necessary to make the code repository complementary rather than duplicative.

Identifying and refactoring common software elements across domains reduces unnecessary code duplication. Proper class design, however, should avoid overgeneralizations, which leads to large abstractions that are hard to maintain. As capabilities and modules grow, one would refactor classes and split modules into smaller ones to maintain a manageable level of complexity.

Example: The Matrix class

To keep data structures maintainable, the Matrix class in *CLARAty* includes only basic operations such as matrix data management (inherited from N2D_Array), addition, subtraction, multiplication, and scalar operations. It does not include functions such as Lower-Upper decomposition, inverse, pseudo-inverse, Cholesky factorization, and singular value decomposition. These are implemented as separate matrix operations. We separate the *matrix* module that defines the structure and basic operations from the *matrix_op* module that includes these complex and mathematically intensive operations. This keeps the *matrix* module and the Matrix class light-weight and simple and allows an Image class to derive from the Matrix class, thus leveraging its basic matrix operations. So, if someone needs to use the Matrix data structure, they do not have to always incorporate the software that includes these mathematically intensive operations.

3.2 Algorithm Integration

Challenge: Different Programming Paradigms

There are two different programming paradigms that are used to develop intelligent robotic software: declarative programming and procedural programming. Declarative programming has dominated software developed by the artificial intelligence community while procedural programming has dominated software developed by the robotics community.

Declarative Programming	**Procedural Programming**

`Rover.navigate_from_to(Loc1, Loc2)`
 	
`Preconditions:` `near(Loc1,Loc2)`
`rover.has_power(Loc1,Loc2)`
`rover.has_time(Loc1,Loc2)`
`Effects:` `rover.is_at(Loc2)`
	`If` `near(Loc1,Loc2) AND`
	`rover.has_power(Loc1,Loc2) AND`
	`rover.has_time(Loc1,Loc2) AND`
	`Then:` `rover.navigate_from_to(Loc1,Loc2)`

Fig. 2. Declarative vs. procedural programming

Solution: Separate declarative and procedural programming into overlapping layers

These two programming paradigms are quite different for building robotic intelligence. In declarative programming, a programmer explicitly describes the activities, models, and constraints but does not provide any program logic (sequences, conditionals, and loops) that describes the order of execution. The program logic is automatically generated and updated by a search-engine that examines all constraints and maintains a plan to order activities without violating these constraints.

Conversely, procedural programming readily provides the program logic that contains the order of execution using activity sequencing, conditionals and loops. The execution flow is only altered through conditionals, exception, and dynamic binding. While declarative programming has infinite flexibility in ordering activities compared to procedural programming, it requires computational resources to generate the program logic and explicit constraints on all activities. In procedural programming, specifying the sequence of activities implies the order constraints. Fig. 2 shows a simple example using the two programming paradigms.

Because of their fundamental differences in formulating program logic, we recommend to separate and layer the two programming paradigms. The declarative programming portion of the software is often referred to as the decision layer while the procedural programming portion is referred to as the functional layer. Where the two layers meet continues to be an active area of research. An architecture where the two layers overlap provides the developers the flexibility to specify where they draw the line between the two layers. Because declarative programming is ideal for situations where the order of activity is less constrained (i.e. many activities can occur concurrently), it tends to dominate higher levels of the application software. On the other hand, procedural programming dominates mid- and lower-level software controls, which have a more constrained sequence of operations.

Challenge: Loosely- versus Tightly-Coupled Integration Model

Selecting the appropriate model for integrating algorithms largely depends on the nature of the algorithms to be integrated. There are several models for integrating robotic algorithms. Some models promote a looser integration where algorithms are encapsulated and wrapped into a framework, while other models promote a tighter model where algorithms are refactored to share a common infrastructure.

Solution: Support different levels for algorithm integration

Using a tightly-coupled integration model, where data structures are consistent, is more efficient and scalable than using a loosely-coupled model. In a loosely-coupled model where algorithms are encapsulated, data structures have to be converted to the format used by the algorithms. This is particularly difficult when memory is managed differently between the framework and the algorithms. Encapsulating algorithms results in redundancy and inconsistency in data representations among components, which can be subject to misinterpretations. It also leads to larger code bases, which can be harder to debug. However, using this model, algorithms are easier to develop at remote institutions since the model imposes fewer constraints on developers who can integrate subsystems fairly quickly for proof-of-concept demonstrations. But the resultant system is often fragile, hard to maintain, and does not scale well as the system evolves. Integrating using a loosely-coupled model is appropriate when the interface is small and the information to be exchanged is limited. The more sophisticated algorithms would require a tighter integration model and a shared infrastructure.

Using a tightly-coupled integration model where algorithms are refactored or implemented against a framework leads to more efficient implementations that are internally consistent and easier to debug. However, they require a commitment to a given framework making the algorithms framework dependent. Therefore, such a framework has to be widely accepted and available, mature and stable enough for developers to adopt and use.

Example: Advanced navigation

Consider a navigation algorithm that uses three-dimensional information from its stereovision sensors to select paths that avoid obstacles. An implementation of such an algorithm for indoor navigation would only need kinematic information about the robot (robot dimensions and types of maneuvers the robot is capable of). A more sophisticated version of this algorithm for use in rough outdoor terrain would also need to incorporate real-time dynamic information from the system. So, as algorithms increase in sophistication, the information and its flow get more sophisticated requiring a richer interface.

Challenge: Redundant Data Structures

Because robotics brings together many domains, software packages are often developed for these domains independently. Bringing together domain packages results in a duplication of data structures.

Solution: Maximize reuse through a cross-domain perspective on data structure classes

Take a global perspective to share common data structures to reduce unnecessary code duplication and reduce overall complexity. Common structure can be refactored into base classes that are cross-domain. Robotics software covers different domains such as motion control, locomotion, manipulation, vision, estimation, planning, scheduling, resource management, and health monitoring. The locomotion, manipulation, vision and estimation domains require similar math and coordinate transformation infrastructure.

Data structures can be general-purpose or domain-specific. General-purpose structures are reusable beyond the scope of robotics applications. The Standard Template Library provides an example of general-purpose data structures. Domain specific data structures include math (matrices, vectors, points), rotation matrices, quaternions, transformations (homogeneous and quaternion transforms), point clouds, paths, fuzzy sets, and so on. Decomposing software into entities that share a common infrastructure enables a more integrated, efficient, and consistent data flow across the system.

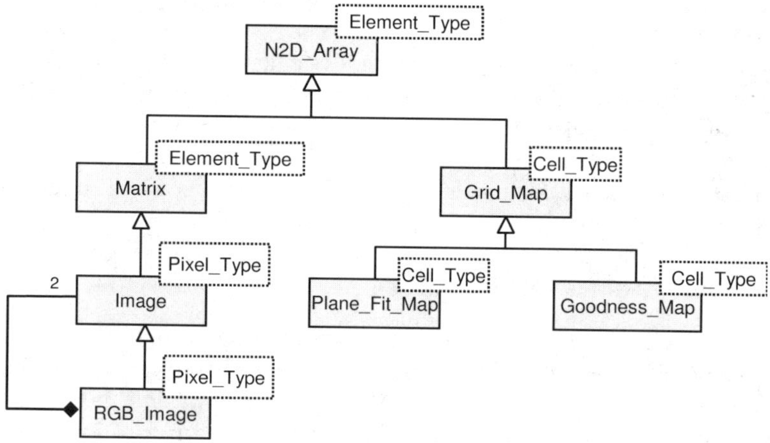

Fig. 3. Data structure abstractions

Example: The Array hierarchy

Fig. 3 shows the array hierarchy that is used in *CLARAty* . Several data structures use two-dimensional arrays. As a result, we define a base N2D_Array class to manage the contiguous storage of two-dimensional information. It provides functions to allocate, resize, retrieve rows and columns, access individual elements, manage sub-arrays, serialize and display the contents of the array. Array elements are not assumed to be numeric. A Matrix class extends the Array class to include numerical operations. Managing the storage of the Matrix elements is handled by the N2D_Array class. The Image class extends the Matrix class to include image specific functions such as pixel interpolation. Another set of classes that extend the Array class is the Grid and Plane_Fit_Map classes that are used for navigation. These classes are two-dimensional arrays whose elements are complex types: the Grid_Cell and the

Plane_Fit_Cell respectively. This example shows how the vision package, the navigation package, and the math package all share a common data type.

Challenge: Different Representations of Information

In robotics, there are multiple ways of representing the same information. Algorithms developed in isolation will most likely use different representations of information. Without agreement on these representations, algorithms will be required to deal with these conversions in an *ad hoc* manner, leading to loosely integrated and inefficient software.

Solution: Use generic programming with templates to support multiple representations efficiently

Because a framework integrates multiple algorithms from multiple sources, it needs to support different representations of information in its data structures. Data structures have to be efficient in dealing with multiple representations, so using polymorphism through inheritance to handle the different representation is neither efficient nor sufficiently flexible for mathematical types. Virtual inheritance adds the overhead of a pointer to the virtual table for every object and cannot support inline functions, which are critical for efficient operations. Instead, in these instances, use template binding to provide both the efficiency and flexibility. Façade classes [GHJV95] can then be used to provide a simple interface to the user to develop against. The principle here is to provide a wrapper façade class to simplify usage while keeping the underlying foundation flexible.

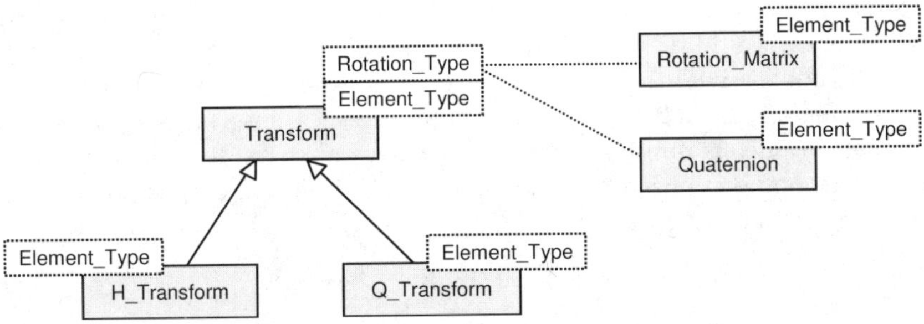

Fig. 4. Coordinate transformation classes

Example: Transforms

We can describe rigid body orientations using multiple mathematical representations such as: Euler angle, rotation matrices (*i.e.* direction cosine matrices), or quaternions. Euler angles are described by three floating-point numbers;

rotations by 3×3 matrices; and quaternions by a scalar and a 3×1 vector. These representations have different characteristics in terms of efficiency and ease of use/understanding. Conversion between them is both inefficient and error prone. This is especially true when dealing with their covariances.

Orientation is an integral part of a coordinate transform that consists of a translation and an orientation. Transforms can use any orientation representation. To represent that in software, we develop a transform as a template class of two types: rotation type and element type. The latter defines whether the elements use single- or double-precision floating-point numbers. Fig. 4 shows the relationship between the Transform class and the Rotation_Matrix and Quaternion classes.

A transform that uses a rotation matrix for its orientation is called a homogeneous transform. A transform that uses a quaternion is then called a quaternion transform. A homogeneous transform is mathematically represented as 4×4 matrix with the 3×3 rotation matrix and a 1×3 translation vector. Quaternion transforms do not have an equivalent mathematical representation. However, thanks to operator overloading, we can think of quaternion transforms as mathematical equivalents to homogeneous transforms. These specialized transform classes are derived from the template transform class by binding their rotation types to the corresponding orientation class: the quaternion transform binds to the Quaternion class and the homogeneous transform binds to the Rotation_Matrix class. This way both types of transforms share a flexible templatized core without exposing users to the complexities of the template implementation.

Challenge: Generalized Interfaces

Once the proper decomposition of modules and classes has been defined, the next challenge would be the design of the class interfaces. Defining proper interfaces is not a trivial task. Generic interfaces need to be stable and mature, but that can only be accomplished by exercising these interfaces across a number of heterogeneous robotic platforms over several years. It is the maturity of the class decomposition, inter-object communication and class interfaces that define the system's architecture.

Solution: Define comprehensive interfaces

Each class needs a complete set of interfaces for application developers to effectively use it. Minimal interfaces are often insufficient. Common base classes for certain types simplify the interface definition of many functions. Devices that represent physical components such as motors, sensors, instruments, and manipulators derive from a base Device class. This enables the connection of multiple instruments on a robotic arm where the end effector is of Device type.

When crafting a generalized interface, it is often the case that neither the union of all possible capabilities nor the intersection of such capabilities (least common denominator) is satisfactory. The solution often lies somewhere in between. In some cases, it is necessary to split the interface into two distinct units and lose the ability to interoperate between the two. This occurs when it is necessary to highlight the differences between platforms rather than their commonality. Trying to find the single unified interface can sometimes lead to undesirable over generalizations.

Example: Locomotor and manipulator interfaces

It is insufficient to provide only *move* functions for a wheeled locomotor class. Additional functions are necessary to control the overall speed and acceleration of the moves as well as stop the robot under normal and emergency conditions. We also need to have functions that access the locomotor's state (moving, stopped, or goal accomplished) and query for the actual speed and acceleration. All these functions are necessary to support the *move* capability. Further extensions may also include selecting the point on the robot's rigid body to control (as opposed to the default center of mass). They also include defining functions to move the rover along paths as opposed to goal-directed moves. Therefore, it is necessary to provide a complete set of interfaces to support a single functionality such as *move*.

One limitation of generic interfaces is the ambiguity they may yield. Consider the example of a four degree-of-freedom (DOF) arm where the three-dimensional position of the end effector can be specified but with only one degree of orientation. A generic manipulator interface would have to support the more general pose $(x, y, z, roll, pitch, yaw)$. If the generic interface is used to pass information to a limited degree-of-freedom arm, then a scheme has to be developed to handle the additional degrees of freedom. One may chose to report an error, ignore these extra terms or achieve the best that the limited DOF arm can do.

Challenge: Unstable Interfaces

Changes to the generic class interfaces reduce the stability of a framework and impact all applications that use the interface. Semantic changes can be checked at compile time, but behavioral changes are harder to manage.

Solution: Use complex data types to stabilize interfaces

While a complete stabilization of interfaces in a generic robotic framework may not be feasible, there is a strong need to minimize the impact of software changes on the generic interfaces. Using complex data types as opposed to primitive types significantly improves the stability of the interfaces. Complex data types hide the details of the implementation allowing the interface to be described using higher-level abstractions. The design of the data types, therefore, becomes of critical importance. The hierarchy of the data types determines what types can be used interchangeably.

There is no single data structure that dominates in a generic framework. While overgeneralizing interfaces to use a single common type provides the most flexibility, it defeats the purpose of a strongly-typed system. Strong type-checking is critical to ensure compatible interfaces and eliminate errors that result from the mapping of different primitive types. Without strong type-checking at the interfaces, changes to data structures cannot be properly managed.

Example: Camera interface

A Camera class has an interface to acquire an image. The interface can use the raw image data as follows:

```
camera.acquire(char* data, int nrows, int ncols)
```

or it can use an Image class to hide the details of the image implementation as follows:

```
camera.acquire(Image & image);
```

In the former case, adding a field to the image, such as image offset, impacts the acquire interface causing it to be changed to:

```
camera.acquire(char* data, int nrows, int ncols,
                           int srow,  int scol)
```

However, such a change does not impact the latter implementation because that change would be encapsulated in the Image class and will not be visible in the camera acquire interface.

3.3 Architectural Mismatches

Unless developed against a framework, robotic software components are likely to have architectural mismatches with the frameworks into which they will be eventually integrated. Consider, for instance, a framework that does not time-stamp measurements collected from various devices. Now consider an algorithm that collects data asynchronously and requires time-stamped measurements. If the underlying framework does not support time-stamped measurements, we have an architectural mismatch. Similar situations occur when an algorithm requires high bandwidth information that may not be available on some platforms.

Another issue is with components that integrate orthogonal functionality into a single modular unit. This introduces artificial coupling of functionalities driven by a specific implementation. While such coupling may optimize local performance, this often comes at the expense of global optimality.

Challenge: Mixing Units

Errors that result from mixing units can lead to catastrophic failures. When integrating heterogeneous algorithms from multiple institutions, it is important to pay careful attention to the inter-mixing of units.

Solution: Use a consistent representation of units

One possible solution is to develop algorithms and models that strictly use the *Systeme Internationale* (SI) units. This will certainly reduce the overall complexity, which is quite important in robotics. Allowing for a mix of units would otherwise require the use of tools and techniques to ensure proper unit conversion.

Several packages have been developed that use template-based classes to do unit checking at compile time. However, the use of unit conversion tools has some limitations and may give a false sense of assurance. This is particularly evident when dealing with larger data structures.

Another consideration for units is to keep the internal representation of units consistent, while supporting mixed units in the input and output files. This localizes the portions of the code that have to deal with mixed units to those handling file I/O operations, which frees up the rest of the software from having to deal with mixed units.

Example: The Image class

Consider a range image. It is desirable to have a single unit attached to an entire image as opposed to units for individual pixels, which would double the size of an image. However, when doing pixel-based operations such as projecting a pixel to its three-dimensional world location, that pixel needs to ensure that it retrieves the proper units from the image prior to any geometric computation. The complexity of adding compile-time unit checking and the risk that will result from the misuse should be weighed against the benefits of unit checking. But whether one chooses to adopt such a framework or use a units standard, that decision has to be made explicit.

Challenge: Representing Uncertainty

One of the primary challenges robots face is dealing with the uncertainty of the environment they operate in. Representation and reasoning about uncertainty is of primary importance. However, because different algorithms may use different representations of uncertainty, interpreting uncertainty becomes quite complex.

Another challenge relates to the assumptions and approaches that algorithms use. An algorithm for an industrial robotic arm, where precise motion and machined fixtures are expected, may handle uncertainty differently from that of an arm mounted on a rover operating in an unknown environment.

Solution: Use templates to represent uncertainty

To handle estimates and their uncertainties properly, use special template classes where the template arguments represent both value and its uncertainty as: `Estimate<Type,Uncertainty_Type>`. While this provides a data structure to capture the data representation, it does not address how this data is interpreted.

Functions that reason about system uncertainty, which are primarily used in estimation, need to use these template constructs to pass information around. Even though the majority of robotic applications use a Gaussian-based distribution, there are other distributions that algorithms use.

Example: Pose uncertainty

In its simplest form, a single variable such as a rover's heading can be represented by an estimate with a mean and a variance. This assumes a probability density function with a Gaussian distribution. Both the heading and the variance can be represented as single floating-point numbers. A rover's (x, y) position with cross correlation between its x and y values will have a 2×2 covariance matrix to represent the cross-correlation terms of x and y. So, an `Estimate<double,double>` is used for a single-valued estimate, and an `Estimate<Vector,Matrix>` is used for a vector and its covariance.

Challenge: Different Time Representations

Algorithm implementations may have different representations of time. They may also make different assumptions about information flow, which have to be reconciled with the framework.

Solution: Use abstract clocks and timers

To manage time, interface the software to clock abstractions as opposed to interfacing directly to the real-time clock. While it may not be quite obvious why managing time is critical for robotic applications, this becomes quite evident when robotic control software is interoperated between real and simulated platforms. On real platforms, it is natural to tie the system clock to the real-time clock. However, when the same software is run against a simulator, a number of options become available. Robotic control software has now the option to run faster or slower than real time. It can also control the stepping of a simulator's clock, enabling the use of otherwise computationally prohibitive control algorithms.

Timers attach to clocks and provide the ability to set, reset and advance time. Tying timers to Clock abstractions keeps time management consistent within a robotic system. Timers are used for measuring time intervals for trajectory generators and planners.

In robotic control applications, barring some exceptions[1], time can be largely used in a relative rather than an absolute sense. A robot needs to know the duration of its functions and activities. Absolute time comes into play when planning a day's worth of activities.

Representing time-values also require some attention. To ensure precise time representation, people use integers to represent absolute time. A single 64-bit integer is insufficient for absolute time measurements of microsecond precision over decades. Therefore two such integers are necessary. However, mathematical operations will require conversion of such integral values to floating-point numbers. Since the majority of time in robot control software is relative, a single integer or a double precision floating-point representation may be sufficient.

To avoid inconsistencies in time representation and to avoid improper mixing of real-time and simulated time, a framework needs to develop a consistent set of these time-based abstractions.

Challenge: Managing Periodic Activities

A single runtime model is often insufficient to handle all aspects of a robotics system. Robotic systems require both synchronous and asynchronous execution of different activities. Some activities have to run periodically at different frequencies. Activities need to be synchronized with other activities. Managing periodic activities can be challenging especially as its execution depends on the underlying operating system.

Solution: Use Periodic_Thread and Periodic_Object abstractions

Using Periodic_Thread and Periodic_Object abstractions provides a consistent way of handling periodic execution across a system. The Periodic_Thread (task) uses the aforementioned time-based abstractions in its implementation. It is responsible for creating a thread and scheduling the periodic execution of registered activities. It also checks the duration of each activity in order not to exceed the assigned interval. The Periodic_Thread class can only register active objects. Active objects inherit from the Periodic_Object base class, which defines a single pure virtual execute method. Active objects then implement this execute method.

The periodic thread uses the time-based abstractions and the thread abstraction to be independent from the underlying operating system. However, if the operating system is a soft real-time, then the periodic thread will only achieve the desired frequency on average. Recording execution statistics about the duration of the activities and the number of skipped activities provide necessary inputs on the performance of the system.

[1] Sun sensors are used to determine a rover's absolute heading using the sensed location of the sun in the sky together with ephemeris data and knowledge of the absolute time of the day.

A robotic system typically has a number of concurrent periodic threads. These include periodic updates to motor controllers, trajectory generators, closed-loop locomotor controllers, and pose estimators. These periodic threads may run at different frequencies depending on the application and hardware configuration.

Challenge: Runtime Models Are System Dependent

Runtime models define the processes and threads that are concurrently running in the system. Because of the variability in hardware architectures and because functionality can be migrated to embedded processors, the runtime models for software change with each platform. Furthermore, both the content and pathways of the information flow change with various device and system configurations, as well as with different application programs.

Solution 1: Separate generic from specific runtime models

It is necessary to separate the generic runtime models from the specific ones. Specific runtime models are system dependent. Therefore, it is necessary to encapsulate these models. The generic and specific runtime models should complement one another to maintain the same abstract behavior.

Example: Motor control

Some systems use motion control chips to servo the motor and generate local trajectories while others do so using software (Fig. 8). The behavior of the Motor class after attaching to either adaptation should remain the same. However, the runtime model for each will be different. In the first case, where the servo loop and trajectory generation are done in hardware, the motor adaptation communicates with the controller chip to send trajectory parameters, set control parameters, and retrieve motor information. In the second case where the motor control is done in software, an additional thread is necessary to periodically compute the trajectory set-points and servo control outputs. The control rate is set by the user.

While these two systems have different threading models to match their control architectures, their abstract motion control behavior should remain the same. The implementation details would be hidden from a person trying to develop a manipulator or a locomotor.

Some runtime models, however, can be generic. Consider the motor state machine shown in Fig. 5. This state machine is generic for all controlled motors. It is a hierarchical state machine with parallel states for the motor. The state diagram shows that the motor can either be in a MOVING or NOT MOVING state and at the same time be in a SERVOING or NOT SERVOING state. For instance, the motor can be in the NOT SERVOING state and the MOVING state at the same time if the motor moves as a result of external forces such as gravity pulling on a robot placed on a slope. From any

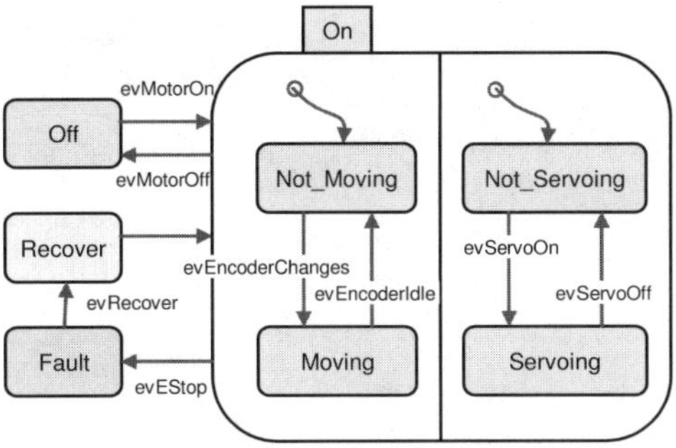

Fig. 5. Motor state machine

of these states, the motor can go into a fault state, which is followed by a transition to the recovering state before resuming normal operations. Such an implementation resides in the generic Motor class and describes the operation modes of the controlled motors. Adaptation of the generic Motor class can extend the state machine to include the specialized modes for the particular hardware component.

Solution 2: Encapsulate runtime models

It is not uncommon to off-load the processing of a computationally intensive algorithm to a separate processor. So, what does this mean for the software system? Since the component interactions are through class interfaces and since the run-time models are encapsulated, we can relatively easily off-load the implementation without impacting the rest of the software.

Example: Remote terrain analyzer

Fig. 6 shows the navigator class, which is one of the robot's behaviors. This class aggregates a locomotor and an action selector. The action selector can either use a grid-based representation of the world or a vector-based one. The grid-based representation that uses a terrain evaluator is computationally intensive because it operates on large data sets. As a result, the grid-base selector is a good candidate to off-load to a separate processor. The framework then implements a specialized grid-based selector that is a proxy to the real implementation. This proxy class is responsible for transferring the terrain and mechanism information to a remote grid-based selector, which in turn computes the data and returns the information to the proxy. The proxy and remote objects can use any communication protocol to exchange the data. However, the implementation is encapsulated such that the grid-based selector and the rest of the navigator classes do not need to know where the processing is taking place. Migrating functionality from one processor to another will be seamless. This is true with the exception of the initialization step that will require a definition of where computation occurs.

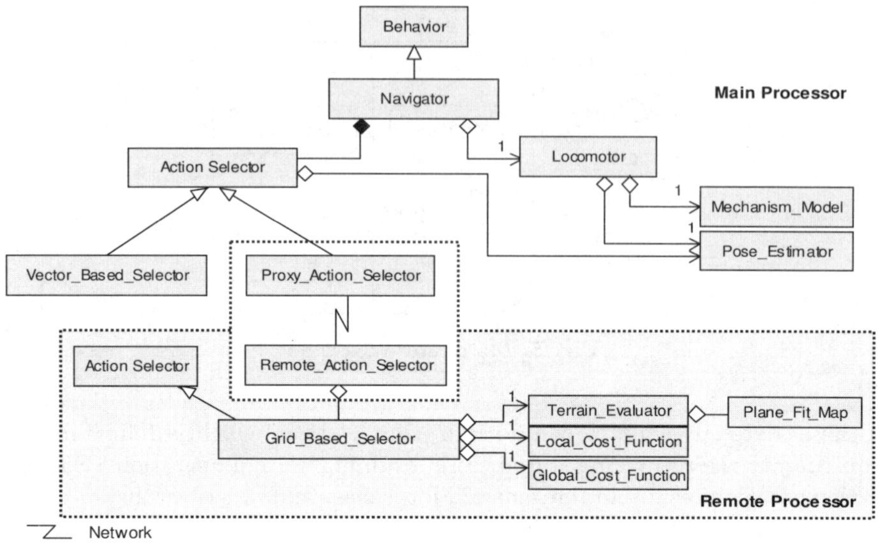

Fig. 6. Navigation abstractions

Challenge: Synchronous and Asynchronous Models

The primary challenge that arises between centralized and distributed systems is in the nature of information flow. In a centralized system, information flow is easily synchronized, while in a distributed peer-to-peer network, information flow and processing is asynchronous. A distributed system in a master/slave configuration is closer to a synchronous system.

Solution: Enable forward processing with constructs to synchronize activities

To ensure a responsive system, device objects must be thread safe and functions must support non-blocking modes. Move functions that do not block enable resumption of processing (*i.e.* forward processing) during a motion. However, it is necessary to provide control over the forward processing. Therefore, we need a function that can block on conditionals such as the percent completion of a move.

Example: Motor thread model

Fig. 7 shows two possible runtime models for the Motor class. The function `wait_until_done(%)` is used to control when the rest of the sequential processing can continue. In the first scenario, the motor is asked to change its position immediately, then wait until the motion is 45% complete before continuing to process that thread. At that instance, the thread blocks until that condition is met and then resumes execution. The motor continues to move while the motor class reads the current position. In the second scenario, there

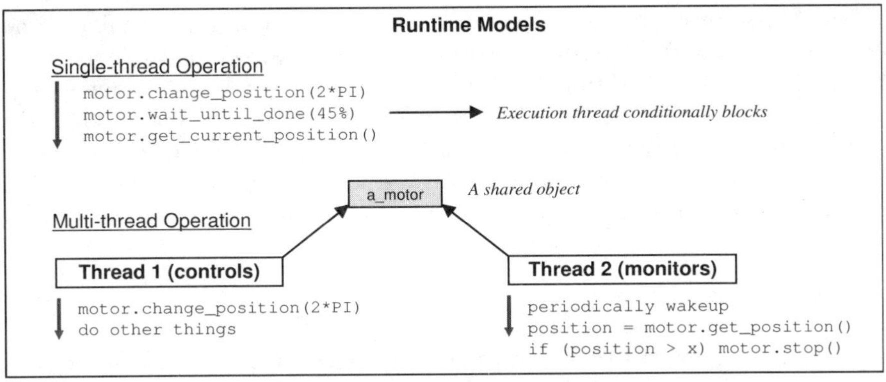

Fig. 7. Motor run-time models

are two threads that communicate with a single instance of the class. The first thread issues a change in position and continues processing other functions. The second thread, which runs in parallel, queries the same Motor instance for the current position. The Motor class supports this parallel interaction and manages the communication link that ties it to the physical hardware.

Challenge: Shared Resources

A robotic system has a number of shared resources that need to be managed. Different resources may need to be managed differently depending on the urgency of the request for these resources. Shared resources in a robotic system include all sensors, actuators, hardware buses, digital and analog I/O, power, memory, and computational time. Multiple applications (clients) require concurrent access to shared resources for controlling various aspects of a robot.

Solution: (1) Protect data integrity using guards; (2) use reservation tokens to manage devices; and (3) manage activities with a global planner

Concurrent access to data requires the proper use of guards to protect data integrity. One can use several mechanisms for multi-threaded inter-object communication. Critical sections are easy to implement but can only be used when the protected operations are very limited in scope and are time bound. Critical sections disable system interrupts and can impact the responsiveness of the operating system. For small data structures, message queues are used. They require a copy of the data, but for a single consumer, such an overhead is acceptable for relatively small data structures. However, this can be quite costly for larger data structures such as images. In such instances, private member data are read and written using proper thread-safe read and write guards.

The above mechanisms can be used to manage resources at small time-scales that require fast context switching. However, using a resource for larger time-scales requires additional flexibility for managing these resources. For large time-scales, one can use reservation tokens where only one client can control a particular device. Only the client that has the token can command that device. However, multiple clients can query the device for information. Overriding reservation tokens is also necessary for handling emergency conditions.

Example: Shared camera

Consider a rover with two stereo camera pairs: one mounted on the front of the rover body while the second is mounted on an articulated pan/tilt mast head (see Rocky 8 in Fig. 1). The rover can acquire images and track targets from the articulated mast head independent of what the rover navigator does. These two applications use different cameras and mechanisms (mobility vs. mast pan/tilt). However, both algorithms share a FireWire bus and a motion control bus. Managing these shared resources efficiently provides a system where the two applications can operate in parallel.

Additionally, algorithms that operate in high-speed robotic system running concurrent activities must address computational latency to ensure correct behavior. A high-speed mobile robot running visual pose estimator in continuous mode needs to handle the latency between the time the data is acquired and the time the estimate is computed. In such situations, an additional step is necessary to compensate for the changes in rover state to produce the most accurate information about the robot's pose.

3.4 Software Efficiency

Challenge: Software Efficiency

Developing efficient algorithms using a generalized framework can be quite challenging. No one wants to trade performance for generality. Generality and flexibility may seem at odds with performance and memory efficiency. Despite the continued increase in available computational power and communication bandwidth, it is necessary to keep the performance of the generic software as close as possible to a custom solution.

Solution: (1) Use common data structures; (2) use templates for math; and (3) use inline functions

An application framework must pay particular attention to avoiding unnecessary copying of data when exchanging information among modules. This is particularly important when using component/connector style interfaces. The framework can also avoid accidental complexities that arise from isolated developments, which require transforming data from one form to another. Many techniques such as common data structures, the use of templates for low-level classes, and the use of inline functions can provide abstractions without the run-time overhead.

3.5 Multiple Operating Systems and Tools

Challenge: Hard versus Soft Real-Time

Many challenges stem from the differences in the runtime architecture of processes and threads that are used in current operating systems. Furthermore, some operating systems provide hard real-time scheduling guarantees while others provide soft real-time performance.

Solution: Use standard tools for operating system independence

To make the robotic software portable, it is necessary to build the software for different hardware architectures, operating systems, and compilers. This process also improves software reliability by eliminating nuances specific to an architecture, an operating system or a compiler. Third party packages such as POSIX and ACE (Adaptive Communication Environment) develop standards that cope with differences in operating systems. Proper handling of little and big-endianness, and the use of ANSI standards for compiler support ensure robust and portable software. However, the cost of developing and maintaining portable software has to be taken seriously into account. Another important aspect of this is the ability to transition from hard real-time operating system such as VxWorks, RTAI Linux, or Integrity, to soft real-time systems such as Linux, Solaris, Mac OS X, or Windows.

4 Hardware Variability

4.1 Hardware Architectures

Challenge: Different Hardware Architectures

While at first, it may seem that adapting abstractions directly to hardware components is all that is necessary to interoperate across robotic platforms, this assumes that the hardware on the various platforms is similar in architecture. This is rarely the case. Most robotic systems have different hardware architectures. On one end of the spectrum, there are robots that use a central processor to generate the motor control laws, motor trajectories, and run the application software. Such systems have their analog and digital signals mapped to memory registers on the central processor, which makes the development of software relatively easy (see Fig. 8). They are very flexible for changing the control laws and coordinating motors, but they lack hardware modularity and can be hard to extend and repair. On the other end of the

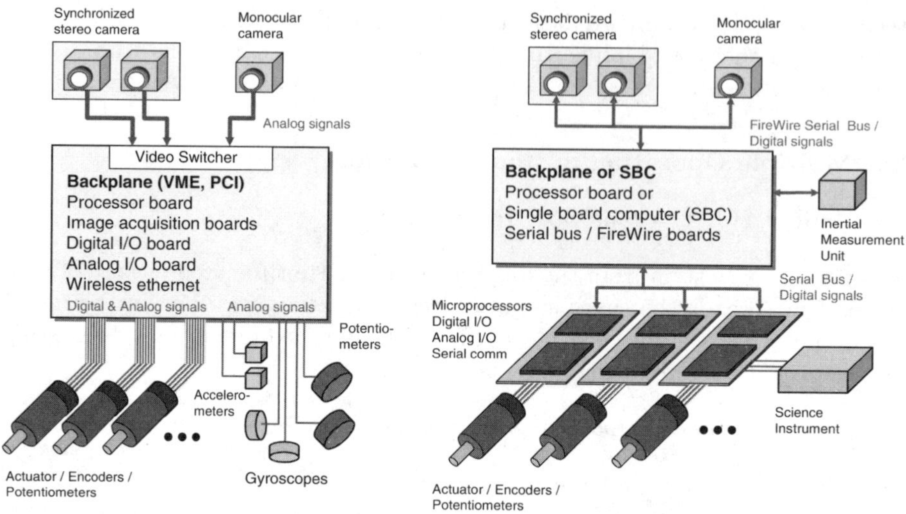

Fig. 8. Centralized (left) and distributed (right) hardware architectures

spectrum, there are systems that migrate much of their control to firmware in distributed nodes in order to improve modularity and reduce the load and real-time requirements on the central processor. They communicate with each other via formatted data streams sent over serial buses. Other systems fall somewhere within this spectrum.

There are similar differences in image acquisition and inertial sensing hardware. Some systems use high-quality analog cameras with centralized image acquisition boards (frame grabbers) while other systems use digital cameras connected through a serial bus (FireWire or USB). Inertial sensors can either be analog sensors integrated through a central processor or an integrated unit that communicates to a processor through a serial interface. A general framework must be sufficiently flexible to handle these variations in the hardware control architectures, which has significant impact on information flow and synchronization.

Solution: Use hierarchical multi-level abstractions

Develop multi-level abstractions to allow adaptations to interface to hardware at different levels. Different robots have different levels of hardware sophistication where some local intelligence may be embedded in distributed microcontrollers. Others may have inaccessible embedded processors limiting access to only higher-level interfaces for controlling the robot. Simulations also need similar multi-level access since each simulation may only support certain fidelity levels. Some may have full dynamics and device simulation capabilities while others may only simulate kinematics and approximate device models. In some cases, it may also be desirable to lower the fidelity level to speed up the simulation. For example, in a rover simulation, one may interface at the

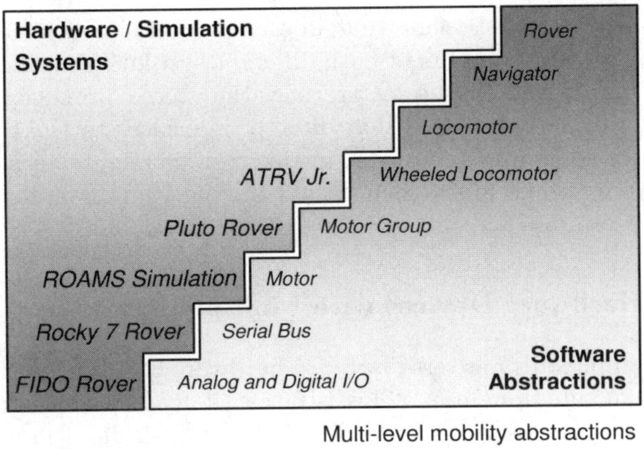

Fig. 9. Multi-level software access

locomotion level to simulate rover kinematic motions bypassing the need to simulate wheel dynamics.

Examples: (1) Multi-level access to a locomotor; and (2) migrating functionality to hardware

Fig. 9 shows a locomotion example that provides access to hardware and simulation at various levels of abstraction. At the lowest levels, the control software interfaces to hardware or simulation using basic analog and digital I/O signals. A higher level would be to interface through a serial bus where information is exchanged through formatted data packets as opposed to toggling I/O signals. At an even higher level, the software interfaces to a motor adaptation that understands motor commands. Higher levels include an interface to a group of coordinated motors, a wheeled locomotor, or a general locomotor.

Such examples not only occur with custom-built robots, but also when interfacing to commercial-off-the-shelf robots. Consider a commercial robot that does not provide access to individual motors. The manufacturer may provide a software layer with an interface to only control the robot's motion. Thence, the concept of a motor is hidden in a layer that is inaccessible to the user. Such a system will have adaptations at the locomotor level providing only access to the locomotor interface as opposed to the motor or I/O level interfaces.

To illustrate the point of migrating functionality into hardware, consider a robot with the stereo camera pair. This robot can acquire images synchronously from these cameras. A stereo processor class takes these camera images as input and outputs a depth map. This algorithm is implemented on the main processor and uses the images acquired by the cameras to generate a third depth image. While the stereo processor component in this example is a purely software component that does not interact with hardware, another stereo processor component on a different system can achieve the same capability in hardware. In the latter case, the stereo processing capability is migrated to a microprocessor that directly interfaces to the cameras. The stereo processor software component, in this case, is merely an adaptation to a hardware device while in the former case the hardware adaptation is at the camera level as opposed to the stereo processor level.

Challenge: Hardware Devices with Multiple Functions

What often complicates matters is when a hardware component provides multiple orthogonal functionalities. This is often the case when a robot system uses distributed processors. A capable micro-controller as the one shown in Fig. 8(right) not only provides motion control but also provides general purpose analog and digital I/O. To complicate things further, some of these analog and digital channels can be connected to other devices and instruments in the system. What we end up with is a system where devices that are logically decoupled become physically coupled. In such cases, we have a shared resource between multiple devices but which exists as a result of a given hardware architecture. We have to handle the physical dependency in a transparent way to keep the logical operation of devices independent.

Solution: Separate the logical architecture from the physical architecture

Decouple the logical architecture of the system from the physical hardware architecture. The physical architecture describes how the hardware works, while the logical architecture defines what the generic abstractions expect to have. Instead of combining these two into a single hierarchy, we recommend to separate them. An adaptation of the logical hierarchy would then bridge logical hierarchy to the physical hierarchy making the mapping between the two unambiguous. This adaptation class often contains little code but clearly shows the logical to physical mappings of functions. Mixing the two can lead to a single adaptation that is hard to understand, difficult to maintain, and hard to use in a specialized application that does not need the generic interface. Developing a hierarchy that is purely hardware specific keeps these drivers independent of any logical hardware architecture. It also makes the hardware specific code easily testable and more portable. This pattern applies to motors, cameras, digital and analog I/O, instruments, sensors and other actuators.

Even though the hardware architecture imposes constraints that result from shared resources (e.g. sharing a bus or a processor), the software has to manage these resources such that the overall behavior results in an independent logical architecture. To clarify this point, consider an operating system that provides a multi-tasking environment. The operating system has a shared resource: the processor. However, it provides the logical functionality of parallel tasks even though at its core it manages a shared resource.

Constraints that result from shared resources are managed locally by specialized classes. This decouples higher-level software that controls logically independent devices from the physical classes. The coupling is handled by the specialized classes, which allocate and free up these resources.

The assumption in the above model is that these resources are needed for only short durations. There are situations, however, where the robotic hardware may have severe constraints and cannot provide a logically decoupled architecture. In such situations, the software can compensate to the best of its ability for the missing or limited hardware functionality. This is similar to graphics acceleration cards where software compensates for missing hardware acceleration functions. Even though the overall performance will degrade, the software continues to operate. A similar example occurs in robotic motion control. If a motor controller cannot generate a motor trajectory to satisfy the requirements of an application, instead of sending trajectory parameters to the controller, the application software sends lower-level motor set-points to generate the desired trajectory.

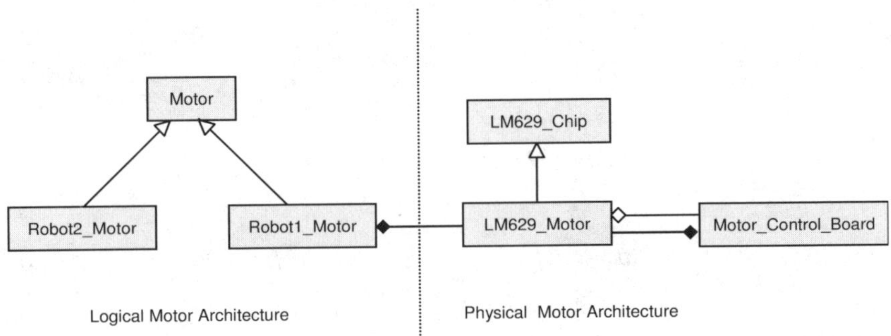

Fig. 10. Logical architecture vs. physical architecture

Example: Motor logical vs. physical architecture

Fig. 10 shows the two class hierarchies. On the left hand side is the logical hierarchy, which defines the generic motor functionality. While some functions will be pure virtual, others will provide a default implementation. On the right hand side is the physical architecture. In this example, the robot uses the LM629 motion control chip. The software driver for this chip can be made generic by keeping the functions that read from and write to the chip pure virtual. This enables the chip class to be specialized for different hardware boards. The LM629_Motor class that uses this chip in a given hardware board defines the communication interface. The Robot1_Motor class, which inherits from the Motor class, aggregates the LM629_Motor controller. The Robot1_Motor maps the physical LM629_Motor class to the generic Motor class. Because the LM629_Motor class does not depend on the Motor class, it can be tested independent of the generic hierarchy.

4.2 Hardware Components

Challenge: Class Design

Defining the proper classes, the interaction between these classes, and the organization of these classes into modules is difficult because different hardware components exhibit different behaviors.

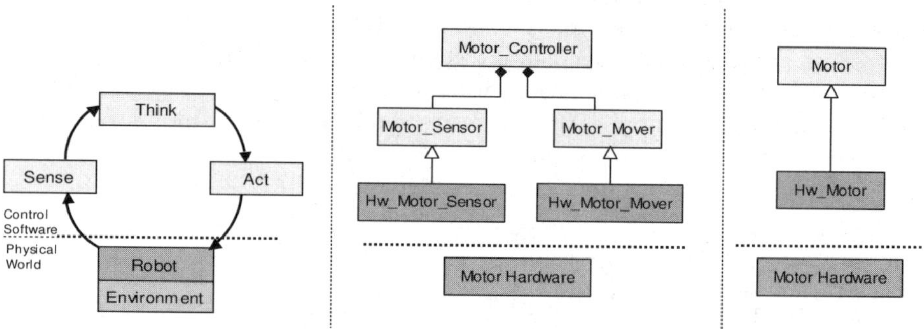

Fig. 11. Motor abstractions

Solution: Separate the specification of "what" to do from the "how" to do it

The classical approach in a robotic design is to separate the actuation from the sensing (see Fig. 11). Using this approach, a framework would have a Motor_Mover class and a Motor_Sensor class. The Motor_Mover would command the motor to move while the corresponding Motor_Sensor would read the encoder/potentiometer feedback and convert the motor shaft position. Another approach is to combine the two abstractions into a single Motor class that defines the capabilities and behavior of a controlled motor. These two approaches are fundamentally different. The first decomposition does not abstract a capability but exposes elements of the motor control. But the second abstracts the capability for controlling motors with a well-defined interface to enable the specialization of this capability. The Motor class provides only a partial implementation of its functionality and behavior. The remaining functionality is implemented inside specialized motor classes that adapt the Motor class to hardware components or to device drivers. Adaptations of motor cover different types of controllers including servo and stepper controllers, hardware and software controllers, and controllers with different sensory feedback such as optical encoders, magnetic encoders, or potentiometers. What is abstracted is not how the motor is doing the control, but rather what functionality a controlled motor provides.

Challenge: Generalize Low-Level Hardware Components

The main challenge in generalizing low-level hardware elements is maintaining the efficiency of a custom implementation.

Solution: Generalize hardware classes for greater flexibility

Even some hardware specific components can be properly generalized to provide the necessary flexibility to replace individual hardware components for low-level interoperability. One may consider digital and analog I/O as hardware specific components that cannot be generalized or, if generalized, become inefficient. Quite to the contrary, a large portion of digital and analog I/O control code can be generalized with proper abstractions, inline functions and templates to ensure a flexible and efficient implementation. Much of the digital I/O software involves bit manipulation and masking, which is independent of the hardware.

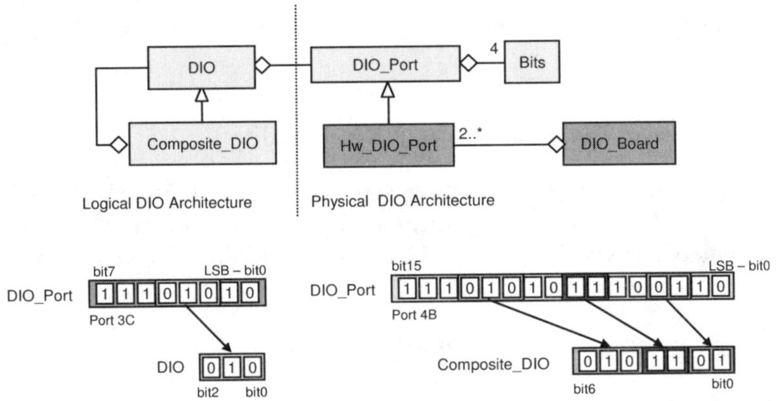

Fig. 12. Digital I/O abstractions

Example: Digital I/O class Hierarchy

Digital I/O provides another example where we separate the logical from the physical architecture. Generic digital I/O classes provide the logical mapping of I/O lines to the various instruments. The mapping of the I/O lines to the physical I/O ports is done through the hardware-based DIO_Port class. The DIO_Port class defines methods for input, output, configuration, and masking that are pure virtual. This class serves as a base class from which the hardware specialized ports are derived. Digital I/O boards often consist of a number of hardware ports. These ports contain a number of I/O lines that can be configured individually or as a group depending on the particular hardware board. These I/O lines can be configured as input, output, or tri-stated as both input and output. Fig. 12 shows the main DIO class, which aggregates a single DIO_Port. This pattern is known as a bridge pattern [GHJV95], which enables both the DIO class and the DIO_Port class to be specialized. This allows inheritance along two axes: the functional (logical) axis by extending the DIO class and the hardware axis by extending the DIO_Port class. Higher-level software now has the flexibility to control the digital I/O without being tied to any given hardware board or driver. With this design, we can replace any digital I/O board and re-map the I/O lines without changing to the application software.

The DIO class handles only contiguous bits. There are situations, however, where non-contiguous groups of bits need to be grouped and controlled synchronously. We handle these using the Composite_DIO class that both inherits from and uses the DIO class. The Composite_DIO class treats disparate DIO lines as a contiguous block of bits. It manages the splitting and grouping of bits to reflect the physical mapping of these I/O lines to hardware ports. Inheriting from DIO forces the Composite_DIO to be of the same DIO type making their objects interchangeable. Aggregating the DIO class enables multiple DIO objects in the Composite_IO. This pattern is known as a composite pattern [GHJV95].

Challenge: Class Flexibility and Extendibility

Because we are proposing to develop a generic framework, the classes and their interfaces need to be sufficiently flexible and rich in functionality to support the different use cases. This can lead to more complex class hierarchies that can be hard to extend and maintain.

Solution 1: Balance flexibility with maintainability

Despite the need for generality and flexibility, it is sometimes necessary to sacrifice flexibility for simplicity and improved maintainability. The challenge is in defining the proper level of flexibility to match the requirements without adding complexity.

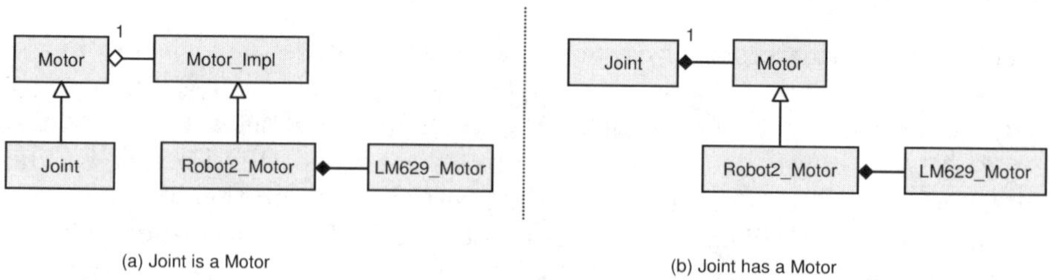

Fig. 13. Extending motor class functionality

Example: Joint specialization of Motor

Consider the motor example that we have presented earlier. The generic motor class gets specialized to a hardware adaptation. By specializing the motor to hardware, we can no longer extend its functionality. If we derive a joint from a motor, we then have to specialize the joint to the same hardware adaptation, thus duplicating the motor hardware adaptation. We can solve this problem using the bridge pattern [GHJV95]. Using this pattern, a new implementation class called Motor_Impl is created as shown in Fig. 13(a). The Motor class aggregates the Motor_Impl class. Now both Motor and Motor_Impl can be extended. A joint can now be derived from Motor to provide limit checking on the joint motions. This way a robot can use the unrestricted Motor objects for its drive wheels and the motion-constrained Joint objects for its steering.

While this seems like a reasonable solution to the problem, it has some drawbacks. First, all state information has to reside in the Motor_Impl class to be accessible to its adaptations. The Motor class would then become an abstract class. Second, every time we add a new function to the Motor_Impl class, we also have to add the same function to the Motor class. This structure can be hard to maintain if all devices in the system use this pattern. Third, this becomes particularly difficult when the Motor class derives from a generic Device class and the Motor_Impl class derives from the Device_Impl class.

In such cases, it may be easier to restrict device classes, such as motor, camera, and IMU from functional extensions. In this case, a Joint class has to aggregate a Motor object as opposed to inherit from it as shown in Fig. 13(b). What is lost here is that a Joint object is no longer of type Motor because it does not inherit from the Motor class. The Joint class would have to then redefine the functions that need to override the motor moves with ones that contain limit checking. The Joint class can also return a reference to the Motor to access to the rest of the functionality.

Solution 2: Group hardware devices into hardware maps

Because different deployments of generic robot software require different hardware maps, it is important to group the hardware components into a separate class that manages the system's configuration. An abstract factory class [GHJV95] can serve as a robot device map. A specialization of the device map creates the appropriate objects using the hardware specialized classes. For example, if the software is deployed on the physical hardware, then the motor, cameras, digital and analog I/O classes use their physical hardware components. If the same robot software is deployed in a simulation, then these devices would use their simulated counterparts.

Challenge: System State

System state is unique to each robot. In most complex systems, state is distributed throughout the system in its various microprocessors. The state information in these controllers can be retrieved, but there is often a limitation on the rate and the latency of this information. Some internal states might not be directly accessible.

Solution: Encapsulate states in the hierarchies

To support heterogeneous robotic platforms, state information should be handled in a hierarchical fashion. State would then be retrieved only through accessor functions as shown in Fig. 14. Similar to states, state machines should also be hidden in class abstractions. Breaking this encapsulation introduces system specific dependencies that reduce the interoperability of the software.

The states of a legged system differ from those of a wheeled system. However, the reason why software can be interoperated across heterogeneous systems is the presumption that there is a level of abstraction at which the generic algorithm can use abstract states to control the robot.

Example: Legged vs. wheeled locomotion

Consider two robots: a wheeled rover with a passive suspension and a legged robot (see Fig. 15). These robots can both be commanded to follow a path in rough terrain. However, the two robots will move differently along the path. The wheeled rover will conform to the terrain while the legged robot will articulate its legs to move its body along the path. While both robots follow the same path, they achieve that using different body motions. So at the generic locomotor level, both systems are commanded to follow a path even though they achieve that goal in different ways. The common states at the locomotor level are the robot's pose, the notion of a path, and the notion of

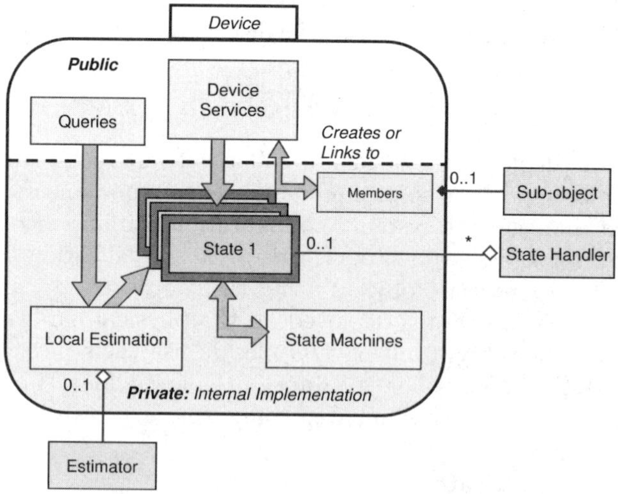

Fig. 14. Dealing with state and state machines

how far along the path a robot is. However, at the lower levels, the states differ. The wheeled locomotor keeps track of the state of each wheel (distance and steering angle) while the legged system keeps track of the joint angles for each leg. But the robot's pose, a higher-level state that is derived from these lower-level states, is common to both.

4.3 Sensor Configuration

Challenge: Different Sensor Configuration

The major challenge comes from differences in sensor configurations that are on similar physical robots. Similar robots may use different sensor configurations that produce similar information. However, different sensor configurations often have different physical constraints.

Solution: Use multi-level generic abstractions for the sensors in the system

By providing various levels of device abstractions, we can cope with the variability in the sensor configurations. This is particularly important as there are multiple implementations that may achieve the same result. Some may be available in hardware while others through software. This is best illustrated by the examples below.

Example: Stereo cameras vs. lidar; sun sensor vs. compass

Consider sensors that generate terrain data that is represented as three-dimensional point clouds. To generate this data, one can either use a lidar or a stereo camera pair. While both devices eventually generate point clouds, these two devices operate with different constraints and have different qualities. A lidar requires a longer time to scan a scene but less time to generate the depth information, while the opposite is true for stereo. These behavioral differences generate constraints on the operation of the robot. A stereo processing algorithm uses two images and their corresponding camera models to generate a three dimensional map. If what we are trying to do is to get a three dimensional map, then our algorithms should not depend directly on a stereo processor but rather on the three dimensional map or a point cloud source. A three-dimensional point cloud can also be generated from a lidar. So, by depending on the point cloud source as opposed to a stereo processor, our algorithms become generalized enabling them to work with a wider range of sensors. If a navigation algorithm that uses this data to find obstacles was interfaced to stereo cameras as opposed to point clouds, then it will not be possible to use this algorithm on rovers that use a lidar sensor in lieu of stereo cameras.

Another example is with algorithms that use a compass. A more general algorithm would substitute its use of a compass class with the generic absolute heading sensor class. That way, a sun sensor, which also computes absolute heading, can be used interchangeably with a compass.

4.4 Different Mechanisms

Challenge: Different Capabilities

Different mechanisms exhibit different capabilities, which have major implications on how the robots move and act. Controlling and maneuvering wheeled robots is very different from controlling legged platforms. Even among wheeled robots, fully-steerable (omni-directional) rovers can move laterally (crab) while partially-steerable (car-like) robot can achieve the same result only via a parallel parking maneuver (Fig. 15(a) to (d)). Mobile robots with passive suspension conform to the terrain with no control over their tilt while those with active suspension have control over their tilt (Fig. 15(e) and (f)). Robotic manipulators have similar nuances. Joint configuration and degrees of freedom result in different constraints on the maneuverability of the end effector. Limited degree-of-freedom arms can only achieve certain poses, while redundant arms can achieve all poses with infinite possibilities.

Solution: Use multi-level model abstractions and separate models from control

Develop a generic model representation for mechanisms. Develop generalized kinematic and dynamic algorithms that use a generic model representation. Provide a means to override the generic capabilities with efficient specialized ones such as specialized forward and inverse kinematics.

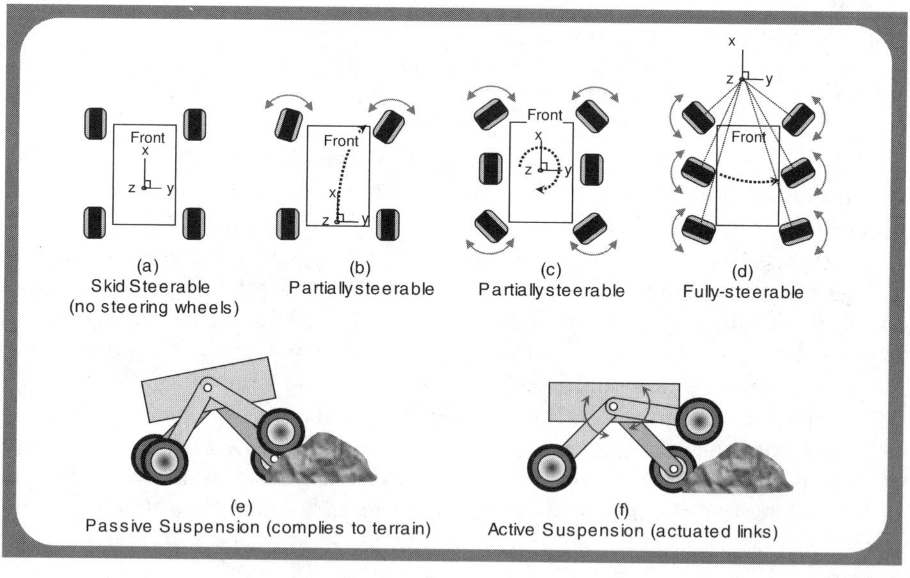

Fig. 15. Different mechanisms for wheeled robots

Separate mechanism models from controls to enable their use for resource and impact predictions. Embedding models into the same software structure as control makes their use outside that context very difficult.

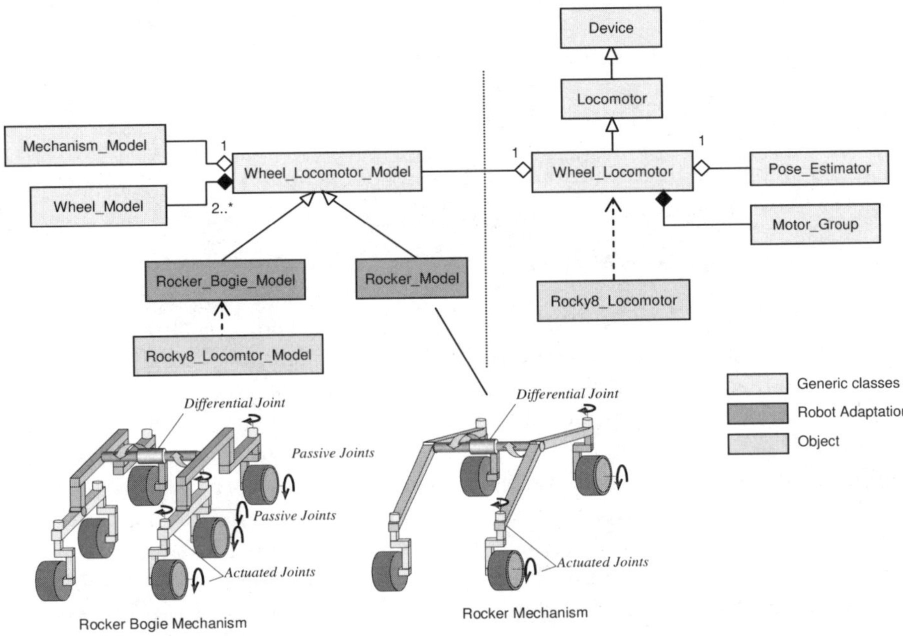

Fig. 16. Separating model from control

Example: Manipulator and locomotor kinematics

Fig. 16 shows the relationship between a generic Mechanism_Model and a Wheel_Locomotor_Model. The Wheel_Locomotor_Model aggregates a Mechanism_Model. The Wheel_Locomotor_Model provides generic capabilities for forward and inverse kinematics of all wheeled robots. It also includes wheel models classes. The Wheel_Locomtor_Model gets specialized to specific mechanisms such as a six-wheel rocker-bogie mechanism or a four-wheel rocker mechanism, which provide specialized kinematic solutions for their respective mechanisms. The Wheel_Locomotor control class aggregates the Wheel_Locomotor_Model class, which keeps the model hierarchy separate from the control hierarchy. Because of this separation, a navigator that requires information about the maneuverability of a vehicle now only relies on the Wheel_Locomotor_Model as opposed to the Wheel_Locomotor control class.

To illustrate the importance of separating models from control, consider a generic manipulator control class that does not separate the mechanism model from its control class. Rather, this class defines the forward and inverse kinematic interfaces in a specialized class that implements the closed-form inverse and forward kinematic equations. Using the forward and inverse kinematic algorithms now requires the instantiation of the arm with its motors and controllers even when only the kinematic portions of that class are desired for analyzing planned arm moves.

Challenge: Coordinate Transformations

Transformations cannot be defined in isolation and require a context that defines the relationships between coordinate frames. This is particularly challenging as some transformations go through articulated joints as shown in Fig. 17.

Solution: Unify mechanism model

Without a uniform representation of the robot mechanism, sharing mechanism information among sub-systems becomes difficult, inefficient and error-prone. A unified mechanism model is the backbone for managing coordinate frame transformations. It handles both fixed transformations and ones that go through articulated or passive joints. It also ensures the integrity of the mechanism information that is used by the multiple algorithms providing a consistent representation of kinematic, dynamic and geometric information.

A mechanism model will reduce code duplication when modeling robotic arms or mobility mechanisms. It also allows the development of generic algorithms for forward, inverse, and differential kinematics. In the absence of

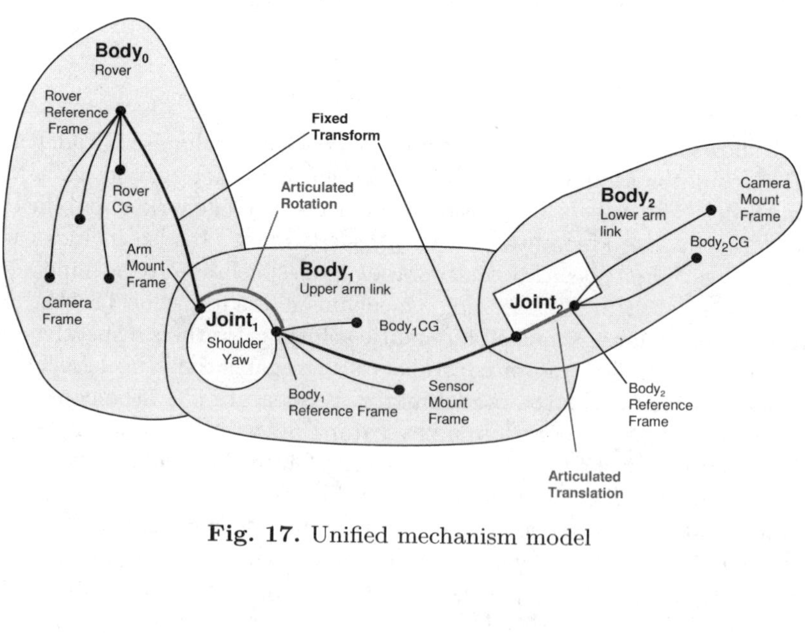

Fig. 17. Unified mechanism model

specialized versions, the generic algorithms provide out-of-the-box functionality. However, the architecture should support specific implementations to override generic algorithms whenever appropriate for optimal performance.

A mechanism can be represented using a tree topology in which an arbitrary number of rigid bodies are connected to one another via joints. The tree topology captures the geometric relationships between all elements in the mechanism such as sensors and bodies, and serves as a repository of mechanical model information. To support multiple clients the tree representation has to be stateless: position, velocity, and acceleration information relative to an inertial frame is not stored in the mechanism model. This enables various system states to be updated at different rates and enables the use of different parts of the tree at a time. It also allows algorithms to use the mechanism model tree to predict future states for any given input state. The trade that is made here is the cost of re-computing derived states vs. making copies of the mechanism model for each client application and keeping all their internal states up to date.

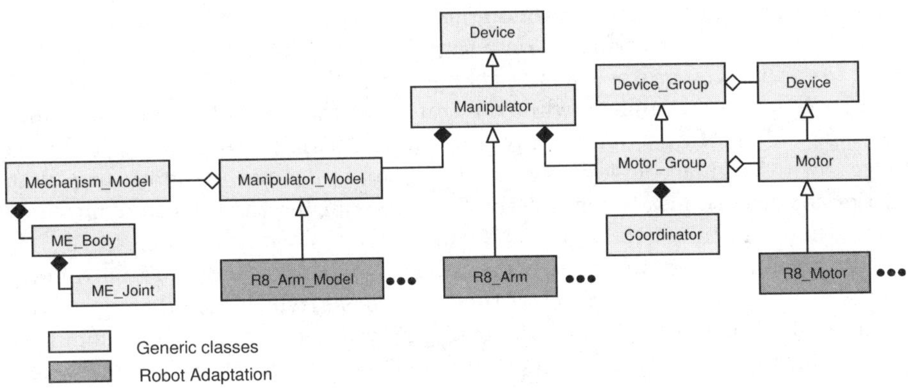

Fig. 18. Manipulator model and control classes

It is desirable for some applications to treat the mechanism as a whole body (for instance when dealing with rover-arm coordination algorithms). Some other applications may require the treatment of arms or legs as separate elements.

A mechanism model should support serial manipulators, closed-chains, wheeled mechanisms (Fig. 15), legged mechanisms, and composite mechanisms.

Example: The Manipulator classes

Fig. 18 shows the relationships between the bodies/joints and the generic mechanism model that are used by *CLARAty* . The manipulator model aggregates the manipulator portions of the complete mechanism model. A more detailed description of mechanism models can be found in [DCNKN06].

5 Conclusion

Many of these recommendations were adopted in the development of the *CLARAty* reusable framework that is used by NASA. Given the heterogeneity of the NASA research rovers, it was incumbent upon us to provide a framework that did not require the redesign of existing hardware. Additionally it was necessary to support legacy algorithms with significant investments.

Algorithms were developed at various institutions and have been integrated and tested on NASA developed robots. Capabilities that were integrated and demonstrated include motion control and coordination, wheeled and legged locomotion, stereo vision, visual tracking, visual odometry, science analysis, pose estimation, continuous trajectory following, path planning, autonomous navigation and obstacle avoidance, and general activity planning. We have also demonstrated autonomous end-to-end capabilities such as placing a rover-mounted instrument on a target selected from a 10 meter distance. Such capability integrates visual tracking of the designated target using multiple rover mounted cameras while navigating to the target location; assessing the safety of the target region; properly positioning the rover relative to the target for instrument deployment; deploying and placing the robotic arm that carries the science instrument on the target; acquiring the scientific data and simulating a downlink to Earth.

We have deployed and have been using *CLARAty* on half a dozen robotic platforms. Fig. 1 shows a subset of these platforms, which include the custom *Rocky 8*, FIDO, *Rocky 7*, and K9 rovers, as well as the ATRV Jr. commercial platform. These platforms have different mobility mechanisms and wheel configurations as well as different sensor suites, manipulators, end effectors, processors, motion control architectures and operating systems. In addition to these real-platform adaptations, we have also adapted *CLARAty* to operate with the high-fidelity ROAMS rover and terrain simulator [Je04].

Developing reusable robotic software presents many challenges. These challenges stem from variability in robotic mechanisms, sensor configurations, and hardware control architectures. They also stem from integrating new capabilities that use different representations of information or that have architectural mismatches with the reusable framework. We found that multi-level abstraction models, object-oriented methodologies and design patterns go a long way to address the extensive variability that is encountered in today's robotic platforms. We have learned that overgeneralizing interfaces makes them harder to understand and use. There is a delicate balance between flexibility and simplicity. Performance cannot be compromised for the sake of flexibility and the least common denominator solution is often unacceptable. It is necessary to have flexible development environments, tools, and regression tests. Reusable software products and processes have to be well-documented. It would be highly desirable to standardize robotic hardware but that may not be feasible today.

There are many challenges in software engineering that any generic framework for robotics will have to address. No matter what approach is used in the design, issues related to the effectiveness of the framework can only be judged over time. The challenge is to find a delicate balance among flexibility, efficiency, scalability, maintainability, and extendibility.

Acknowledgement

The author would like to acknowledge the contributions of current and former members of the *CLARAty* team: Gene Chalfant, Caroline Chouinard, Dan Clouse, Antonio Diaz-Calderon, Tara Estlin, Daniel Gaines, Ron Garrett, John Guineau, Won Kim, Richard Madison, Michael McHenry, Michael Mossey, Darren Mutz, Hari Nayar, Richard Petras, Mihail Pivtoraiko, Gregg Rabideau, Babak Sapir, I-Hsiang Shu, and Marsette Vona from the Jet Propulsion Laboratory; Clayton Kunz, Lorenzo Fluckeiger, Susan Lee, Eric Park, Randy Sargent, Hans Utz, and Anne Wright from NASA Ames Research Center; Reid Simmons, David Apfelbaum, Kam Lasater, Nik Melchoir, Chris Urmson, and David Wettergreen from Carnegie Mellon; and Stergios Roumeliotis, Anastasios Mourikis, and Nikolas Trawny from the University of Minnesota. He would also like to acknowledge the contributions of former principal investigator Richard Volpe and former lead engineers Anne Wright and Max Bajracharya as well as the numerous contributors from universities and NASA centers who have provided algorithms into *CLARAty* .

The author would also like to thank NASA's Mars Technology Program and the Robotics management at NASA and the Jet Propulsion Laboratory for their vision and support: David Lavery, Samad Hayati, Paul Schenker, Richard Volpe, and Gabriel Udomkesmalee. The work described in this chapter was carried out at the Jet Propulsion Laboratory, California Institute of Technology, NASA Ames Research Center, Carnegie Mellon, and University of Minnesota under a contract to the National Aeronautics and Space Administration.

References

[ACF98] R. Alami, R. Chautila, S. Fleury, M. Ghallab, and F. Ingrand, *An architecture for autonomy*, The International Journal of Robotics Research **17** (1998), no. 4.

[Alb00] J. Albus, *4-d/rcs reference model architecture for unmmanned ground vehicles*, IEEE Internation Conference on Robotics and Automation (San Francisco), April 2000.

[AML87] J. Albus, H. McCain, and R. Lumia, *Nasa/nbs standard reference model for telerobot control system architecture (nasrem)*, NBS Technical Note 1235, National Bureau of Standards, Gaithersburg, Maryland, July 1987.

[cla06] *http://claraty.jpl.nasa.gov*, 2006.

[DCNKN06] A. Diaz-Calderon, I.A. Nesnas, W. Kim, and H. Nayar, *Towards a unified representation of mechanisms for robotic control software*, International Journal of Advanced Robotic Systems **3** (2006), no. 1, 61–66.

[GHJV95] E. Gamma, R. Helm, R. Johnson, and J. Villisides, *Design patterns: Elements of reusable object oriented software*, Addison-Wesley, 1995.

[Hat03] M. Hatig, *Robotic engineering task force*, IEEE Int'l Conference on Intelligent Robots and Systems (Las Vegas, Nevada), October 2003.

[HP86] V. Hayward and R. Paul, *Robot manipulator control under unix rccl: A robot control c library*, International Journal of Robotic Research, 1986.

[jau06] *http://www.jauswg.org*, 2006.

[Je04] A. Jain and et.al., *Recent developments in the roams planetary rover simulation environment*, IEEE Aerospace Conference (Big Sky, Montana), 2004.

[KT98] C. Kapoor and D. Tesar, *A reusable operational software architecture for advanced robotics*, CSIM-IFTiMM Symposium on Theory and Practice of Robots and Manipulators (Paris, France), 1998.

[NSG06] I.A. Nesnas, R. Simmons, D. Gaines, C. Kunz, A. Diaz-Calderon, T. Estlin, R. Madison, J. Guineau, M. McHenry, I. Shu, and D. Apfelbaum, *Claraty: Challenges and steps toward reusable robotic*, International Journal of Advanced Robotic Systems **3** (2006), no. 1, 23–30.

[NWB03] I. A. Nesnas, A. Wright, M. Bajracharya, R. Simmons, T. Estlin, and W. Kim, *Claraty: An architecture for reusable robotic software*, SPIE Aerosense Conference (Orlando, Florida), April 2003.

[PCSCW98] G. Pardo-Castellote, S. Schneider, V. Chen, and H. Wang, *Controlshell: A software architecture for complex electromechanical systems*, International Journal of Robotics Research **17** (1998), no. 4.

[SVK97] D. Stewart, R. Volpe, and P. Khosla, *Design of dynamically reconfigurable real-time software using port-based object*, IEEE Transactions on Software Engineering **23** (1997), no. 12.

[VG06] R.T. Vaughan and B.P. Gerkey, *Reusable Robot Software and the Player/Stage Project.* In D.Brugali (Ed.) Software Engineering for Experimental Robotics, STAR, Springer Verlag, 2006.

Simulation and Testbeds of Autonomous Robots in Harsh Environments

Richard S. Stansbury[1], Eric L. Akers[1], Hans P. Harmon[2], and Arvin Agah[1]

[1] University of Kansas richss@ku.edu, eakers@ku.edu, and agah@ku.edu
[2] Pinnacle Technology, Inc. hharmon@pinnaclet.com

1 Introduction

Software development for autonomous field robots can be quite challenging due to the difficulties associated with testing and evalution of the robot and its components. Software for the mobile robot must undergo extensive testing throughout its lifecycle. However, testing a robot for harsh environments requires access to the field, logistic support, maintenance of the robot, and adherence to safety standards. Time, budgetary, and logistics constraints often limit the amount of testing, which in turn adversely affects the quality of the software developed for the field mobile robot.

A harsh environment is one that is very difficult and possibly dangerous to traverse because of obstacles or difficult terrain, dangerous climate, or a location that is difficult to reach by human or robot. For the purposes of this chapter, the harsh environments referred to are the polar regions. These locations have extreme cold temperatures, are very windy, sastrugi and crevasses make traversal difficult and dangerous, and they are difficult to reach.

Through simulation and robot testbeds, software for field robots can be designed, developed, and tested, while effectively addressing many of these challenges. Virtual prototypes can be used to realistically simulate the field robot for hardware and/or software design. Testbed platforms can be used to test and refine the robot software within a real environment. The resulting software can then be implemented on the field robot. It should be noted that although this approach is of great benefit to software development, it does not eliminate the need for eventual testing in the field because the virtual prototype and testbed are only an approximation of the real system. This methodology should, however, reduce the time required to physically test the robot in the field and improve the chances of success. The transition from simulation to physical robot includes iteration of testing and evaluation, modifying the simulation, and retesting and reevaluating, and it is this step that is the most difficult when dealing with harsh environments.

D. Brugali (Ed.): Software Eng. for Experimental Robotics. STAR 30, pp. 71–92, 2007.

For the evolution of software from a virtual prototype to the real robotic system, a software architecture for the mobile robots must exist. The architecture must allow the reusability and portability of software between virtual and physical platforms. The architecture must also support the interaction between the robotic components both locally and in a distributed fashion.

Virtual prototypes can be used to produce a fully simulated and realistic representation of the robot and its environment. The simulated robot can be controlled within the simulator in order to test its software, as well as to determine various constraints and design parameters that could affect the software or hardware of the robot.

Smaller mobile robots may also act as testbeds by providing analogous sensors and actuators. Higher-level control software can be developed and refined on a testbed within a laboratory environment, reducing the need for field experiments. Software development on an analogous robot provides further opportunity to develop software functionality and to refine robot behaviors.

In this chapter, a polar robot developed as part of the Polar Radar for Ice Sheet Measurement (PRISM) project at the University of Kansas [AHSA04, SAHA04] will be presented as a successful case study for the application of the proposed methodology. The PRISM project will be briefly introduced as well as the autonomous field robotic systems used as part of this research. A software architecture for mobile robots will be presented. Next, the virtual prototype of the PRISM mobile robot is described. The transitions between the prototype, the testbed, and PRISM mobile robot is discussed. Finally, the performance of the robot in Greenland is presented.

1.1 Polar Radar for Ice Sheet Measurement Project

The PRISM project is currently underway at The University of Kansas [KU 04], with the goal to develop radar systems to measure polar ice sheet properties in order to accurately determine their mass balance and other characteristics. Such data will help scientists to determine and to better model the contributions of polar ice sheet melting to global climate change and its effects on the rising sea levels.

In order to accommodate a pair of bistatic synthetic aperture radars (SAR), two vehicles have been employed, carrying the transmitter and the reciever components of the SAR, respectively. The first is a manned tracked vehicle that will tow an antenna array for the bistatic/monostatic SAR and will also carry an on-board dual-mode radar. The second is an autonomous robot that will be utilized to carry the second SAR. When operating in bistatic mode, the robot's SAR will act as the receiver. The two vehicles will move along side one another in a coordinated pattern. The autonomous robot will traverse a wide area along side the tracked vehicle collecting data to build the radar image.

Challenges of Polar Traversal

In June of 2001, a workshop analyzed the feasibility and resourcefulness of using mobile robots similar to planetary rovers for collection of scientific data on the ice sheets [CSB01]. This group defined several tasks with which an autonomous rover could aide. These tasks include: traverses with detailed and tedious paths, extremely remote and/or inhospitable environments, data collection parallel to a manned traversal, and data collection at slow speeds. In 2002, researchers at NASA's Jet Propulsion Laboratory proposed that utilizing mobile robotics for polar exploration can provide future benefits toward planetary exploration [BCW02].

Both groups discussed the numerous challenges that exist with polar traversals. These challenges include deep and blowing snow, crevasse detection and avoidance, sastrugi detection and avoidance, and limited supervision. Operating in the polar regions can be more challenging than in more hospitable environments. Equipment may be damaged as the result of sub-zero temperatures, high wind speeds, and blowing snow.

Several unique terrain-based obstacles present themselves in the polar environments. For instance, sastrugies which are snow drifts that form as the result of wind erosion, have heights of up to one meter. In addition, crevasses in the ice sheet are encountered that remain invisible because they are covered by snow. Vision can be limited due to the lack of contrast on the icy surface. With this lack of depth perception, natural obstacles can be overlooked.

PRISM Robotics Testbed

A Nomadic Scout mobile robot [Nom99a] acts as a testbed for development of higher-level software applications. This platform, also known as Bob, acts as a scaled down version of the PRISM mobile robot. Similar to its larger counterpart, it possesses a position sensor, a heading sensor, and obstacle detection sensors. It utilizes differential drive which is similar to the skid-steering of the PRISM robot. Therefore, control software such as waypoint navigation and the precise motion required by the PRISM mobile robot can be tested within the laboratory first, under controlled conditions.

The Nomadic Scout mobile robot base [Nom99a] has a drive system that is comprised of two motors driving two wheels. Each wheel motor is equipped with shaft encoders. The shaft encoders are used to approximate both the heading and the position of the mobile robot. Obstacles are detected using an array of sonar transducers. Communicating via a simple ASCII language allows outside programs to query sensors for data, receive sensor data, and control the actuators. A Sony Vaio Picturebook laptop [Son04] executes Bob's control software and drivers. A variety of sensors utilized by the field robot were also utilized when performing software tests with Bob. Bob operating in the laboratory is presented in Figure 1.

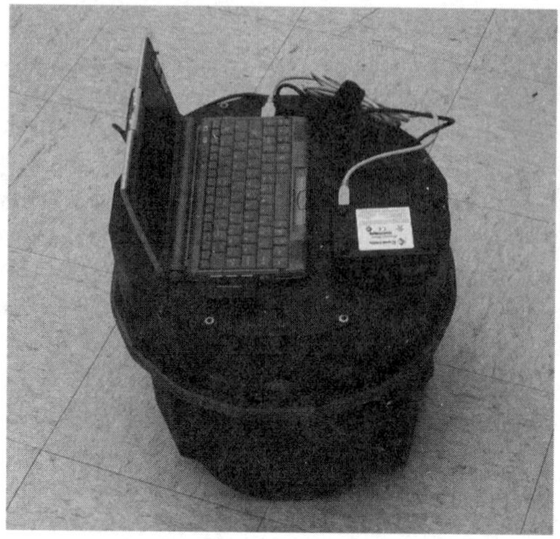

Fig. 1. Bob: a test platform for the PRISM Mobile Robot software.

Mobile Autonomous Robot with Intelligent Navigation (Marvin)

In order to meet the requirements of the PRISM project, the Mobile Autonomous Robot with Intelligent Navigation (Marvin), shown in Figure 2, was designed and built. This mobile robot is the field robot for which the software techniques discussed within this chapter are utilized (referred to as the PRISM Mobile robot, or Marvin, interchangeably).

The MaxATV Buffalo [Rec04], an amphibious six-wheeled ATV, was selected for the mobile base platform of Marvin. With tracks installed, the surface pressure of the Buffalo is greatly reduced, allowing it to better travel across snow and ice. In order to automate the platform, three components had to be placed under actuator control. The left and right brakes levers had to be controlled as they allow the turning of the vehicle. The throttle had to be controlled using an actuator. Magnetically driven linear actuators were used to automate the platform.

The sensor suite selected for the PRISM mobile robot included a Topcon Legacy-E RTK GPS system that was selected to support the robot's scientific objectives, as well as its navigation and localization. The system was composed of a fixed base station and a mobile rover system, and provided centimeter-level position accuracy [Mar02]. The LMS221 laser range finder by Sick AG was used to detect obstacles, and it was also equipped with an internal heater for operating in temperatures below 0°C [SIC98]. An inertial measurement unit provided heading information while the robot turned, as well as monitored its acceleration. It acted as three sensors in one, namely, a three-axis gyroscope, a 3-axis accelerometer, and an internal temperature sensor.

Fig. 2. Marvin operating in Greenland at the Summit Camp

2 A Software Architecture for Autonomous Mobile Robots

A software application programming interface (API) was developed for PRISM mobile robotics research. The API was designed with an emphasis on portability and robustness. As the underlying hardware of the mobile robot changes, the overall structure of the software system should remain intact and only require minor changes to adapt to the new hardware. The software libraries should also be general enough so that they could be utilized for a variety of mobile robots. In order to accomplish these goals, careful design of the software API was required to emphasize abstraction of the higher-level logic from the underlying hardware. The Java [Sun04b] programming environment was utilized to develop the system. It facilitates object-oriented design, which is necessary to provide the abstraction of higher-level control software from the lower-level hardware drivers.

In this section, the design of the software API is discussed. In general, a mobile robot is composed of actuators, sensors, and a higher-level control system that produces actuator output based on sensor input and outside control parameters. Figure 3 illustrates the general design of the software API and how it interfaces with more complex software systems.

2.1 Hardware Abstraction

Software portability was a major goal for designing the API for controlling the PRISM mobile robot. This would allow much of the software developed for one mobile platform to be usable with others. Such portability can greatly aide the development of a mobile robot for field applications. Using this approach, various aspects of the field mobile robot can be tested indoors on smaller mobile robots.

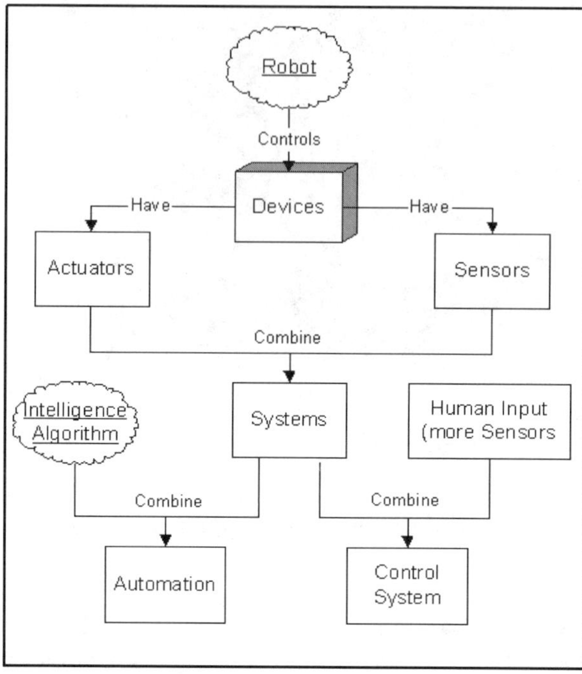

Fig. 3. The PRISM Robotics Software API.

Accomplishing hardware abstraction makes it necessary to write device drivers, or wrapper code, that implement the interfaces necessary for other software components to interact with the hardware. The robot API defines these interfaces such that the hardware can be viewed both generically as its overall purpose (sensor or actuator) and more specifically based on its actual function (e.g., bump sensor, linear actuator, etc.).

Sensors Interfaces

The *Sensor Interface* defines the most elementary representation of a sensor. This class is implemented subsequently by all other sensor interfaces. Its purpose is to define all sensors under common terms. These interfaces are implemented as device drivers or wrappers for a uniform interface for accessing data that are produced by the actual sensor hardware.

A sensor may either poll for new data, or wait for an Input/Output (I/O) event to specify that new data are available. Data from an instantiated sensor class may be accessed in one of two ways. The requesting application may query the sensor for the current data. Alternatively, the application may register itself with the sensor such that whenever the sensor data are updated, the application will receive a data update event. The Sensor interface of this API allows data access through either method.

Interfaces that extend this basic sensor interface are also defined. These new interfaces provide additional abstraction of the sensor data from the underlying sensor. The *Position Sensor Interface* defines sensors that help the mobile robot navigate over a two-dimensional plane such as a dead reckoning system, or the longitude and latitude of a GPS receiver. The *GPS Receiver Interface* extends the *Position Sensor Interface* by adding elevation and time. The *Heading Sensor Interface* defines a variety of sensors that measure the current heading or yaw orientation. The *Tilt Sensor Interface* facilitates orientation sensors that measure the vehicle's orientation in the pitch and roll (tilt) directions. The *Distance Sensor Interface* defines sensors that measure the distance in meters from nearby objects, i.e., the obstacle detector.

Several other sensors are specified by the robotics API. For each sensor type, a unique interface is created to define the means of accessing the available data. Other sensor types include: temperature, bump, weather, force, current, level, etc.

Actuator Interfaces

Actuators allow the mobile robot to act upon the world. Similar to sensors, some actuators are capable of providing feedback regarding their current state. However, often the actuators are unaware of their current state and simply perform the requested action.

For the robotics API, two basic actuator interfaces are defined, namely, the *Motor Interface* and the *Switch Interface.* Each of these extends a more general *Actuator Interface.* Other categories of actuators do exist. However, for PRISM, these two categories encompassed all of the actuator classes required by polar mobile robot. For each category of actuator, abstract classes provide a general definition of more specific actuators. Figure 4 shows the actuator hierarchy for the API.

A motor causes an object to move in the real world. The *Motor Interface* defines the most basic requirements of a motor such as the ability to tell it to stop all motion. However, to define more specific motor type, the motor type actuators were divided into three sub-types of wheel motors, servo motors, and linear motors.

Switches are capable of changing the current state of the mobile robot without the use of motion. Switches are useful for controlling circuits and the transfer of current through robotic subsystems. A binary switch has only two states, on and off, such as momentary switches, toggle switches, and relays. Linear switches control devices that have a range of input values. For instance a dimmer switch controls the rate of current flow to a light bulb over a given range.

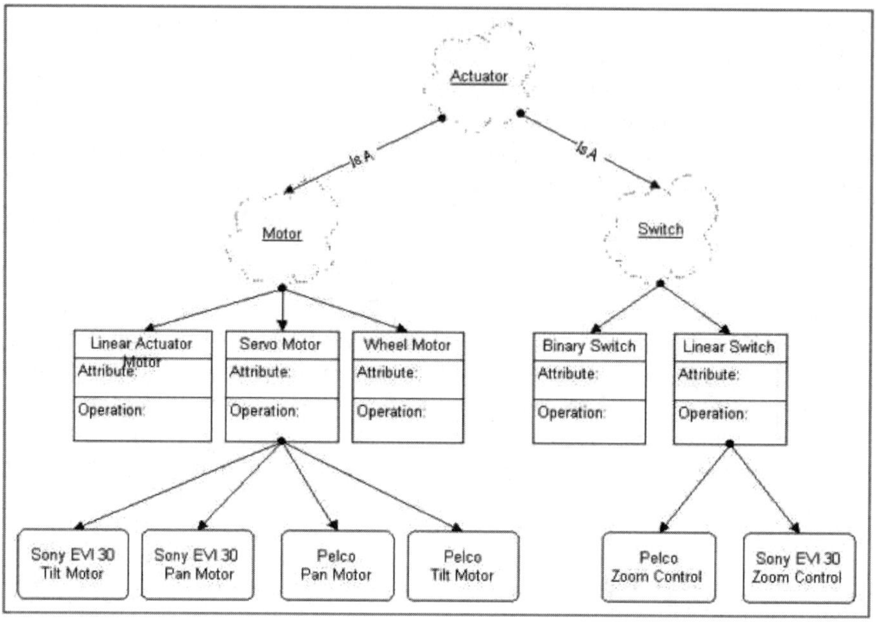

Fig. 4. PRISM Mobile Robot Actuator API.

2.2 Movement2D Interface

The *Movement2D Interface* defines an extension of the Robotics API for more intelligent movement and abstracts away from the underlying mechanism for motion. The extension to the API is shown in Figure 5. This gives a common interface to control the mobile robot regardless of the underlying hardware. The controller sets the left and right velocities for the vehicle's drive system. This defines three potential drive systems which can be utilized for mobile robots.

The *Skid Steer* defines control of vehicles that utilize skid-steering such as the PRISM field robot. In hardware, this system is made up of three linear actuators to control the left brake, the right brake, and the throttle. Given an input of left or right velocities, the linear actuators are sent control signals to define their appropriate position. The *Differential Drive* controls vehicles whose drive system is comprised of two independently driven wheel motors. For this system, the left and right velocities directly translate to wheel motor speeds. The Nomadic Scout utilizes this system. The final drive system, which was not implemented, is the *Steering Drive*. This represents drive systems in which the front wheels are steered in order to direct the heading of the vehicle such as in cars. Such a system can be implemented using a servo for steering and linear motors for throttle and brake.

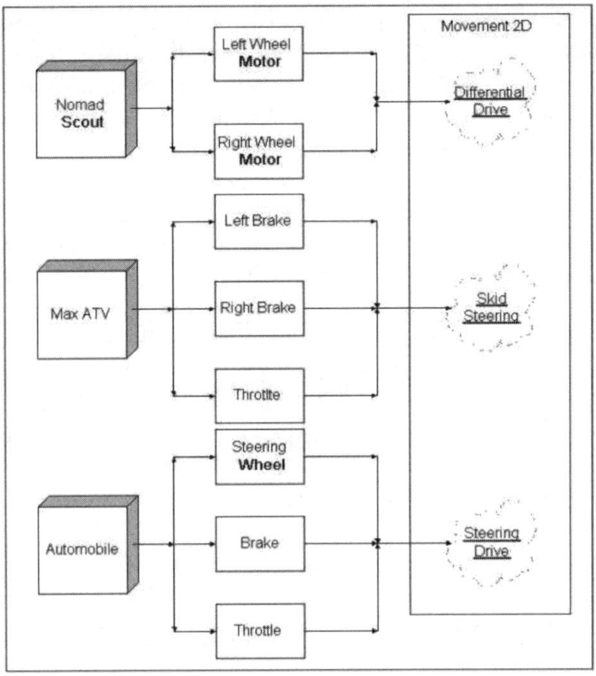

Fig. 5. Extension to the API for two-dimensional movements.

2.3 Challenges and Limitations

Using the object-oriented paradigm, the PRISM Robot API supports distributed computation, control, and sensing. The software architecture is portable and reusable, and through careful abstraction, software can be transitioned between systems without the need to re-implement the control system. However, there are some limitations to the proposed architecture. While the system is portable across platforms, it does limit the developers primarily to Java and Java-based libraries. Java Native Interfaces allows developers to access libraries written with other programming languages. However, depending upon the complexity of the external software, the development of such interfaces may be more difficult.

2.4 Other Software Architectures for Autonomous Mobile Robots

A number of research efforts have focused on software design for mobile robot systems. A common goal of software systems for autonomous robots is to create a system that can be utilized on multiple different platforms with little or no modification. The University of Florida created such a system that can be moved between platforms without modification [ACNW00]. That system divided up the various robotics tasks into more abstract modules. These included the Mobility Control Unit (MCU), Path Planner (PLN), Position System (POS), Detection and Mapping System (DMS) and a Primitive Driver (PD). This division of the robotic components helped inspire the Join Architecture for Unmanned Systems (JAUS).

JAUS, lead by the Department of Defense, seeks to define robotic control systems for a variety of physical platforms. Such platforms include Uninhabited Air Vehicles (UAVs), Uninhabited Ground Vehicles (UGVs), Uninhabited Underwater Vehicles (UUVs), etc. Software for JAUS is abstracted into a four-level hierarchy. For each layer, the software abstracts further away from the actual hardware. These are the component, node, subsystem, and system levels. The component level provides a direct interface with the hardware. The node level abstracts from the component level by hiding the details of the underlying hardware and accounting for redundancies in component level systems. The subsystems level represents software components that control nodes (e.g., a single uninhabited system). Finally, the systems level acts as a planner that controls the subsystems to achieve higher-level goals [JAU02].

The proposed software architecture shares similar goals with the Player architecture developed by the University of Southern California [GVH03, VGH03]. Unlike the PRISM robotics architecture, Player's primary focus was the control of multi-robot systems. Therefore, more concern was placed upon lowering overhead in order to increase scalability for a large groups of collaborative robots. Similar to the PRISM software architecture, Player relies upon the interface/driver model for interaction with the robot's hardware while abstracting away the underlying details of the robot's actual hardware. Similar to Unix-based systems, for each hardware driver, Player treats the device as a file that can be assigned read or write access for communication. Player also utilizes TCP sockets for communication between software modules within a system.

3 Virtual Prototyping

The PRISM mobile robot's operating conditions in Kansas are quite different than that of the polar regions in Greenland and Antarctica. Properties such as the robot's maximum turn speed, how the robot would react when it impacts an obstacle, or how the robot would handle climbing various slopes are difficult to obtain outside of the field. These properties can strongly influence the design and construction of the mobile robot. Factors such as weight distribution and center of gravity of the robot can also affect these properties. In order to design the PRISM mobile robot and test its abilities, it was first modeled and constructed using the MSC.visualNastran 4D [MSC04b] simulation package. Numerous experiments were performed using the simulated robot.

3.1 Virtual Prototyping of Mobile Platforms

Virtual prototyping involves the development of a realistic model (kinematics and/or dynamics) to represent the mechanical system that is to be developed. An environment can also be simulated within which the mechanical model functions. Certain simulation software packages allow the model to be controlled within the environment and utilize a physics engine to simulate the dynamics of the environment. Within the robotics domain, a few simulation packages also provide interfaces so that the robot can be controlled externally, and the data from within the environment can be sensed through virtual sensors. Virtual prototyping is not unique to robotics as it has been utilized in many other research and commercial domains such as mechanical engineering.

Researchers at the University of Perugia in Italy [BCO04] have focused on the analysis and the design of snowmobiles and methods to improve their design. The researchers used the MSC.ADAMS [MSC04a] software to build the model of the snow track vehicle and to test it. They presented a working model and the analysis of all the different parts of the snow track vehicle. The components that were modeled and tested included the track, suspension, frame, upper structural components (cabin and motor), auxiliary rope traction system (winch), front snow shovel, and the rear snow-crushing device.

The model and its components were tested on plane ground, rough ground, a 15-degree slope, and a 30-degree slope. The results included the histories of data such as the track force, gear angular velocity, the velocity of the vehicle, and much more. This work illustrates how different components can be modeled and tested, and the data that can become available by using such methodology.

A research project at Helsinki University of Technology [AKS00], describes the simulations used to study the load balancing and stability of arobot. Two models were used, a kinematics model and a dynamic model. The dynamic model was the same as the kinematics model, except that dynamic properties such as mass, inertia, and ground contact forces were added to the model. The kinematics model was used only for locomotion visualization and monitoring purposes. The dynamic model was used for torque and stability analysis.

Several software applications exist that explicitly support the simulation of robotic systems. As discussed in [KH04], virtual prototypes for mobile robots may be utilized for the co-design of the robot's hardware and software systems. Webots [Mic04] is a commercial application for simulating and controlling mobile robots within a virtual environment. It provides a flexible interface so that it can be controlled using programs written in C++ or Java. Two open source tools have features similar to Webots. Gazebo [KH04] generates 3Ds environment and models to represent mobile robots and their environment. It simulates a variety of sensor types such as position, ray proximity (obstacle detection), and a virtual camera. It is designed such that software can be written to control the model using the Player [GVH03] robot software architecture. A controllable simulation package was not available at the start of the PRISM project.

3.2 Modeling the PRISM Mobile Robot

Prior to the selection of a specific mobile platform, the Nastran simulation package provided the means for virtual prototyping a variety of mobile platforms to determine their performance in polar-like conditions.

The MaxATV Buffalo which was selected as the mobile platform for the PRISM mobile robot had an optional track kit. It was of interest to compare the performance of the vehicle when driven by six-wheels versus with tracks. A model of the new vehicle was constructed, with extra care given to distribute the vehicle's weight accurately among each component. The tracks were realistically modeled such that they covered the appropriate surface area on the snow and ice. The models for these configurations are shown in Fig. 6.

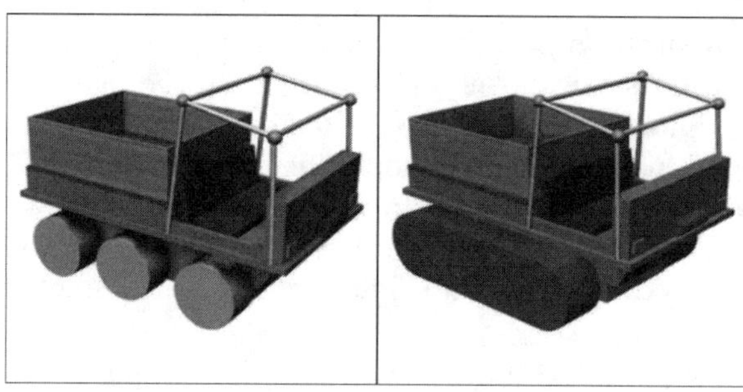

Fig. 6. Modeling the base platform with and without tracks.

In order to simulate Marvin, various robotic components had to be modeled. The rackmounts and generator were modeled as rectangular boxes. The roll cage and truck bed were removed from the existing MaxATV model as these components were replaced by a winterized enclosure. The winterized enclosure was modeled with the use of lexan for the windshield and an aluminum/plastic composite material for the sides and the roof. The simulator allowed the modification of the enclosure so that it could contain the various components onboard the robot. The configuration was modified until the best load balance and lowest center of gravity was developed given all the onboard components. The resulting model is shown in Figure 7, along with a picture of the actual robot.

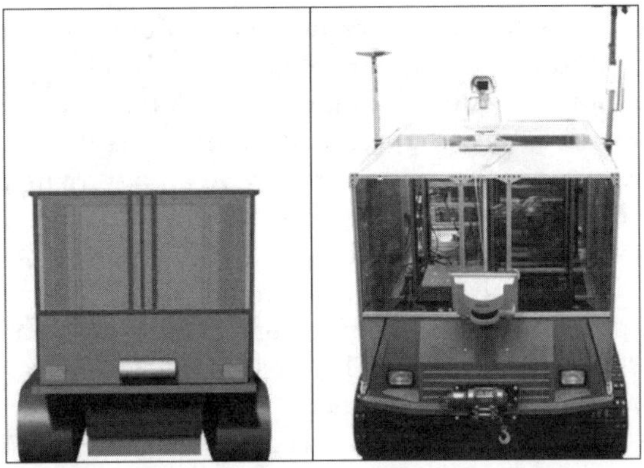

Fig. 7. Modeling the robot and its structure.

4 From Prototype to Testbed

Using the virtual prototype and software architecture, the software for the PRISM mobile robot evolved as transition between a prototype simulation, software development and testing using a testbed platform, and the final implementation of the software for the field robot. In this section, the simulation of the PRISM robot is discussed in further detail with a focus on how these results influenced the software development. Next, several robotic behaviors implemented on the testbed platform are presented. Software reuse and portability is demonstrated as the software is seamlessly evolved to control the field robot.

4.1 Simulation of PRISM Mobile Robot

As discussed previously, the simulation package used did not provide an adequate interface to create a virtual prototype that could be controlled using software for the PRISM mobile robot. Instead, initial software development was performed using the testbed platform. However, the simulation experiments produced results that influenced not only the hardware design of the robot, but also helped define several requirements of the controller.

The simulation experiments were intended to determine how well the robot performs basic tasks such as towing an antenna while turning, and what slopes the robot could climb. This information was necessary to design and build the robot, as it have to be able to traverse potentially harsh terrain. The tests estimated a number safe running parameters such as at what point the robot would tip over when driving over uneven surface. Without these tests, much more experimentation would have to be performed at the actual field site which is difficult due to the logistical limitations.

The experiments performed were designed to answer some specific questions about the performance of the robot. One objective was to decide the starting point be for the safety parameters. The term starting point is used because the model is only an approximation of the real world, therefore, the answers from the experiments should only give an approximation. The safety parameters include the maximum slope the robot could climb (the pitch angle) and the point at which it might roll over (the roll angle). Another objective was to determine how the robot would handle while performing basic movements and with different load configurations. The experiments were grouped into three series of experiments. The first series of experiments generated a baseline of how the model performed. These experiments were performed without any load on the vehicle. The second series of experiments were performed with the antennas and towing mechanisms added into the model. The antenna experiments were performed with different antennas, different towing mechanisms, and at different speeds. The third series of experiments were performed with a single antenna and weights added to the inside of the model to simulate a fully loaded vehicle.

4.2 Software Implementation for Testbed Mobile Robot

The Nomadic Scout platform is a single system treated as six individual components. It is composed of a left motor, right motor, bump sensor, position sensor, sonar sensor, and heading sensor. Input and output for all of these sensors occur through a single communication interface. Drivers were written for each of the robot's sensors and actuators. Each of the wheel motors was passed to the differential drive movement class in order to simplify the control of the mobile robot's drive system.

Some typical robot behaviors were developed using the testbed such as waypoint navigation and obstacle avoidance using a wall-following algorithm. Using the testbed, each of these behaviors could be tested within the controlled environment of the robotics laboratory without many of the risks involved with testing a much larger field robot. These algorithms were developed using class definitions for abstract robot components to facilitate the transition between robot platforms. This approach produced reusable control software for the mobile robots.

Once these behaviors were refined for the testbed, more complex behaviors were developed to meet the requirements of the PRISM project. In particular, a movement pattern was defined for the collection of radar data using the synthetic aperture radar. This movement pattern, shown in Figure 8 will be referred to as the SAR path. The SAR path directs the robot to make parallel swaths of a specified length at a fixed width apart.

The code section presented in Figure 9 configures the SAR Navigator to operate on the testbed using its suite of sensors and differential drive movement. This code will later be compared with an equivalent section of code from the field robot's controller in order to demonstrate the transition requirements

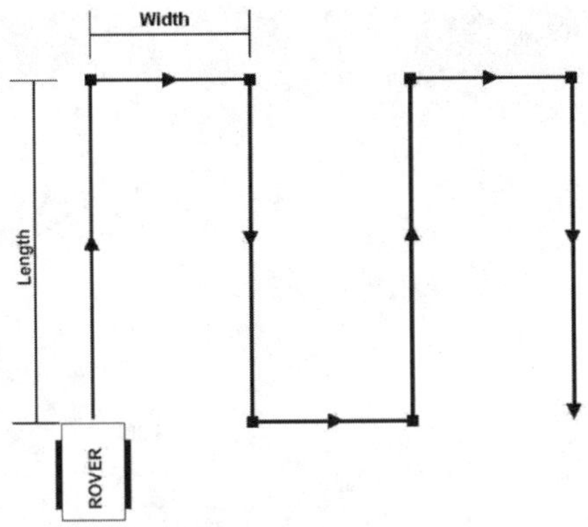

Fig. 8. The SAR path required for radar data collection.

```
From scout.java

//Connect to Nomadic Scout
NomadScoutAsDevice nc = new NomadScoutAsDevice(SCOUT_PORT_KEY,
                                               UPDATE_RATE);

//Sensors Setup
Sensor [] sensors = nc.getSensors();
PositionSensor position = (PositionSensor) sensors[nc.POSITION];
HeadingSensor heading   = (HeadingSensor)  sensors[nc.HEADING];

//Drive System Setup
Actuator [] motors  = nc.getActuators();
WheelMotor left     = (WheelMotor) motors[nc.LEFT];
WheelMotor right    = (WheelMotor) motors[nc.RIGHT];
Movement2D movement = new DifferentialDriveSystem(left, right);


//Build SAR Navigator given sensors and drive.
Navigator navigator = new SARNavigator(movement,
                                       position,
                                       heading);
```

Fig. 9. Code: Configuration of the testbed robot for SAR path navigation.

between platforms. Figure 10 shows the testbed following the SAR path within the laboratory.

Fig. 10. Approximate path followed by the testbed robot navigating the SAR path.

4.3 Software Implementation for PRISM Mobile Robot

Once software development was completed for the testbed robot, the software was ported to the real field robot. Testing and refinement of the control software then began for the new platform. However, use of the testbed did not cease; any major changes to the control software were first validated using the testbed before they were transitioned to field robot. This approach preserves the requirement of backwards compatibility between the testbed's and the field robot's control behaviors.

In order for the transition between platforms to be performed, drivers were written for each of the robot components utilized by the field robot (and not by the testbed robot). Each driver was developed independently using the software API's available driver interfaces and then tested prior to integration. With the goal of ensuring portability, drivers for the field robot's hardware were implemented such that they were backward compatible with more generic components. For instance, the Topcon GPS receiver's drivers implemented both the *GPSReceiver* interface as well as the *PositionSensor* interface such that it would operate seamlessly with the SAR path navigator.

Figure 11 illustrates the simplicity of the software transition from the testbed to the field robot, i.e., its reusability and portability. Similar to the previous code section, the SAR Navigator receives as input a movement controller, a position sensor, and a heading sensor. Analogous to the *DifferentialDriveSystem* used for the testbed robot, the *SkidSteeringDriverSystem* implements the *Movement2D* interface. Therefore, the navigator can command the robot to perform the same set of actions. Similarly, the drivers for the Top-

```
From Marvin.java,

//Sensor Configuration
PositionSensor position  = new TopconGPSReceiver(GPS_PORT);
HeadingSensor heading    = new Motionpak2(GYRO_PORT);

//Actuator Configuration
LinmotControlBox lcb4000 = new LinmotControlBox(STEERING_PORT,
                                                CONTROLLER_ID,
                                                NUM_ACTUATORS);
Motor [] motors       = lcb4000.getMotors();
Movement2D movement = new SkidSteeringDriveSystem(
                        (LinearActuatorMotor) motors[LEFT],
                        (LinearActuatorMotor) motors[RIGHT],
                        (LinearActuatorMotor) motors[THROTTLE]);

//Build SAR Navigator given sensors and drive
Navigator navigator = new SARNavigator(movement,
                                        heading,
                                        position);
```

Fig. 11. Code: Configuration of the field robot for SAR path navigation.

con GPS Receiver and the MotionPak II Gyroscope implement the *Position* and *HeadingSensor* interfaces, respectively.

Figure 12 plots the movement of the field robot along the SAR path in a field at the University of Kansas. Through slight modifications to the controller, the robot makes smoother turns toward the next waypoint in order to protect the radar array that is being towed.

By utilizing the PRISM robotics API, software reuse and platform independence emerges. This is demonstrated by porting the software from a testbed robot to a field robot. Not all of the software could be ported, but through abstraction, all higher-level control is reusable. Only drivers for the platform dependent hardware must be re-implemented with every platform change.

4.4 Testbeds for Mobile Robot Design and Testing

A common practice for mobile robot design and development is the utilization of a testbed system so that the designers can focus on the development of software and the control system on a stable mobile platform. Performing experiments to test the software becomes easier and more efficient because the tests can be performed in a controlled environment, thus eliminating limitations such as resource availability, space, and logistics. Many theoretical concepts may be discovered and revised prior to being incorporated into the production system.

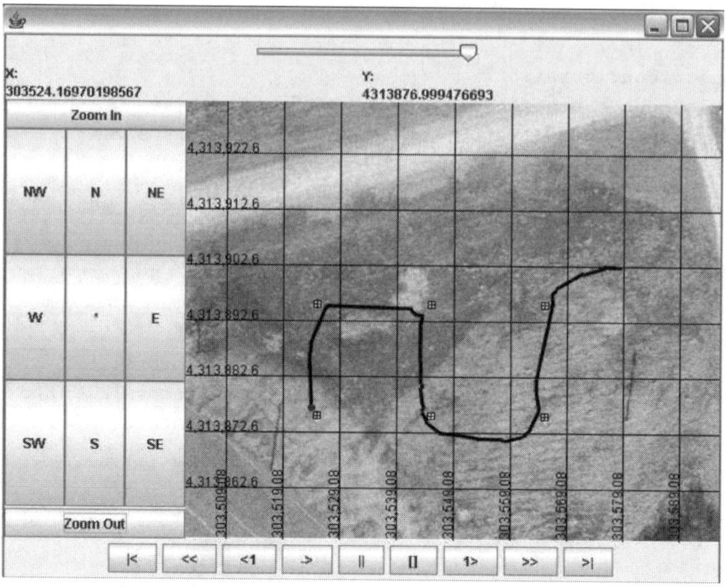

Fig. 12. The precise path followed by the field robot while performing SAR path navigation.

Carnegie Mellon University's (CMU) Robotics Institute has conducted research in the area of autonomous land vehicles. The goal of this research was to develop a mobile robot that was capable of navigating along sidewalks, or along curved roads. To accomplish these tasks, two mobile platforms were employed. The Terregetor mobile robot acted as a testbed [GS87]. It was a custom-built mobile robot platform small enough to safely travel along sidewalks. Its larger counterpart was NAVLAB, an automated delivery truck. This is an example of an application for which testing near the university campus was much easier and safer whenever a testbed mobile robot was utilized, instead of the full sized system.

As part of DARPA's Tactical Mobile Robotics Program, researchers at the Georgia Institute of Technology have worked to develop a mobile robot capable of urban warfare and reconnaissance applications. Their software system is developed so that it may execute on multiple platforms [ACE99]. One such platform is the Pioneer AT developed by ActivMedia Robotics [Act04]. This testbed is designed for both indoor and outdoor robotic applications and is capable of transporting several mobile robot sensors.

The utilization of testbed systems is also common on much larger projects such as the Mars rovers. These testbed systems much more closely resemble the rovers that travel to Mars. However, they provide stable platforms for which new hardware and software components can be tested and integrated. For instance, the Athena Software Development Model (SDM) rover at NASA's Jet Propulsion Laboratory represents prototypes for the planetary rovers, Spirit and Opportunity, that successfully traversed the Martian surface [SHP03, BMM01].

5 Conclusion

Development of autonomous mobile robots for field applications presents many unique challenges. These challenges are compounded for robots operating in harsh environments such as polar regions. Testing the mobile robot within these environments can be both difficult and costly.

In order to meet these challenges, a software development methodology for mobile robots has been presented that results in portable and reusable software for such robots. A virtual prototype is used to simulate the mobile robot to develop various design parameters for the software system and the hardware system. Some simulators provide virtual prototypes with interfaces such that they can also be used for initial software development. Testbed platforms such as small indoor (or outdoor) mobile robots are then used to develop the robot's higher-level behaviors such as obstacle avoidance and waypoint navigation. Finally, these behaviors are transitioned to the field robot and tested both locally and in the field.

The robotics API was developed as a platform independent architecture for mobile robots. The API can be expanded for other robotic endeavors and othersensor/actuator modalities for robots. Robot control software was developed using the API to produce autonomous mobile robots for the PRISM project. It was first developed and tested using a testbed mobile robot, and the software was then transitioned to control the field robot.

During the summer field seasons of 2004 and 2005, field experiments were conducted at Greenland's North GRIP and Summit camps, respectively. The first field experiments provided the proof of concept that the virtual prototype had successfully produced a mobile robot capable of operating within the polar environment. During the second field experiments, the mobile robot was automated using the robotics API. The robot was integrated with the SAR radar equipment and autonomous data collection tasks were performed. Using the proposed methodology, a viable mobile robot for field applications was produced.

This research is not without some limitations. Improvements could be made to both the simulation and the implementation of the software architecture. The simulator was capable of modeling the robot and general properties of the environment, but is not yet sufficient to simulate more specific hazards of the environment. A simulation environment with additional features would also include hazards such as rough terrain. Virtual prototypes may also be used for initial development of software for robot control and autonomy.

Similar to other software development endeavors, the implementation of the software architecture can always be expanded. The robotics API can be augmented to include additional sensor and actuator modalities. The architecture includes components (sensors and actuators) that were encountered and utilized within the PRISM robotics laboratory. As the software architecture grows and matures, it is envisioned that more formal techniques may be applied to the design of software for mobile robots.

Acknowledgements
The work described in this chapter was supported by the National Science Foundation (grant #OPP-0122520), the National Aeronautics and Space Administration (grants #NAG5-12659 and NAG5-12980), the Kansas Technology Enterprise Corporation, and the University of Kansas.

References

[ACE99] Ronald C. Arkin, Thomas R. Collins, and Yoichiro Endo, *Tactical mobile robot mission specification and execution*, Mobile Robots XIV **3838** (1999), 150–163.

[ACNW00] David G. Armstrong II, Carl D. Crane III, David Novick, and Jeffrey Wit, *A modular, scalable, architecture for unmanned vehicles*, Association for Unmanned Vehicle Systems International (AUVSI) Unmanned Systems 2000 Conference (Orlando, Florida), July 2000.

[Act04] Active Media Robotics, *Robots and Robot Accesories*, URL: http://www.activrobots.com, 2004.

[AHSA04] Eric L. Akers, Hans P. Harmon, Richard S. Stansbury, and Arvin Agah, *Design, fabrication, and evaluation of a mobile robot for polar environments*, IEEE International Geoscience and Remote Sensing Symposium (IGARSS 2004) (Anchorage, Alaska), vol. I, September 2004, pp. 109–112.

[AKS00] P. Aarnio, K. Koskinen, and S. Salmi, *Simulation of the hybtor robot*, 3rd International Conference on Climbing and Walking Robots (Madrid, Spain), October 2000, pp. 267–274.

[BBD00] Greg Bollella, Ben Brosgol, Petter Dibble, Steve Furr, James Gosling, David Hardin, Mark Turnbull, and Rudy Belliardi, *The real-time specification for java*, Addison-Wesley, 2000.

[BCO04] C. Braccesi, F. Cianetti, and F. Ortaggi, *Modeling of a snow track vehicle*, Institute of Energetics, Faculty of Engineering, University of Perugia, Italy, 2004.

[BEI01] BEI Systron Donner, *Bei motionpak ii: Multi-axis inertial sensing system*, BEI Systron Donner Inertial Division, Concord, 2001.

[BMM01] Jeffrey Biesiadecki, Mark Maimone, and Jack Morrison, *The athena sdm rover: A testbed for mars rover mobility*, 6th Intl. Symp. on AI, Robotics and Automation in Space (Montreal, Canada), no. AM026, June 2001.

[Bot04] Per Bothner, *Kawa, the java-based scheme system*, URL: http://www.gnu.org/software/kawa/, 2004.

[CSB01] F. Carsey, P. Schenker, J. Blamont, S. Gogineni, Kenneth Jezek, Chris Rapley, and William Whittaker, *Autonomous trans-antarctic expeditions: an initiative for advancing planetary mobility system technology while addressing earth science objectives in antartica*, Sixth International Symposium on Artificial Intelligence, Robotics and Autonomy in Space (Montreal, Canada), no. AM017, June 2001.

[FH04] Ernest Friedman-Hill, *Jess, the rule engine for the java platform*, URL: http://herzberg.ca.sandia.gov/jess/, 2004.

[GS87] Y. Goto and Anthony (Tony) Stentz, *Mobile robot navigation: The cmu system*, IEEE Expert **2** (1987), no. 4, 44 – 55.

[GVH03] Brian P. Gerkey, Richard T. Vaughan, and Andrew Howard, *The player/stage project: Tools for multi-robot and distributed sensor systems*, Proceedings of the International Conference on Advanced Robotics (ICAR 2003) (Coimbra, Portugal), July 2003, pp. 317–23.

[Han04] Chris Hanson, *The scheme programming language*, http://www.swiss.ai.mit.edu/projects/scheme/, 2004.

[HSAA04] Hans P. Harmon, Richard S. Stansbury, Eric L. Akers, and Arvin Agah, *Sensing and actuation for a polar mobile robot*, International Conference on Computing, Communications and Control Technologies (Austin, Texas), vol. IV, August 2004, pp. 371–376.

[JAU02] JAUS Work Group, *Joint architecture for unmanned systems*, United States Department of Defense, September 2002.

[KH04] Nathan Koenig and Andrew Howard, *On device abstractions for portable, reusable robot code*, Proceedings of the IEEE/RSJ International Conference on Intelligent Robots and Systems (IROS 2004) (Sendai, Japan), September 28 - October2 2004, pp. 2149–2154.

[KU 04] KU PRISM Team, *Prism home: Polar radar for ice sheet measurements*, URL: http://www.ku-prism.org, 2004.

[Lai99] Sheng Laing, *The java native interface: Programmer's guide and specification*, Addison-Wesley, Reading, MA, 1999.

[Lin04] Linmot Inc., *Linmot data book 2003-2004*, URL: http://www.linmot.com/datasheets/LinMot Data Book 2003-2004.pdf, 2004.

[Mah01] Qusay H. Mahmoud, Sun Microsystems: http://http://java.sun.com/developer/technicalArticles/ALT/sockets/, 2001.

[Mar02] Jim Martin, *Personal communications. griner and schmitz, kansas city, mo*, June 2002.

[Mic04] Olivier Michel, *Cyberbotics ltd - webotstm: Professional mobile robot simulation*, International Journal of Advanced Robotic Systems **1** (2004), no. 9, 39–42.

[MSC04a] MSC Software, *Msc.adams*, URL: http://www.mscsoftware.com/ products/products detail.cfm?PI=413, 2004.

[MSC04b] MSC Software, *Msc.visualnastran desktop tutorial guide*, URL: http://www.mscsoftware.com, 2004.

[Nie04] Douglas Niehaus, *Kurt-linux: Kansas university real-time linux*, URL: http://www.ittc.ku.edu/kurt, 2004.

[Nom99a] Nomadic Technologies Inc., *Nomadic scout user's manual*, URL: http://nomadic.sourceforge.net/production/scout/scout_user-1.3.pdf, 1999.

[Nom99b] Nomadic Technologies Inc., *Nomadic scout user's manual*, URL: http://nomadic.sourceforge.net/production/scout/scout_langman-1.3.pdf, 1999.

[Pre04] Precise Navigation Inc, *Tcm2-50 feature sheet*, URL: ftp://www.pnicorp.com/technical-information/pdf/TCM2-50productsheet.pdf, 2004.

[Rec04] Recreative Industries, *Buffalo all terrain truck*, URL: http://www.maxatvs.com/buffalo-interior.htm, 2004.

[Ril04] Gary Riley, *Clips: A tool for building expert systems*, URL: http://www.ghg.net/clips/CLIPS.html, 2004.

[SAHA04] Richard S. Stansbury, Eric L. Akers, Hans P. Harmon, and Arvin Agah, *Survivability, mobility, and functionality*, International Journal of Control, Automation, and Systems **2** (2004), no. 3, 334–353.

[Sel03] Denish Selvarajan, *Implementation of real-time java using kurt*, Master's thesis, University of Kansas, Lawrence, KS, 2003.

[SHP03] P.S. Schenker, T. L. Huntsberger, P. Pirjanian, E. T. Baumgartner, and E. Tunstel, *Planetary rover developments supporting mars exploration, sample return and future human-robotic colonization*, Autonomous Robots **14** (2003), 103–126.

[SIC98] SICK, *Laser management systems: Technical description*, SICK AG, Germany, 1998.

[SIC04] SICK, *Collision prevention: Active area monitoring with lms laser scanner*, URL: http://www.sick.dk/lasscan/picolli.pdf, 2004.

[Son04] Sony, *Using your sony viao picturebook computer*, URL: http://www.docs.sony.com/release/PCGC1MV.PDF, 2004.

[SS96] Leon Sterling and Ehud Shapiro, *The art of prolog*, 2nd ed., MIT Press, Cambridge, MA, 1996.

[Sun04a] Sun Microsystems, Inc., *Java remote method invocation*, URL: http://java.sun.com/products/jdk/rmi/, 2004.

[Sun04b] Sun Microsystems, Inc., *Java technology*, URL: http://java.sun.com, 2004.

[TOHC01] A. Trebi-Ollennu, T. Huntsberger, Yang Cheng, E.T. Baumgartner, B. Kennedy, and P. Schenker, *Design and analysis of a sun sensor for planetary rover absolute heading detection*, IEEE Transactions on Robotics and Automation **17** (2001), no. 6, 939–947.

[VBSH00] R. Volpe, E. Baumgartner, P. Schenker, and S. Hayati, *Technology development and testing for enhanced mars rover sample return operations*, Proceedings of the 2000 IEEE Aerospace Conference (Big Sky, MT), March 2000, pp. 247 – 257.

[VGH03] Richard T. Vaughan, Brian P. Gerkey, and Andrew Howard, *On device abstractions for portable, reusable robot code*, Proceedings of the IEEE/RSJ International Conference on Intelligent Robots and Systems (IROS 2003) (Las Vegas, Nevada), October 2003, pp. 2421–2427.

[VGH06] Richard T. Vaughan, Brian P. Gerkey, and Andrew Howard, *Reusable Robot Software and the Player/Stage Project*, In Brugali D. (Ed.) Software Engineering for Experimental Robotics, Springer STAR series, 2006.

[BCW02] A. Behar, F. Carsey, and B. Wilcox, *Polar Traverse Rover Development for Mars, Europa, and Earth*, Proceedings of the IEEE Aerospace Conference (Big Sky, MT), MArch 2002, pp. 389–396.

Writing Code in the Field: Implications for Robot Software Development

William D. Smart

Department of Computer Science and Engineering
Washington University in St. Louis, United States wds@cse.wustl.edu

1 Introduction

Running robots on a deployment in the "real world" is very different from running them in a laboratory setting. Environmental conditions, such as lighting, are often completely beyond your control. There is often a limited amount of time on-site to test and tune the robot system, and deadlines are generally very inflexible. Finally, the public is notoriously unforgiving of failure. In this chapter, we identify some key design issues that can affect real world robot deployments. We discuss these issues and propose some solutions, based on our own experiences, designed to ease the pain of sending a robot out into the real world.

1.1 The Deployment Scenario

In this chapter we will consider a particular type of real world deployment, characterized by six main features:

Environment: The deployment takes place in an environment where we do not have complete control over the environmental conditions, such as the lighting level. We also assume that we cannot completely replicate the environmental conditions in the laboratory before arriving at the deployment.
We cannot alter the environment in any substantial way for the deployment. This includes removing furniture and obstacles that the robot has difficulty with, adding brightly-colored navigation aids, and forbidding the public from areas of the environment.

Schedule: The robot is not the most important thing in the environment, and existing schedules cannot be altered to accommodate our deployment. For instance, the opening time of a building cannot be delayed if the robot is not ready on time.

D. Brugali (Ed.): Software Eng. for Experimental Robotics. STAR 30, pp. 93–105, 2007.

We often have a limited amount of time on-site to make sure that the robot behaves as expected, and to tune it to the specifics of the environment. This limitation generally comes from the policies of the deployment site which may, for example, restrict after-hours access to the venue.

Task: The robot is performing some autonomous task during the deployment, and is not just a static exhibit. It is expected to interact with the environment, and possibly the general public, in an "intelligent" and purposeful manner.

Duty Cycle: The robot must perform its task for a set period of time, often several hours, without intervention. All maintenance and repairs must happen outside of this time period (after the building has closed, for example).

The Public: The public will be observing the robot as it performs its task, and might even be interacting with it directly. We cannot assume that the public knows anything about robots, computer science, algorithms, or the like. Although human "interpreters" can be used to explain the robot's behavior, they should be used sparingly.

Reputation: The public has a low tolerance of failure, and will associate poor performance with our lab, institution, and students. Failures in the laboratory, in front of a knowledgeable audience, can be (somewhat) explained away at a technical level. No such easy escape exists in the real world. If the robot seems broken, or is acting erratically, the public will lose interest and move on. More importantly, they will remember that the robot didn't work as expected, and will tell their friends.

As a concrete example, consider the deployment of a mobile robot in a science museum. The museum has fixed hours, which we have no control over. The robot *must* be ready when the doors open at 9am, and we cannot perform any obvious maintenance until the museum closes at 5pm. The robot is likely to be one of a large number of exhibits. If someone comes by while the robot is not working, they will simply move on to the next exhibit. This means that it is important to keep the system working for the entire duty cycle. This is especially true if it has been advertised, since people might have made a special trip to see the robot in operation.

It is unlikely that we can replicate the conditions of a typical science museum in our lab, although we might be able to approximate some of them. This means that we will need to spend time tuning our system to operate in the museum. The amount of time we can spend doing this will be limited by museum policies. It also must happen at an inconvenient time, in the evenings and very early mornings.

1.2 Typical Deployment Problems

For any non-trivial deployment of the type described above, we will have to adapt our systems on-site. This adaptation will involve the tuning of software parameters to account for environmental conditions, and will usually also include writing new code to deal with unforeseen circumstances. Often, several subsystems must be tuned, which leads to contention a single shared resource: the robot.

It is a fact of life that hardware and software will fail. This is especially true on deployment. Software written on-site, under time pressure, will not be of the same quality as that developed over time in the laboratory. This means that it will fail more often, often catastrophically. Simply shipping the robot to the deployment site can increase the likelihood of hardware problems. Components and connectors can be shaken loose and gears can be knocked out of alignment. Even the most careful pre-deployment inspection cannot be guaranteed to catch all hardware problems.

We must accept that things will fail, and try to manage this failure so as to minimize its effect on the deployment. We must ensure that any failures that do occur result in a graceful performance degradation, rather than stopping the system dead. The robot should continue to operate, perhaps at a reduced capacity, until the end of the duty cycle, when we can perform the needed maintenance.

1.3 Software Architectures for Deployments

We have deployed our mobile robot in a number of real world situations, and observed many instances of the problems outlined above [SDM03, BDG03, BDSG04]. This has led us to modify our software architecture [MS04] and development process to ease the deployment experience. Our modifications have been aimed at two things: reducing software development time on-site and increasing robustness to the inevitable hardware and software failures that occur. In this chapter, we describe some of the problems encountered on deployment, and the modifications that they caused us to make in our control architecture.

2 Reducing Software Development Time On-Site

Performing some amount of software development in the field, while on a deployment, is almost inevitable. Thus, we need to make the on-site development process as efficient as possible. Although we should always strive for efficient software development, there are some additional pressures on a deployment, and often severe penalties for missing deadlines. For example, consider a robot that is part of a larger exhibition in a science museum. The museum has a hard deadline for opening in the morning, and will not postpone it if the robot is not ready. If the public arrives to see the robot surrounded by puzzled-looking students, there is a real price to pay. Our public reputation, and that of our institution, can suffer. Perhaps more importantly, our credibility with the museum will be hurt, which can affect plans for subsequent deployments.

It is vital to have an effective development and testing process in place, well before we arrive at the deployment. Although we cannot eliminate on-site tweaking of code, we can certainly minimize it by planning ahead. Arriving at a deployment without a full, working application is simply asking for trouble. Time in the field should be spent adapting a system, not implementing it. Even if we cannot replicate the conditions of the deployment in the laboratory, we can, at least, come armed with a system that works under *some* conditions.

While it seems obvious to say that good software engineering practices are useful, this is often forgotten in a university or research lab environment. A good deal of the code written in such an environment can be considered to be proof-of-concept code, designed to test a theory and then to be discarded. Code is often modified incrementally, without a formal specification, leading to undocumented behavior. We can often get away with this for internal deployments, since the person who wrote the code is probably there to answer questions. However, this can cause severe problems while on an external deployment.

All code that is to be used on a deployment should be maintained under a source code control system, such as CVS. When a new version of some part is written, unit and integration tests should be carried out before it is checked into the official code base. Documentation and test code should also be checked in at the same time as the main code.

We have found that it is extremely useful to have a student act as the configuration manager for the robot. While any student can check code out of the repository, only the configuration manager (CM) may check in new versions. This allows us to easily make sure that the other students do not "forget" to include documentation and test routines with their new code.

The role of configuration manager is doubly important for a robot, since hardware can also be changed. Just as no-one is allowed to add to the code base without the permission of the CM, no-one can alter the hardware configuration. The removal or addition of hardware can cause many problems, and a huge amount of frustration. We have had a student spend several hours trying to debug the speech synthesis software on our robot, only to find that someone had removed the sound card without telling anyone. While one could argue that the hardware should be the first thing that we check, we tend to make assumptions about the sources of failure. Computer science students tend to assume that the software is at fault.

Adding hardware to a robot can be just as bad. Everything on a robot is powered from a shared set of batteries. If we have too many devices, there may not be enough power to go around, leading to strange hardware behavior. We have observed this problem first-hand (with another team) at a major robot competition. A student plugged in a new camera system just before the start of the run. It drew enough power from the batteries to cause the laser rangefinder to reset, and stop returning values. Since the obstacle-avoidance system relied on the laser, the robot was essentially blind, and drive straight into the first table it encountered. A large group had gathered to watch the robot, making it a high-visibility failure.

Perhaps worse than the failure itself was the time needed to debug the problem. All of the symptoms were of a faulty laser range-finder. However, this device was fine, as all testing in isolation confirmed. It is a testament to the experience of the deployment team that the problem was found at all.

Suggestion 1
Designate someone as the Configuration Manager (CM). No code or hardware is changed on the robot without the explicit approval of the CM. The CM is also responsible for enforcing all coding standards and procedures.

Most of time on-site is spent tuning parameters in the robot software, to adapt it to environmental conditions. We will begin by discussing how this process can be optimized. We will then cover how to minimize the effects of any on-site software development that may be necessary. In all of the following, we assume that we have a single robot (or a small number of robots) and multiple developers, each with a computer. The robot is a shared resource, and one of our primary goals is to reduce the contention for this resource.

2.1 Tuning Parameters

The most obvious way to reduce the time spent tuning parameters is to have fewer parameters to tune. Although it is not always possible to remove parameters from an algorithm, it might be possible to set reasonable default values that are "close enough". These default values might be set according to some context given by the developer. For example, there might be one set of vision algorithm defaults for a bright environment with direct sunlight, and another for a window-less room with artificial lighting. The parameters can be properly tuned if there is sufficient time, but the default values will prevent catastrophic failure if we run out of time. If we do tune the parameters for a specific algorithm, the settings should be saved, to be reused in similar environments in the future.

For some parameters, there may be no good default value. In this case, a procedure for determining a good setting should be included in the documentation. If this procedure is straightforward and algorithmic, it means that the tuning can be done by any member of the deployment team, not just those with an intimate knowledge of the algorithm.

Suggestion 2
Reduce the number of parameters to tune, and provide defaults where possible. Provide guidelines for determining a good setting for parameters without a default.

If it is reasonable to have parameters optimized automatically, then this is almost certainly the best approach. Even if this optimization is not complete, it might provide a better starting point for a human programmer. We make extensive use of this approach with the vision system on our robot. We calibrate the stereo vision system by taking a picture of a known target (a checkerboard pattern), and running code to estimate the camera parameters. The stereo vision calibration is not entirely automatic, and requires some human input. However, this requires no skill on the part of the human, and can be done by any of the deployment team. We can also correct for local lighting conditions by taking a picture of a target with a (known) variety of colors.

Suggestion 3
Write code to automatically tune as many parameters as possible. If human input is required for any of these calibrations, have explicit instructions that do not assume any knowledge of the underlying algorithm.

Parameters settings should not be defined in the code. If they are, then every parameter setting change requires that the code be re-compiled, which will cost valuable time (see below for more details). Each subsystem should save its parameter settings in a configuration file, in some human-editable form, such as XML. If the robot has been successfully deployed in a similar setting before, all of the parameter settings for that deployment should be saved in a single place. This will potentially reduce the amount of time spent tuning parameters in similar settings, by providing a better starting point.

In our architecture, parameters are stored in configuration files, but can also optionally be provided by a configuration server. This service, based on a similar system in CARMEN [TMR06] allows us to change parameter values while the robot is running, using a graphical interface. This lets us see the effects of a parameter setting immediately, without having to stop and start the code. This saves time, and we have found that it is often easier to set some parameters, like how close to come to obstacles, interactively. Additionally, being able to change parameter values while the robot is running allows us to recover from some failures transparently (see section).

Suggestion 4
Save parameter settings in a configuration file, in some human-editable format, such as XML. Provide a mechanism for modifying the parameter values while the robot is running.

It is painful to watch programmers waiting in line to be able to use the robot, while someone tunes *their* set of parameters. If we design our applications as monolithic systems, parameter changes to one sub-system can affect other sub-systems. Even if they do not interact at run-time, if we have a monolithic controller, we cannot run two copies of it at the same time, meaning that only one programmer can be occupied.

If, however, the architecture is modular, and allows sub-systems to be run independently, we can parallelize the process of calibration. The vision programmer can tune the color tracking model, while the sensor programmer calibrates the laser range-finder. Better yet is the ability to run sub-systems on a workstation that is not directly connected to the robot, from the command line, using data logged to files. This allows, for example, the vision programmer to capture a sequence of images, then use these to tune the color tracking parameters on a workstation while the robot is being used for other tasks.

To address this, our software architecture is composed from many small, independent modules. Each of these modules is a single process (in a Linux environment), and is designed to be run as a service (the basic compositional unit of our architecture), a pipe element (in the Unix sense, reading from the standard input, writing to the standard output), and a command-line program (reading from and writing to files), depending on the arguments used. This allows us to use the same code for each mode of operation, ensuring that tuned parameter settings for one mode will be valid in another.

This arrangement allows us to quickly prototype prototype simple sequences of processing steps on the command-line. A saved image can be piped into the color blob finder, with the output being saved to a file. Looking at this file allows us to quickly debug problems with the blob finder. Once it is working as expected, we can pipe the output into the face detector, which again outputs to a file that can be examined, and so on.

Suggestion 5
Provide as much modularization as possible in the system. Allow sub-systems to be run independently of each other, and to deal with input and output from a variety of sources.

2.2 Writing New Code

In addition to reducing the time it takes to tune parameters, we must reduce the time it takes to write new code while in the field. Regardless of how well-prepared we are, it is almost inevitable that we will have to write *some* new code.

Instead of writing completely new code in the field, we should aim to compose existing (compiled) software into new applications. This achieves two important goals: it avoids introducing bugs due to hastily-written code, and it avoids having to recompile code. Software written under (often extreme) time-pressure will almost certainly be less reliable than that written and tested in a more controlled environment. Re-compiling code (often many times, as incremental changes are made and tested) is not only time-consuming, but increases the stress level of the programmers. As time becomes short, staring at a long compile is a harrowing thing.

If we are forced to re-compile code, we want to compile (and link) as little as possible. Although a build management system will help with this, we can reduce the size of the compile even further if our system is composed of small, independent modules.

Our approach is to have a large number of small, independent modules. Each module provides a small number of interfaces, over which information flows. Typically, each interface (such as "distance") can be provided by more than one service (laser range-finder, sonar range-funder, stereo vision, *etc.*). Applications request interfaces from a broker agent, which selects appropriate services to provide the interface (which may request interfaces themselves, and so on). There is considerable flexibility in the selection of actual services, which means we can adapt to different environments quickly, by selecting appropriate ones. For instance, we have several services that supply the "obstacle-avoider" interface, each of which use a different algorithm. Depending on the specifics of the deployment, we can easily select the most appropriate algorithm, without having to recompile any code. We can also be sure that it will work with the rest of the system, because all interactions with it are performed through the generic "obstacle-avoider" interface.

Suggestion 6
Build applications by composing small, independent, previously written (and tested) modules.

3 Guarding Against Hardware and Software Failures

Despite our best efforts, hardware and software will fail in a variety of ways. This is especially true when we are running the robot out of the laboratory. Transporting the robot to the deployment site might have caused some hardware components to shake loose. Software written during the deployment is not going to be as reliable as that written and tested in the laboratory. Although we should always try to avoid hardware and software failures, we must also recognize that we cannot realistically eliminate them all. This means that we must plan for these failures, and try to minimize their impact on the deployment.

The most basic safeguard is to have a complete backup of the *entire* baseline system. If the hard drive in the robot fails, we need to be able to replace it with a new one, and restore the baseline system as quickly as possible. On deployments, we actually have two identical disks, so that we can restore the system by physically installing the backup. Any changes to the baseline system made on-site are under source code control on a remote workstation, and can be applied to the baseline system in the case of failure.

Suggestion 7
Have an up-to-date backup of everything that you can. Be able to restore the software to its baseline state quickly in the case of failure. Commit on-site changes to a source code control system, so that they can be easily applied to the baseline system.

When running robot applications in the laboratory, we want to know about problems with the hardware and software immediately. When they occur, we stop the robot, diagnose and fix the problem, and restart the application. However, when we are on deployment in the real world, our goals are different. In this case, if some subsystem fails we want the robot as a whole to continue operating as well as possible until the end of the current duty cycle. At that time, we can intervene to fix the problem. Ideally, members of the public watching the robot will never even be aware that something failed.

Our robots must show graceful performance degradation in the face of hardware and software failures. This means that we must identify failures, and switch to some backup system quickly enough that the overall behavior of the robot is not noticeably affected. Moreover, this failure tolerance should be built into the software architecture (as much as is possible), so that application developers do not have to consciously think about rare failure conditions while writing their code.

To sequence applications at a higher level, we have a scripting language that describes augmented finite state machines. This lets us quickly assemble high-level applications by composing lower-level behaviors. It removes the need to re-compile these low-level behaviors, and also gives us strong encapsulation, and crash protection, which we discuss in the next section.

A key element of our approach is to provide strong encapsulation for subsystems. The failure of a subsystem should not cause other systems to fail. One way of accomplishing this is to use the process encapsulation mechanisms of the underlying operating system (Linux, in our case), and make each subsystem an independent process. If one process fails catastrophically, the others will be insulated from this failure. This is especially relevant on deployment, where we might be running code that was hastily-written, and not well tested.

Suggestion 8
Provide strong, operating system level encapsulation between subsystems where possible.

Our approach to adding robustness to failures draws heavily on ideas from Recovery Oriented Computing (ROC) [PBB02], and in particular from the idea of micro-reboots [CF01]. If a subsystem fails, restarting it will often cause it to work again, for a while at least. If this reboot is quick, the overall performance of the system will not suffer. We extend this idea by allowing new services (that provide the same interface) to be started in place of the one that failed.

Even code that has been well-tested cannot be guaranteed to be free of bugs. However, if a piece of software fails, it can often be successfully restarted. This is a reasonable strategy for keeping a robot active during a deployment, since software becomes more unreliable the longer it runs (due to slow memory leaks, and the like). We can often make quite strong empirical claims about the short-term behavior of software. For example, we can be confident that a piece of code will run for one hour if we repeatedly run it for one hour without it failing. We cannot do so well if we want to claim that it runs for a whole day. However, if we can seamlessly restart the program every hour (or when it fails), we can approach this stronger claim. To make this possible, individual modules in the software architecture should have as little internal state as possible. If we must have internal state, it should be stored in some way that allows it to persist through a process restart (in shared memory, or stored in a file, for example).

Suggestion 9
Have as little internal state as possible in a process. If the subsystem absolutely requires internal state, store in a persistent way, that can survive a restart.

The broker agent in our architecture monitors every service (process) in the system. Those that fail catastrophically are restarted up to a certain number of times. If a service fails too often, it is removed and replaced by another service that provides the same interfaces. If no such interface is found, then the system enters a failsafe mode, and requests human assistance.

Each process also has a watchdog timer that is monitored by the broker agent. If a process seems to be inactive, it is restarted or replaced. The broker can also optionally observe the output from sensor services, trying to identify failure. By their nature, sensor readings have measurement error in them. This means that a healthy sensor will rarely return exactly the same value every time it is read. If a sensor does return the same value every time, this is an indication that the hardware has failed, and that the service should be shut down.

Suggestion 10

Provide automatic monitoring of processes and (where possible) hardware systems to automatically detect and correct failures.

We can perform this replacement without notifying the application because of the abstractions used in the interfaces. For example, to determine the location of objects in the world, an application might request the "distance" interface, with a preference for data generated by a laser range-finder. The broker agent performs the steps necessary to connect the application to the laser range-finder service, and all is well. If the laser fails, however, another service that supplies the "distance" interface, such as the sonar range-finder, can be substituted in, without the application being aware. The data generated by the sonar range-finder will, of course, be different from that generated by the laser. However, the hope is that it is "good enough" to keep the application from failing catastrophically.

There is a price to be paid for using such abstractions. We can no longer ask for data from a particular sensor, such as the laser range-finder. This also means that we cannot make assumptions about, for instance, the failure modes of the data being supplied. Our applications must be written in more generic terms, requesting the specific characteristics of sensors before they are used, rather than making the assumptions that are common in current robot software. This makes writing software more time-consuming, in our experience. However, we believe that the level of added robustness is well worth it. We can, of course, still directly request laser range-finder data, but by doing this we lose any graceful degradation. If we require the laser to be working, and it breaks, then our application must deal with this failure in some intelligent manner.

Abstractions should not be limited to sensor data. Our architecture using abstractions for common tasks such as path-planning. When requesting the "path-planner" interface, the application can request a particular algorithm, with a preference level ("required" to "don't really care"). If a path-planner repeatedly fails, it can be replaced by another (often using a different algorithm). This allows us to provide robustness against poorly-written code.

> **Suggestion 11**
> Provide abstractions for commonly used things like distance measurements and path-planning, and write all applications in terms of these abstractions. This allows backup services to be substituted for primary ones that have failed.

One of our goals on deployments is to hide hardware and software failures from the general public. While the techniques described above have proven to be extremely useful, they are not foolproof. There are still failures that will cripple our system. Some of these, such as the primary computer failing, cannot be covered without implementing multiply-redundant parallel hardware systems. However, for other unforeseen problems we have implemented an option of last resort. If the robot needs an interface that cannot be provided, often due to multiple failures, it launches a graphical interface on a nearby workstation, and asks for human help. A human operator can intervene, extricate the robot from its dilemma, and then give control back to the autonomous system. The operator can also instigate this recovery mode. Although this is technically cheating, and not appropriate in all situations, it has saved us embarrassment in a few situations.

> **Suggestion 12**
> Have an emergency recovery mode that allows a human to take direct control of the robot, extricate it from trouble, and then give control back to the autonomous system without the general public being aware of the intervention.

4 Some Concluding Thoughts

Most robot architectures are designed to make robot programming easier [OC03, MAR06, USEK02], to provide a framework for particular application research [TMR06], or to provide support for experiments [UMK04]. Our architecture is the only one we are aware of that is designed with deployments in mind. Although we have a complete architecture, the ideas discussed in this chapter could be applied to (almost) any existing robot architecture to provide robustness, and to make deployments easier.

Our architecture provides a basic level of fault tolerance. However, explicitly building in fault detection and recovery systems into the application will result in a vastly more robust system [MNPW98, WN96]. This is especially true if advanced, model-based reasoning techniques are used [DWH04, VGST04]. However, this required detailed knowledge of the application, and is hard to make generic.

In this chapter, we have discussed some of the observations that we have made during several real-world deployments of a mobile robot system. These observations have caused us to change our perspective on robot software architectures and the middleware provided by them. The underlying ideas of our current architecture are

1. Support for efficient on-site adaptation of applications.
2. Graceful degradation under various failures.

Our observations are directed at robot deployments where the public are present, and there is a cost to stopping the system to fix a problem. They are not meant to cover every possible deployment, although we believe that many of the lessons will also have value in other deployment scenarios.

We are currently working on evaluating and extending our system. It was successfully demonstrated during the Mobile Robot Competition and Exhibition held at AAAI 2004, and successfully recovered from an unscripted sensor failure during one of the runs in the Challenge event. We are currently focusing on more advanced failure recovery and prediction mechanisms.

References

[MAR06] *Mobile and Autonomous Robotics Integration Environment (MARIE)*, http://marie.sourceforge.net/, 2006.

[TMR06] Sebastian Thrun, Michael Montemerlo, and Nicholas Roy, *Carnegie Mellon robot navigation toolkit (CARMEN)*, http://www.cs.cmu.edu/carmen/, 2006.

[UMK04] Hans Utz, Gerd Mayer, and Gerhard K. Kraetzschmar, *Middleware logging facilities for experimentation and evaluation in robotics*, September 2004.

[USEK02] Hans Utz, Stefan Sablatnog, Stefan Enderle, and Gerhard K. Kraetzschmar, *Miro Middleware for mobile robot applications*, IEEE Transactions on Robotics and Automation 18 (2002), no. 4, 493497.

[SDM03] William D. Smart, Michael Dixon, Nik Melchior, Joseph Tucek, and Ashwin Srinivas, *Lewis the graduate student: An entry in the AAAI robot challenge*, AAAI Mobile Robot Competition 2003: Papers from the AAAI Workshop, 2003, Available as AAAI Technical Report WS-03-01, pp. 4651.

[MS04] Nik A. Melchior and William D. Smart, *A framework for robust mobile robot systems*, Proceedings of the SPIE: Mobile Robots XVII (Douglas W. Gage, ed.), vol. 5609, December 2004, pp. 145154.

[CF01] George Candea and Armando Fox, *Recursive restartability: Turning the reboot sledgehammer into a scalpel*, Proceedings of the 8th Workshop on Hot Topics in Operating Systems (HotOS-VIII), May 2001, pp. 125132.

[PBB02] David Patterson, Aaron Brown, Pete Broadwell, George Candea, Mike Chen, James Cutler, Patricia Enriquez, Armando Fox, Emre Kiciman, Matthew Mertzbacher, David Oppenheimer, Naveen Sastry, William Tetzlaff, Jonathan Traupman, and Noah Treuhaft, *Recovery oriented computing (ROC): Motivation, definition, techniques, and case studies*, Computer Science Technical Report UCB//CSD-02-1175, Department of Computer Science, University of California at Berkeley, March 2002.

[BDG03] Zachary Byers, Michael Dixon, Kevin Goodier, Cindy M. Grimm, and William D. Smart, *An autonomous robot photographer*, Proceedings of the IEEE/RSJ International Conference on Robots and Systems (IROS 2003), vol. 3, October 2003, pp. 26362641.

[BDSG04] Zachary Byers, Michael Dixon, William D. Smart, and Cindy M. Grimm, *Say cheese!: Experiences with a robot photographer*, AI Magazine 25 (2004), no. 3, 3746,

[OC03] Anders Oreback and Henrik I. Christensen, *Evaluation of architectures for mobile robotics*, Autonomous Robots 14 (2003), 3349.

[DWH04] Richard Dearden, Thomas Willeke, Frank Hutter, Reid Simmons, Vandi Verma, and Sebastian Thrun, *Real-time fault detection and situational awareness for rovers: Report on the mars technology program task*, Proceedings of the IEEE Aerospace Conference, March 2004.

[MNPW98] Nicola Muscettola, Pandu Nayak, Barney Pell, and Brian Williams, *Remote agent: To boldly go where no AI system has gone before*, Artificial Intelligence 103 (1998), no. 12, 547.

[VGST04] Vandi Verma, Geoff Gordon, Reid Simmons, and Sebastian Thrun, *Particle filters for rover fault diagnosis*, IEEE Robotics and Automation Magazine (2004), Special Issue on Human-Cenetered Robotics and Dependability (In Press.

[WN96] Brian C. Williams and P. Pandurang Nayak, *A model-based approach to reactive self-configuring systems*, Proceedings of the Thirteenth National Conference on Artificial Intelligence (AAAI-96), AAAI Press / The MIT Press, 1996, pp. 971978.

Software Environments for Robot Programming

Bruce MacDonald, Geoffrey Biggs, and Toby Collett

Robotics Group, Department of Electrical and Computer Engineering, University of Auckland, New Zealand

1 Introduction

Many challenges must be met if the robot programming process is to be improved for human programmers; challenges involving the environment, robots and tasks. These challenges, listed in Fig. 1, result from the complex interactions robots have in real environments and the complex sensors and actuators that robots use. They are present in most robotic systems, however the emphasis on interaction with human users in a human environment amplifies the difficulties for programmers of robotic assistants. Some of the challenges are also exhibited by some traditional, nonrobotic applications, but have generally not been addressed by mainstream development tools.

A final weakness in robot development systems is that concurrency is not well handled by debugging tools. In fact, concurrency is not well handled in many generic debugging environments. This weakness is more problematic in robot development because concurrency is more prevalent in robot problems.

The result of attempts to deal with these difficulties has been a set of tools in which much of the software infrastructure is proprietary and much is targeted at specific hardware, additionally robot software development kits may be limiting, there is a lack of open standards to promote collaboration, code reuse and integration, and there is a lack of techniques that integrate the human into the robot's programming infrastructure. Robot researchers typically program robot applications using an ad hoc combination of tools and languages selected from both traditional application development tools and proprietary robotic tools. Each robotics lab develops its own tools for debugging robot programs, often with different tools for each robot in the lab.

We believe robot programming systems must be targeted more closely to robotics, paying attention to the needs of the robot developer and therefore the nature of typical robot programs, the typical skills of robot programmers, the interactions between humans and robots, and the programming constructs that prevail in robotic applications. The approach must be oriented towards

D. Brugali (Ed.): Software Eng. for Experimental Robotics. STAR 30, pp. 107–124, 2007.

The nature of the robot environment

- Environments are dynamic, asynchronous and operate in real time.
- Unexpected variations cause nonrepeatable behaviour and unexpected conditions.

The nature of the robot being debugged

- Robots are mobile. The target hardware moves away from the programmer!
- A large number of devices are used for input, output and storage. The combination is dramatically more than a programmer is presented with by a common desktop computer. Although the robot is in the same world as the human programmer, the robot's sensory and motor channels may not match human programmers' familiar senses and effectors, and can be difficult for the human programmer to comprehend.
- There are wide variations in hardware and interfaces for different robots, as opposed to the highly commoditized and standardized desktop.

The nature of mobile robot tasks

- Robot tasks emphasize geometry and 3D space.
- Complex data types must be represented.
- Some robot tasks cannot be interrupted without disrupting the task. For example, one robot cannot be interrupted for programming and debugging while it is helping another robot carry an object.
- There is simultaneous and unrelated activity on many inputs and outputs, requiring a programmer to manage multiple activities in all parts of a robot program.

Fig. 1. Challenges facing human programmers of robots that operate in typical human environments.

the human users with which robots will interact, and the human and robot environments in which both live and work.

If we examine the full process of programming a robot assistant, there are two fundamental aspects; the toolchain a human uses for developing robot programs, and the underlying technologies used to instantiate both the toolchain and the robotic system that is the end product. The toolchain consists of a number of components, including:

- Design Tools
- Code Entry Tools
- Compilers
- Test and Debug Utilities

The technology components include:

- Robot Platform
- Programming Languages

- Frameworks and API's, which can be broken into:
  - Library functionality
  - Interface definitions
- Middleware

There is considerable work in some areas, but others are neglected. Some tools, in particular Player/Stage [PSP2005], are emerging as de facto standards for the middleware, framework and API since they support a large variety of hardware and have a growing user community. Some test and debugging facilities are provided. However, the full process involving human developers and the immersive robot environment is not so well addressed. Fundamental in the challenges above is that it is the *robot's interaction with the environment* that makes robot programming different and challenging. One aspect is the *programmer's lack of understanding of the robot's world view* that makes robot programs difficult to code and debug. Standard debugging tools give programmers access only to program data. This makes debugging robot programs difficult because program data is at best an indirect representation of the robot and environment. The interactive systems must enable a human programmer to create, modify and examine programs and system resources, both on and off line. The human programmer may also interact with the framework and API, to examine, monitor and configure resources, and directly with robots as they perform tasks.

Another fundamental aspect is the *expressiveness of the language* that the human programmer uses to describe the robot's behaviour. Although the trend in recent years is to a few common, general purpose programming languages, such as C and C++, with advanced application libraries and tools, we expect some of the robotic programming challenges cannot be overcome without improving the expressiveness of the programming language itself.

In this chapter we focus on case studies of improvements in two somewhat neglected areas related to the human programmer; in section 2 the expressiveness of the robot programming language, and in section 3 the need for human–oriented robot debugging tools.

2 Programming Languages and Environments

This section discusses the nature of robot programming, and presents a case study where a programming language is enhanced so that it is able to manage dimensioned robot data. Robot systems must be able to:

- reason about geometric, temporal, and statistical data to help interpret the input received from sensors and determine outputs to actuators
- operate in real time; the world will not wait for the robot
- react with specified timing constraints to both internal and external events

- use distributed resources and concurrency to provide redundancy and simultaneous execution of components such as localisation, navigation, planning, and image recognition.

However existing programming tools for humans are lacking:

- geometric, temporal and statistical sensor data are not natively supported in programming languages
- real time support is difficult for programmers and relies on the scheduling in the underlying operating system
- specifying events and responses requires complex exception triggering and handling code
- support for concurrency is limited to simplistic threading models and distributed system programming relies on 3rd party tools that are complex and difficult.

The programming process can be improved by:

- reducing development times and costs for robot systems
- allowing more powerful and expressive robot programs
- allowing less skilled programmers to create robot programs with less training.

2.1 Case Study: Programming with Dimensional Analysis

Consider the need for robots to reason about dimensioned data. This provides a case study for how languages can be adapted with features specifically required when programming robot software.

Robots operate in the real world, which is rich in dimensioned data, including:

- odometry readings for tracking translations and rotations
- IR, sonar and laser scanner range data
- motion control for mobile robots and limbs
- navigation systems that input geometric data, estimate geometric quantities such as obstacles, and give geometric commands for movement
- manipulation systems that measure and control forces and torques as well as movements
- intelligent controllers that often use geometric data, for example in path plans and maps
- time constraints for real time processes
- kinetic equations for robot motion.

Existing systems do not directly support dimensioned data, instead using fixed or floating point numeric types and an implied standard of what dimensionality is used in programs, such as those shown in Table 1. It is potentially unsafe to rely on programmers to maintain dimensional consistency, especially in larger projects where many programmers are involved. It is difficult

Table 1. Current methods for dealing with geometric data in some common robot programming systems.

System	Units used	Representation
Player [VGH2003, CMG2005]	Metres, radians	double
CARMEN [MRT2003]	Metres, radians	double or integer
Marie [CLMV2004]	Millimetres, degrees	integer
Orca [BKMWO06]	Metres, radians	double
Pyro [BMK2003]	Varies by robot	Python number
KUKA [KU2005]	Varies by use	REAL type
ABB [ABB2005]	Varies by use	NUM type

to change the implied standard; the recent Player/Stage unit change from millimetres to metres generated much discussion and considerable reengineering of clients. Another, well known, incident was the failure of NASA's Mars Climate Orbiter mission; the root cause was data with incorrect units [STE1999]. In addition, the lack of direct support for dimensioned data makes debugging difficult as dimensionality is not immediately clear from the code.

To enhance programmability, dimensioned data in robot programs should appear and behave as programmers expect. It should look like dimensioned data in program code, allow new dimensions and units to be defined as necessary by the programmer, support unit algebra (the combination of different units to create more complex units, for example speed being metres divided by seconds), and above all emphasise simplicity.

These goals can be achieved by adding support for dimensional analysis to the programming system. The dimensional analysis technique can help make programs more robust against errors in units by verifying the consistency of the units given. It allows a new class of errors to be caught [GE1977] and enables automatic conversions between compatible dimensionality. Dimensional analysis systems have been proposed for Ada [GE1985, HI1988, RO1988, GP1993], Pascal [GE1977, AG1984, DMM1986, BA1987], C++ [CG1988, UMR1994, TUO2005, BR2001], and Java [ACLM2004], among others, from 1977 to 2005. These efforts vary considerably in their flexibility, usefulness and clarity of program code. However, despite the potential in many common areas of programming and numerous proposals over the years, dimensional analysis has still not become a feature of the mainstream languages commonly used in robotics. Our work applies and extends some of these techniques for robotic programming systems. We believe an application specific design in robotics may be more successful. The limited domain of robotics, with its smaller number of common dimensions and simple units, allows for a simple system.

Dimensional analysis can be implemented by one of four methods:

- a new category of run time dimension information, in addition to type and sign information

- a preprocessor system that performs dimensional analysis and outputs standard program code
- object oriented extensions for managing dimensioned data
- a new data type specifically for dimensioned data.

An object oriented extension allows for dimensioned data support without the need to change the language itself. However, it does not provide any special syntax for dimensioned data and so does not significantly improve the clarity of dimensioned data in program code. On the other hand, a new data type can use a new syntax to specify dimensioned data directly in program code, improving the clarity and making debugging easier. Such a data type is more strongly integrated into the language than an object oriented approach, meaning it may be supported by more of the language's constructs. For example, in C++ user defined types are not supported in the `switch` statement. A new datatype would, however, require modification of the language, which can be a difficult task.

Both methods can be implemented at compile time only, run time only or a mixture of both, simply by deciding when the dimensional analysis and consistency checking are performed. Compile time checking has the advantage that it does not introduce any inefficiencies into the run time program, and allows for earlier program validation. On the other hand retaining the dimension information for use at run time can provide more flexibility, for example by allowing data to be output with the dimension information or allowing input data to be checked rather than requiring it be in the correct dimension already.

The design presented here uses a new primitive data type aimed at supporting dimensional analysis for robotics. The semantics of the data type are designed in such a way as to be implementable in any object oriented language.

The design considers dimensions to be categories into which units are assigned by compatibility. Each dimensioned value is associated with a unit expression, which is made up of one or more units. For example, `1m` is a value of `1` associated with the unit `m` (metres), and `2.5m/s` is a value of `2.5` associated with the unit expression `m/s` (metres per second), which is in turn made up of the unit `m` divided by the unit `s` (seconds) using unit algebra. The `m` unit falls into the `distance` dimension, while the `s` unit falls into the `time` dimension, meaning that values measured in metres cannot be converted to values in seconds. If a programmer attempts to do so, an error will result. However, conversions between values with equivalent dimensionality will be automatic, for example degrees to radians or `m/s` to `km/h`.

Dimensions and units must be defined before they are used, and cannot be modified. Four dimensions commonly used in robotics are predefined (distance, angle, time and mass) along with several units in each. Each dimension is defined as a name. Units require a name, dimension, a ratio relative to other units in the dimension, the numerical representation used (floating or fixed point), and any physical range limits. Aliases, for example speed aliasing metres per second, are not allowed, to maintain simplicity and clarity.

A simple formalisation of the semantics is shown below, based on the formalised arithmetic of [Ken96]. Assume the following definitions:

$\mathbb{R}$	set of real numbers
r	member of $\mathbb{R}$
$d(v, u)$	a piece of dimensioned data with numeric value v and unit expression, u
u_a, u_b	unit expressions a and b
u_f	unit expression with floating point representation
u_i	unit expression with fixed point representation
$=_D$	dimensional equivalence relation
$C(d, u)$	conversion operator, convert d to units u
$\times$	operation

The conversion operator can be defined as allowing conversions between units of equivalent dimensionality while disallowing conversions between units of inequivalent dimensionality. The conversion factor between units is used to alter the value as necessary when performing the conversion.

$$C(d(v, u_a), u_b) \rightarrow \begin{cases} d(v, u_b) & \text{if } u_a =_D u_b \\ \text{error} & \text{otherwise} \end{cases}$$

The operations between two pieces of dimensioned data can then be defined as follows.

$$\begin{array}{lrcl}
+, - & d(v_1, u_a) \times d(v_2, u_b) & \rightarrow & \begin{cases} d(v_1, u_a) \times C(d(v_2, u_b), u_a) & \text{if } u_a =_D u_b \\ \text{error} & \text{otherwise} \end{cases} \\
 & d(v_1, u_a) \times d(v_2, u_a) & \rightarrow & d(v_1 \pm v_2, u_a) \\
 * & d(v_1, u_a) \times d(v_2, u_b) & \rightarrow & d(v_1.v_2, u_a.u_b) \\
 \div & d(v_1, u_a) \times d(v_2, u_b) & \rightarrow & d(v_1.v_2^{-1}, u_a.u_b^{-1}) \\
 <, \leq, ==, \geq, > & d(v_1, u_a) \times d(v_2, u_b) & \rightarrow & \begin{cases} d(v_1, u_a) \times C(d(v_2, u_b), u_a) & \text{if } u_a =_D u_b \\ \text{error} & \text{otherwise} \end{cases} \\
 & d(v_1, u_a) \times d(v_2, u_a) & \rightarrow & \text{bool}
\end{array}$$

The conversion operator is used in some operators to reduce dimensioned data to the units needed for the simpler form of the operation.

Similarly, the operations between dimensioned data and a numeric value can be defined:

$$\begin{array}{lrcl}
+, - & d(v, u_a) \times r & \rightarrow & \text{error} \\
 * & d(v, u_a) \times r & \rightarrow & d(r.v, u_a) \\
 \div & d(v, u_a) \times r & \rightarrow & d(r^{-1}.v, u_a) \\
 <, \leq, ==, \geq, > & d(v, u_a) \times r & \rightarrow & \text{error}
\end{array}$$

Finally, the interaction between floating point and fixed point values is defined, governed by the units used for each piece of dimensioned data.

For all except $*, \div$

$$\begin{aligned} d(v,u_f) \times d(v,u_i) &\rightarrow d(v,u_f) \\ d(v,u_i) \times d(v,u_f) &\rightarrow d(v,u_i) \end{aligned}$$

For $*, \div$

$$\begin{aligned} d(v,u_f) \times d(v,u_i) &\rightarrow d(v,u_f) \\ d(v,u_i) \times d(v,u_f) &\rightarrow d(v,u_f) \\ d(v,u_i) \times d(v,u_i) &\rightarrow d(v,u_i) \end{aligned}$$

The general rule is to convert the right hand operand to the left hand operand's units. So, for example, adding a floating point value and a fixed point value will give a result with floating point units. However, for multiplication and division no conversion is performed, as the units are combined via unit algebra and the unit expression becomes floating point if one of the involved units is floating point.

A prototype, RADAR, has been implemented in the Python language[1] to test the semantics and syntax [BM2005]. While it is arguable that interpreted languages are not suited to robotics, Python was chosen because it has a clean syntax and the interpreter is simple to modify. Since Python is an interpreted language, RADAR uses a run time checking approach to the dimensional analysis. The RADAR syntax for creating dimensioned data is shown below.

```
dimensioned ::= (integer | float)~unit(('*' | '/')unit)*
unit ::= [['-'] digit+] letter+ ['^' ['-'] digit+]
```

Note the overloading of the `*` and `/` operators in unit expressions. This could be considered confusing, for example since `s` might be a variable name. However, we expect most robot programmers will be comfortable with the common mathematical symbols for units.

The *operators* available for the primitive dimensioned types are illustrated in Table 2. Table 3 shows the attributes of dimensioned values and some examples.

The sample code in Fig. 2 illustrates the use of units in the sonar avoidance example from the Player/Stage project.

```
dimensioned ::= (integer | float)~unit(('*' | '/')unit)* unit ::=
[['-'] digit+] letter+ ['^' ['-'] digit+]
```

A built in module `dimension` provides the functionality for managing dimensions and units. As an example, consider a programmer who wishes to use a light intensity sensor. A new dimension, "lumens", would be required, as well as a unit of measure in that dimension. If the programmer wishes to ensure the intensity value is within certain limits, then range limits would be set, as shown in Fig 1.

The primitive data type can be used to create higher level data types representing more complex dimensional structures. For example a tuple of dimensional values can represent a robot pose. A list of these can represent a path. Add a time value to each tuple and the path becomes a motion plan

[1] http://www.python.org

Table 2. Examples of operations on dimensioned data.

Operator	Example	Result
Addition	5m + 2cm	5.02m
	5m + 0.02	Error
Subtraction	5m − 2cm	4.98m
	5m − 2rad	Error
Multiplication	5m × 2	10m
	5m × 2m	$10m^2$
Division	5m ÷ 2	2.5m
	5m ÷ 2s	2.5m/s
	5m ÷ 2m	2.5
Comparison	2m < 3cm	False
	2m > 3cm	True

Table 3. Attributes of dimensioned values and some examples.

Attribute	*Examples*		
Value	43.0	5	1.5
Unit	cm	cell	radians
Representation	float	fixed	float
Unit Range Limits	-10 to 120	0 to 10	0 to 2π
Code	43~cm	5~cell	1.5~rad

Listing 1. New dimensions can be created using the `dimension` module.

```
dimension.DefineDimension ('lumens') dimension.DefineUnit
('lumens',
                          'intensity',
                          UnitPrec_Floating,
                          1.0,
                          UnitRange_Include, 0,
                          UnitRange_Exclude, 10 )
```

that can be iterated over and interpolated based on the time value. Similar to the consistency checking of dimensionality in terms of units of measure, higher level structures could provide support for checking consistency of dimensionality in terms of spatial dimensions. For example, ensuring that a programmer does not inadvertently combine the X and Y dimensions of a robot's pose could be done using a vector structure that uses dimensioned data types for its individual components.

The dimensional analysis system makes the management of real world data in robot software easy and clear, improving the programmability and maintainability of robotic systems. This illustrates just one aspect of how programming languages can be improved specifically for human programmers of robots.

```
if motorsOn:
    position.set_cmd_vel(0~m/s, 0~m/s, 0~rad/s, 1)
    position.enable(1)
else:
    position.set_cmd_vel(0~m/s, 0~m/s, 0~rad/s, 0)
    position.enable(0)

newSpeed = 0~m/s newTurnRate = 0~rad/s while 1:
    robot.read()

    if len (laser.Ranges()) == 0:
        print 'No laser data'
        continue
    leftRanges = laser.Ranges()[:len(laser.Ranges())/2]
    rightRanges = laser.Ranges()[len(laser.Ranges())/2:]
    minLeft = min(leftRanges)
    minRight = min(rightRanges)
    l = (minLeft * 10) / 5 - 1~m
    r = (minRight * 10) / 5 - 1~m
    if l > 1~m:
        l = 1~m
    if r > 1~m:
        r = 1~m
    newSpeed = ((r + l) / 1~s) / 10
    newTurnRate = ((r - l) / 1~m) * 1.57~rad/s
    newTurnRate = min (newTurnRate, 40~deg/s)
    newTurnRate = max (newTurnRate, -40~deg/s)

    position.set_cmd_vel(newSpeed, 0~m/s, newTurnRate, 1)
```

Fig. 2. The Player laser obstacle avoidance example.

3 Robot Debugging Tools

This section presents a case study aimed at improving the process of debugging robot systems. It begins with an analysis of developer–robot interaction and then presents an Augmented Reality (AR) system for visualising robot data.

A shared perceptual space which both the human user and robot understand is a requirement for effective interaction [BEFS2001]. Both the human user and the robot have a perceptual space where communication is meaningful, and it is the overlap of these two spaces that is the interaction space of the human–robot coupling (Fig. 3). The interaction space for input may be different from the output space. For example a robot that knows how to represent an emotion using facial expressions, but cannot recognise the facial expressions of others, will have differing input and output spaces. Any mismatch between the perceptual input and output space may put additional cognitive load on the user, who will have to perform additional translations before communicating with the robot.

This case study focuses on providing a shared perceptual space for the robot–developer interaction. By allowing the developer to understand the robots world view we are effectively enhancing the shared space as we are allowing the developer to view the world in the same way as the robot.

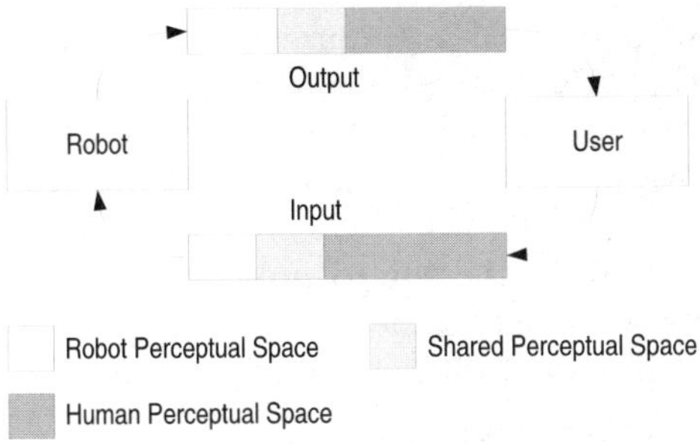

Fig. 3. Shared human–robot perceptual space.

3.1 AR for Developer Interaction

AR provides an ideal presentation of the robot's world view; it displays robot data layered within the real environment. By viewing the data in context with the real world the developer is able to compare the robot's world view against the ground truth of the real world image without needing to know the exact state of the world at all times. This makes clear not only the robot's world view but also the discrepancies between the robot's view and reality. The main challenge in AR is to accurately track the human user so that the overlaid data is accurately positioned when the human viewer changes position and orientation. However this is generally a simpler problem than the alternative, which is to track the entirety of a potentially large and dynamic real world environment and explicitly calculate a comparison between the robot view and the actual world view.

Related Work

Player [PSP2005] has a built in tool, playerv, for visualising its interfaces. However this tool provides the visualisation in isolation. By providing an AR interface for the visualisation we show the data in context with the real objects, making understanding the data more intuitive.

Shikata *et al* [SGNI2003] present an algorithm evaluation system that uses virtual robots in a virtual world to test the performance of avoidance algorithms when human users are involved. The main advantage is that a real human is used rather than a scripted one, thus giving more accurate behaviour while at the same time the virtual robots mean there are no safety issues. In particular the system is used to test avoidance algorithms with no danger to the human user from collisions occurring when the algorithm fails. Milgram *et al* used AR in creating a virtual measuring tape and a virtual tether to assist inserting a peg in a hole [MZDG1993, MRG1995]. Freund *et al* [FSR2001] use AR to more simply display complex 3D state information about autonomous systems. Raghavan *et al* [RMS1999] describe an interactive tool for augmenting the real scene during mechanical assembly. Pettersen *et al* [PPS2003] present a method to improve the teaching of way points to a painting robot, using AR to show the simulated painting process implied by the taught way points. Brujic-Okretic *et al* [BGHMP2003] describe an AR system that integrates graphical and sensory information for remotely controlling a vehicle.

Daily *et al* [DCMP2003] use AR to present information from a swarm robotics network for search and rescue. Recently KUKA began to investigate the use of AR to visualise data during training [BK2004].

Amstutz and Fagg [AF2002] describe an implementation for representing the combined data of a large number of sensors on multiple mobile platforms, using AR and VR to display the information.

3.2 AR Toolkit Implementation

Given the rapidly changing nature of available AR hardware and software techniques, to remain relevant any system must be flexible and independent of any specific AR implementation.

To be practically effective an AR toolkit must:

- Be modular and flexible in order to make use of core technology advances.
- Be robust, and able to run as a permanent installation in a robotics laboratory.
- Provide sufficient accuracy in the underlying AR system for developer interaction.

All are implemented in our toolkit.

The software architecture has been designed in a highly modular fashion allowing for individual components to be replaced. While the implemented toolkit is Player based, the modular nature of the AR toolkit allows for the easy addition of support for any other device interface system.

The rendering process for the toolkit is broken into four basic stages: capture, preprocessing, rendering and postprocessing. The rendering stage is further broken down into three rendering layers and a ray trace step is optionally performed to enhance stereo representations. Each step is described below.

1. Capture: the background frame, orientation and position of the camera are captured.
2. Preprocessing: such as blob tracking for robot registration.
3. Render - Transformation: the position of the render object is extracted from its associated list of position objects, and appropriate view transforms are applied.
4. Render - Base: invisible models of any known 3D objects are rendered into the depth buffer. This allows for tracked objects such as the robot to obstruct the view of the virtual data behind them. The colour buffers are not touched as the visual representation of the objects was captured by the camera.

5. Render - Solid: the solid virtual elements are drawn.
6. Render - Transparent: transparent render objects are now drawn while writing to the depth buffer is disabled.
7. Ray Trace: to aid in stereo convergence calculation, the distance to the virtual element in the centre of the view is estimated using ray tracing. This is of particular relevance to stereo AR systems with control of convergence, and optical see through stereo systems.
8. Postprocessing: once the frame is rendered any post processing or secondary output modules are called. This allows the completed frame to be read out of the frame buffer and, for example, encoded to a movie stream.

The toolkit architecture is centred around an output device, each of which contains a capture device for grabbing the real world frame, and a camera device for returning the camera parameters, including pose. The capture device could be a null object for optical see through AR, or for a purely virtual environment. Also, the output device maintains three component lists: secondary outputs, which monitor the interface and provide movie capture; preprocessing objects used for image based position tracking; and finally a list of render item pairs. Each *Render Pair* consists of a render object that performs the actual rendering of a virtual element and a chain of position objects that define where it should be rendered.

Fig. 4 shows the software structure and each of the components is summarised in Table 4. The first four items (Capture, Camera, Secondary Output and Preprocessing) are all unique to the output object. The render objects and position objects can be used in multiple combinations, potentially with different output objects. For example a Stereo head mounted display (HMD) needs the same laser data (render object) on both displays (output objects), while for a single output the same origin (position object) could be used to render both laser and sonar data.

To create a permanent laboratory setup it was important to detect the presence of robots. This presence is in two forms, first the system must cope with operational presence, if a robot is switched on or off the system must detect this and if required start (or stop) displaying appropriate data. Secondly, if a robot is physically removed from the system, i.e. is outside the active field of view, or is unable to be tracked then the data for that robot must not be displayed, this is particularly relevant for optically tracked robots.

The toolkit has been tested with two core AR configurations. A magnetically tracked video see–through HMD (Trivisio ARVision3D) provides an immersive, stereo environment for a single user. A fixed overhead camera and an AR display on a large wall mounted screen near the robots provides a communal environment where humans and robots may interact together.

The toolkit has been tested with three different Player compatible robots, the B21r robot from iRobot, the Pioneer 3DX from ActivMedia and our own in house Shuriken robot, representing a rich set of capabilities and a range of scales. Sample output of these systems is shown in Fig. 5. The visualisation

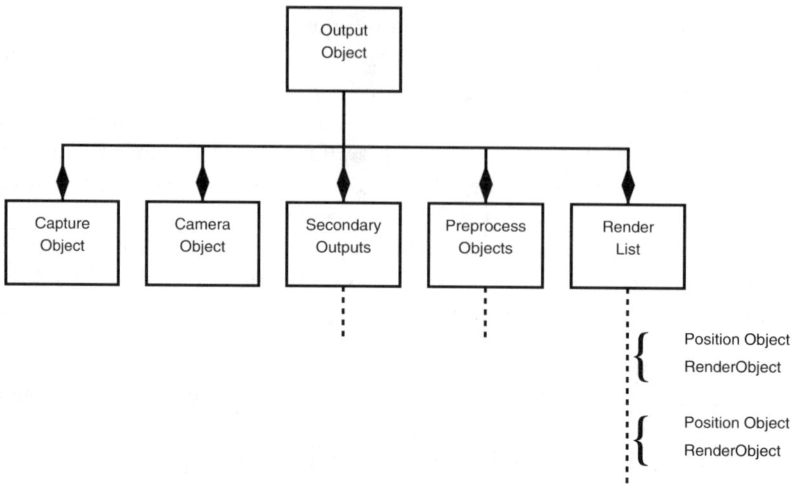

Fig. 4. The ARDev software architecture.

Table 4. Core objects.

ObjectName	Description
OutputObject	Handles the actual rendering of the information and its display
CaptureObject	Supplies the real world input frames, for example from a Video for Linux device, or a blank frame for optical see through AR
CameraObject	Provides the geometry and optical properties of the camera
SecondaryOutputObject	Allows a secondary output stream to exist, such as output to video, or capturing stills
PreProcessObject	Handles processing of a frame before rendering (allows for tracking of markers using image processing)
PositionObject	Returns the position of a tracked object, relative to an arbitrary base
RenderObject	Handles the rendering of a single augmented entity

system has run continuously in the lab for 7 days, at which point it was stopped for an upgrade.

The AR toolkit provides intuitive representations of geometric data from Player interfaces for the robot developer. The use of AR for the visualisations gives the developer an immediate understanding of the robot data in context with the real world baseline, allowing any limitations in the robots world view to be understood, and leading to much faster solutions to related software

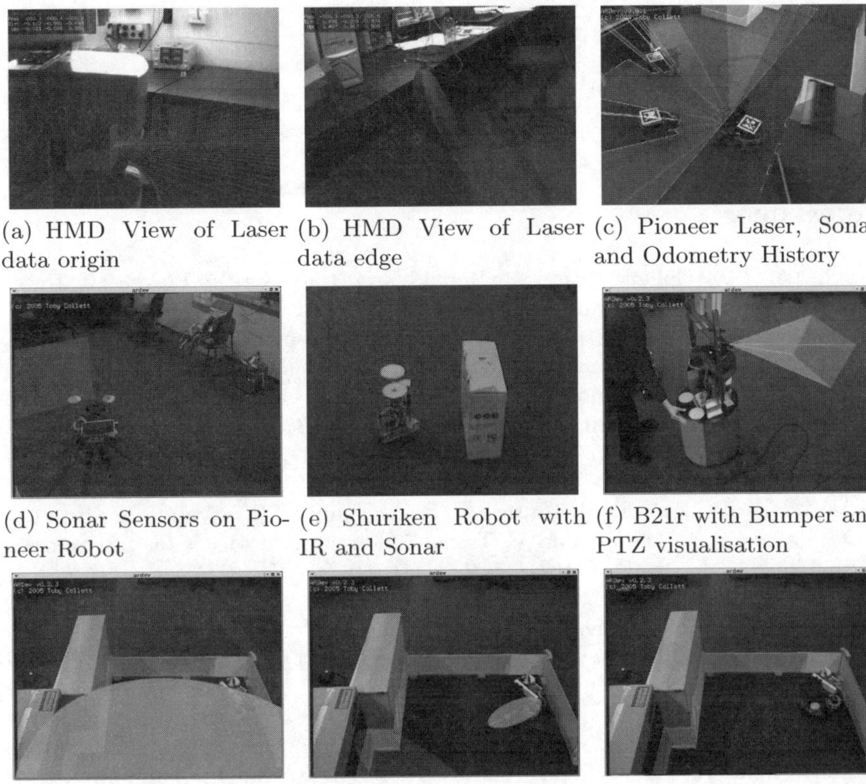

(a) HMD View of Laser data origin (b) HMD View of Laser data edge (c) Pioneer Laser, Sonar and Odometry History

(d) Sonar Sensors on Pioneer Robot (e) Shuriken Robot with IR and Sonar (f) B21r with Bumper and PTZ visualisation

(g) AMCL Initial Distribution, High Covariance (h) AMCL Improved Estimate (i) AMCL Converged Result

Fig. 5. AR system output.

issues. Our initial implementation of these concepts has shown useful and promising results for robot software development.

4 Conclusion

The three case studies promote an important goal that designers of robotic systems include the human developer as a significant role. Robot development tools should provide expressive programming languages and frameworks that enhance the opportunity for human developers to describe robot behaviour. Debugging tools should be human–oriented and improve the immediate visualisation of robot data for the human developers.

Acknowledgements

Toby Collett is funded by a top achiever doctoral scholarship from the New Zealand Tertiary Education Commission.

References

[CMG2005] Toby Collett, Bruce MacDonald, and Brian Gerkey. Player 2.0: Toward a practical robot programming framework. In *Proc. Australasian Conference on Robotics and Automation*, Sydney, Australia, December 5–7 2005.

[MRT2003] Michael Montemerlo, Nicholas Roy, and Sebastian Thrun. Perspectives on standardization in mobile robot programming: The Carnegie Mellon Navigation (CARMEN) toolkit. In *Proc. of the IEEE/RSJ Intl. Conf. on Intelligent Robots and Systems (IROS)*, volume 3, pages 2436–2441, Las Vegas, NV, October 2003.

[CLMV2004] C. Côté, D. Létourneau, F. Michaud, J.-M. Valin, Y. Brosseau, C. Raïevsky, M. Lemay, and V. Tran. Code reusability tools for programming mobile robots. In *Proc IEEE Intl. Conf. on Intelligent Robots and Systems*, volume 2, pages 1820–5, Sendai, Japan, Oct 2004.

[BMK2003] D. Blank, L. Meeden, and D. Kumar. Python robotics: An environment for exploring robotics beyond legos. In *Proceedings of the Thirty-Fourth SIGCSE Technical Symposium on Computer Science Education*, Reno, Nevada, February 2003. ACM Press.

[KU2005] KUKA Automatisering + Robots N.V. http://www.kuka.be/, June 2005.

[GP1993] Dean W. Gonzalez and Tim Peart. Applying dimensional analysis. *Ada Lett.*, XIII(4):77–86, 1993.

[ABB2005] The ABB group. http://www.abb.com/, June 2005.

[GE1985] N. H. Gehani. Ada's derived types and units of measure. *Softw. Pract. Exper.*, 15(6):555–569, 1985.

[GE1977] Narain Gehani. Units of measure as a data attribute. *Computer Languages*, 2(3):93–111, 1977.

[HI1988] Paul N. Hilfinger. An Ada package for dimensional analysis. *ACM Trans. Program. Lang. Syst.*, 10(2):189–203, 1988.

[PPS2003] T. Pettersen, J. Pretlove, C. Skourup, T. Engedal, and T. Lokstad. Augmented reality for programming industrial robots. In *Proceedings of the Second IEEE and ACM International Symposium on Mixed and Augmented Reality*, pages 319–20, 7–10 Oct 2003.

[PSP2005] Player/Stage. The player/stage project. http://playerstage.sourceforge.net/, January 2005.

[RMS1999] V. Raghavan, J. Molineros, and R. Sharma. Interactive evaluation of assembly sequences using augmented reality. *Robotics and Automation, IEEE Transactions on*, 15(3):435–449, 1999.

[RO1988] P. Rogers. Dimensional analysis in Ada. *Ada Lett.*, VIII(5):92–100, 1988.

[SGNI2003] R. Shikata, T. Goto, H. Noborio, and H. Ishiguro. Wearable-based evaluation of human-robot interactions in robot path-planning. In *Proc. IEEE International Conference on Robotics and Automation (ICRA 03)*, volume 2, pages 1946–1953, 2003.

[STE1999] A. G. Stephenson and group. Mars Climate Orbiter Mishap Investigation Board Phase I Report. ftp://ftp.hq.nasa.gov/pub/pao/reports/1999/MCO_report.pdf, November 10 1999.

[TUO2005] The units of measure library. http://tuoml.sourceforge.net/, 2005.

[UMR1994] Zerksis D. Umrigar. Fully static dimensional analysis with C++. *SIGPLAN Not.*, 29(9):135–139, 1994.

[VGH2003] R.T. Vaughan, B.P. Gerkey, and A. Howard. On device abstractions for portable, reusable robot code. In *Proceedings of the 2003 IEEE/RSJ Intl. Conference on Intelligent Robots and Systems (IROS03)*, volume 3, pages 2421–2427, Las Vegas, Nevada, October 2003.

[BHGC06] Antonio C. Domínguez-Brito, Daniel Hernández-Sosa, Josep Isern-González and Jorge Cabrera-Gámez , *CoolBOT: a Component Model and Software Infrastructure for Robotics*, In Brugali D. (Ed.) Software Engineering for Experimental Robotics, Springer STAR series, 2006.

[BKMWO06] Alex Brooks, Tobias Kaupp, Alexei Makarenko, Stefan Williams and Anders Orebäck, *Orca: a Component Model and Repository*, In Brugali D. (Ed.) Software Engineering for Experimental Robotics, Springer STAR series, 2006.

[AG1984] Mukul Babu Agrawal and Vijay Kumar Garg. Dimensional analysis in Pascal. *SIGPLAN Not.*, 19(3):7–11, 1984.

[DMM1986] A Dreiheller, B Mohr, and M Moerschbacher. Programming Pascal with physical units. *SIGPLAN Not.*, 21(12):114–123, 1986.

[BA1987] Geoff Baldwin. Implementation of physical units. *SIGPLAN Not.*, 22(8):45–50, 1987.

[CG1988] R.F. Cmelik and N.H. Gehani. Dimensional analysis with C++. *Software, IEEE*, 5(3):21–27, 1988.

[BR2001] Walter E. Brown. Applied template metaprogramming in siunits: the library of unit-based computation. Technical report, Computational Physics Department, Computing Division, Fermi National Accelerator Laboratory, August 2001.

[ACLM2004] Eric Allen, David Chase, Victor Luchangco, Jan-Willem Maessen, and Guy L. Steele, Jr. Object-oriented units of measurement. In *OOPSLA '04: Proceedings of the 19th annual ACM SIGPLAN Conference on Object-oriented programming, systems, languages, and applications*, pages 384–403. ACM Press, 2004.

[BM2005] Geoffrey Biggs and Bruce MacDonald. A design for dimensional analysis in robotics. In *Third International Conference on Computational Intelligence, Robotics and Autonomous Systems*, Singapore, December 2005.

[BEFS2001] C. Breazeal, A. Edsinger, P. Fitzpatrick, and B. Scassellati. Active vision for sociable robots. *IEEE Trans. Syst., Man, Cybern. A*, 31(5):443–453, 2001.

[BK2004] Rainer Bischoff and Arif Kazi. Perspectives on augmented reality based human-robot interaction with industrial robots. In *Proc. IEEE/RSJ International Conference on Intelligent Robots and Systems (IROS 04)*, pages 3226–3231, 2004.

[DCMP2003] M. Daily, Youngkwan Cho, K. Martin, and D. Payton. World embedded interfaces for human-robot interaction. In *Proc. 36th Annual Hawaii International Conference on System Sciences*, pages 125–130, 2003.

[BGHMP2003] V. Brujic-Okretic, J.-Y. Guillemaut, L.J. Hitchin, M. Michielen, and G.A. Parker. Remote vehicle manoeuvring using augmented reality. In *International Conference on Visual Information Engineering. VIE 2003.*, pages 186–9, 7–9 July 2003.

[FSR2001] E. Freund, M. Schluse, and J. Rossmann. State oriented modeling as enabling technology for projective virtual reality. In *Proc. IEEE/RSJ International Conference on Intelligent Robots and System (IROS 01)*, volume 4, pages 1842–1847, 2001.

[MRG1995] P. Milgram, A. Rastogi, and J.J. Grodski. Telerobotic control using augmented reality. In *Proceedings., 4th IEEE International Workshop on Robot and Human Communication. RO-MAN'95*, pages 21–9, Tokyo, 5–7 July 1995.

[MZDG1993] P. Milgram, S. Zhai, D. Drascic, and J. J. Grodski. Applications of augmented reality for human-robot communication. In *Proc. IEEE/RSJ International Conference on Intelligent Robots and System (IROS 93)*, volume 3, pages 1467–1472, 1993.

[AF2002] P. Amstutz and A.H. Fagg. Real time visualization of robot state with mobile virtual reality. In *Proc. IEEE International Conference on Robotics and Automation (ICRA 02)*, volume 1, pages 241–247, 2002.

[Ken96] Andrew John Kennedy, *Programming Languages and Dimensions*, Technical Report St. Catherine's College, March 1996.

Sidebar – Programming Commercial Robots

José María Cañas[1], Vicente Matellán[2], Bruce MacDonald[3], and Geoffrey Biggs[4]

[1] Universidad Rey Juan Carlos, Madrid, Spain jmplaza@gsyc.escet.urjc.es
[2] Universidad Rey Juan Carlos, Madrid, Spain vicente.matellan@urjc.es
[3] University of Auckland, New Zealand b.macdonald@auckland.ac.nz
[4] University of Auckland, New Zealand g.biggs@auckland.ac.nz

Lozano–Pérez [LP1986] divided robot programming into methods for guiding, robot–level programming, and task–level programming. A more useful distinction for modern methods is between manual programming and automatic programming, based on the actual method used for programming as this is the crucial distinction for users and programmers.

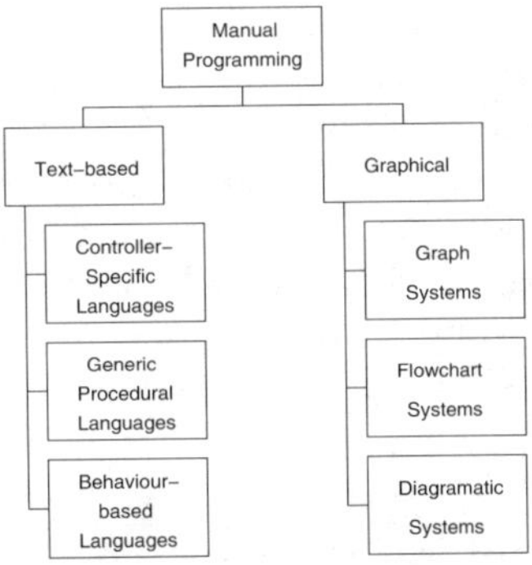

Fig. 1. Categories of manual programming systems. A manual system may use a text-based or graphical interface for entering the program.

Manual systems require the user/programmer to directly enter the desired behaviour of the robot, usually using a graphical or text-based programming language, as shown in Fig. 1. Text-based systems are either controller-specific languages, generic procedural languages, or behavioural languages, which typically differ by the flexibility and method of expression of the system. Graphical languages [BKS2002, BI2001] use a graph, flow-chart or diagram based graphical interface to programming, sacrificing some flexibility and expressiveness for ease of use.

D. Brugali (Ed.): Software Eng. for Experimental Robotics. STAR 30, pp. 125–132, 2007.

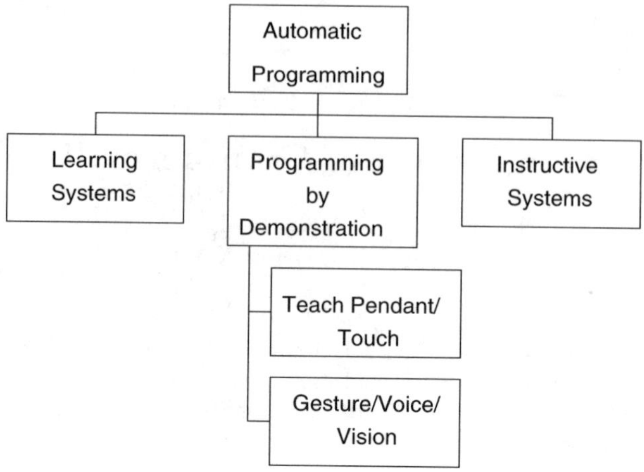

Fig. 2. Categories of automatic programming systems. Learning systems, programming by demonstration (PbD) and instructive systems are all methods of teaching robots to perform tasks.

The user/programmer has little or no direct control over the robot code in an automatic programming system, which may acquire the program by learning, programming by demonstration (PbD), or by instruction, as indicated in Fig. 2. Often automatic systems are used "online," with a running robot, although a simulation can also be used.

In this sidebar we will focus on the characteristics of commercial programming environments. Simple robots can be programmed directly using their own operating systems. More sophisticated robots include SDKs to simplify the programming of their robots. Mobile robots programming environments vs. industrial manipulators are also presented.

1 Industrial Manipulators

Programming systems for industrial manipulators include both manual and automatic methods of programming. Initially manual programming tools were common, in the form of text-based controller-specific languages. Controller–specific languages are designed for a single robot system, for example the system provided by KUKA, shown in Fig. 3.

Coupled with touch screens, graphical languages can enable rapid configuration of industrial robots.

PbD was developed for industrial robot manipulators, using a teach pendant or similar method to move the manipulator to each position in a task, where the robot's joint positions are recorded for later playback. Recent work in PbD has focussed on creating more flexible robot programs by segmenting demonstrations to identify key actions [EZRD2002, CM1998, CM2000,

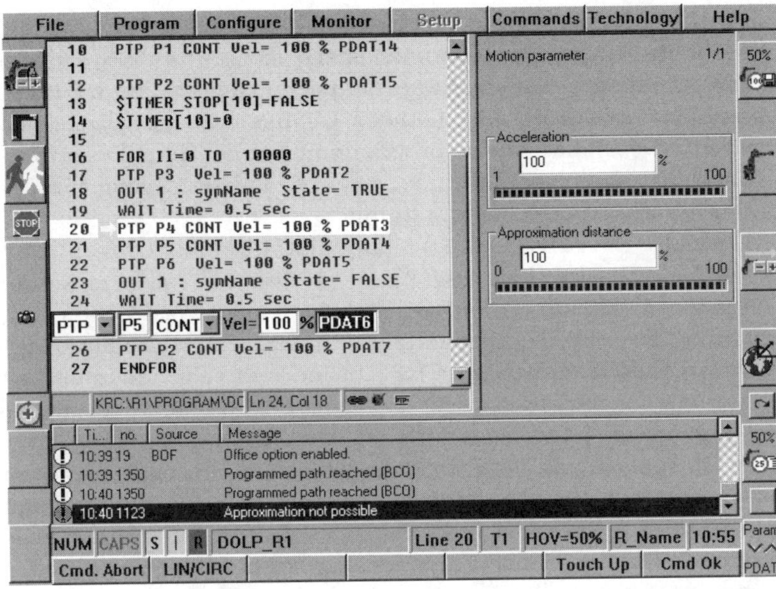

Fig. 3. The KUKA programming environment. [From [KU2005]]

CZ2001, OKKI2002], and on using more natural interaction methods such as voice and touch to perform the demonstrations [GSAH2001, YKY2002, TOKI2002]. Other work includes virtual environments for PbD [OSK2002], finger sensors to detect fine manipulations [ZRDZ2002], and graphical display of demonstration results [FHD1998].

2 Mobile Robots

Mobile robot programming has evolved significantly in recent years, and two approaches are currently found, both manual programming methods. On one hand, application programs for simple robots obtain readings from sensors and send commands to actuators by directly calling functions from the drivers provided by the seller. On the other hand, we have identified many common features across commercial SDKs.

First, they offer a simple and more abstract access to sensors and actuators than the operating systems of simple robots. For example, in a Pioneer with a laser rangefinder, the applications can obtain readings using ARIA or directly through a serial port. Using ARIA, one need only invoke a method and ARIA will take charge of refreshing the variables. Using the operating system directly, the application must request and periodically read the data from the laser through the serial port, and must identify the protocol of the device to compose and analyze the low level messages correctly. The abstract access is also offered for actuators.

Second, the software architecture of the SDK sets the way the application code obtains sensor data, commands the motors, or uses a developed functionality. There are many software options: calling to library functions, reading variables, invoking object methods, sending messages via the network to servers, etc.. Depending on the programming model the robot application can be considered an object collection, a set of modules talking through the network, an iterative process calling to functions, etc.

Third, usually the SDK includes simple libraries and common use functionality, such as robust techniques for perception or control, localization, safe local navigation, global navigation, social abilities, map construction, etc. The robot manufacturers sell them separately or include them as additional value with their own SDK. For example, ERSP includes three packages in the basic architecture: one for interaction, one for navigation and another for vision.

There are several advantages of using the SDKs. First, they favor the portability of applications between different robots. Second, they promote code reuse, shortening the development time and reducing the programming effort needed to code the application as long as the programmer can build the program by reusing the common functionality, keeping herself focused in the specific aspects of her application. And third, the software architecture offers a way to organize code, allowing the handling of code complexity when the robot functionality increases.

The next sections present some case studies for different mobile robot environments. Most of them are based on *libre*[5] software because it lets us freely explore the underlying technologies.

2.1 LEGO RCX and BrickOS

The RCX[6] in Fig. 5 is sold as a creative and educational toy. It has a central processor, the RCX brick, and a set of LEGO pieces that are assembled to build the body. There are many ways to program it. LEGO offers a graphical programming environment oriented to children, named RCX-code (Fig. 4). Another possibility is NQC [Bau00], a variation of C which includes instructions to access to the sensors and actuators.

The open-source operating system BrickOS[7], developed by Markus L. Noga [Nie00], allows programming the brick in C. And the LeJOS operating system allows the creation of Java applications. These operating systems, including the original LEGO, offer multitasking. Since its sensors and actuators are simple, an SDK is not necessary and applications can be developed without difficulty, programming directly over the API of the operating system.

[5] We use the Spanish term "libre" to avoid the common misunderstanding between free as in "free beer" and free as in "free speech."

[6] http://www.legomindstorms.com

[7] http://brickos.sourceforge.net/

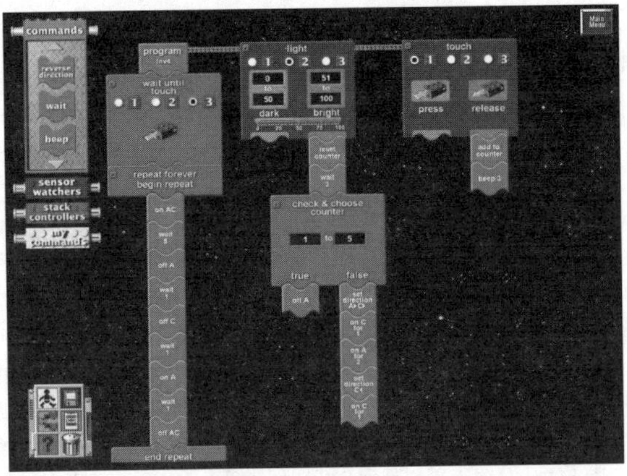

Fig. 4. The Lego Mindstorms graphical programming environment, used to create simple programs for Lego robots.

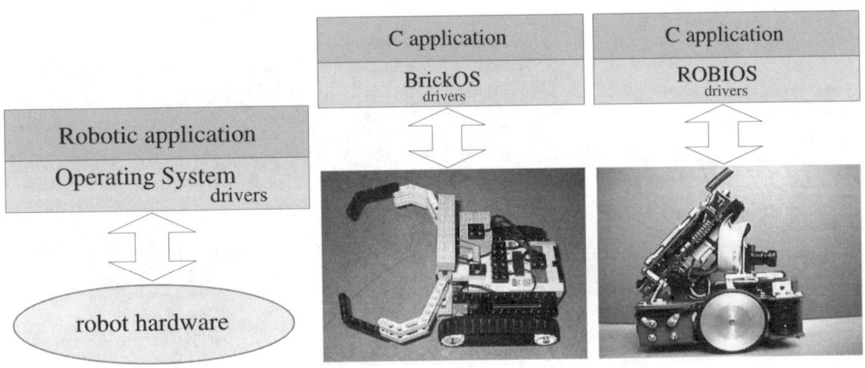

Fig. 5. LEGO (center) and EyeBot (right) are programmed over their Op. Systems

2.2 EyeBot and ROBIOS

The EyeBot [8] [Bra03] is a small robot. Its operating system, ROBIOS, is a meaningful example of an ad-hoc operating system. It allows programs to be loaded through a serial port and executed by pushing buttons. The ROBIOS API includes functions to read the infrared sensors, capture images from the camera and move the motors at a certain speed. It also includes two functions to send and receive bytes through the radio link to other EyeBots or a PC. ROBIOS includes primitives to monitor whether a button is being pushed, and to display images and text on the screen. Concerning multitasking, ROBIOS has primitives to create, pause or kill threads. It offers two ways to share the processor time between threads: with and without preemption. If also offers

[8] http://robotics.ee.uwa.edu.au/eyebot/

locks for the coordination of concurrent execution of these threads, and access to shared variables.

2.3 Aibo and OPEN-R

The Aibo robot[9] in Fig. 6) is sold as a pet, with a program that governs its movements to exhibit dog behaviors (follow a ball, look for a bone, dance, etc.) and learn. Since the summer of 2002 it has been possible to program it, and so the AIBO is useful for research. The operating system in charge of controlling the hardware devices is Aperios, a real-time operating system based on objects. On top of this, Sony provides the OPEN-R SDK [Cor03], which includes many specific C++ objects to access the Aibo hardware. There are objects for basic access to camera images, joint positions, management of the TCP/IP stack and management of the microphone and speaker. Additionally, OPEN-R allows multitasking and event oriented programming.

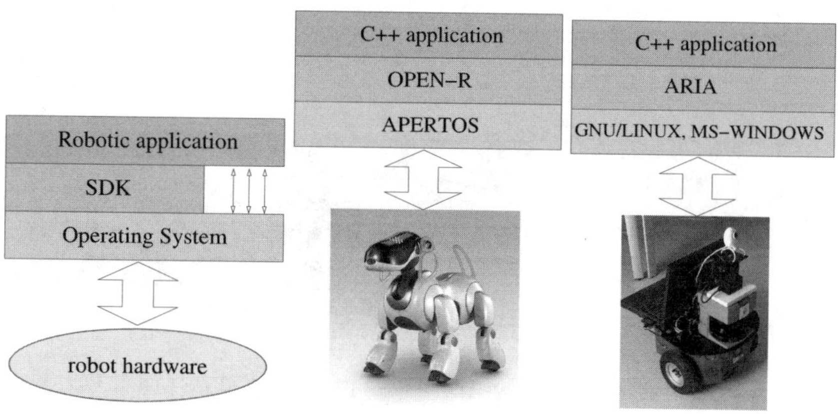

Fig. 6. Two robots programmed using an SDK: Aibo (center) and Pioneer (right)

2.4 Pioneer and ARIA

The Pioneer[10] is a medium size robot, shown in Fig. 6. Its onboard PC is connected to the base microcontroller via a serial port. ARIA (ActivMedia Robotics Interface for Applications) [Rob02] is the manufacturer SDK for the Pioneer robot. ARIA is supported by ActivMedia Robotics, but it is distributed with a GPL license. It offers an object-oriented programming environment, which includes support for multitasking programming and network

[9] http://www.aibo.com/
[10] http://www.activrobots.com/

communication. Applications must be written in C++, and since ARIA runs both on Linux and MS-Windows, the same ARIA application can control robots from either operating system.

For hardware access, ARIA offers a collection of classes, which setup an object based API. The main class *ArRobot* has many relevant methods. There are classes for range sensors and their objects have methods which allow the application to access the data from the proximity sensors. The objects of ARIA are not distributed, ARIA allows the programming of distributed applications using *ArNetworking* to manage remote communications.

Concerning multitasking, the application on ARIA can be programmed as mono-threaded or multi-threaded. In the latter, ARIA offers infrastructure for both user threads and kernel threads, which are a wrapper of the native Linux-pthreads or Win32-threads. For concurrency and synchronization ARIA offers resources such as *ArMutex* and *ArCondition*. It also has basic behaviors such as safe navigation and obstacle avoidance, but it does not include map construction or localization functionalities, which are sold separately. Recently, it has included the open-source simulator Stage (renamed as MobileSim), which has been adapted to work with ARIA. It is a clear example of code reuse in robot applications, involving a private company.

References

[Bau00] Dave Baum, *Dave baum's definitive guide to lego mindstorms*, Apress, 2000.

[Bra03] Thomas Braunl, *Embedded robotics*, Springer Verlag, 2003.

[Cor03] Sony Corporation, *Open-r sdk, programmers guide*, Technical Report 20030201-E-003, 2003.

[Nie00] Stig Nielsson, *Introduction to the legos kernel*, Technical Report, 2000.

[Rob02] ActivMedia Robotics, *Aria reference manual*, Technical Report (1.1.10), 2002.

[USEK02] Hans Utz, Stefan Sablatnog, Stefan Enderle, and Gerhard Kraetzschmar, *Miro – middleware for mobile robot applications*, IEEE Transactions on Robotics and Automation, Special Issue on Object-Oriented Distributed Control Architectures **18** (2002), no. 4, 493–497.

[LP1986] Tomás Lozano-Pérez. Robot programming. Technical Report Memo 698, MIT AI, December 1982, revised April 1983 1982. Also published in Proceedings of the IEEE, Vol 71, July 1983, pp.821–841 (Invited), and IEEE Tutorial on Robotics, IEEE Computer Society, 1986, pp.455–475.

[BKS2002] R. Bischoff, A. Kazi, and M. Seyfarth. The MORPHA style guide for icon-based programming. In *Robot and Human Interactive Communication, 2002. Proceedings. 11th IEEE International Workshop on*, pages 482–487, 2002.

[BI2001] A. Bredenfeld and G. Indiveri. Robot behavior engineering using DD-Designer. In *Robotics and Automation, 2001. Proceedings 2001 ICRA. IEEE International Conference on*, volume 1, pages 205–210, 2001.

[KU2005] KUKA Automatisering + Robots N.V. http://www.kuka.be/, June 2005.

[EZRD2002] M. Ehrenmann, R. Zollner, O. Rogalla, and R. Dillmann. Programming service tasks in household environments by human demonstration. In *Robot and Human Interactive Communication, 2002. Proceedings. 11th IEEE International Workshop on*, pages 460–467, 2002.

[CM1998] J. Chen and B. McCarragher. Robot programming by demonstration-selecting optimal event paths. In *Robotics and Automation, 1998. Proceedings. 1998 IEEE International Conference on*, volume 1, pages 518–523, May 1998.

[CM2000] J.R. Chen and B.J. McCarragher. Programming by demonstration - constructing task level plans in hybrid dynamic framework. In *Proceedings of the IEEE Intl. Conf. on Robotics and Automation (ICRA '00)*, volume 2, pages 1402–1407, apr 2000.

[CZ2001] J.R. Chen and A. Zelinsky. Programming by demonstration: removing sub-optimal actions in a partially known configuration space. In *Proceedings of the IEEE Intl. Conf. on Robotics and Automation (ICRA '01)*, volume 4, pages 4096–4103, May 2001.

[OKKI2002] K. Ogawara, J. Takamatsu, H. Kimura, and K. Ikeuchi. Generation of a task model by integrating multiple observations of human demonstrations. In *Proceedings of the IEEE Intl. Conf. on Robotics and Automation (ICRA '02)*, volume 2, pages 1545–1550, May 2002.

[GSAH2001] G. Grunwald, G. Schreiber, A. Albu-Schaffer, and G. Hirzinger. Touch: The direct type of human interaction with a redundant service robot. In *Robot and Human Interactive Communication, 2001. Proceedings. 10th IEEE International Workshop on*, pages 347–352, 2001.

[YKY2002] Y. Yokokohji, Y. Kitaoka, and T. Yoshikawa. Motion capture from demonstrator's viewpoint and its application to robot teaching. In *Proceedings of the IEEE Intl. Conf. on Robotics and Automation (ICRA '02)*, volume 2, pages 1551–1558, May 2002.

[TOKI2002] J. Takamatsu, K. Ogawara, H. Kimura, and K. Ikeuchi. Correcting observation errors for assembly task recognition. In *Intelligent Robots and System, 2002. IEEE/RSJ International Conference on*, volume 1, pages 232–237, 2002.

[OSK2002] H. Onda, T. Suehiro, and K. Kitagaki. Teaching by demonstration of assembly motion in vr - non-deterministic search-type motion in the teaching stage. In *Intelligent Robots and System, 2002. IEEE/RSJ International Conference on*, volume 3, pages 3066–3072, 2002.

[ZRDZ2002] R. Zollner, O. Rogalla, R. Dillmann, and M. Zollner. Understanding users intention: programming fine manipulation tasks by demonstration. In *Intelligent Robots and System, 2002. IEEE/RSJ International Conference on*, volume 2, pages 1114–1119, 2002.

[FHD1998] H. Friedrich, J. Holle, and R. Dillmann. Interactive generation of flexible

Part II

Component-Based Robotics

Trends in Component-Based Robotics

Davide Brugali[1], Alex Brooks[2], Anthony Cowley[3], Carle Côté[4], Antonio C. Domínguez-Brito[5], Dominic Létourneau[4], Françis Michaud[4], and Christian Schlegel[6]

[1] Universitá degli Studi di Bergamo, Italy brugali@unibg.it
[2] University of Sydney, AUSTRALIA a.brooks@cas.edu.au
[3] Federal University of Minas Gerais, MG, Brasil chaimo@dcc.ufmg.br
[4] Université de Sherbrooke, Department of Electrical Engineering and Computer Engineering, Sherbrooke (Québec), CANADA {Dominic.Letourneau, Carle.Cote, Francois.Michaud}@USherbrooke.ca
[5] Universidad de Las Palmas de Gran Canaria, Spain adominguez@iusiani.ulpgc.es
[6] University of Applied Sciences Ulm schlegel@fh-ulm.de

1 Introduction

Component-Based Software Engineering (CBSE) is an approach that has arisen in the software engineering community in the last decade. It aims to shift the emphasis in system-building from traditional programming to composing software systems from a mixture of off-the-shelf and custom-built components [Cas00, HC01, Szy02, DW98, CC01]. Component-Based Software Engineering is said to be primarily concerned with three functions [HC01]:

1. Developing software from pre-produced parts
2. The ability to reuse those parts in other applications
3. Easily maintaining and customizing those parts to produce new functions and features

As robotics systems are becoming more complex and distributed there is the need to promote a similar environment, where systems can be constructed as the composition and integration of reusable building blocks. System modularity and interoperability are key factors that enable the development of reusable software. If a system is modular, its functionalities can be customized by replacing individual components. When two or more systems are interoperable, they can be (re)used as components of more complex systems.

As an analogy, in the electronics domain re-usable off-the-shelf electronic components have been available for many years in the form of integrated chips (ICs) which can be bought and deployed in other parts of the world. This is possible because each IC packages a clear set of functionality and provides a

D. Brugali (Ed.): Software Eng. for Experimental Robotics. STAR 30, pp. 135–142, 2007.

well-defined external interface. Furthermore, numerous standard tools exist to design electronic devices based on the composition, assembly and combination of these electronic components.

Out of the multiple definitions for software components we can find in literature, we adopt the one given in [Szy02] as it illustrates clearly the concept:

"A software component is a unit of composition with contractually specified interfaces and explicit context dependencies only. A software component can be deployed independently and is subject to composition by third parties".

According to the previous definition, a software component is a piece of software that has been independently developed from where it is going to be used. It offers a well-defined external interface that hides its internals, and it is independent of context until instantiation time. In addition, it can be deployed without modifications by third parties. Also, it needs to come with a clear specification of what it requires and provides in order to be able to be composed with other components.

In an attempt to clarify the definition, we offer four properties of components identified by Collins-Cope [CC01]:

1. A component is a binary (non-source-code) unit of deployment
2. A component implements (one or more) well-defined interfaces
3. A component provides access to an inter-related set of functionality
4. A component may have its behaviour customized in well-defined manners without access to the source code.

Components can be put together in various configurations to form a system. This flexibility comes from modularity, enforced by the contractual nature of the interfaces between components. Interfaces and their specifications are viewed in isolation from any specific component that may implement or use the interface. In addition, the contractual nature of interfaces allows components on either side of the interface to be developed in mutual ignorance. Designing each component and interface in isolation allows a component that implements an interface to be seamlessly replaced with a different component having the same interface.

The component approach improves software development by reducing the amount of code that has to be written by the application designer. In particular, assembling previously-existing components greatly reduces time to test new applications. When a component is used in a large number of systems by different developers, the knowledge about the component usage, robustness and efficiency is available in the user community. The more a component is used the less it costs.

From the component developer's viewpoint, domain analysis is the most critical phase. In order to be broadly reusable, individual components have to be written to meet the requirements of a variety of applications in a specific domain. The current trend is toward a "product line" approach: the developer designs and implements a complete set of components for a family of similar applications.

From the application developer's viewpoint, the user requirements specification should be made by having in mind which components are already available. According to Boehm [Boe95], "in the old process, system requirements drove capabilities. In the new process, capabilities will drive system requirements...it is not a requirement if you can't afford it". In the early stages of an application development, the designer has to take into consideration which components are available, which integration effort they require, and whether to reuse them as they are or build new components from scratch.

2 Opportunities to Exploit CBSE in Robotics

General purpose component approaches require far too much software engineering knowledge, in particular if one has to address all the challenges of robotics software development and integration. Therefore, there is a need for a methodology and technology that tailor general purpose component approaches such that one can take advantage from the progress made in the software engineering community without being forced to become a software engineering expert as robotics researcher.

The following domain characteristics make experimental robotics particularly suited to a component-based development approach:

Inherent Complexity Even a fairly simple robotic system, a single vehicle working in isolation, is complex. Correct operation requires the interaction of a number of sensors, actuators and algorithms. Robotics software always requires to cope with the inherent complexity of concurrent activities, the deployment of software components on networked computers, ranging from embedded systems to personal computers, a bunch of different platforms, operating systems and programming languages and even with different real-time requirements. Typically, only some parts require hard real-time and for many parts soft real-time or even no real-time requirement exist.

Requirements for Flexibility Researchers need systems that are flexible enough to allow them to experiment with one particular aspect or algorithm, without their experimentation having large repercussions for the rest of the system. The possibility of the development of software systems based on interoperable components would be extremely beneficial to the research community since every research group, even the smallest, would have the possibility to concentrate its efforts on a small piece of the robotics puzzle. For example, experts in automated planning could experiment new path planning algorithms for a mobile robot relying on the obstacle avoidance and self-localization functionalities encapsulated in off-the-shelf components. Reuse of consolidated and shared components would allow different teams to test their algorithms on common benchmarks in order to assess performance.

Distributed Environments Robotic systems are inherently distributed. When dealing with single vehicles it's often convenient to develop on workstations, controlling robots remotely. Often the processing power on a single robot is insufficient, requiring the use of off-board processing power. Code should not have to change if the off-board components are moved on-board. Increasingly, complex single vehicles are becoming multi-processor [UAC04]. Multi-robot systems and sensor networks have many more distribution issues. Location transparency allows components to be distributed readily and easily, according to processing and bandwidth constraints.

The chapters in the second Part of this book advocate Component-Based Software Engineering (CBSE) as an valuable approach for facing the challenges listed above. When the application domain, like Robotics, has rapidly evolving requirements, a computational model which allows a higher adaptability of the reusable software components and flexibility in their pattern of interactions is mandatory.

The Sidebar *Software Architectures* by Patricia Lago and Hans van Vliet reports on the role that software architectures play in the software development process.

In the following sections we identify four issues that need to be addressed in order to define a computational model of reusable components for robotic software applications. Finally Section 3 draws the relevant conclusions.

2.1 Specifying a Component Internal Behavior

Certain level of uniformity in behavior and structure in software components, in spite of the functionality they have individually, is critical to allow easy integration and replacement of interchangeable components.

Components have an internal state, which reflects their knowledge. What are the specific requirements of a component model for robotic applications with respects to the definition of its internal state? In other words, which requirements in terms of internal structure should components have in order to apply CBSE to robotics? At first sight, a question comes naturally out from a development approach based on integrating software components. Is the whole the joining of its parts?, or, in other words, is a system just the joining of its components?

Components have a behavior which is determined by the set of admissible operations on their internal state. What are the specific requirements of a component model for robotic applications with respects to the definition of its behavior?

Chapter *CoolBOT: a Component Model and Software Infrastructure for Robotics* by Antonio Dominguez-Brito et al. will present a component model for robotic applications that faces the above issues on internal structure and behavior uniformity. The proposed component model captures two key aspects that enable component integration: the component behavior should be *observable* and *controllable* by an external supervisor (e.g. another component).

By *observable* we mean that components have some internal aspects about their internal state which could be observed and monitored by an external entity. Examples of observable aspects of a component are its execution priority and its error handling behavior in the occurrence of any faulty situation.

As to *controllable* we mean that an external supervisor should be provided with means and mechanisms to control and to affect some aspects of a specific component during its execution. Examples of some of these controllable aspects might be, for instance, forcing the component to transit to specific states, changing the priority at which a component is executed at runtime, injecting artificially an exception in a component in order to originate a faulty situation during its execution, interrupting, suspending or resuming its execution.

2.2 Specifying a Component External Interface

Components have a message-based interface, which define the set of messages that the component can interpret and process. What are the specific requirements of a component model for robotic applications with respects to the definition of its external interface?

While hardware platform designers benefit from reusable device components with standard interfaces, application developers benefit from the existence of tested, documented components that implement common algorithms and techniques. This composition approach requires that each component must be strictly limited in scope and designed such that it functions entirely in its own context, completely independent of other components, or any container application. A clear definition of a component's interface facilitates modular robot design by allowing an incremental construction system wherein software capabilities may be added to a platform in step with additional hardware capabilities.

To aid this process, it is often helpful to consider the potential explicit coupling between components. This is the familiar process of interface definition, but with the requirement that any transaction between components be entirely self-contained and independent of context. This context-free, explicit coupling can be used as a blueprint for the functional scope of the component: all functionality should follow directly from the explicitly defined component interface. In this way, the component's general identity is immediately established by how it may be coupled with other components.

Chapter *ROCI: Strongly Typed Component Interfaces for Multi-Robot Teams Programming* by Luiz Chaimowicz et al. presents the ROCI approach for the definition of component interfaces that relies on the use of strongly typed, self describing data structures. The Components built using this type of interface can be coupled in a simple and bug proof manner, making the task of the application developer easier. The ROCI approach has been applied to the programming teams of robots.

2.3 Enforcing Components Interoperability

Robotics software always requires to cope with the inherent complexity of concurrent activities, the deployment of software components on networked computers, ranging from embedded systems to personal computers, a bunch of different platforms, operating systems and programming languages and even with different real-time requirements. Typically, only some parts require hard real-time and for many parts soft real-time or even no real-time requirements exist.

The basic idea of the approach presented in Chapter *Communication Patterns as Key Towards Component Interoperability* by Christian Schlegel is to provide a small set of generic communication patterns that support easy integration of robotic components. All component interactions are squeezed into those predefined patterns. Using communication patterns with given access modes prevents the user of a component from puzzling over the semantics and behavior of both component interfaces and usage of component services.

Communication patterns relieve the component builder from error-prone details of distributed and concurrent systems by providing approved and reusable solutions for inter-component interactions. They handle complex synchronization problems in distributed systems and decouple the component internals from the externally visible and standardized behavior.

Communication patterns provide the basis for providing dynamic wiring. Dynamic wiring supports a context and task dependent assembly of components as is needed in nearly all robotics architectures.

2.4 Supporting Components Integration

Many existing software components can be found in the robotics community and it would be beneficial to reuse them in an integrated fashion using their initial implementation, saving time and avoiding introducing errors when reimplementing everything from scratch. Unfortunately, integration of existing software components is difficult knowing that they are typically developed independently, following their own set of requirements (e.g. timing, communication protocol, programming language, operating system, objectives, applications). Moreover, those components are often available "as is", with not much support from their developers and with minimal documentation. Reusability in this context is challenging but crucial for the evolution of the field, avoiding becoming experts in all the related areas that must be integrated.

Chapter *Using MARIE for Mobile Robot Component Development and Integration* by Carle Côté et al. presents MARIE, a software architecture that addresses three important issues in component-based robotic software development:

1. Being able to interconnect heterogeneous software components, from legacy to novel "state of the art" software components.
2. Being able to support a wide range of communication protocols and mechanisms to cope with the fact that there is no unified protocol available, and no consensus has yet emerged from the robotic software community.
3. Being able to support multiple sets of concepts and abstractions within the integration frameworks, to help multidisciplinary team members to collaborate without having them to become experts in every aspect of robotics software development.

MARIE's layered software architecture provides integration tools to help wrapping existing and new application functionalities in reusable software blocks. Each layer abstracts and manages different aspects needed to integrate heterogeneous software components together, such as communications, data handling, shared data types, distributed computing, low-level operating system functions, component architecture, reusable software blocks and system deployment, etc. With this approach, MARIE offers a flexible and extendable software architecture suitable in various integration situations, and allows software developers to work at the required abstraction level to craft reusable software blocks or to build robotic systems.

2.5 Deploying Components in Large-Scale Robotic Systems

A particular issue which is often overlooked concerns work that is required after a set of interoperable components have been developed, downloaded or purchased. There are several steps which must occur before reliable operation of the system can be achieved. Components must be connected to form a system, configured with appropriate parameters, and deployed onto the hardware hosts. To ensure reliable operation and to detect and correct any faults, the components must be monitored effectively.

For small systems operating over a short period of time, these extra steps can be and often are performed manually. For more realistic systems with more components, distributed over more hosts, or required to operate continuously for long periods of time, these steps can come to dominate the total effort required. In addition to the failures caused by components' internals, larger systems are more likely to exhibit communication failures and errors due to the interactions between components. The mean time between failures in the system can become far smaller than the mean time between failures of the least reliable constituent component.

Chapter *Orca: a Component Model and Repository* by Alex Brooks et al. identifies some of the more problematic issues and discusses ways in which a CBSE framework can assist developers and system integrators in overcoming them. The discussion is based on experience using Orca to implement systems ranging from single indoor robots to teams of heterogeneous outdoor vehicles. Details of several of these systems are presented.

3 Conclusions

The different solutions presented are giving hints concerning what a complete component-based approach should provide in order to support the development of reusable robotic software building blocks:

- A component model that allows the observation and control of its internal behavior.
- Well defined component interfaces and data structures that clearly define its usage modalities.
- Communication patterns that enable the interconnection of reusable components.
- Communication abstractions that support the interoperability of heterogeneous components.
- Well identified components repositories that simplify documentation, retrieval, and deployment of a large number of reusable components.

The combination of all these concepts is by itself a hot topic for future research.

References

[Boe95] B. Boehm, *A technical perspective on systems integration*, Proceedings of the SEI/MCC Symposium on the Use of COTS in Systems Integration (1995).

[Cas00] Castek, *Component-based development: The concepts, technology and methodology*, Castek Company's white paper, available at www.castek.com, 2000.

[CC01] M. Collins-Cope, *Component based development and advanced OO design*, White paper, Ratio Group Ltd., 2001.

[DW98] D. D'Souza and A. Wills, *Objects, components, and frameworks with UML: The catalysis approach*, Addison-Wesley, 1998.

[HC01] G. T. Heineman and W. T. Councill (eds.), *Component-based software engineering : putting the pieces together*, Addison-Wesley, Boston, 2001.

[Szy02] C. Szyperski, *Component software - beyond object-oriented programming*, Addison-Wesley / ACM Press, 2002.

[UAC04] C. Urmson, J. Anhalt, M. Clark, T. Galatali, J. P. Gonzalez, J. Gowdy, A. Gutierrez, S. Harbaugh, M. Johnson-Roberson, H. Kato, P. L Koon, K. Peterson, B. K. Smith, S. Spiker, E. Tryzelaar, and W. Whittaker, *High speed navigation of unrehearsed terrain: Red team technology for grand challenge 2004*, Tech. Report TR-04-37, Robotics Institute, Carnegie Mellon University, Pittsburgh, PA, June 2004.

CoolBOT: A Component Model and Software Infrastructure for Robotics*

Antonio C. Domínguez-Brito, Daniel Hernández-Sosa, José Isern-González, and Jorge Cabrera-Gámez

IUSIANI - Universidad de Las Palmas de Gran Canaria (ULPGC), Spain
{adominguez,dhernandez,jisern}@iusiani.ulpgc.es,jcabrera@dis.ulpgc.es

1 Introduction

In general, we face recurrently some common problems when programming robotic systems: multithreading and multiprocessing, distributed computing, hardware abstraction, hardware and software integration, multiple levels of abstraction and control, the development of a programming tool for a group of users which may become wide and diverse, etc. In this document we will introduce CoolBOT, a component oriented programming framework implementing primitives and mechanisms aimed to support the resolution of some of these common problems. This framework allows building systems by integrating "off-the-shelf" software components following a port automata model [SVK97] that fosters controllability and observability. Next section, Section 2, will introduce some of the recurrent problems we can find when developing the software infrastructure aimed to control a robotic system. In Section 3 we introduce a component oriented programming framework for programming robotic systems called CoolBOT which is the main subject of this chapter. In Section 4 we will outline the use of CoolBOT for building a real example. Finally in Section 5 we outline some of the conclusions we have drawn along the way of building and using CoolBOT.

2 Recurrent Problems

There are some questions and problems that appear repeatedly and recurrently in every robotic system which is not strange to solve again and again

* This work has been supported by research projects *PI2003/160* and *PI2003/165* funded by the Autonomous Government of Canary Islands (Gobierno de Canarias - Consejera de Educacin, Cultura y Deportes, Spain), by the MEC (Ministerio de Educacin y Ciencia, Spain) and FEDER research project *TIN2004-07087*, and by the ULPGC research projects *UNI2004/11*, *UNI2004/25* and *UNI2005/18*.

D. Brugali (Ed.): Software Eng. for Experimental Robotics. STAR 30, pp. 143–168, 2007.

when we face the design of the system control software, and its implementation for each new system or project. Our own experiences and necessities created at our lab the real demand of investing resources and research effort in building a common software infrastructure implementing primitives and mechanisms to facilitate the resolution of some of these recurrent problems. In the next subsections we enumerate the problems we had in mind when developing the software infrastructure which is the subject of this contribution: CoolBOT, a C++ component-oriented programming framework for programming robotic systems.

2.1 Uniformity, Controllability, Observability and Robustness.

Certain level of uniformity in internal structure and external interface in software components is critical to allow a basic uniform treatment of components in spite of their individual functionality. This is a wide-spread and well-know principle in the operating systems community. In fact, one of the main challenges an operating system developer must face is how to model processes and threads for a specific operating system, and which services and primitives it provides for such a model. All operating systems impose on the programs they can execute a uniform internal structure and a uniform external interface in order to make them uniformly treatable, administrable and executable, independently of the functionality these programs have.

Moreover, a uniform internal structure for components facilitates its observability and controllability, i.e. the possibility of monitoring and controlling the inner state of a component. At the same time, component's internal uniformity sets a real basis for the development of debugging and profiling tools, and makes component design and implementation less error-prone.

Furthermore, robustness is a permanent design goal in every robotic project. Having systems integrated by components which can be observed and controlled externally constitutes a fundamental feature that facilitates the design of robust systems.

2.2 Concurrent/Parallel Programming

Concurrency and parallelism are serious issues to which is necessary to dedicate many efforts when programming robotic systems. Simultaneous and non-synchronized accesses to the same resources by multiple units of execution are in the origin of a wide set of problems. The use of multiple operating systems with different thread models introduces even a higher level of difficulty. A developer should not be worried about simultaneous non-synchronized accesses to shared resources or critical sections, the programming infrastructure should care about all of it behind the scenes.

2.3 Real Time Requirements

In general, it is possible to establish a taxonomy dividing robotic systems in two big groups attending to their real time requirements. On one side there are systems with *hard real time* constraints. Their requirements correspond usually to the strict observation of deadlines in the achievement of different and/or multiple tasks in run-time, because otherwise, the safety of the system and its environment during operation can not be guaranteed. On the other side, we have systems with *soft real time* requirements. In these systems it is possible to miss spuriously deadlines for tasks without compromising system safe operation.

A software infrastructure aimed to programming robotic systems should support at least soft real time system implementations providing means to know when missed deadlines have happened in a system (*cognizant failures* [Gat92]). Supporting mechanisms and primitives for guaranteeing hard real time requirements would need a tighter coupling between the software infrastructure and the mechanisms and primitives it provides, and the underlying operating system where it runs.

2.4 Distributed Programming

Components should be integrable and reachable in a system with independence of the machines where they run. From a developer's point of view, whether components were in the same computer, or in a distinct one residing in the same computer network, should be transparent in terms of component integration and interconnection.

3 CoolBOT

CoolBOT [DB03] [DBHSIGCG04] is a C++ component-oriented framework for programming robotic systems that allows designing systems in terms of composition and integration of software components. Each software component [Szy99] is an independent execution unit which provides a given functionality, hidden behind an external interface specifying clearly which data it needs and which data it produces. Components, once defined and built, may be instantiated, integrated and used as many times as needed.

CoolBOT has been mainly originated by very practical reasons. While developing several robotic systems we got to a point where the necessity of some common software infrastructure was a clear demand. This infrastructure had to be generic enough to design any imaginable architecture for the projects we were involved at that moment. At the same time, it had to allow us to integrate software more easily. As one of the results of these projects we developed an agent-based software framework called *CAV* (**C**ontrol **A**rchitecture for **A**ctive **V**ision Systems) [DBHTCG00]. An active vision system termed *DESEO* (**DE**tección, **SE**guimiento y Reconocimiento de **O**bjetos)

[HTCGCS+99] aimed to detect, track and recognize faces, and an entertainment museum robot called *Eldi* [DBCGHS+01] were developed using CAV as software infrastructure. CAV was a tool that allowed modelling software as networks of interconnected software agents. It provided mechanisms for intercommunication between agents, whether residing in a remote computer or in the local machine. It lacked many primitives and resources we consider were also necessary, like a more rich set of intercommunication mechanisms, and a support to make multithreading less error-prone and more systematic. Specially, it lacked mechanisms to facilitate software integration that still remained being an important problem for us. Further work and experiences using CAV has driven us to design and build CoolBOT, whose evolution can be tracked along different documents [DBAC00], [CGDBHS02], [DB03] and [DBHSIGCG04], being [HSDBGACG05] one of last ones. CoolBOT, the framework we will present in the rest of sections of this chapter, is the current result of this line of work, and its name does not stand for any acronym, it only reflects our desire of disposing of a robotic framework to make software easier and "cool" to integrate and reuse.

3.1 Port Automata Model for Components

In CoolBOT, components are modelled as *Port Automata* [SVK97]. This concept establishes a clear distinction between the internal functionality of an active entity, an automaton, and its external interface, sets of input and output ports. Fig. 1 displays the external view of a component where the component itself is represented by a circle, input ports (the i_k's) by the arrows oriented towards the circle, and output ports (the o_k's) by the arrows oriented outwards. Fig. 2 depicts an example of the internal view of a component, concretely the automaton that models it, where circles are states of the automaton, and arrows, transitions between states. Transitions are triggered by events (the e_k's) caused either by incoming data through an input port, or by an internal condition, or by a combination of both. Double circles indicate automaton final states.

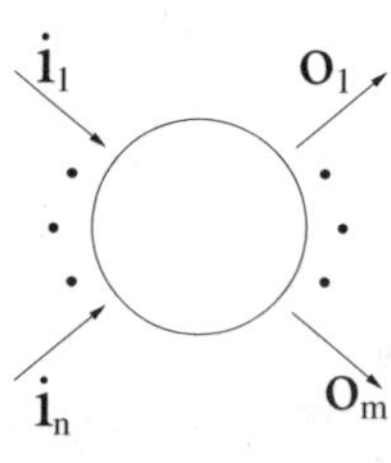

Fig. 1. Component external view.

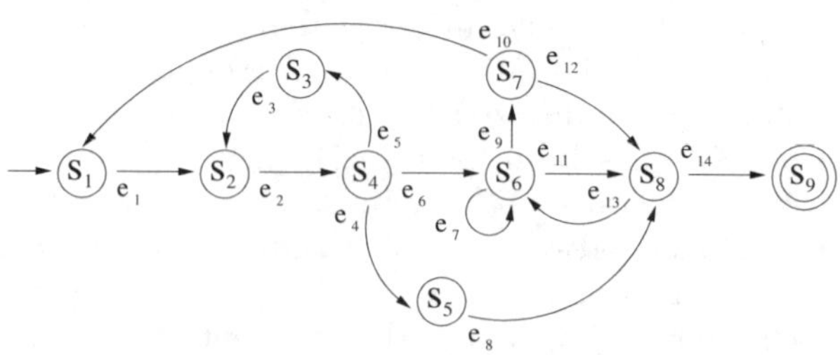

Fig. 2. Component internal view.

CoolBOT components interact and inter communicate each other by means of *port connections* established among their input and output ports. Data are transmitted through *port connections* in discrete units of information called *port packets*. *Port packets* are also classified by their type, and usually each input and output port can only accept a specific set of port packet types.

3.2 Observability and Controllability

Components should be observable enough to know whether they are working correctly or not, and in that case, they should be controllable enough to make some adjustment in their internal behavior to regulate and adjust their operation. CoolBOT introduces two kinds of variables as facilities in order to support monitoring and control of components.

- *Observable variables*: Represent features of components that should be of interest from outside, they are externally observable and permit publishing aspects of components which are meaningful in terms of control, or just for observability and monitoring purposes.
- *Controllable variables*: Represent aspects of components which can be externally controlled, i.e., modified or updated. Thence, through them the internal behavior of a component can be controlled.

Additionally, to guarantee external observation and control, CoolBOT components provide by default two important ports: the *control port* and the *monitoring port*, both depicted in Fig. 3.

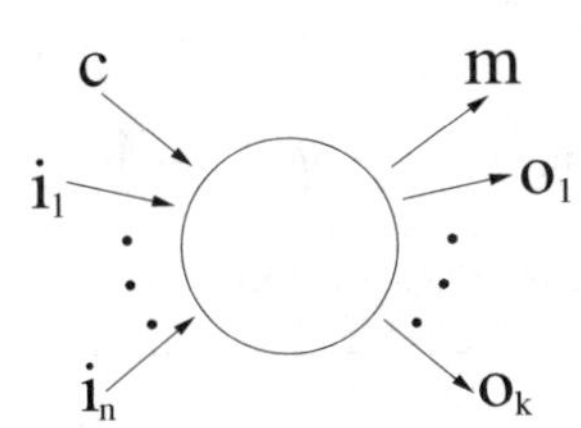

Fig. 3. The *control* port, *c*, and the *monitoring* port, *m*.

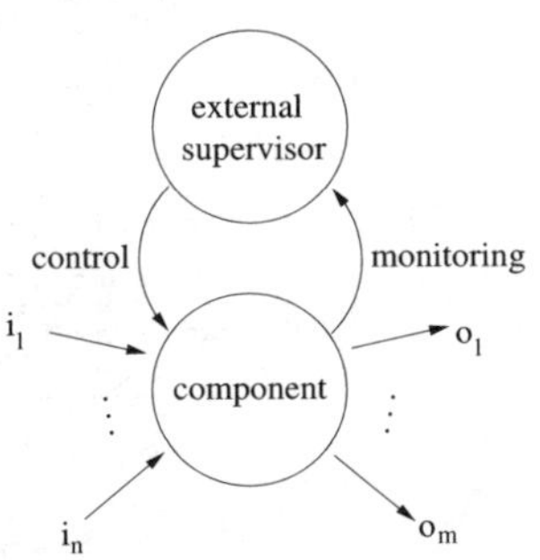

Fig. 4. A typical component control loop.

- The *monitoring port*: This is a public output port by means of which component *observable variables* are published. Through this port, an external observer or supervisor can observe and monitor a component.
- The *control port*: This is a public input port through which component *controllable variables* are modified and updated. An external controller or supervisor can control a component by means of it.

CoolBOT provides components with several default observable and controllable variables, in addition to the observable and controllable variables specific to each component. A brief description of these default variables is shown in Tables 1 and 2 (the symbols representing each variable are given beside variable names in parenthesis).

Table 1. Default observable variables.

Default Observable Variables	
Name	Brief Description
state (s)	Automaton state where the component is situated.
priority (p)	Current component execution priority.
config (c)	Requests a supervised configuration change, or confirms configuration commands.
result (r)	Result of execution.
error description (ed)	Error description indicating a locally unrecoverable exception.

Fig. 4 illustrates graphically a typical execution control loop for a component using another component as *external supervisor*. This is possible thanks to the default *control* and *monitoring* ports CoolBOT imposes on all components.

Table 2. Default controllable variables.

Default Controllable Variables	
Name	Brief Description
new state (ns)	Desired automaton state where the component is commanded to go.
new priority (np)	Desired execution priority the component is commanded to have.
new exception (nex)	Externally induced exception.
new config (nc)	Component's configuration can be modified and updated during execution through this controllable variable.

3.3 Default Automaton

Internally all components are modelled using the same default state automaton, the *default automaton*, shown in Fig. 5 that contains all possible control paths that a component may follow. In the figure, the transitions that rule the automaton are labelled to indicate the event that triggers each one. Some of them correspond to internal events: *ok*, *exception*, *attempt*, *last attempt* and *finish*. The other ones indicate default controllable variable changes: ns_r, ns_re, ns_s, ns_d, *np*, *nc* and *nex*. Subscripts in ns_i indicate which state has been commanded: *r* (*running* state), *re* (*ready* state), *s* (*suspended* state), and *d* (dead state). Event *np* happens when an external supervisor forces a priority change; event *nc* when a change of configuration is commanded from the external supervisor; and event *nex* occurs when the supervisor injects “artificially the occurrence of an exception.

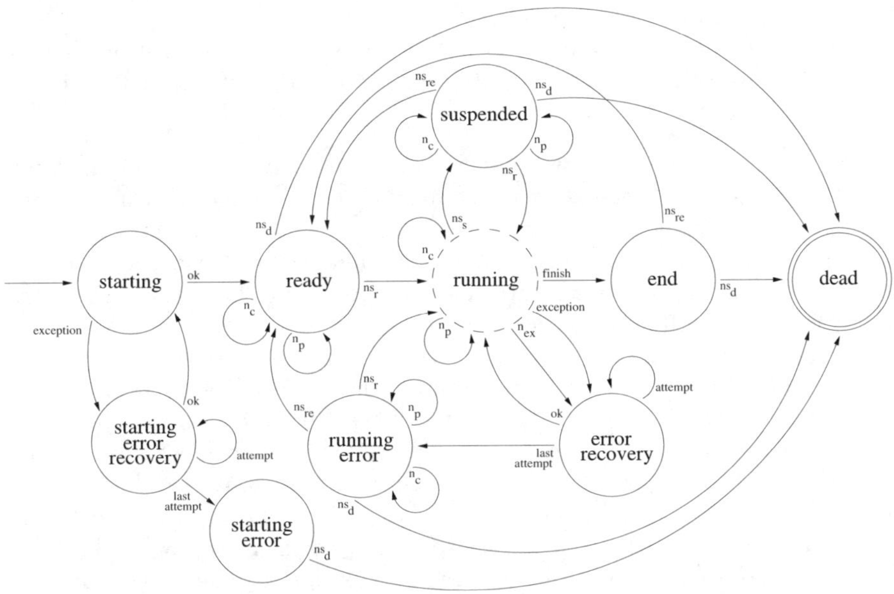

Fig. 5. The Default Automaton.

The *default automaton* is said to be *“controllable* because it can be brought in finite time by an *external supervisor* by using the *control port*, to any of the controllable states of the automaton, which are: *ready*, *running*, *suspended* and *dead*. The rest of states are reachable only internally, and from them, a transition to one of the controllable states can be forced externally.

The *running* state, the dashed state in Fig. 5, constitutes or represents the part of the automaton that implements the specific functionality of the component, and it is called the *user automaton*. The *user automaton* varies among components depending on their functionality, and it is defined during component design and development. The initial state of a *user automaton* constitutes its *entry state*.

Having a look to Fig. 5 we can see how CoolBOT components evolve along their execution time, since they are launched until they finish their execution. Basically, the *default automaton* organize the life of a component in several phases which correspond to different states:

- *starting*: This state is devised to allocate resources for a correct task execution.

- *ready*: In this state the component is ready to execute, and waits for an external command to start (ns_r).
- *running*: In this pseudo-state the component is carrying out its specific task running its *user automaton.*
- *suspended*: In this state the component is suspended and remains idle until ordered to transit to other state.
- *end*: In this state the component has just finished a task execution. Getting to this state the component publishes the result of its execution, if any, through its *monitoring port.*

Furthermore, there are two pair of states conceived for handling faulty situations during execution. One of them devised to face errors during resource allocation (*starting error recovery* and *starting error* states), and the other one though to deal with errors during task execution (*error recovery* and *running error* states). These states are part of the support CoolBOT provides for error and exception handling in components.

In fact, in the same manner that exceptions constitute a useful concept which is present in numerous programming languages (C++, Java, etc.) to separate error handling from the normal flow of instructions in a program., we may define and use exceptions in CoolBOT components. Thus, a component may define a list of *component exceptions* to signal and handle erroneous, exceptional or abnormal situations during execution, where each component exception is defined using the pattern shown in Fig. 6. In the figure `<exceptionId>` is a number identifying the exception; `<description>` is a description of the exception; `<handler>` is an optional handler to try a recovery procedure, optionally `<retries>` indicates the number of recovery attempts, and `<period>` specifies the period in milliseconds between attempts; `<onSuccessHandler>` is an optional handler to be executed in case of a successful recovery; and finally, `<onFailureHandler>` is also an optional handler which is executed when all recovery exception attempts have failed.

```
On Exception: <exceptionId>
    <description>
    [<handler> [<retries> <period>]]
    [<onSuccessHandler>]
    [<onFailureHandler>]
```

Fig. 6. Exception Pattern.

All in all, if an exception comes up during runtime in a component, the component is driven to one of the recovery error states (*starting error recovery* or *error recovery*) where the handler for the exception is tried several times. Only when all recovery attempts have been unsuccessful the component is took to a situation of unrecovered exception (*starting error* or *running error*). In this situation, an error description is published through the monitoring port of the component, and it remains idle waiting for external supervision.

3.4 Multithreading

CoolBOT components are *weakly coupled entities* that execute concurrently or in parallel, on their own initiative, in order to achieve their own independent objectives. Thence, components are not only data structures, but execution units as well. In fact, CoolBOT components are mapped as *threads* when they are in execution; Win32 threads in Windows, and POSIX threads in GNU/Linux, which are the two operating systems where CoolBOT is supported in its current version.

```
void Component::kernel()
{
  PortPacket* packet;

  initialization();

  while(true)
  {
    packet=waitForSomething(inputPorts);

    if(!processPortPacket(packet)) break;
  } // end of while

  finalization();
}
```

Fig. 7. A component in execution: A continuous loop processing port packets. Simplified C++ kernel code.

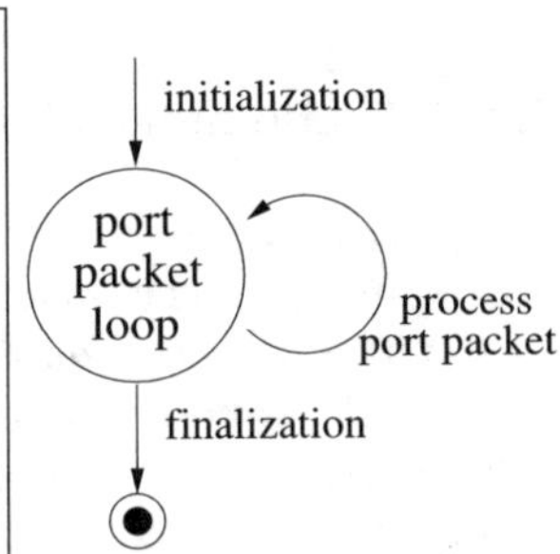

Fig. 8. Component simplified kernel. Graphical representation.

At runtime a CoolBOT component executes a continuous loop processing port packets where the component carries out different actions depending on which input port has received each port packet, and in which state of its automaton the component is. In general, incoming port packets trigger the transitions which form the components automaton. Figures 7 and 8 shows respectively a simplified C++ version and a graphical representation of the kernel code which drives a component at runtime. Observe that according to this code, depending on at which rate or frequency a component is receiving port packets, the component would be blocked waiting for incoming port packets for most of its execution time. Thus, components could be described as *data-flow-driven machines*, processing when they dispose of data in their input ports, and otherwise, keeping idle, waiting for processing new input port packets.

The loop for processing incoming packets which is located at the kernel of a component can be multithreaded or not, as Fig. 9 illustrates. In general, a component needs for its execution at least a thread in the underlying operating system, called the *main thread.* Furthermore, in order to make a component more responsive, it is possible to distribute the attention of a component on different input ports using different threads of execution. These threads are called *port threads*, and they are responsible for disjoint sets of component's input ports, and similarly to the *main thread*, they are also *data-flow driven* through their incoming port packets. The *main thread* executes the automaton of the component, and controls and observes any other components *port threads*, if any. The *main thread* is also responsible for maintaining the consistency of the internal data structures that constitute the internal state of the whole component. As a result, in CoolBOT any component corresponds with at least an underlying thread, concretely the *main thread*, guaranteeing in this way that if the code of a component hangs completely the systems will not collapse.

3.5 Inter Component Communications

Analogously to modern operating systems that provide *IPC* (*Inter Process Communications*) mechanisms to inter communicate processes, CoolBOT provides *Inter Component Communications* or *ICC* mechanisms to allow components to interact and communicate among them. CoolBOT *ICC* mechanisms are carried out by means of input ports, output ports, and ports connections. Communications are one of the most fragile aspects of distributed systems. In CoolBOT, the rationale for defining standard methods for data communications between components is to ease inter operation among components that have been developed independently, offering optimized and reliable communication abstractions.

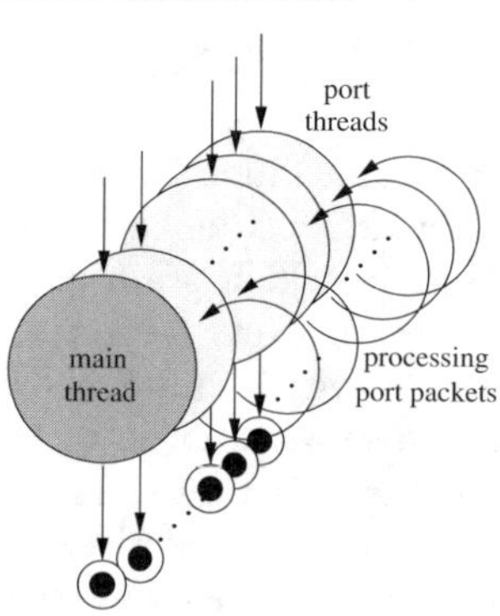

Fig. 9. Multiple threads.

There are several types of output and input ports supported by CoolBOT which combined adequately in the form of port connections implement different protocols of interaction between components. Fig. 10 shows all the different types of output and input ports supported by CoolBOT (on the right and on the left respectively), and all the possible combinations between them to form port connections. Below the arrows, the cardinality of each type of port connection is also indicated. Remember that a *port connection* between an output port and an input port is only possible whether both ports match the type of port packets they accept, besides, it is necessary that they also constitute a compatible pair of output and input ports. Each type of port connection implements a different protocol of interaction between components, Table 3 resumes the protocol implemented by each one (the letter in parenthesis beside each port name is its corresponding symbol).

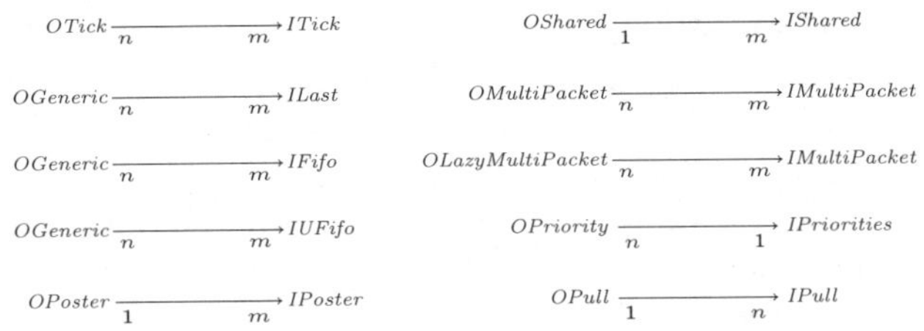

Fig. 10. Port connections ($\forall\, n,m \in \mathbb{N}; n,m \geq 1$).

Table 3. Port connections.

Output Port	Input Port	Brief Description
OTick (t)	*ITick (t)*	*Tick Connections*: Implements a protocol to signal events between components
OGeneric (g)	*ILast (l)* *IFifo (f)* *IUFifo (uf)*	*Last, Fifo and Unbounded Fifo Connections*: There is a queue (fifo) of port packets in the input port
OPoster (p)	*IPoster (p)*	*Poster Connections*: There is a "master copy" of port packets in the output port, input ports keep local copies
OShared (s)	*IShared (s)*	*Shared Connections*: Components share a "shared memory" residing in the output port. Implements a protocol of shared memory
OMultiPacket (mp) *OLazyMultiPacket (lmp)*	*IMultiPacket (mp)*	*Multi Packet Connections*: Accepts multiple types of port packets through the same port connection
OPrioriy (pr)	*IPriorities (pr)*	*Priority Connections*: Implements a protocol of sending with priority
OPull (pu)	*IPull (pu)*	*Pull Connections*: Implements a protocol of request/answer between components

In CoolBOT there are two basic communication models for port connections. The first one is the *push model*, where the initiative for sending port packets relies on the output port part of a connection, the data producer, which sends port packets on its own, completely uncoupled from its consumers. The second model for communications is the *pull model*, where packets are emitted when the input port part of a connection, the consumer, demands new data to process. In this model the consumer keeps the initiative, sending a request to the producer whenever it need a new incoming port packet. In all cases consumers connect to producers *by subscription*, so if a component connects one of its input ports to the output port of another component it gets subscribed to this port conforming a port connection (this is indicated by the cardinality in Fig. 10). Thus, depending on the nature and typology of the connected ports the cardinality of the relationship between components can be either "many to many, "one to many, "many to one, etc.

All types of port connections in Table 3 observe a push communication model that allows for uncoupled and asynchronous interactions among components. The only one type of port connection that utilizes a pull communication model is the *pull connection* (an *OPull*/*IPull* pair of ports).

Additionally to the connections shown in Fig. 10 there are other possible pairs of port connections which are collectively called *single multi packet* connections. All the possibilities appear in Fig. 11. Their main characteristic is that they combine a type of output or input port that accepts only one type of port packet (a *single packet* port) with a type of input or output port that accepts several types of port packets (a *multi packet* port).

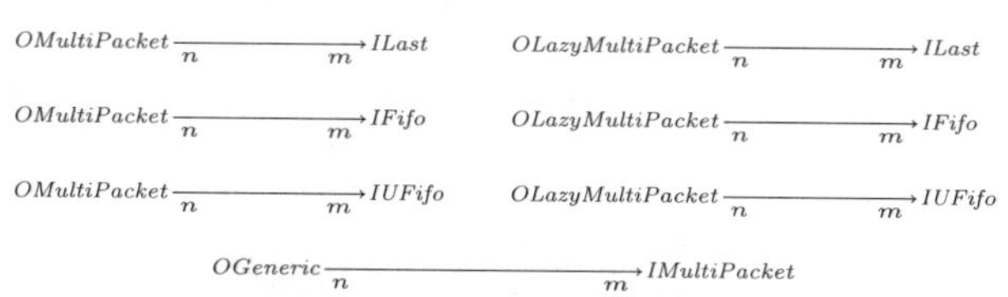

Fig. 11. Simple multi packet connections ($\forall\, n,m \in \mathbb{N}; n,m \geq 1$).

All in all, CoolBOT provides a wide and rich set of output and input ports for inter component communications that frees developers of an important workload in software design and development, what implies a less error-prone system programming. In addition, the typology of different ports offers a number of communications patterns wide enough to allow a broad variety of component interactions.

3.6 Distributed Components

CoolBOT components reside in specific computers when they are instantiated and executed. Frequently, the functionality of a component which runs in one computer may be needed in other machine elsewhere in a computer network. CoolBOT provides a mechanism based on *"mediator"* components [GHJV95] called *proxy components*, and network processes called *CoolBOT servers*, to allow components to be accessible through a computer network.

3.7 Real Time Features

Currently, CoolBOT is supported under the Windows family of operating systems, and GNU/Linux. Given that those operating systems are not real-time, and that CoolBOT relies on the thread model of the underlying operating systems to map its support for multithreading, the framework can not guarantee any realtime requirement by itself. However, CoolBOT provides some mechanisms and resources which usually are characteristic of real time operating systems (timers, watchdogs, etc.), and it can keep soft real-time requirements, since the operating systems where it runs actually can keep such requirements [RS98] [Gop01].

3.8 Modularity and Hierarchy

CoolBOT components are classified into two kinds: *atomic* and *compound components. Atomic components* have been mainly devised in order to abstract low level hardware layers to control sensors and/or effectors; to interface and/or to wrap third party software and libraries, and to implement generic algorithms.

Compound components are compositions of instances of several components which can be either atomic or compound. The functionality of a compound component resides in its *supervisor*, depicted in Fig. 12, which controls and observes the execution of its *local components* through the control and monitoring ports present in all of them. The *supervisor* of a compound component concentrates the control flow of a composition of components, and in the same way that in atomic components, it follows the control graph defined by the default automaton of Fig. 5. All in all, *compound components* use the functionality of instances of other atomic or compound components to implement its own functionality. Moreover, they, in turn, can be integrated and composed hierarchically with other components to form new compound components.

3.9 Developing Components

The process of developing CoolBOT components and systems is resumed in six steps on Fig. 13:

(1) *Definition and Design*: In this step the component is completely defined and designed. This comprises deciding whether it is atomic or not, its functionality – *user automaton* –, thread use, resources, output and input ports, port packets, observable and controllable variables, exceptions, timers and watchdogs.

(2) *Skeleton Generation*: There is already a small set of developed components, and component examples in the form of C++ classes illustrating the most common patterns of use. It is possible to start from one of them

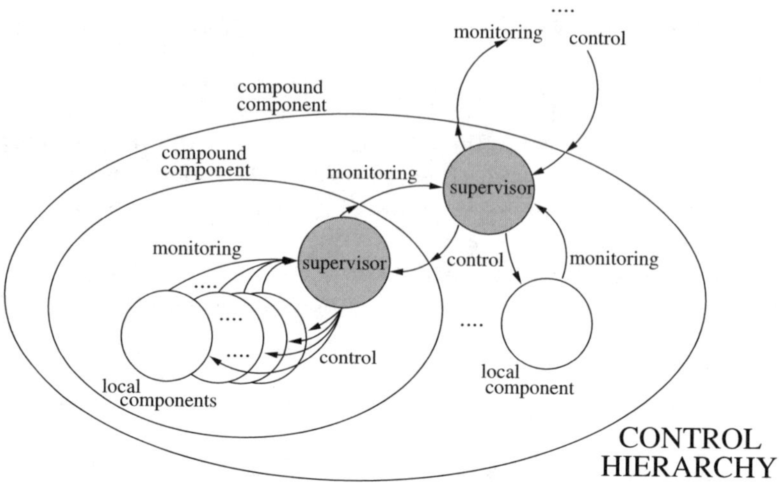

Fig. 12. Compound components.

as skeleton, or generate a new one from a component skeleton description file.

(3) *Code Fulfilling*: Using the components skeleton obtained in the previous step we complete the component fulfilling its code.

(4) *Library Generation*: Then the component is compiled obtaining a library.

(5) *System Integration*: Next the component may be integrated in a system alone or with other components.

(6) *System Generation*: And finally, the system gets compiled and an executable system is obtained. With it, we can already test the whole integration with our component.

4 Using CoolBOT

CoolBOT is a C++ framework, and every CoolBOT component is a C++ class. Next subsections will introduce briefly how it is the process of building components, and how to integrate them to have a working system following the process illustrated in Fig. 13.

4.1 Building a System

CoolBOT has been conceived to promote integrability, incremental design and robustness of software developments in robotics. In this section, a simple demonstrator will be outlined to illustrate how such principles manifest in systems built using CoolBOT. The first level of this simple demonstrator is shown in Fig. 14 and it is made up of four components: the *Pioneer*

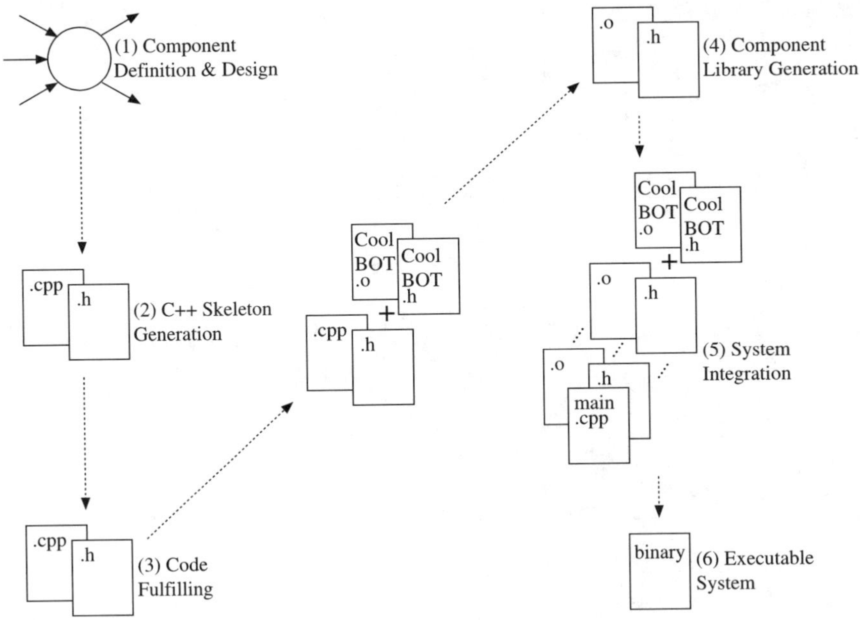

Fig. 13. The development process.

component which encapsulates the set of sensors and effectors provided by an ActivMedia Robotics Pioneer robot; the *PF Fusion* component which is a potential field fuser; the *Strategic PF* component that transforms high level movement commands into combinations of potential fields; and finally, the *Joystick Navigation* component which allows controlling the robot using a joystick. The integration shown in the figure makes the robot able to avoid obstacles while executing a high level movement command like, for example, going to a specific destination point.

CoolBOT maps all components as C++ classes. As an example, Fig. 15 shows the skeleton of the C++ mapping for the component *Pioneer* of Fig. 14. In general, all components inherits a default external interface and a default internal structure through superclass `CoolBOT:Component` as illustrated in Figures 15 and 16. In this manner the framework imposes on components a default minimum external and internal behavior.

As mentioned in subsection 3.9, in the process of developing a component, once we have generated the C++ skeleton for a specific component (step *"(2) C++ Skeleton Generation"* of Fig. 13), we have to start fulfilling its code (step *"(3) Code Fulfilling"* in the same figure). Mind that the C++ skeleton obtained in step *(3)* only has structure but it is empty in terms of functionality. This functionality should be coded in the transitions, and in the entry and exit sections of each one of the states of the component's automaton, which are

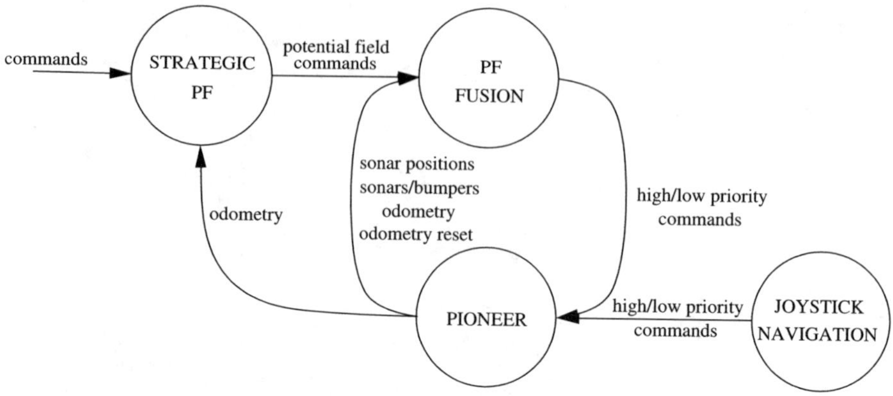

Fig. 14. The avoiding level

```
...
#include "coolbot.h"
using namespace CoolBOT;
...
namespace PioneerSpace
{
    ...
    class Pioneer: public Component
    {
        public:
            ...
        private:
            ...
    };
    ...
}
```

Fig. 15. Component *Pioneer*: class skeleton.

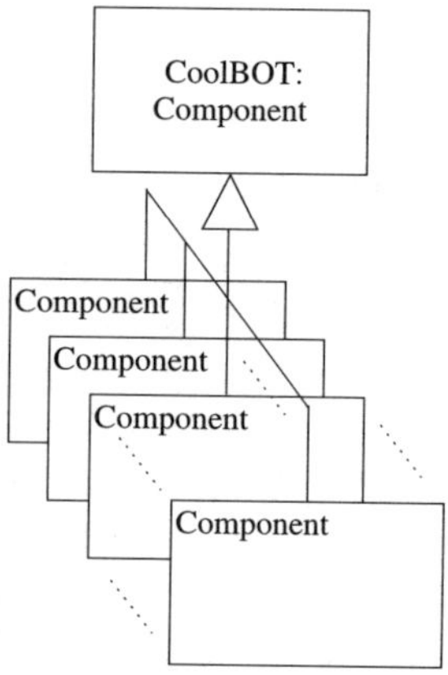

Fig. 16. The superclass `CoolBOT:Component`.

mapped as function members in the C++ skeleton class. An example appears in Fig. 17, where a transition for the component*Pioneer* of Fig. 14 is shown, concretely it maps a transition in a state called `main` which is triggered when an incoming packet reaches the input port called `commands`.

In order to clarify a bit more the code, in Fig. 18 we show the external public interface of input and output ports for this component. Output and input port types are indicated by their symbols in parentheses (consult Table 3 for the symbols associated with each type). Tables 5 and 4 describes respectively the public interface of output and input ports for the component *Pioneer*.

```
...
// state main: begin
...
int Pioneer::_mainCommands_(int port, Component* pComp) {
    Pioneer* pThis=static_cast<Pioneer*>(pComp);
    pThis->_debugPortTransition_(port);

    ...

    return pThis->_currentState_;
}
...
// state main: end
...
```

Fig. 17. Component *Pioneer*: a transition.

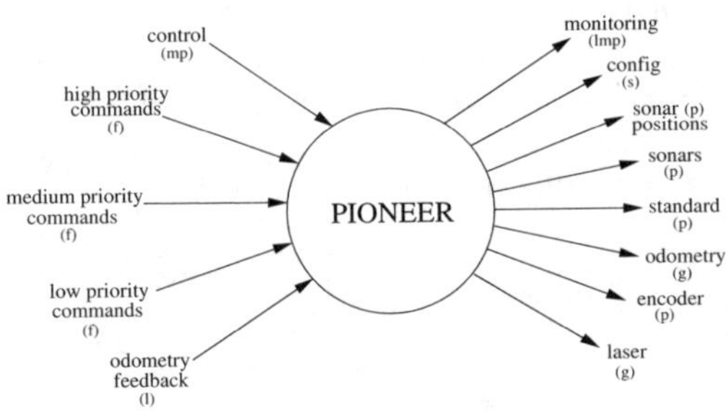

Fig. 18. Component *Pioneer*: external interface.

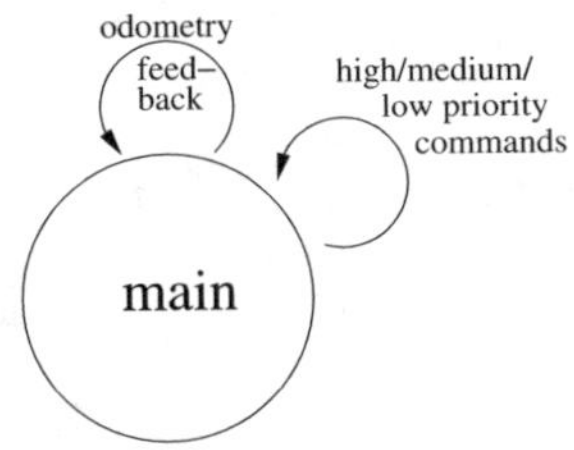

Fig. 19. Component *Pioneer*: *user autoamaton.*

Besides, in Fig. 19 appears the internal *user automaton* for this component which corresponds to the super-state *running* in Fig. 5. The behavior of the *Pioneer* component is not complex. As it is shown in the figure, the components user automaton has only one state named *main* state. In this state, the component is listening continuously the serial port to publish the information that is periodically sent through the port by the robot. The component only formats conveniently the information which it receives, and publishes it through its different output ports (figure 18). At the same time, in this state, the component is attending all its input ports. Thus it sends to the robot any command that it receives through its command ports, and it corrects internally the odometry that is published when a correction is received through its *odometry feedback* input port. Once in *main* state, the component is running there until, either it is commanded by means of the *control* port to go to one of the *default automaton* states, or, on the contrary, it gets into *error recovery* state because an exception has been raised.

Table 4. Component *Pioneer*: public output ports.

Public Output Ports	
Name	Brief Description
monitoring	This is the default *monitoring* output port by means of which the component publishes all its observable variables. It is an *OLazyMultiPacket*.
config	This is an *OShared* output port. It publishes configuration data of the robot to which the component is attached.
sonar positions	Once connected to a physical robot this *OPoster* output port publishes the coordinates of the different sonars the robot provides.
sonars	By means of this *OPoster* output port the component provides robots sonar readings.
standard	Some additional information related to the robot is published in this *OPoster* output port: bumper status, motor stall, ...
odometry	Through this *OGeneric* output port the component publishes periodically the robots odometry.
encoder	This *OPoster* output port allows direct access to the motor encoders of the robot.
laser	This *OGeneric* output port resends information related to a **SICK** laser range scanner, if the robot is equipped with one.

Table 5. Component *Pioneer*: public input ports.

Public Input Ports	
Name	Brief Description
control	This is the components default *control* port through which *Pioneers* controllable variables may be modified and updated. This component does not add any new controllable variable. It is a *IMultiPacket* input port.
high priority commands	*IFifo* input port that accepts commands to be sent to the robot the component is connected to. High priority input port.
medium priority commands	*IFifo* input port that accepts commands to be sent to the robot the component is connected to. Medium priority input port.
low priority commands	*IFifo* input port that accepts commands to be sent to the robot the component is connected to. Low priority input port.
odometry feedback	It is a *ILast* input port, by means of it the odometry data produced can be affected by a correction.

In Fig. 14 all components are *atomic*, the system is simple enough to have no necessity of using a hierarchy of *compound* components. To continue with our small system. We have added new components as a second control level to our little demonstrator of Fig. 14, this is depicted in Fig. 20. Observe that the system has now two new components, the *Sick Laser* component which controls a Sick laser range finder and the *Scan Alignment* component that performs map building and self localization using a SLAM (Simultaneous Localization And Mapping) algorithm.

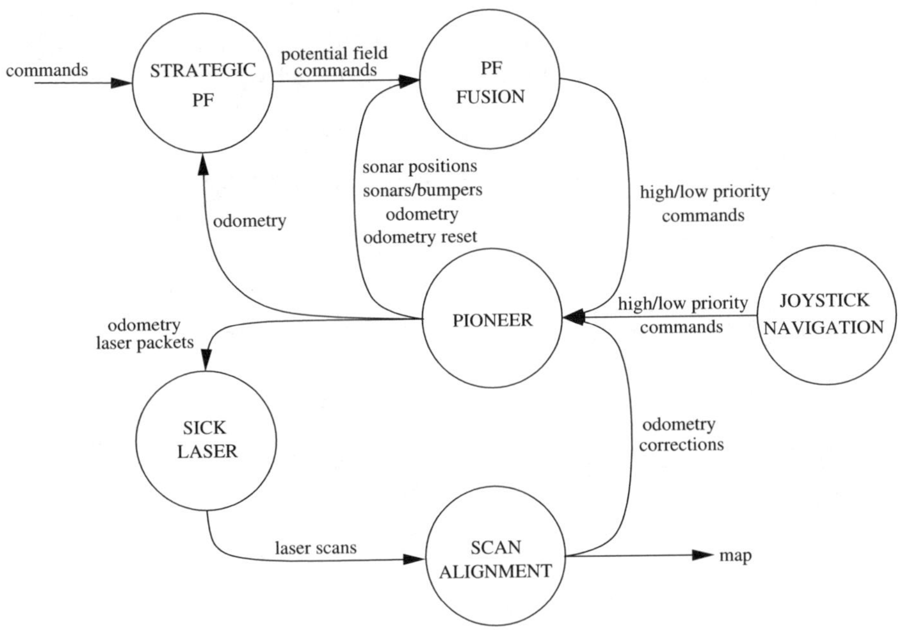

Fig. 20. The whole system

In particular, the integration depicted in Fig. 20 works as follows. The *Sick Laser* component executes a continuous loop, periodically reading the laser device and publishing the data by means of a poster output port. This component receives odometry packets from the *Pioneer* component in order to be able to stamp every scan with the robot pose at the time the scan was taken. The *Scan Alignment* component communicates with the *Pioneer* and *Sick Laser* components. It reads scans and preforms pose corrections which are sent to the *Pioneer* component that in turn will correct the pose and its associated covariance. The component does this by means of a simultaneous localization and mapping algorithm which produces a map of the environment [LM94][LM97]. This map is also published in an output port to be used by any other component. So, now, the system allows a robot to make self-localization and map building, at the same time that it avoids obstacles.

4.2 Control Interface

In CoolBOT, as we have mentioned in section 3, every component presents a uniformity of external interface and internal structure. This is suitable for implementing a control system that allow us to connect to several components and control them. Such a system could be able to control and observe the behavior of a component by means of its *control* and *monitoring* default ports. At the same time it could be able to "sniff" port connections between components. This is very helpful when trying to debug such systems or to check its normal operation.

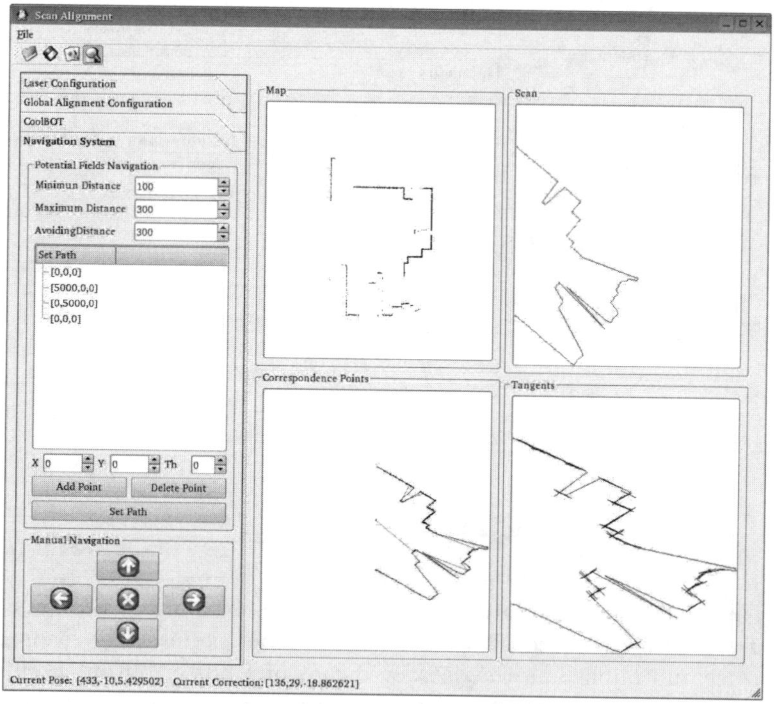

Fig. 21. The control interface.

For the system integration of Fig. 20 a graphical front-end was designed to control and observe the system during operation, a snapshot appears in Fig. 21. As we can see the figure shows the map, the current scan and the result of the corresponding points and tangent calculations. On the left side we can see part of the controlling interface which allow us to change the state of every component, and to connect to different ports of the system to have a look to the different data and results available (robot odometry, odometry corrections underdone, laser scans, the map as it gets built, etc.). Moreover, it is possible to change some operational *controllable* variables (frequency of laser scans, *Scan Alignment* component parameters, etc.). Additionally, it is possible to establish different paths the robot should follow during map building, or, as we have commented previously, to control the robot movements directly with a joystick device.

4.3 Test and Results

Robustness is an important aspect in robotics, and CoolBOT has aimed some of its infrastructure to facilitate robust system integration. As every component works in an autonomous and asynchronous fashion, the system does not have to stop running whenever any of its components enters into an non-recoverable error state. If this occurs the system can keep working with part of its functionality (if possible), or waiting for the faulty component to restart its normal execution.

Focusing on the robustness of the system we prepared several tests that show the behavior of the the system previously presented upon the malfunction of any of its components.

In the system of Fig. 20, the most important components are: the *Pioneer*, *Sick Laser* and the *Scan Alignment* components. Based on these three components, two tests were made. The first of them shows the way the system works whenever any of the components hangs, and the second one is related to the degradation of the system when the *Sick Laser* component stops running.

- **First Test**. The main goal of this test is to overcome a situation provoked by a faulty component without collapsing. The system should be able of keeping operation, either running, or waiting for the faulty component to restore the erroneous situation. As result of this test we found that the system did not collapse for any combination of its components in an non-recoverable error state. Table 6 summarizes the results derived from the test for every possible combination of error states in the components. The first three columns indicate which component is in a faulty situation (R stands for *Running* and E for *Error*), and the fourth one describes briefly the system behavior for each case. Any of the errors shown in Table 6 does not lead the system to a fatal situation, except if we consider the last case. The asynchronicity of execution, initiative and communications between components allow them to keep the system working until the faulty one is recovered, or the whole system is explicitly stopped.
- **Second test**. In Table 6 there are some situations during system operation that lead to odometry degradation. If the *Sick Laser* or the *Scan Alignment* component gets into erroneous operation, the SLAM algorithm responsible for the pose error, will not work. In this situation, the *Pioneer* component will keep working but as the robot moves, its pose will

degradate. For showing this degradation and its subsequent recovering we have made a real experiment where the *Sick Laser* is provided with a mechanism for reinitializing the connection with the laser device upon abnormal disconnection. Thus, for example if the serial connection to the laser is physically broken the component will keep trying to recover communications with the device by restarting the connection periodically. Therefore, when serial connection is eventually reestablished, the component restarts its normal operation. Once the laser is working properly, the *Scan Alignment* component will receive again new scans and will produce new corrections which, in turn, will be sent to the *Pioneer* component, finishing in this way, odometry degradation.

Table 6. Error States

Pioneer	Sick Laser	Scan Alignment	Description
R	*R*	*R*	Normal state, everything is right.
E	*R*	*R*	The *Pioneer* component does not work. The *Sick Laser* component publishes every scan with the last received pose. Notice that if the *Pioneer* component does not run, the robot also stops because of a watchdog event in such a way that the scans are published with the actual pose of the robot. The *Scan Alignment* component keep reducing the error of the map built so far.
R	*E*	*R*	The *Scan Alignment* component reduces errors in the current map, this is done by means of an iterative process to achieve consistent maps. But it stops integrating new scans to the map as the *Sick Laser* component is not working. In this state the robot pose degradates (explained in more detail in the second test).
E	*E*	*R*	The *Scan Alignment* component reduces errors of the current map built so far, but it does not add new scans to the map.
R	*R*	*E*	In this case the *Scan Alignment* component does not work. This leads to a degradation of the pose of the robot, but the rest of the system still works. This case is also the basis of the second test.
E	*R*	*E*	The *Sick Laser* component publishes scans with the last robot pose received.
R	*E*	*E*	The odometry of the robot degradates.
E	*E*	*E*	The system here does not collapse itself, but it can do nothing.

5 Conclusions

This chapter outlines a first operating version of CoolBOT, a component-oriented C++ framework where the software that controls a system is viewed as a dynamic network of units of execution modeled as port automata interconnected by means of port connections. CoolBOT is a tool that favors a programming methodology that fosters software integration, concurrency and parallelism, asynchronous execution, asynchronous inter communication and data-flow-driven processing.

CoolBOT imposes some uniformity on the units of execution it defines, CoolBOT components. This uniformity makes components externally observable and controllable, and treatable by the framework in an uniform and consistent way. CoolBOT also promotes a uniform approach for handling faulty situations, establishing a local and an external level of exception handling. Exceptions are first handled locally in the components where the exceptions come out. If they can not be handled at this local level, they are deferred to an external supervisor, in turn, this handling may scale going up in a hierarchy of control loops.

In developing CoolBOT we have collected and reused many design ideas which can be found in the literature. Out of the bibliography we can find in the field, CoolBOT owes a big deal of inspiration from mainly three works in the area:

- **G^{en}oM**: a work developed at CNRS-LAAS (Laboratoire d'Analyse et d'Architecture des Systmes) in Toulouse. G^{en}oM is a software framework that was built to support the software architecture for the mobile robots available at LASS. For a general introduction consult [FHC97], [MFB02] and for a deeper look there is a more detailed user manual [FHM]. Recently G^{en}oM has become an open source project available at `softs.laas.fr/openrobots` being part of a bigger initiative from this labaratory called *"LAAS Open Software for Autonomous Systems.* G^{en}oM has been a strong source of inspiration for modelling CoolBOT's components and, not surprisingly, they share many features. The main is that components in both frameworks have a well defined internal architecture organized around a state automaton. This characteristic, that does not compromise the functionality of the component, seems to be central in order to achieve properties like reactivity, uniform error recovery mechanisms and generic control and monitoring procedures.
- **Smartsoft**: [Sch06] developed by Schlegel and Wrst at FAW (Research Institute for Applied Knowledge) at the University of Ulm in Germany. For more details you can consult [SW99], [Sch03], and [Sch04]. Like G^{en}oM, SmartSoft is a software framework developed to support a software architecture for mobile robots. SmartSoft has some strong points of coincidence with CoolBOT. Both frameworks conceived a perception-action system as a network of components whose functionality and capabilities arise from the coordinated interaction of individual components. And for that purpose both frameworks provide a reduced set of communication patterns, that essentially implement a basic set of communication protocols with an

emphasis on asynchronicity. Both frameworks also hide the difficult issues of concurrency and synchronization to the programmer.
- **Chimera**: [Ste94] [SVK97] is a real-time operating system developed at CMU (Carnegie Mellon University) aimed to control robotic systems. Chimera is a multiprocessor real-time operating system designed specifically to support the development of dynamically reconfigurable software for robotic systems. The main inspiration we used from Chimera was the concept of *port automaton* given by Steenstrup in [SAM83]. In Chimera, the units of execution handled by the operating system are defined as software modules called *port-based objects* which are units of execution defined formally as *port automata*.

For us, CoolBOT is a long-term experimental tool of which this work is only a first design and development effort. Contrarily to other approaches in the literature, it needs still further experimentation. Only the development of complex demonstrators out of a wide-enough set of robust ready-to-use components will allow us to determine how long CoolBOT scale, and in general to validate this programming tool in the long-term.

References

[CGDBHS02] J. Cabrera-Gámez, A. C. Domínguez-Brito, and D. Hernández-Sosa, *Coolbot: A component-oriented programming framework for robotics*, Lecture Notes in Computer Science, Sensor Based Intelligent Robots: International-Workshop Dagstuhl Castle, Germany, October 15-20, 2000 Selected Revised Papers, vol. 2238, pp. 282–304, Springer Berlin / Heidelberg, January 2002.

[DBHTCG00] A. C. Domínguez-Brito, F. M. Hernández-Tejera, and J. Cabrera-Gámez, *A Control Architecture for Active Vision Systems*, Frontiers in Artificial Intelligence and Applications: Pattern Recognition and Applications, M.I. Torres and A. Sanfeliu (eds.), pp. 144-153, IOS Press, Amsterdam., 2000.

[DBAC00] A. C. Domínguez-Brito, M. Andersson, and H. I. Christensen, *A Software Architecture for Programming Robotic Systems based on the Discrete Event System Paradigm*, Technical Report CVAP 244, Centre for Autonomous Systems, KTH - Royal Institute of Technology), S-100 44 Stockholm, Sweden, September 2000.

[DBCGHS+01] A. C. Domínguez-Brito, J. Cabrera-Gámez, D. Hernández-Sosa, M. Castrillón-Santana, J. Lorenzo-Navarro, J. Isern-González, C. Guerra-Artal, I. Pérez-Pérez, A. Falcón-Martel, M. Hernández-Tejera, and J. Méndez-Rodríguez, *Eldi: An Agent Based Museum Robot*, Systems Science, ISSN 0137-1223 **27** (2001), no. 4, 119–128.

[DB03] A. C. Domínguez-Brito, *CoolBOT: a Component-Oriented Programming Framework for Robotics*, Ph.D. thesis, Dpto. Informática y Sistemas, Universidad de Las Palmas de Gran Canaria (mozart.dis.ulpgc.es/Publications/publications.html), September 2003.

[DBHSIGCG04] A. C. Domínguez-Brito, D. Hernández-Sosa, J. Isern-González, and J. Cabrera-Gámez, *Integrating Robotics Software*, IEEE International Conference on Robotics and Automation, New Orleans, USA, April 2004.

[FHC97] S. Fleury, M. Herrb, and R. Chatila, *$G^{en}oM$: A Tool for the Specification and the Implementation of Operating Modules in a Distributed Robot Architecture*, Proceedings of the IEEE/RSJ International Conference on Intelligent Robots and Systems (IROS) (Grenoble, Francia), September 1997, pp. 842–848.

[FHM] S. Fleury, M. Herrb, and A. Mallet, *$G^{en}oM$: User's Guide*, softs.laas.fr/openrobots/distfiles/genom-manual.pdf, -.

[Gat92] E. Gat, *Integrating Planning and Reacting in a Heterogeneous Asynchronous Architecture for Controlling Real-World Mobile Robots*, Proceedings of the Tenth National Conference on Artificial Intelligence (San Jose, CA, USA), July 1992, pp. 809–815.

[GHJV95] E. Gamma, R. Helm, R. Johnson, and J. Vlissides, *Design patterns: Elements of reusable object-oriented software*, Addison-Wesley Professional Computing Series, Addison-Wesley, 1995.

[Gop01] K. Gopalan, *Real-Time Support in General Purpose Operating Systems*, ECSL Technical Report TR92, Experimental Computer Systems Labs, Computer Science Department. State University of New York at Stony Brook, 2001.

[HTCGCS+99] M. Hernández-Tejera, J. Cabrera-Gámez, M. Castrillón-Santana, A. C. Domínguez-Brito, C. Guerra-Artal, D. Hernández-Sosa, and J. Isern-González, *DESEO: an Active Vision System for Detection, Tracking and Recognition*, vol. 1542, pp. 376–391, International Conference on Vision Systems, Las Palmas de Gran Canaria, Spain. Springer-Verlag Lecture Notes on Computer Science, January 1999, ISBN 3-540-65459-3.

[HSDBGACG05] D. Hernández-Sosa, A. C. Domínguez-Brito, C. Guerra-Artal, and J. Cabrera-Gámez, *Runtime self-adaptation in a component-based robotic framework*, IEEE/RSJ International Conference on Inteligent Robots and Systems (IROS 2005), August 2-6, Edmonton, Alberta, Canada, 2005.

[LM94] F. Lu and E. Milios, *Robot pose estimation in unknown environments by matching 2d range scans*, Proc. IEEE Comp. Soc. Conf. on Computer Vision and Pattern Recognition, Seattle, USA, 1994.

[LM97] F. Lu and E. Milios, *Globally consistent range scan alignment for environment mapping*, Autonomous Robots **4** (1997), no. 4, 333–349.

[MFB02] A. Mallet, S. Fleury, and H. Bruyninckx, *A specification of generic robotics software components: future evolutions of $G^{en}oM$ in the orocos context*, IROS 2002 Conference, 2002.

[MH01] R. Monson-Haefal, *Enterprise JavaBeans*, O'Reilly, September 2001.

[Rof01] Jason T. Roff, *Ado: Activex data objects*, O'Reilly, 2001.

[RS98] K. Ramamritham and C. Shen, *Using Windows NT for Real-Time Applications: Experimental Observations and Recommendations*, IEEE Real-Time Technology and Applications Symposium (merl.com/people/shen/pubs/rtas98.pdf), 1998.

[SAM83] M. Steenstrup, M. A. Arbib, and E. G. Manes, *Port automata and the algebra of concurrent processes*, Journal of Computer and System Sciences **27** (1983), 29–50.

[Sch06] C. Schlegel, *Communication Patterns as Key Towards Component Interoperability*, In D.Brugali (Ed.) *Software Engineering for Experimental Robotics*, Spinger STAR series, 2006

[Sch03] C. Schlegel, *Overview of the OROCOS::SmartSoft Approach*, http://www1.faw.uni-ulm.de/orocos/, 2003.

[Sch04] C. Schlegel, *Navigation and execution for mobile robots in dynamic environments: An integrated approach*, Ph.D. thesis, University of Ulm, 2004.

[Ste94] D. B. Stewart, *Real-Time Software Design and Analysis of Reconfigurable Multi-Sensor Based Systems*, Ph.D. thesis, Carnegie Mellon University, Dept. Electrical and Computing Engineering, Pittsburgh, 1994.

[SVK97] D. B. Stewart, R. A. Volpe, and P. Khosla, *Design of Dynamically Reconfigurable Real-Time Software Using Port-Based Objects*, IEEE Transactions on Software Engineering **23** (1997), no. 12, 759–776.

[SW99] C. Schlegel and R. Wörz, *Interfacing Different Layers of a Multilayer Architecture for Sensorimotor Systems using the Object Oriented Framework SmartSoft*, Third European Workshop on Advanced Mobile Robots - Eurobot99. Zürich, Switzerland, September 1999.

[Szy99] C. Szyperski, *Component software: Beyond object-oriented programming*, Addison-Wesley, 1999.

ROCI: Strongly Typed Component Interfaces for Multi-robot Teams Programming

Anthony Cowley[1], Luiz Chaimowicz[2], and Camillo J. Taylor[1]

[1] GRASP Laboratory – University of Pennsylvania, PA, USA
{acowley,cjtaylor}@grasp.cis.upenn.edu
[2] Computer Science Department - Federal University of Minas Gerais, MG, Brazil
chaimo@dcc.ufmg.br

1 Introduction

The software development process in robotics has been changing in recent years. Instead of developing monolithic programs for specific robots, programmers are using software components to build all kinds of robotic applications. As discussed in Chapter *Trends in Component-Based Robotics* of this book [Bru06], component based development offers several advantages such as reuse of code, increased robustness, modularity and maintainability.

Considering this basic guideline, we have been developing ROCI – *Remote Objects Control Interface* [CCS2003, CHT2004b]. ROCI is a software platform for programming, tasking and monitoring multi-robot teams. In ROCI, applications are built in a bottom-up fashion from basic components called ROCI modules. A module encapsulates a process which acts on data available on its inputs and presents its results as outputs. Modules are self-contained and reusable, thus, complex tasks can be built by connecting the inputs and outputs of specific modules. We can say that these modules create the language of the ROCI network, allowing task designers to abstract from low level details and focus on high level application semantics. ROCI is specially suited for programming and monitoring distributed ensembles of robots and sensors, since modules can be connected across local networks in a transparent way and data can be transmitted and visualized in different formats.

One key characteristic of ROCI is its approach for developing robust interfaces to connect individual modules. In component based development, external interfaces should be clearly defined to allow an incremental and error resistant construction of complex applications from simpler, self-contained components. By making interfaces explicit and relying on strongly-typed, self-describing data structures, ROCI allows the development of robust applications. Moreover, ROCI's modularity supports the creation of parallel data flows which favors the development of efficient applications.

D. Brugali (Ed.): Software Eng. for Experimental Robotics. STAR 30, pp. 169–182, 2007.

In this chapter we present an overview of the ROCI framework and discuss in more detail its modular, strongly-typed design. The chapter is organized as follows: the next section describes the ROCI framework, giving an overview of its architecture and main features. Section 3 describes in more details some of ROCI's benefits for programming distributed robots. Specifically, we discuss modularity, strongly typed interfaces, parallelization, logging and web accessibility. In Section 4, we present a couple of examples that illustrate some these benefits and, in Section 5, we conclude the paper discussing some of ROCI's main contributions.

2 ROCI Overview

ROCI is a dynamic, self-describing, object-oriented, strongly typed programming framework for distributed robot teams. It provides programmers with a network transparent framework of strongly typed modules - assemblies of metadata, byte code, and machine code that can consume, process and produce information. ROCI modules are injectable (they can be automatically downloaded and started on a remote machine), reusable, browseable, and support automatic configuration via XML. These features, coupled with a dynamic database of available nodes and network services, allows a programmer to write code that utilizes networks of robots as resources instead of independent machines.

2.1 Modules and Tasks

The building blocks of a ROCI application are ROCI modules. Each module encapsulates a process which acts on data available on its inputs and presents its results on well defined outputs. In ROCI, there is no specification in the code relating to where input should come from or where output should go; the only specification is the type of data each computational block deals with. The key design principle of ROCI is to keep modules simple, with well designed interfaces and no feature overlap. This minimalist approach allows modules to be easily tested, composed and reused in different scenarios [CCT2006].

Complex tasks can be built by connecting the inputs and outputs of specific modules. A ROCI task is a way of describing an instance of a collection of ROCI modules to be run on a single node, and how they interact at runtime. It is defined in an XML file which specifies the modules that are needed to achieve the goal, any necessary module-specific parameters, and the connections between these modules. A good analogy is to view each of these modules as an integrated circuit (IC), that has inputs and outputs and does some processing. Complex circuits can be built by wiring several ICs together, and individual ICs can be reused in different circuits.

As will be explained in the next section these connections are made through a pin architecture that provides a strongly typed, network transparent communication framework. Pins can connect modules within the same task on

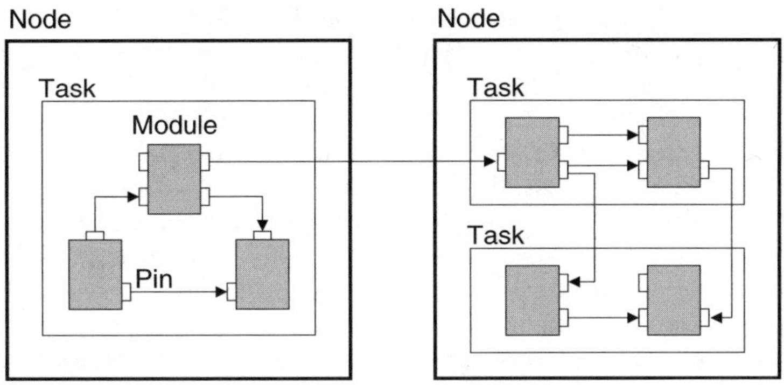

Fig. 1. ROCI Architecture: tasks are composed of modules and run inside nodes. Communication through pins can be seamlessly done between modules within the same task, modules in different tasks, or in different nodes.

the same computer, between tasks on the same computer, or between tasks on different computers. Figure 1 shows an example of this organization.

2.2 Pin Based Communication

The wiring that connects ROCI modules is the pin communications architecture. Basically, a pin provides the developer with an abstract communications endpoint. Pins in the system are nothing more than strongly typed fields of a module class, and thus connecting a producer pin to a consumer pin is as simple as setting a reference to the producer in the consumer's field. One can set a pin's input in two distinct ways that are each useful in different situations. On a lower level, pins can be connected dynamically during program execution. This can be accomplished by querying nodes on the ROCI network for available pins – usually with a type constraint – and may involve dynamically creating local pins to bind to the discovered remote pins. However, the simpler way of binding statically declared pins together is via the XML descriptions that define ROCI tasks.

As the producer generates data it assigns an updated reference to the latest data to its pin. This assignment causes the pin to fire messages to all of the registered consumers in the network, alerting them to the availability of fresh data. A typical usage of this system is for a module to make a blocking call to one of its input pins which returns once the input pin has gotten new data from the output pin it is registered to. Alternatively, a module can ask its input pin to copy over whatever data is available immediately, whether it is new or not. Pin data is time stamped, allowing the consumers to determine how current their data is. Together with logging mechanisms, the time stamps also allow the creation of sensor databases and the execution of distributed queries [CHT2004].

It is important to stress the advantages of such strongly typed communication channels. In ROCI, the data exchanged between modules are highly structured entities and not simply byte streams. This ensures that valid connections are made between different modules, increasing robustness. Also, it allows the development of specific "translator" modules based on these types that, for example, automatically format the information for an human operator. Additionally, the system supports what we call *blind polymorphism* in which the data being exchanged inherits from some base class and the upcast is made transparently on the source node, not on the destination. In this way, a producer module that exports a very complex data type that inherits from a more basic one will transmit an instance of the relevant, smaller, ancestor class to any consumer module that only needs the basic information. The blind polymorphism guarantees that only the necessary data is transmitted, and that the receiver need not be updated to parse new data types that it does not directly make use of.

2.3 ROCI Kernel

The kernel is the core control element of ROCI. It can be considered a high level OS that mediates the interactions of robots in a team. The kernel manages the Remote Procedure Call (RPC) system, the real-time network database, module and task allocation and injection, and a RESTful web service interface for remote monitoring and control. There is a copy of the kernel running on every entity that is part of the ROCI network (robots, remote sensors, etc.).

The system management functionality provided by the kernel focuses on resource discovery, usage, and prioritization. Peer-to-peer file discovery and sharing are included to facilitate automatic distribution of injected tasks. Usage statistics are monitored, and controllable, for both CPU and network load. These loads are monitored at the scale of modules and pins, respectively, thus allowing for an appropriately fine grain of control over resource allocation that is completely dynamic and modifiable at runtime.

For example, when the user requests a task be started on a given node, the kernel running locally on that node first ascertains whether or not the proper versions of component modules are available locally. If they are not, it queries the network for a node that does have the modules in question and downloads them automatically. Once the byte code of the modules that comprise the task all exists locally, the task is loaded. Resource allocation priorities may be set at startup, but they may also be modified at runtime by any module, thus allowing for reactive, context-aware, resource allocation. An example of such a dynamic behavior is one in which CPU resources are largely devoted to obstacle avoidance while a robot moves quickly, but shifted to target detection as the robot's rate of motion decreases.

The ability to support this type of flexibility is characteristic of ROCI's data-driven design. On the far end of the flexibility-spectrum, all the data that flows through ROCI is accessible as XML through a web browser. Towards the efficiency end of things, this same data is internally shared through more compact, event-based binary serialization. The pervasive use of metadata throughout the system allows for this style of multiple views of the same data. There is very little effort required to keep, for example, binary and XML serialization code paths in sync because translation between these forms, as well as HTML, Email, IM, SMS, etc., occurs automatically based on the metadata. The ultimate win is that the minimal metadata necessary to effect basic translation and formatting of data structures is present in the type declarations themselves. The reflective type system means that there is never an unidentifiable byte stream; instead, the programmer effort put into creating data structures is exploited again on the front-end to structure and label live data.

2.4 ROCI Browser

The job of presenting this network of functionality to the user falls upon the ROCI Browser. The browser's main job is to give a user command and control over the network as well as the ability to retrieve and visualize information from any one of the distributed nodes, and make informed decisions about network operations. The browser displays the multi-robot network hierarchically: the operator can browse nodes on the network, tasks running on each node, the modules that make up each task, and pins within those modules.

Using the browser, the user can start, stop and monitor the execution of tasks on the robots remotely, change task parameters, send relevant control information for the robots or even to tap into and display the outputs of pins for which display routines exist. Since modules and pins are self-describing entities, when the user browses through the tasks, he or she can immediately have a complete description of the modules and pins in use. Given this information, the browser can automatically start appropriate modules locally to tap into the remote data for visualization or processing purposes. This can be very useful for debugging purposes during development and for situational awareness during deployed execution. Figure 2 shows a snapshot of the browser during the execution of a multi-robot mission. In this mission, a team of robots should search an urban environment for a specific target while maintaining network connectivity and send a picture back to a base station [CCG2005].

Elaborate missions may also be constructed within the browser using *scripts*. ROCIScripts echo the domain language foundations of the ROCI task development process. In task development, semantically meaningful constructs such as modules and pins are reasoned about. Similarly, ROCIScript is based on the notion that the language in which a human commands a robot should be a conversation at an appropriately high level of abstraction. Furthermore, the specific abstractions relevant to a particular application may be used to define a compact working vocabulary. Rather that pre-determine a fixed-syntax

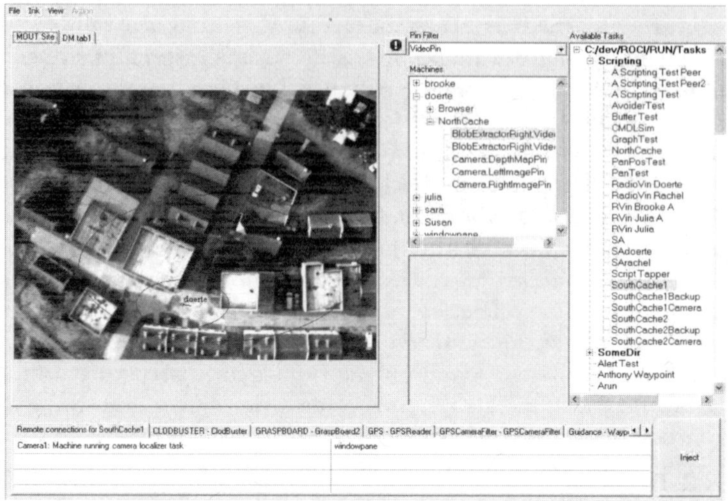

Fig. 2. ROCI Browser during the execution of a multi-robot mission: It is possible to monitor the status of each node, the tasks and modules being executed and even see the real time location of each robot on a map.

scripting language for robotics, ROCI extracts the scripting commands a robot supports from the modules that make up a task. In this way, the vocabulary used to script a particular task is always relevant to that task: the user is given a meaningful set of commands to work with for a given robot, and the robot runtime only need handle commands that are actually relevant to its capabilities. The scripting system lets developers create high level behavioral scripts that may be re-parameterized by any software package capable of editing XML. This parameter assignment may be human operator-driven through the browser, or it may be accomplished by another piece of software that can then inject the new script into the robot. The script interface thus represents a very high level of robot interfacing: commands such as *Search Room* or *Go To Waypoint* are implemented as scripts that may be invoked from a wide variety of programming languages, environments and GUIs.

3 Programming the ROCI Platform

The components of the ROCI framework presented so far make up a system that encourages a modular design style based on the construction of loosely coupled components. While the benefits of such design techniques are manifold, we have found that the interaction between software and well-designed robot hardware provides a fresh perspective that clarifies the benefits of this technique while simultaneously making the path to implementation more obvious.

3.1 Modular Design

ROCI modules, written by users, create the language of the ROCI network. This short statement touches upon two key issues: users are better positioned to establish their working language than a central ROCI development authority, and the domain language established can be far more powerful than any fixed system language. The intuition of the latter statement motivates the former. The desire to have non-kernel developers determine the primitives of the framework demands that the system be built around structures and standards that aid this effort. To this end, ROCI defines a module development model where the boundaries of an object, its interfaces, place bounds on its functionality. In other words, no module should contain functionality that could not be generally inferred from an inspection of the module's pins. Furthermore, we encourage developers to minimize the number of moving parts within a single module. This method of focusing the utility of any given module actually makes it far easier to describe what a module does to other developers, encourages reuse by not including unwanted functionality, minimizes functional overlap with other modules, eases task refactoring, and makes code maintenance simpler.

As discussed in [CCT2006], the watchwords in ROCI module design are "Keep It Simple". Developers should focus on creating small, stand-alone processing loops that do one thing and do it well. The benefits of such a design are many, but primarily we wish to avoid feature overlap and component complexity. Feature overlap occurs when multiple components are capable of doing the same thing. The danger here is not only confusion when overlapping components appear in the same project, but also a duplication of development effort and a greater chance for something we refer to as interface divergence. Redundant component functionality is an obvious inefficiency, but interface divergence is often the more dangerous of the two problems. This phenomenon occurs as basic functionality is expanded upon in different places in parallel. The end result is that there exist multiple ways of doing very similar things, but the different methods are not entirely compatible. This leads to correct usage patterns being applied incorrectly due to a change in which component provides the service; a change which can go unnoticed due to the overall similarity in functionality. Component complexity rises when a developer continues adding functionality to a single module in order to supposedly accomplish a near-term goal more rapidly. The result is that the component often becomes brittle and difficult to test. The brittleness comes from unchecked intra-component dependency growth, wherein each part of the component is dependent upon one or more other parts; a problem typical of monolithic designs. Testing difficulty is related to the number of unit tests required to adequately confirm a component's correctness, a number that rises combinatorially as functionality is added.

3.2 Strongly Typed Interfaces

Abstractly, one can view a ROCI task as being built on top of two distinct levels of protocol design. There is a high level application protocol defined by how the various modules are connected and ordered in the parallel data flows described above. This protocol is created by the task designer in simple declarative fashion in the task XML file. The low level protocol is established by pin type declarations that define how connected modules communicate, but say nothing for how end-to-end data flow is accomplished. Critically, the details of the low-level protocol are transparent from the high-level, and the high-level data flow protocols are invisible from the module level. This separation allows for the easy creation of application-level protocols, i.e. data flow patterns, which can be distributed to take advantage of whatever parallelism is offered by the unique IO patterns and priorities of the given application.

Viewed another way, semantic significance is created at the application protocol level. It is at this level that developers connect, for example, Position Pins or Video Pins. These constructs define a working set of ideas that are meaningful in the specific context of the application, as opposed to any sort of pre-defined language. Furthermore, they may properly be reasoned about at the application level. That is, the task developer need not be concerned with bits on the wire. Instead, the task designer works only with high-level, complex data structures and terminology specific to the desired behavior. ROCI pins, on the other hand, are concerned with establishing the bit-level protocol; how data is passed between functional elements. This protocol is derived from the structure of the pin data itself. The programmer defines various structures that allow for a clean, single-argument method invocation syntax, and this data structure definition work is exploited to define wire protocols. ROCI determines if there is any type ambiguity involved in a transaction between two connected pins, and, if so, is able to insert the necessary type metadata into the byte stream. For example, value types in the .NET framework are transferred by direct binary serialization while instances of reference types that are used polymorphically at runtime may be serialized, but must also be tagged with the metadata necessary for the receiver to parse the incoming data. This transaction protocol is automatically created by ROCI through type reflection.

3.3 Parallel Computing

Modular design promotes a style of application architecture that scales well, and is a natural fit for distributed computing. Whether the distribution is across a network or across a multi-core CPU, breaking an application into discrete, potentially-parallelizable parts is an important technique given modern hardware. Currently, many parallelization efforts focus on a particular application and a single target platform rather than more general techniques. Such ad-hoc efforts typically involve breaking a sequential application design into two or three minimally synchronized flows. Instead of focusing on such a top-down approach, we advocate a bottom-up approach that does not necessarily consider any particular hardware or application. Instead, we consider the potentially-parallelizable parts in isolation with the focus on functional integrity rather than parallelization.

The difficulty in actually determining the potential threading granularity of an application is, unfortunately, considerable. Indeed, significant portions of many applications have such rigid data dependencies that there is no way to avoid inherent linearity in processing. The field of robotics, however, offers new insight into the question of when individual application building blocks may execute in parallel via its characteristic utilization of multi-channel, variable rate IO.

Data input is, intrinsically, a blocking operation. It is an opening in the loop of a program that the calling thread may not have control over. Thus it makes sense that when one thread can not proceed for want of input, another thread that is not starved of data should proceed. The alternative of polling each sensor once per loop iteration limits all processing to the rate of the slowest sensor. Furthermore, such an architecture imposes constant latency across all data processing activities. Instead, we can easily address the natural desire to better utilize processing resources by addressing the flow that has data ready to be processed. Once we have broken the application into multiple parallel streams, we can also control, to a certain extent, the rate of iteration for each flow to better address latency requirements of the system. For example, Figure 3 shows a diagram of a typical robotic application: a robot has to maintain a particular heading until a color target is spotted. In this case, we can take advantage of multiple processing units and conform to the naturally asynchronous pattern created by the input devices: a high-frequency magnetometer and a low-frequency imaging sensor. Importantly, the only module that needs to be at all thread-aware is the motion planner that implicitly constructs an application-specific compound data type out of bearing and image data. As long as this fusion module maintains some imaging state, the motors can be controlled at the rate of the higher-frequency sensor.

Parallel data flows are characteristic of ROCI tasks. Each task can be seen as a collection of data flows that always progress forward, sometimes joining to form new streams. The joining points of these data flows define an application-specific protocol that, ultimately, describes what the application does but is entirely absent at the module level. In this fashion, we have effected a clean separation between application-level architecture and building block construction.

3.4 Logging

The same idea behind the creation of application-protocols through the composition of low-level type-based protocols can be seen in the way data logs are implemented in ROCI. Rather than jumping straight into a compound

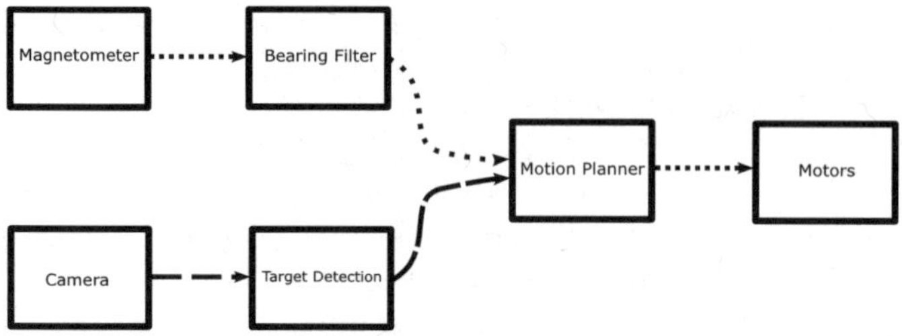

Fig. 3. Parallel data streams present a natural separation scheme for multi-threading. The dotted lines in this figure denote high-frequency data transactions, while the dashed lines represent relatively slow data flow.

logged data type, each data flow remains simple until a more complex type is required. By efficiently cross-indexing multiple simple data logs, we can provide compound data types that do not require a fixed schema, or any changes to the providers of the low level data. The compound data type returned is defined by the query itself, thus cleanly separating such high-level data mining from the low-level code that generates the data.

This system is again built around the reflective type system. A universal logging mechanism can tap into any pin connection, ascertain what type of data is being transmitted, and log all the data with type-appropriate indexing. These loggers can then describe the data that has been logged so that complex queries can be constructed that find data stores on the ROCI network and combine them in novel ways. Such a logging mechanism is completely reliant on all data being self-describing. Without this metadata, loggers for each data type would need to be written. The smart, type-aware logging process lets the development team add features to all type logging at once while also making the code maintenance job tractable.

3.5 Web Accessibility

As mentioned earlier, ROCI's data-driven internals are completely accessible to network peers. The deep extent of data accessibility is a manifestation of the desire to integrate ROCI with the broadest possible spectrum of other software packages as well as hardware devices. To this end, ROCI supports the standards and protocols of the Internet: HTTP and XML. A human operator may navigate a ROCI node with a web browser as the data content is converted from raw XML to formatted HTML through the use of XSLT and CSS. Machine peers, meanwhile, may consume the raw XML documents using standard HTTP methods.

This data accessibility is entirely provided by ROCI itself, and not dependent on developer support. While individual formatting methods may be overridden, developers can benefit from various Pin Exporters capable of operating on any ROCI pin. This creates an environment of universal accessibility, and allows the development team to support new interoperation technologies without having to update every already-defined interface. Instead, new translators may be developed that bring all ROCI data to a new format or protocol.

4 Examples

4.1 Module Reuse: Indoor/Outdoor Navigation

As discussed in this chapter, ROCI's modular design and strongly typed interfaces make component reuse a simple job. Here, we present an example in which a trajectory controller module was used to navigate different robots in different scenarios. Basically, this module receives two inputs: the robot pose (x, y, θ) and a list of waypoints (X_i, Y_i) and outputs velocities (v, ω) to the robot's low level controller. Since this trajectory controller is completely self-contained, it can be used in combination with different modules that export and import the correct pins.

Figure 4(a) shows an implementation where the waypoint controller receives input from a GPS and a waypoint planner and outputs velocities to a clodbuster robot. This configuration was used to navigate teams of robots in outdoor environments [CCG2005]. The waypoint controller (the exact same module) has also been used for some indoor demos in the GRASP Lab in the configuration shown in Figure 4(b). In this case, instead of a GPS, an overhead camera is used for localization: the robot is tagged with a colored blob and a color blob extractor is used to compute its position. The waypoint list is given by a user interface running in the ROCI Browser and the robot being controlled is a Segway. It is important to mention that some of the modules are running in different machines but, since the pin architecture is network transparent, the module programmers are insulated from this usage issue.

4.2 Scheduling and Data Translation: The Networked Robot

In this example, a robot searches for a colored blob and, upon spotting, it dynamically changes the schedule to track the target and sends an e-mail to a cellular phone. This will exemplify several ROCI features such as allocation of priorities, strongly-typed interfaces, multiple data views, and general modular development. Importantly, there is minimal coupling between individual modules at the system code level. In fact, many of these modules are used without change or recompilation in many other tasks (for example, the Blob Extractor from the previous example). From another point of view, any one module in this task may be replaced to reflect a different hardware platform

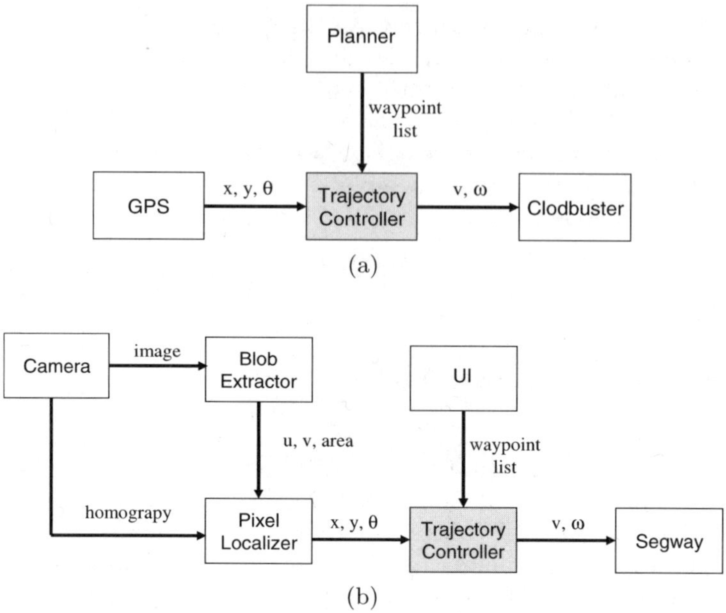

Fig. 4. Diagrams of two tasks using the same waypoint controller module: (a) outdoors with a GPS and a planner; (b) indoors with an overhead camera and an user interface.

or different functionality requirements without necessarily affecting any other module. Such robustness is achieved through the usage of the common, defined interfaces embodied by the ROCI pins in use.

A diagram of the relevant components (ROCI Modules) that make up this task is shown in Figure 5. Data is brought in from the outside world through a stereo camera module that interfaces, via native system code, with a camera driver. Color images from each camera are fed into blob extractors whose outputs are combined by a stereo blob localizer that computes the 3D position of the blob relative to the camera. This range and bearing information is fed into a motion controller that directs a mobile robot, a clodbuster, to follow the blob.

In this particular task, when no target has been sighted the robot pans its camera back and forth in a search mode. While in this state, the robot may be involved in processing data from other sensors or devoting a large portion of its CPU resources to long-term goals such as map building. When the blob extractors find a target, this data is fed to the blob localizer, but it is also fed to a schedule monitor module. This is the one module in this task that is specifically designed for one purpose: when a viable target is detected, processing resources are reallocated more heavily to the vision processing modules. In the actual experiment, the robot was able to process video data at approx-

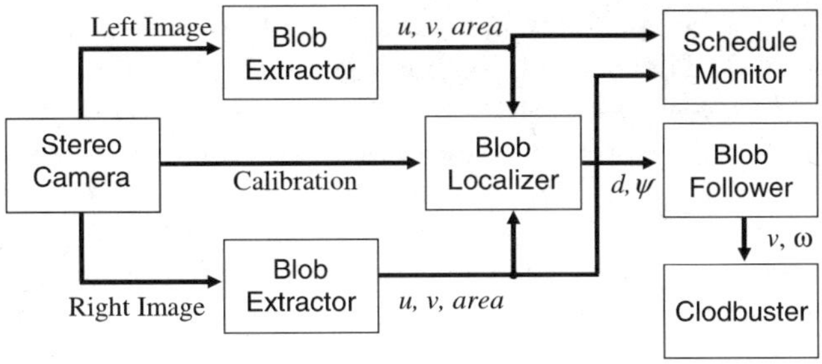

Fig. 5. Block diagram of modules involved in the blob following behavior.

imately 2.5 Hz while searching for a target. Once a target was detected, this rate jumped to 5 Hz, while the other CPU intensive processing jobs' rates slowed. This re-prioritization is effected by a change in the task schedule that specifies the relative rates each module iterates at. When the target is lost, the original schedule is replaced.

An additional feature of this experiment was an addition to the task file that specified that schedule change alerts should be messaged to a cell phone. This change did not require any change to the code for the modules or pins involved in the task. Instead, a single line of XML caused the system to automatically translate the internal binary data to a human-readable message, with labeled data, that was then sent to the cell phone at the number specified in the XML. In other experiments, robot status messages were forwarded to mobile devices. These status messages actually included the most recent views of a particular target being tracked. Such images appear in the status message as hyperlinks which, when followed, lead to images served up by the robot itself via HTTP.

5 Conclusion

The single theme that runs through all core ROCI development is the minimization of redundancy. In many cases, this involves the automation of as much programming effort as possible through the use of pervasive metadata. Since the system can always inspect what the programmer has done, it can generate code, invoke methods, and, in general, take action upon recognition of supported design patterns as simple as pin type declarations. This type of proactive system frees the developer from having to write the same glue code more than once, thus increasing efficiency and decreasing the likelihood of programmer error.

By focussing on the needs of the programmer, we have attempted to make component development as natural and clean as possible. As a result of freeing the developer from work that can be automated, and assisting in programmer cooperation through the use of self-describing objects, we are offering a system wherein the lone developer may concentrate on the particular problem he or she wishes to solve. The enabling technology here is the modularity of development. Each developer may contribute his or her own high-level primitive to an already functioning system, and automatically gain from all existing works. In the end, the very early decision to define modules by their interfaces has lead all the way to truly high-level robot programming with all the features and support of a tightly integrated platform, and all the flexibility, specificity, relevancy, and robustness of ad hoc, loosely coupled, domain languages.

References

[Bru06] Brugali, D. and Brooks, A. and Cowley, A. and Côté, C. and Domnguez-Brito, A.C. and Létourneau, D. and Michaud, F. and Schlegel, C. *Trends in Component-Based Robotics*, In D.Brugali (Ed.) *Software Engineering for Experimental Robotics*, Spinger STAR series, 2006

[CCS2003] Chaimowicz, L., Cowley, A., Sabella, V. and Taylor, C. J. (2003) ROCI: A distributed framework for multi-robot perception and control. In *Proceedings of the 2003 IEEE/RJS International Conference on Intelligent Robots and Systems*, pp. 266–271, 2003.

[CCG2005] Chaimowicz, L., Cowley, A., Gomez-Ibanez, D., Grocholsky, B., Hsieh, M. A., Hsu, H., Keller, J. F., Kumar, V., Swaminathan, R., and Taylor, C. J.(2005) Deploying Air-Ground Multi-Robot Teams in Urban Environments. In *Proceedings of the 2005 International Workshop on Multi-Robot Systems*, pp. 223-234, 2005.

[CCT2006] Cowley, A., Chaimowicz, L. and Taylor, C. J. (2006) Design Minimailsm in Robotics Programming. *International Journal of Advanced Robotics Systems.* To appear, 2006.

[CHT2004] Cowley, A., Hsu, H. and Taylor, C. J. (2004) Distributed sensor databases for multi-robot teams. In *Proceedings of the 2004 IEEE International Conference on Robotics and Automation*, pp. 691–696 2004.

[CHT2004b] Cowley, A., Hsu, H. and Taylor, C. J. (2004) Modular programming techniques for distributed computing tasks. In *Proceedings of the 2004 Performance Metrics for Intelligent Systems (PerMIS) Workshop*, 2004.

Communication Patterns as Key Towards Component Interoperability

Christian Schlegel[1]

University of Applied Sciences Ulm
Fakultät Informatik
Prittwitzstr. 10, D-89075 Ulm, Germany
schlegel@hs-ulm.de

1 Introduction

A component based software engineering approach does not per se ensure that independently developed components finally fit together. The reason is that general purpose component based approaches still provide far too much freedom with respect to defining and implementing component interfaces. There is always a gap from a general purpose software component approach towards reusable and easily composable components in a specific application domain.

The challenge of component based software approaches for robotic systems is to address the numerous functional and non-functional requirements of this application domain. As argued in Chapter *Trends in Component-Based Robotics*, the major characteristics are related to the *inherent complexity* of robotics systems, the *requirements for flexibility* and the challenges of *distributed environments.*

Of particular importance in the robotics domain are the following aspects that need to be addressed explicitly when tailoring the general idea of component based software engineering to the needs of the robotics domain:

- Components at different levels of a robot system can follow completely different designs. Thus, roboticists ask for as few restrictions as possible with regard to component internal structures.
- *Dynamic wiring* of components can be considered as *the* pattern in robotics. It is the basis for making both, the control flow and the data flow configurable at run-time. That is the basis for situated and task dependent composition of skills to behaviors.

Furthermore, a framework is accepted by roboticists if it provides an obvious surplus value, requires a flat learning curve only, is a software framework and not a robotic architecture, addresses middleware and synchronization issues and provides lots of device drivers and components for widespread sensors and platforms.

D. Brugali (Ed.): Software Eng. for Experimental Robotics. STAR 30, pp. 183–210, 2007.

2 The Approach

The basic idea to address the above requirements is to master component interfaces since these are crucial with respect to component interoperability. Mastering the component interfaces also allows to abstract from middleware aspects and to ensure decoupling to address the complexity issue. That idea is implemented by means of a small set of generic and predefined communication patterns. All component interactions are squeezed into those predefined patterns. Thus, component interfaces are only composed out of the same set of well-known patterns with a precise and predefined semantics. This ensures decoupling of components, enforces the appropriate level of abstraction at the externally visible component interfaces and thereby results in components that are composable to form complex robotics applications.

2.1 Communication Patterns to Define Component Interfaces

Components: A component can contain several threads and interacts with other components via predefined communication patterns that seamlessly overcome process and computer boundaries. Components can be dynamically wired at runtime.

Communication Patterns assist the component builder and the application builder in building and using distributed components in such a way that the semantics of the interface is predefined by the patterns, irrespective of where they are applied. A communication pattern defines the communication mode, provides predefined access methods and hides all the communication and synchronization issues. It always consists of two complementary parts named *service requestor* and *service provider* representing a *client/server*, *master/slave* or *publisher/subscriber* relationship.

Communication Objects parameterize the communication pattern templates. They represent the content to be transmitted via a communication pattern. They are always transmitted *by value* to avoid fine grained intercomponent communication when accessing an attribute. Furthermore, object responsibilties are much simpler to manage with locally maintained objects than with remote objects (see [HV99] for nifty details of *CORBA* lifecycle issues). Communication objects are ordinary objects decorated with additional member functions for use by the framework. Genericity of the approach is achieved by using arbitrary and individual communication objects to instantiate communication patterns.

Service: Each instantiation of a communication pattern provides a service. Generic communication patterns become services by binding the templates by types of communication objects.

The service based view comes along with a specific granularity of a component based approach. Services are not as fine grained as arbitrary component interfaces since they are self-contained and meaningful entities. Major characteristics are as follows.

- Communication patterns provide the only link of a component to the external world and can therefore ensure decoupling at various levels. Communication patterns decouple structures used inside a component from the external behavior of a component. Decoupling starts with the specific level of granularity of component interfaces enforced by the communication patterns which avoids too fine-grained interactions and ends with the message oriented mechanisms used inside the patterns.
- Using communication patterns with given access modes prevents the user from puzzling over the semantics and behavior of both component interfaces and usage of services. One can neither expose arbitrary member functions as component interface nor can one dilute the precise interface semantics and the interface behavior. Given member functions provide predefined user access modes and hide concurrency and synchronization issues from the user and can exploit asynchronicity without teasing the user with such details.
- Arbitrary communication objects provide diversity and ensure genericity with a very small set of communication patterns. Individual member functions are moved from the externally visible interface to communication objects.
- *Dynamic wiring* of intercomponent connections at runtime supports context and task dependent assembly of components. Reconfigurable components are modular components with the highest degree of modularity. Most important, they are designed to have replacement independence. Many component approaches only provide a deployment tool to establish component connections before the application is started.

2.2 The Set of Communication Patterns

Restricting all component interactions to given communication patterns requires a set of patterns that is sufficient to cover all communicational needs. One of course also wants to find the smallest such set for maximum clarity of the component interfaces and to avoid unnecessary implementational efforts. On the other hand, one has to find a reasonable trade-off between minimality and usability. The goal is to keep the number of communication patterns as small as possible without ignoring easy usage. Table 1 shows the set of communication patterns. A service provider can handle any number of clients concurrently.

Communication patterns make several communication modes explicit like a *oneway* or a *request/response* interaction. Push services are provided by the *push newest* and the *push timed* pattern. Whereas the *push newest* pattern can be used to irregularly distribute data to subscribed clients whenever updates are available, the latter distributes updates on a regularly basis. The *event* pattern is used for asynchronous notification if an event condition becomes true under the activation parameters and the *wiring* pattern covers dynamic wiring of components at runtime.

Table 1. The set of communication patterns

Pattern	Relationship	Communication Mode
send	client/server	one-way communication
query	client/server	two-way request/response
push newest	publisher/subscriber	1-to-n distribution
push timed	publisher/subscriber	1-to-n distribution
event	client/server	asynchronous conditioned notification
wiring	master/slave	dynamic component wiring

2.3 Reference Implementation

The *CORBA* based reference implementation called SMARTSOFT uses the *TAO ORB* [Schb] and can be found at [Sma05]. The operating system abstractions are provided by the *ACE* package [Scha][SH02][SH03] so that interoperability and usage across most operating systems is ensured.

Using communication patterns as means to master the otherwise inevitably exploding diversity of component interfaces has been proposed first in [SW99b]. The approach matured over several projects and a most complete description is contained in [Sch04]. Further aspects are described in [Sch06] and [SW99a].

3 The User View

The user view comprises the application builder view and the component builder view. The main focus is on guiding the component builder without enforcing a particular robot architecture and thus to allow the application builder to compose off-the-shelf components due to replacement independence.

3.1 Examples of Deployed Robots

Figure 1 shows some robots based on components with communication patterns as component interfaces. A small part of the overall components is shown in figure 2. All services are instantiations of the small set of generic communication patterns. Components are wired dynamically at runtime according to the executed task and can thus be rearranged dynamically.

The interaction with higher layers like task sequencing for a discrete description of execution sequences is coordinated via events. These not only report successfully reached intermediate goals, for example, but also non-default events like *got stuck*, *no path* and others. This allows the task sequencing layer to decide on the next configuration and to monitor and coordinate the task execution progress.

Fig. 1. Different robots composed out of standardized navigation components.

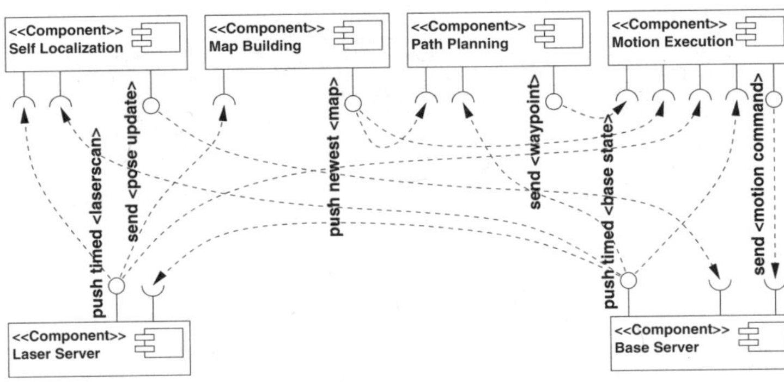

Fig. 2. Composability of components due to standardized component interfaces.

3.2 User View on the Query Pattern

The user visible interface of the communication patterns is illustrated by means of the *query* pattern. Figure 3 shows the client side user interface that provides both synchronous and asynchronous access modes that can be used in any order and even interleaved without requiring any further access coordination. This also includes the proper handling of blocking calls in case of dynamically performed rewirings of the client to another server.

The server side user interface is shown in figure 4. All incoming requests are forwarded to a handler and answers are returned via the *answer* method. The used handler class defines the processing model like thread per request, thread pool or single thread, for example.

3.3 Code Example

The basic usage of the communication patterns to build a component is shown in figure 5. The first component uses a service of the second component implemented by the *query* pattern transmitting a laser scan object. Of course, components can provide and require any number of services and can be interweaved arbitrarily. In this example, the *laserQueryClient* of component *first* is exposed as port *laserPort*. The connections can then even be rearranged from outside the component.

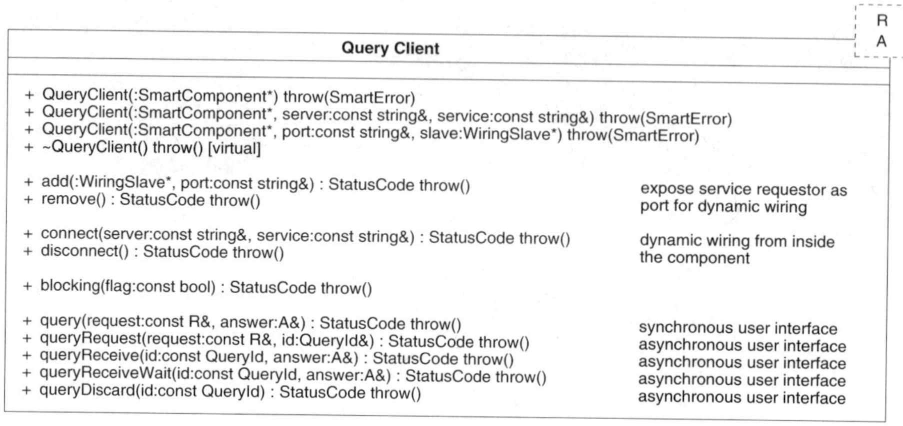

Fig. 3. The *query* pattern and its client side user interface.

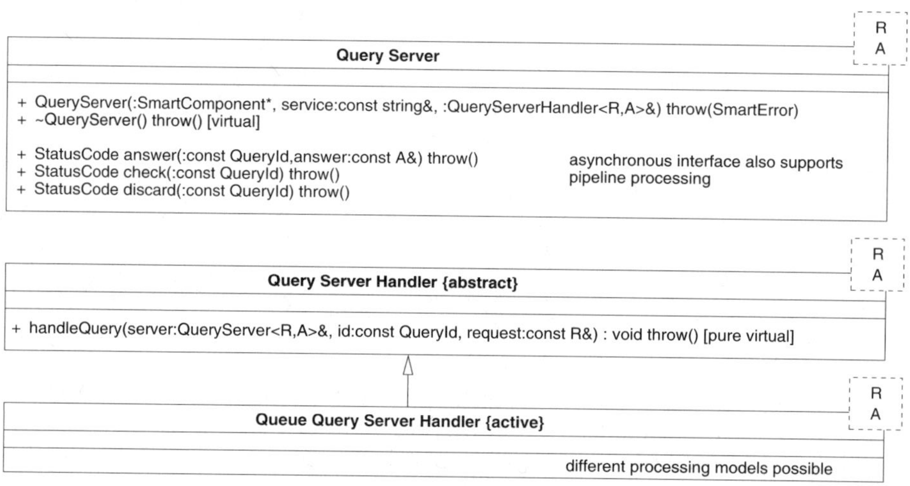

Fig. 4. The *query* pattern and its server side user interface.

3.4 Communication Objects

Communication objects are transfered *by value*. The relevant data structures are (de)marshalled via the *set/get* methods that form the framework interface together with the *identifier* method. Those methods are the only user visible ones that depend on the underlying communication mechanism. Since these methods can be overloaded, one can implement communication objects such that they work with different middleware systems. The structure of the *CORBA* based implementation is shown in figure 6.

A major advantage is that all user access methods are independent of the middleware data types. In particular, one can use *STL* classes [MDS01] or heap

```
#include "smartSoft.hh"                    // standard include
#include "commVoid.hh"                     // a communication object
#include "commMobileLaserScan.hh"          // a communication object

CHS::QueryClient<CommVoid,CommMobileLaserScan> *laserQueryClient;

class UserThread : public CHS::SmartTask {   // separate thread for user activity
public:
    UserThread() {};
    ~UserThread();
    int svc(void);
};

int UserThread::svc(void) {
    CommVoid request1, request2;
    CommMobileLaserScan answer1, answer2;
    CHS::QueryId id1, id2;
    ...
    status = laserQueryClient->connect("second","laser");
    status = laserQueryClient->queryRequest(request1,id1);
    status = laserQueryClient->queryRequest(request2,id2);
    ...
    status = laserQueryClient->queryReceiveWait(id2,answer2);
    status = laserQueryClient->queryReceiveWait(id1,answer1);
    ...
}
...
int main(int argc,char *argv[]) {
    ...
    CHS::SmartComponent component("first",argc,argv);
    CHS::WiringSlave wiring(component);
    UserThread user;

    laserQueryClient = new CHS::QueryClient<CommVoid,CommMobileLaser>
                           (component,"laserPort",wiring);
    user.open();
    component.run()
    ...
}
```

```
#include "smartSoft.hh"                    // standard include
#include "commVoid.hh"                     // a communication object
#include "commMobileLaserScan.hh"          // a communication object

// this handler is executed with every incoming query
class LaserQueryHandler
  : public CHS::QueryServerHandler<CommVoid,CommMobileLaserScan> {
public:
  void handleQuery(
          CHS::QueryServer<CommVoid,CommMobileLaserScan>& server,
          const CHS::QueryId id,
          const CommVoid& r) throw()
  {
      CommMobileLaserScan a;
      // request r is empty in this example, now calculate an answer
      server.answer(id,a);
  }
};

int main(int argc,char *argv[])
{
   ...
   // the component management is mandatory in all components
   CHS::SmartComponent component("second",argc,argv);
   // the following implements a query service for laser scans
   // with an active handler
   LaserQueryHandler laserHandler;
   CHS::QueueQueryHandler<CommVoid,CommMobileLaserScan>
                 activeLaserHandler(laserHandler);
   CHS::QueryServer<CommVoid,CommMobileLaserScan>
                 laserServant(component,"laser",activeLaserHandler);
   ...
   // the following call operates the framework by the main thread
   component.run();
}
```

Fig. 5. Two example components named "first" and "second".

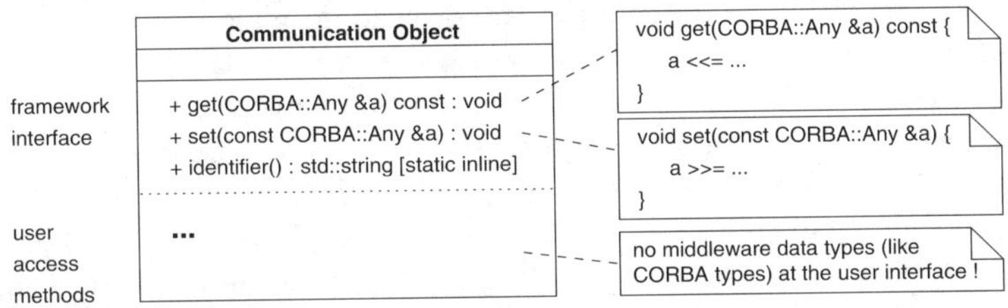

Fig. 6. Structure of a communication object.

memory even if these cannot be represented via the *CORBA IDL*, for example. Since all conversions are handled inside the *set/get* methods, one can easily flatten *STL* lists by iterating through them, for example. Furthermore, local extensions of the user access methods have no influence on the transmitted content such that other components relying on the same set of communication objects are not affected. Introducing additional user access methods has purely local effects.

4 The Framework Builder View

This section presents the pattern internal structures and protocols. These are independent of any specific communication approach and they mediate between the characteristics of the user interface and the characteristics of the communication system. These details are neither seen by the application builder nor the component builder.

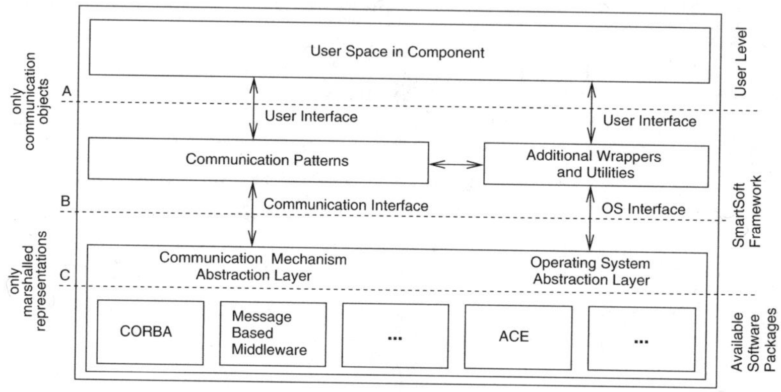

Fig. 7. The communication patterns provide the glue logic between the user interface and the underlying communication mechanism.

Communication patterns provide the glue logic which coordinates everything on top of a communication mechanism to enforce the required behavior of the user interface. As shown in figure 7, communication patterns provide the glue logic between the user interface and the underlying communication mechanism. The dotted stripline labeled with C furtheron designates the framework internal interface between the actual communication mechanism and the communication system abstraction. B denotes the framework internal interface between the communication system abstraction and the communication patterns. The interface that separates the framework internal level from the user visible part is labeled with A.

Communication patterns provide a standardized semantics and behavior independently of the underlying communication mechanism and make sure that the characteristics of the used communication mechanism do not influence the behavior of the user interface. An important demand on the communication patterns is to decouple the service requestor from its service provider including an asynchronous operation of both sides. That is a nontrivial task since not all communication mechanisms provide all the required features. For example, implementing an asynchronous user interface on top of a synchronous communication mechanism requires carefully chosen protocols. Otherwise, the asynchronous user interface behaves like a synchronous one and becomes obsolete due to the introduced dependencies with respect to execution orders. The challenge is to provide both in parallel synchronous and asynchronous user interfaces even if only synchronous two-way or asynchronous one-way interactions are supported by the communication system. Both the implementation of a synchronous two-way user interface on top of asynchronous one-way interactions as well as the implementation of an asynchronous user interface on top of synchronous interactions requires additional glue logic respectively well chosen protocols to achieve the expected user interface semantics.

4.1 Interaction Patterns

Communication patterns provide different user access modes like a synchronous invocation, an asynchronous invocation or a handler based processing of incoming requests or answers. From the user level perspective, user access modes provided at the service requestor and those provided at the service provider of a communication pattern can be used in any combination.

From the framework builder view, all reasonable combinations of a user access mode at service requestors and service providers form an *interaction pattern*. The semantics of the user interfaces of the communication patterns and the user access modes assignable to the communication patterns are independent of the underlying communication system as soon as all interaction patterns are implementable on top of the selected communication system.

Interaction patterns each consist of a service invoking and a service providing part named *client* and *server*. Both parts of a communication pattern, the service requestor and the service provider, can invoke operations on each other and can thus contain both clients and servers of interaction patterns. For example, the client part of the interaction pattern beneath the *put* method of the server side user interface of the *push* communication patterns is located inside the service provider of the *push* communication pattern.

The Client Part Characteristics

The characteristics of the client part of the interaction patterns are shown in figure 8. These characteristics are motivated by the access modes of the communication patterns and can be characterized with respect to the *direction* and the *invocation* mode. The client part interfaces are always based on member function calls. They are also expected to be thread safe, that is no further user level synchronization is needed with concurrent access.

The direction mode can be either *one-way* or *two-way* and determines the transmission direction of arguments. *In* arguments are transmitted from the client to the server and can be modified there but without affecting the client side. An *out* argument cannot provide an initial value to the server but it is returned to the client. An *inout* argument provides an initial value to the server and all modifications are returned to the client. In one-way mode, interactions can have *in* arguments only since there is no back channel and

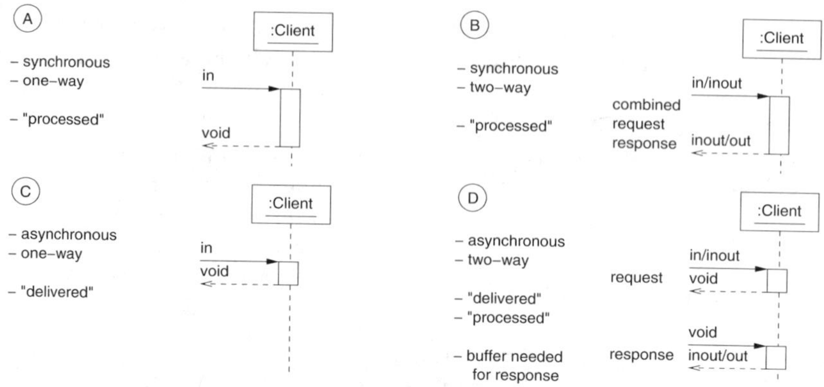

Fig. 8. Overview on the characteristics of the client part of the interaction patterns.

arguments can be transmitted solely from the client to the server. In two-way mode, interactions can have all three types of arguments since a two-way interaction provides the required back channel from the server to the client.

The invocation mode can be either *synchronous* or *asynchronous*. A *synchronous* invocation blocks and returns *after* the server side processing is completed according to a *processed* policy whereas an *asynchronous* invocation returns *before*. The invocation mode solely characterizes the client side behavior and does not characterize the communication mechanism used between the client and the server. Asynchronous invocations allow to benefit from concurrent calculations and thus normally result in better reactivity and shorter answer times if calculations can be invoked in parallel.

A client part *synchronous one-way* characteristic (A) accepts *in* arguments only, blocks and returns either after the server side processing is finished or with an error. A feedback channel is still needed to return the *processed* acknowledgment for the synchronous client side interface. However, it is just empty and does not carry any arguments to be returned.

A client part *synchronous two-way* characteristic (B) is invoked with the *in* and *inout* arguments, blocks and either returns with the *inout* and *out* arguments after the server part processing is completed or with an error. The only difference to the synchronous one-way characteristic is that the feedback message is not just empty. The feedback message again corresponds to a *processed* acknowledgment.

A client part *asynchronous one-way* characteristic (C) also transmits *in* arguments only but returns *before* the server part processing is finished or even before it is started. Asynchronous one-way interfaces can be distinguished with respect to the level of guarantees they provide. The *unreliable send* policy returns as soon as the message is delivered to the client part transportation layer and accepts that messages might get lost. The next level is the *reliable send* policy that guarantees the delivery of a message in case the recipient exists. Otherwise, no feedback is given on the whereabout of the message. Even if no message gets lost, one cannot be sure that it is delivered successfully and that it is going to be processed. The recipient might be in the process of destruction or might have disappeared after the message was sent and just before the message is tried to be delivered. Both policies provide no acknowledgment and user level protocols have to take additional precautions to make sure that communicating partners never hold wrong assumptions about the states of their opponents due to messages that never reached their destination. In contrast, the *delivered* policy returns after the message has been delivered to the server part but before it is processed there. However, being delivered guarantees that it is going to be processed and that the recipient cannot get destroyed as long as there are pending requests. Thus, that policy provides an acknowledgment that the message arrived at the server and is going to be processed for sure. The client part *asynchronous one-way* interface (C) is always meant to implement the *delivered* policy.

A client part *asynchronous two-way* characteristic (D) splits the two-way interface into a request and a response method. The request method provides the *in* and *inout* arguments and the response method returns the deferred answer consisting of the *inout* and *out* arguments. The asynchronous two-way characteristic follows a *delivered* policy with respect to the request and a *processed* policy with respect to the response. Since the user decides on when to fetch responses, a buffer is needed to store the answers meanwhile.

The Server Part Characteristics

The server part characteristics of the interaction patterns are shown in figure 9. The server part can be categorized with respect to the *initiation* and the *invocation* mode. The initiation mode can be either *pattern* or *user* and describes the responsibility for initiating the request processing. In *pattern* mode, it is the interaction pattern that makes sure that every incoming request initiates its processing. Thus, one has to consider server part processing models for the upcall to decouple several concurrent requests or to separate the communication from the request processing. In *user* mode, it is the user who invokes the processing of requests. Thus, the processing models are not in the scope of the server anymore. However, a buffer is needed for incoming requests that are not yet fetched by the user.

The invocation mode can be either *synchronous* or *asynchronous* and specifies whether a request has to be processed within a single call or is splitted into an invocation and a completion part. A *synchronous* invocation has a single interface method only and interprets the completion of that method as having processed the request. An *asynchronous* invocation is based on an *invocation* and a *completion* interface method. The invocation method provides the *in* arguments and in case of a two-way interaction the *in* parts of the *inout* arguments but does not return any arguments. The completion method has

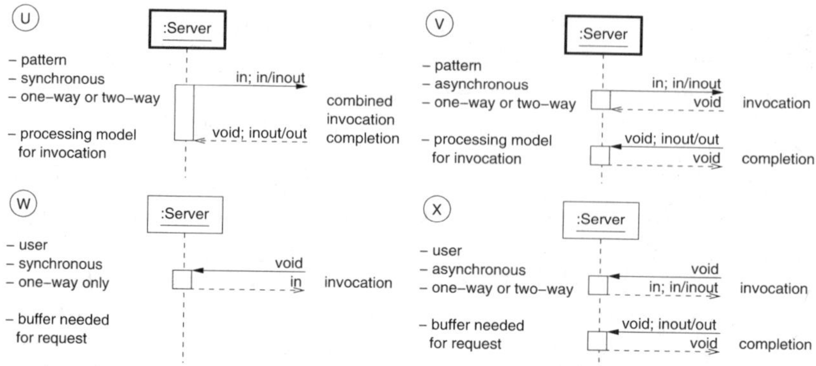

Fig. 9. Overview on the characteristics of the server part of the interaction patterns.

to be called from the user level to provide the *out* parts of the modified *inout* arguments and the *out* arguments that have to be returned to the server. Returning from the invocation method to the server does not indicate that the request is already processed. On completion, the server can send back any designated arguments respectively a *void* argument. The completion method indicates when a *processed* policy considers the processing as completed.

The advantage of the asynchronous invocation mode is the flexibility with respect to user level processing models. For example, one can easily propagate a request through a pipeline of processing steps and return the result from a thread that is completely different to the one that handled the invocation method. In contrast thereto, a synchronous invocation mode is much easier to implement by the server since it does not require the glue logic to correctly assign provided responses to open requests. However, pipeline processing models with a synchronous upcall block the upcalling thread until the pipeline returns the answer.

The server part *synchronous pattern* characteristic (U) invokes the processing from the interaction pattern and processes the request within a single upcall. A request is completed with returning from the upcall. The server part *synchronous user* characteristic (W) implements a downcall from the user level to get a request for processing. The actual processing is moved out of the scope of the server and there is no back channel from the user level to the server. Thus, the characteristic (W) is unsuited for interaction patterns that require a *processed* acknowledgment or expect arguments to be returned. The characteristic (W) can be used only in combination with a client part asynchronous one-way characteristic.

The server part *asynchronous pattern* characteristic (V) splits the request into a pattern invoked upcall for the request and a user invoked downcall for the response. Of course, the completion method can also be invoked from inside the invocation method since an asynchronous interface is expected to be organized such that selflocks are impossible. The request is considered as being completed as soon as the completion method is called and independently of any subsequent activities even if these are executed inside the invocation method. In case of one-way arguments, one has to check carefully when to call the completion method since any subsequent activities are not considered as belonging to the scope of the request processing. That behavior is desired with the communication patterns since the user can already complete a request with respect to the client even if there are still some server part housekeeping activities to be executed afterwards. However, some asynchronous interfaces send back an answer earliest after both the invocation method and the completion method returned to the server. Finally, the server part *asynchronous user* characteristic (X) is different to (V) only with respect to the invocation method.

The Remaining Interaction Patterns

Since the client part characteristic (A) can easily be emulated by (B) and since the server part characteristic (X) is not justified by reasonable use cases, only six different interaction patterns remain on top of which all reasonable combinations of user access modes of the communication patterns can be implemented. Figure 10 shows the remaining interaction patterns.

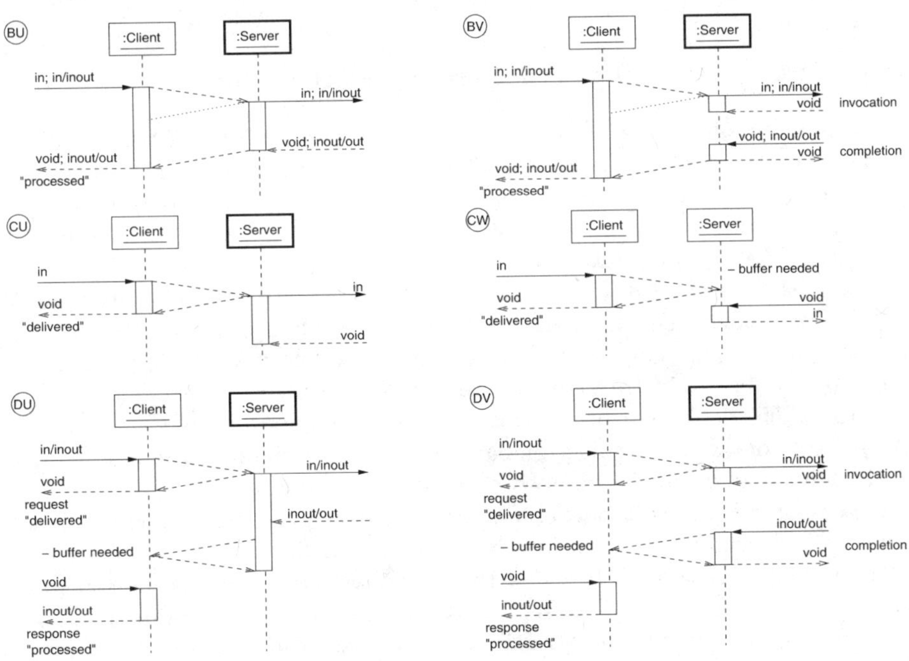

Fig. 10. The remaining interaction patterns.

The interaction patterns (B/U) and (B/V) represent the standard synchronous interactions that can be used in either one-way or two-way mode. The client part returns after the server part processing is completed. In case of (U) this corresponds to the return of the server part upcall and in case of (V) to calling the completion method. The dotted lines correspond to a *delivered* policy with respect to the request and mark the point of time after which the client part invocation has to be abortable. That is in particular important to enable the client part to react before the currently running request is completed since being forced to first await the completion could take far too much time with long running calculations. The completion method of (V) has to be callable from inside the invocation method.

The interaction patterns (C/U) and (C/W) represent the standard asynchronous one-way interaction with either a pattern or a user invoked server part processing and a *delivered* policy. The client part returns before the server part processing is completed or even before it is started but after the request has been delivered at the server part. Being delivered is tantamount to getting processed for sure.

The interaction patterns (D/U) and (D/V) represent the standard asynchronous two-way interaction. The client part invocation has to implement a *delivered* policy. The server part characteristics (U) and (V) behave exactly like the one in the (B/U) and (B/V) interaction patterns. The client part response methods block if they are called before the response is available and if they are allowed to block. Thus, they have to be abortable.

Mapping Interaction Patterns onto Interaction Models

Various communication and middleware systems provide different interaction models. For example, the standard CORBA model provides synchronous remote method calls (B/U), oneway messages (C/U) or asynchrononous method invocations (AMI, D/U). In contrast thereto, a mailbox system or TCP sockets behave like a (C/W) interaction pattern.

The goal is to find a generic mapping between interaction patterns and interaction models such that the modifications that are required in case of migrating to another communication system are reduced to a minimum. The idea is to emulate all interaction patterns by solely using the (C/U) interaction pattern. Then the (C/U) interaction pattern is the only one that is visible at the interface of the communication patterns to the communication middleware system and that has to be mapped onto an interaction model.

The main advantage is the lean interface between the communication patterns and the communication system as shown in figure 11. *B* denotes the interface of the communication patterns to the communication system abstraction. The emulation of interaction patterns is part of the communication patterns. Communication patterns provide and accept user level data in marshalled form only. Messages are sent by invoking the client part of an appropriate (C/U) interaction pattern and messages are received via callbacks that

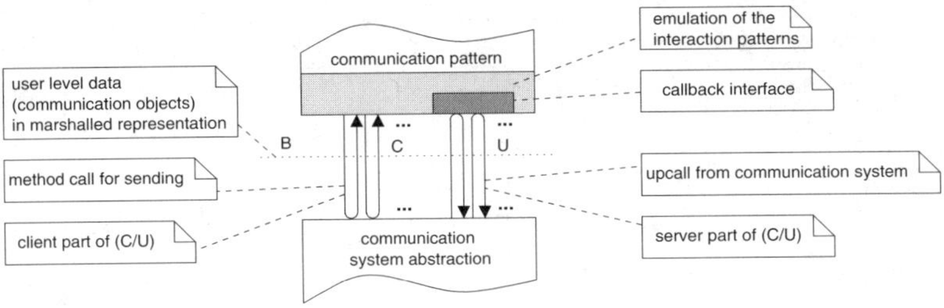

Fig. 11. The interface between the communication patterns and the underlying communication mechanism.

are invoked by the upcalling server part of a (C/U) interaction pattern. Of course, this protocol requires appropriate glue logic inside the communication patterns to coordinate the independent (C/U) interactions. The price to pay is the increased complexity of the implementation of the interaction patterns since these have to be emulated on top of the (C/U) interaction pattern inside the communication patterns. However, that additional complexity has to be mastered only once with the implementation of the communication patterns and not with every single migration to a new communication middleware system. It is much better to cope with these demanding details only once and inside the communication patterns rather than with every middleware migration. The additional complexity of the communication patterns that host the emulation of the interaction pattern thus pays off very soon.

It is important to note that there are two features of the (C/U) interaction pattern that predestinate it as interface pattern. Due to its *one-way* characteristic, it can be mapped onto communication systems that provide a one-way interaction only and due to the *delivered* policy, it already provides the decoupling between the client and the server part as required by the interaction patterns.

In particular, the *delivered* policy is mandatory and is exploited by the emulation of the other interaction patterns. Even if it looks like the (C/U) interaction pattern could be mapped easily onto all interaction models, it is challenging to achieve the *delivered* policy on top of the various interaction models.

The naive usage of synchronous interactions to emulate the (C/U) interaction pattern can make asynchronous user interfaces obsolete as shown in figure 12. The request returns only after the server side processing has been completed and even after the server has already returned the results. On top of a synchronous interaction model, one has to convert the *processed* policy into a *delivered* policy without ending up at a *reliable send* policy only.

In case of a *reliable send* policy messages without a regular recipient get dropped silently by the communication system. A recipient can disappear due to destruction, for example. If one cannot be sure that the recipient still exists, one has to use timeouts when awaiting an acknowledgment message. An

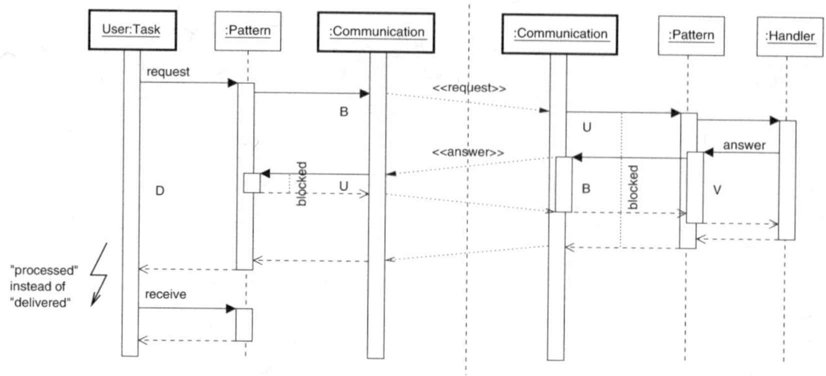

Fig. 12. A passive handler and a (D/V) interaction pattern emulated by two (B/U) interaction patterns.

absent acknowledgment can now also mean that the message has been delivered but the sent back acknowledgment returned too late. Then, the message is processed at the recipient even if the sender experienced a timeout and the acknowledgment arrives anyhow. Thus, in case of a timeout, assumptions on the state of the recipient could get out of sync which is in contrast to the *delivered* policy of the (C/U) interaction pattern. Thus, one cannot simply map the (C/U) interaction pattern onto an interaction model with a *reliable send* policy only. On top of a *reliable send* policy, one has to achieve the guarantee of the *delivered* policy that is requests get processed for sure once the sending activity got completed successfully.

4.2 The Communication Protocol of the Communication Patterns

Another option is to handle disappeared recipients at the level of the protocol used above the interaction patterns that is inside the communication patterns. This shift allows to map the (C/U) interaction pattern onto the weaker *reliable send* policy without requiring further glue logic at the interface to the communication system. However, that is possible only if the interaction of both parts of a communication pattern are protected by an appropriate protocol as it is the case for the *connection oriented* protocol presented subsequently.

The *connection oriented* protocol that is used inside the communication patterns also requires one-way interactions only and possesses the same lean interface to the communication system as the before introduced protocol. The major extension is the full exploitation of the connection oriented design of the communication patterns. Besides the interactions related to the connection management, all interactions between a service requestor and a service provider part of a communication pattern are monitored by the connection management of a communication pattern. Changes to a connection can be made at any time and the connection management assumes the responsibility for the proper handling of affected interaction patterns. Once a service requestor is connected to a service provider, both inform each other about getting unreachable. Communication patterns always know when their opponent disappears and thus also know which messages reach their destination and which messages can be awaited. The consequence is that the requirements on the interaction pattern that is used as interface between the communication patterns and the communication system can be further alleviated. In principle, the *connection oriented* protocol achieves the features of the *delivered* policy at the level of the communication patterns and relieves the interface to the communication system from that task.

The big step is that a (C*/U) interaction pattern with a *reliable send* policy instead of a (C/U) interaction pattern with a *delivered* policy is now sufficient as interface to the underlying communication system. That makes it much easier to map the generic interface onto interaction models of widespread communication systems. The uniform (C*/U) interface of the *connection oriented* protocol can even be mapped onto a *reliable send* interaction model. One does neither require any support from the communication mechanism to automatically generate acknowledgment messages nor does one run into the pitfalls of timeout procedures. Since a *reliable send* policy is now sufficient, decoupling mechanisms and buffers that convert a *processed* policy into a *reliable send* policy, are not critical anymore. Above all, most of the administrative interactions do not depend on the decoupling properties of the (C*/U) interaction pattern. Due to the to be presented locking strategies, they even work on top of a *processed* policy without requiring any further decoupling. That significantly simplifies the implementation of the *connection oriented* protocol since the decoupling has to take effect for the service related interactions only and can thus be done by and hidden in the user level processing models of the handlers. Standardized handler classes providing active queues, thread pools or other processing models already due to the job.

To summarize, administrative interactions can be mapped onto a *delivered*, a *processed* and a *reliable send* policy without requiring additional mechanisms. Service related data messages on top of a *processed* policy become decoupled by provided user level handler models, in any case work on top of a *delivered* policy and now also work with a *reliable send* policy.

The (C*/U) Interactions of the Communication Patterns

The interactions between both parts of a communication pattern are emulated on top of the (C*/U) interaction pattern. Figure 13 illustrates the (C*/U) interaction patterns of the *send* communication pattern.

Before one can use a service requestor, it needs to be connected to a service provider. The connection management consists of the messages R0, A0 and R1 for the *connect* procedure and the messages R2 and A2 for the *disconnect* procedure. The *server initiated disconnect* message R3 is needed to properly disconnect service requestors in case that a service provider gets destroyed.

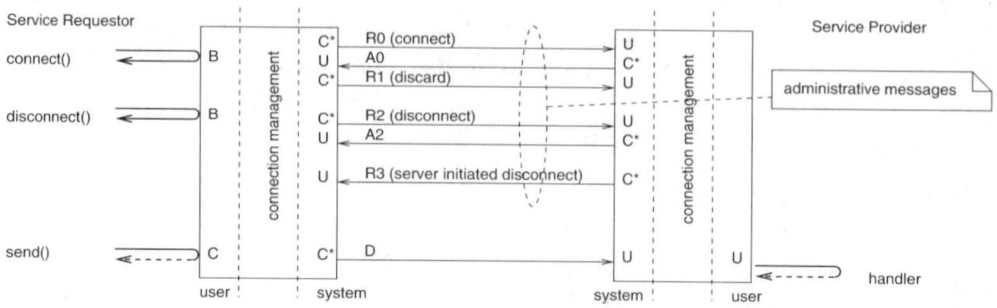

Fig. 13. The (C*/U) interactions of the *send* communication pattern.

The messages R0, R1 and R2 can be emitted by all service requestor parts of communication patterns, the acknowledgment messages A0 and A2 and the message R3 by all service provider parts.

The *connect* message provides the address of the service requestor and a connection identifier to the service provider. The address enables the service provider to inform the service requestors that are connected to it in case it gets destroyed. The connection identifier is generated by the service requestor and uniquely identifies each connect procedure. It is returned by the message A0 so that one can identify outdated acknowledgments that are possible due to an underlying *reliable send* policy. A *connect* is rejected if the service provider is either not yet ready or is already in the process of destruction. In depth explanations of the connect/disconnect procedures and how they cope with a *reliable send* policy, service provider destructions and concurrent service provider destructions and connects of service requestors are contained in [Sch04].

The address of the service requestor is needed with a *disconnect* to remove the appropriate entry from the list of connected service requestors. Once the service requestor received the acknowledgment, it knows that from now on no further messages that are related to the just closed connection can be on their way towards the service requestor.

The *server initiated disconnect* is invoked if the service provider wants to remove all its service requestors. That is needed if the service provider gets destroyed. The message R3 contains the connection identifier that enables the service requestor to check whether the order to get disconnected is still relevant. The connection identifier prevents a meanwhile newly established connection to another service provider from getting closed in case that a *server initiated disconnect* at the service provider coincides with a *disconnect* at the service requestor that is immediately followed by a *connect* to another service provider.

The Interface Objects and the Communication Pattern Interface

The generic structure of the connection between the communication patterns and the underlying communication system is shown in figure 14. So-called *interface objects* mediate between the communication system and the communication patterns. The task of an interface object is to map the (C*/U) interaction patterns onto the communication system, thereby performing all the required adjustments and conversions. The interface objects are managed by the communication patterns and are typically implemented as part of the communication patterns.

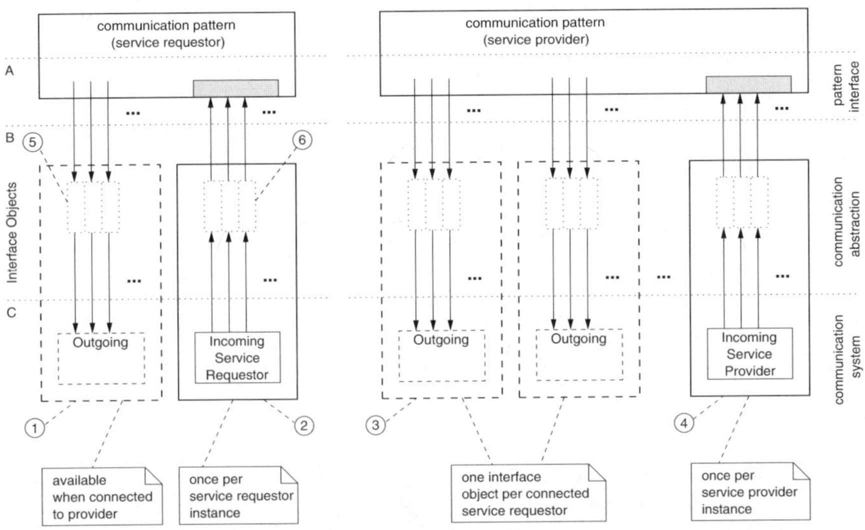

Fig. 14. The generic structure of the connection between the communication patterns and the communication system.

The interface object ② is generated with the creation of the service requestor. It contains an interface to the communication system with a unique address such that the service requestor can receive messages. All method arguments besides the communication objects are first demarshalled and are converted into the format that is used at the callback interface of the communication patterns ⑥. As soon as a service requestor gets connected to a service provider, it generates the interface object ① which performs all the marshalling ⑤ for the outgoing messages. The communication objects are already marshalled when they are forwarded from the communication pattern to the interface object. The interface object ① is destroyed with a disconnect and the interface object ② gets destroyed with a destruction of the service requestor.

The interface object ④ is generated with the creation of the service provider and provides a unique address over which all messages from all service requestors are handled. The service provider generates a separate interface object ③ for each connected service requestor. The interface objects ③ are destructed as soon as the corresponding service requestor gets disconnected. The interface object ④ gets destructed with the shutdown of the service provider.

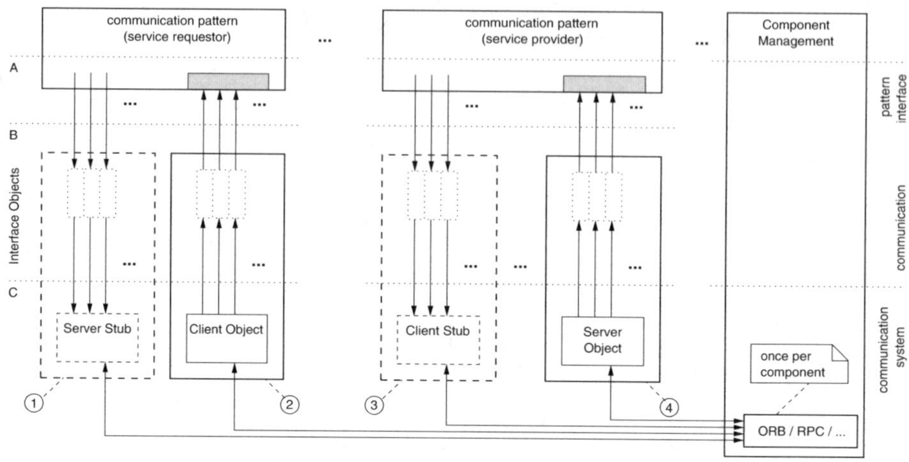

Fig. 15. The interface objects of the communication patterns with an object based middleware.

Figure 15 illustrates the interface objects with an object based middleware like *CORBA* or *remote procedure calls*. Each message is represented by a member function of a remotely callable object. Sending a message is calling a member function of a remote object via the appropriate stub and receiving a message corresponds to executing the corresponding member function at the servant object. The implementations of the methods of the servant objects provide the required adjustments and invoke the upcall interface of the communication pattern. The interface objects for outgoing messages provide the same methods as the stubs, perform the adjustments of the parameters and invoke the corresponding method of the stub. This kind of interface is completely independent of the capabilities of the actual *CORBA ORB* since one does neither depend on *oneway-messages* [SV00] [OSK00] nor on *value-types* or *asynchronous method invocations (AMI)* [AOS00].

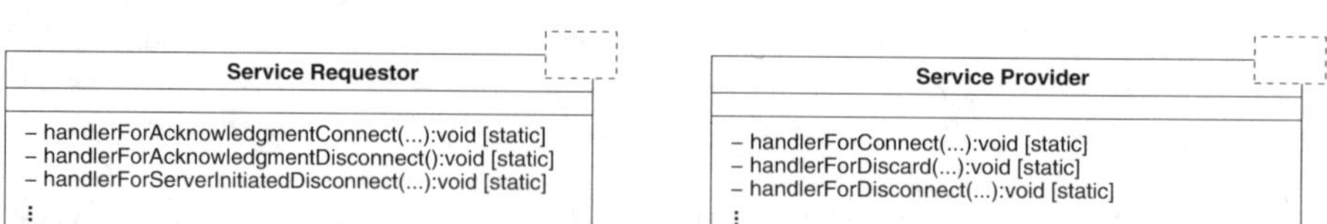

Fig. 16. Implementation details of the upcall interface of the communication patterns.

Figure 16 shows some details of the implementation of the upcall interface of the communication patterns. The communication patterns provide handlers for all incoming messages. The arguments of the handler functions correspond to the arguments of the messages.

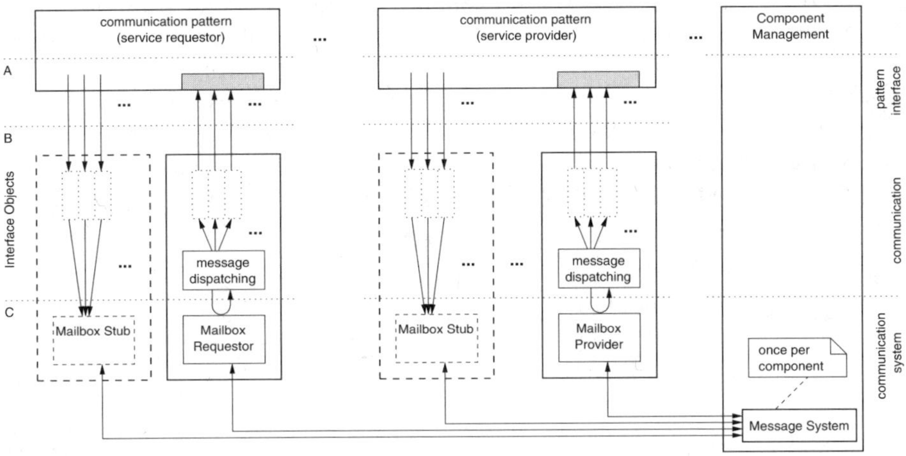

Fig. 17. The interface objects of the communication patterns with a message based middleware.

Figure 17 illustrates the interface objects with a message based communication system that can be either a mailbox based communication system or even a socket based communication mechanism. In contrast to the object based approaches, one now needs a message dispatching mechanism. A message based approach is normally based on a reactor pattern [SSRB00] to accept incoming messages and to dispatch and forward them to the appropriate handlers of the communication patterns. The servant object respectively the mailbox or the socket that is contained in the interface objects ② and ④ is shared by all interaction patterns of a communication pattern.

The Internal Locking Strategy of the Communication Patterns

A crucial role inside the communication patterns is played by the locking mechanisms. These have to ensure that no deadlocks can occur and that all concurrently active interaction patterns never interfere. The internals of an interaction pattern are coordinated by a monitor. However, the *connection oriented* protocol additionally requires mechanisms to coordinate the administrative interactions and to coordinate them with the interaction patterns that implement the actual service. Only the proper interaction of the administrative interaction patterns with the service related interaction patterns ensures that the latter show the specified characteristics even on top of ($C^\star$/U) interaction patterns.

The locking mechanisms are designed such that the administrative interactions from a service requestor to a service provider do not depend on any decoupling. Thus, they even work on top of a *processed* policy and a *delivered* policy needs not to be emulated for them in case that only a *processed* policy is available which makes their implementation very efficient.

The overall locking strategy is explained by means of the administrative (B/U) interaction. All administrative interactions are serialized such that they are never active concurrently. The serialization is mandatory since it makes no sense, for example, to invoke a *subscribe* while an *unsubscribe* or a *disconnect* is still active. The difference between administrative and service related interactions is that always only one administrative interaction is active. Thus, both types of interactions differ with respect to releasing locks.

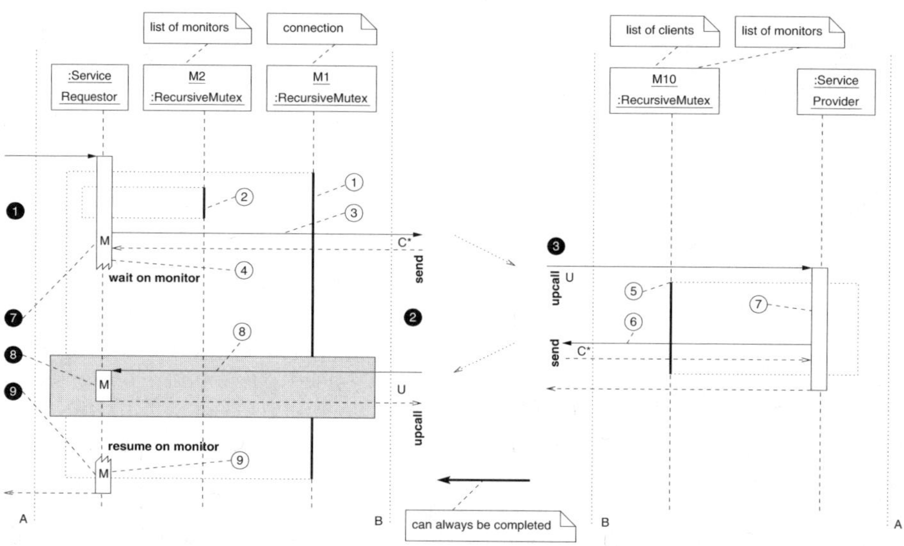

Fig. 18. Administrative (B/U) interaction from the service requestor to the service provider (*delivered* policy).

Figure 18 illustrates an administrative (B/U) interaction from the service requestor to the service provider. An M in the activation box of the service requestor denotes the access to the monitor of the administrative interaction. The top most M denotes the *prepare* method that is invoked prior to the *send* ❼. It sets the state automaton to *await*. The M in the upcall ❽ sets the state automaton inside the monitor according to the received answer and performs a broadcast. Finally, ❾ is the simplified representation of releasing the monitor lock after it is not needed anymore. The monitor lock was acquired automatically with resuming the activity after the *wait*.

At the service requestor part of a communication pattern, the mutex M1 protects the connection related status flags like the *connected* and the *sub-*

scribed flag and it serializes the administrative interactions so that these cannot get interleaved and mess up the connection status. Thus, every administrative interaction must always hold the mutex M1 while being executed. Of course, the mutex M1 can also be acquired to inhibit any connection related changes since no administrative interaction gets executed then. Holding the mutex M1 protects the connection to the service provider from being modified while an interaction pattern of the service requestor accesses the service provider. The mutex M10 performs the analogous task at the service provider part of a communication pattern. It must be hold either if a connected service requestor is accessed or if the list of connected service requestors is modified.

Since always only one administrative interaction can be active, the administrative (B/U) interactions are based on a static monitor instance that could even be shared by all administrative interactions. In contrast thereto, a service related interaction can be invoked by any number of concurrent threads. For example, each not yet completed *query* can be seen as a separate instance of the appropriate interaction pattern. Of course, all *queries* are handled by the same interaction pattern instance but they all possess their own monitor instance. Thus, the monitor instances of the service related interactions are generated dynamically. The administrative interactions always have to be able to notify the currently active service related interactions about connection related state changes. Therefore, each communication pattern that has to handle dynamically generated monitor instances, possesses a *list of monitors*. At the service requestor the *list of monitors* is protected by the mutex M2. At the service provider, the mutex M10 also undertakes the task of protecting the *list of monitors*. For reasons that get clear with the subsequent explanations on the locking strategy, the mutex M1 cannot be used to protect the *list of monitors* at the service requestor.

The dynamically generated monitor instances have to be managed by *smart pointers* [ED94]. The reason is that several threads can be blocked on one monitor instance and one never knows when all threads are released such that the monitor instance is not needed anymore. That problem is further intensified by the Mesa-style semantics of the condition variables. With *smart pointers*, the dynamically generated monitor instance is destroyed automatically as soon as the last shared pointer pointing to the monitor instance is deleted and pointers to monitors keep valid as long as there is at least one remaining reference.

The administrative (B/U) interaction shown in figure 18 is now explained in detail. At first, the mutex M1 ① is acquired to ensure that no other administrative interaction is executed concurrently and that the interaction starts only after all sending activities of the interaction patterns of the service requestor are completed ❶. While holding the mutex M2 ②, one can iterate through the *list of monitors* to access the dynamically generated monitor instances of the currently active service related interactions. With a *disconnect*, for example, one has to appropriately set the state automatons of the service related interactions to notify the pending interactions about the connection related state changes. In case of a *disconnect*, for example, these have to know that the expected response cannot be received anymore. The next steps are to first call the *prepare* method of the monitor of the administrative (B/U) interaction, to then send the request message ③ and to finally invoke the *wait* ④.

As explained below, it is mandatory that one neither holds the mutex M2 nor the lock of the monitor of the administrative interaction when performing the *send*. Even if the monitor lock is released for the *send*, no administrative interaction can be invoked concurrently so that no other administrative interaction than the currently executed one accesses the monitor. Thus, although the monitor is accessible, its state does not get messed up by concurrent administrative interactions.

At the service provider, the upcall performs some work and acquires the mutex M10 ⑤ as soon as either the list of connected service requestors needs to be accessed or a response is to be sent ⑥. The *list of monitors* again contains the dynamically generated monitor instances of the service related interactions. With a *disconnect*, for example, one needs to iterate through the *list of monitors* to notify affected service related interactions by appropriately setting their state automatons. In case of the *query* communication pattern, for example, the not yet answered requests of a service requestor, that gets disconnected, need not to be processed and answered anymore. In contrast to the service requestor part, even the monitor lock can be hold while performing the *send* ⑥. Again, the reason why this holds true gets clear with the subsequent explanations on the locking strategy.

At the service requestor, the upcall for the response ⑧ accesses the monitor belonging to the interaction pattern. If the identifier of the received response matches the expected one, the blocked administrative (B/U) interaction resumes processing ⑨. The upcall does not need to acquire the mutex M1 since it leaves the modification of any items that are protected by the mutex M1 to the resumed thread. Of course, the upcall ⑧ can acquire the mutex M2. The resumed method can acquire the mutex M2 only after it released the monitor lock since one always has to take into account the locking order.

The ordering of the mutexes is fixed to prevent from deadlocks and one always has to acquire them in the appropriate order. At the service requestor, that order is M1 first, then M2 and finally a monitor lock. Besides the mutex M1, no locks must be hold while performing a *send* or while awaiting a response. The monitor locks are automatically released when invoking the *wait*. Thus, before acquiring a monitor lock that is automatically released by a *wait*, one always has to release the mutex M2 first. At the service provider, the locks are ordered in the same way that is one first has to acquire the mutex M10 and then a monitor lock. In contrast to the service requestor, both the mutex M10 and a monitor lock can be hold while performing a *send*. In contrast to the mutex M1, however, the mutex M10 must never be hold while awaiting a response as is explained now.

In principle, the availability of the mutex M2 never depends on the availability of any communication activities. It is never hold while awaiting a response or while waiting until a message can be sent. Since the mutex M1 at the service requestor is never acquired by any upcall ❷, all upcalls at the service requestor can always be completed even if the mutex M1 is hold due to an administrative (B/U) interaction that is awaiting its response. Thus, all sending activities from the service provider to the service requestor ⑥ can always be completed independently of the interaction they belong to and thus, sooner or later, the mutex M10 gets released. As soon as the mutex M10 is released, the upcalls at the service provider ❸, that require the mutex M10, can proceed. Thus, the upcall of the administrative (B/U) interaction also gets through and invokes the *send* ⑥ that completes the interaction ⑨. Even if the upcalls at the service provider ❸ get temporarily blocked on the mutex M10, that causes no deadlocks since the release of the mutex M10 never depends on the capacity of the service provider to process further upcalls.

Administrative (B/U) interactions always have the client part (B) at the service requestor of the communication pattern and the server part (U) at the service provider. The *connection oriented* protocol never requires an administrative (B/U) interaction from a service provider to a service requestor. That is the crucial factor why the administrative (B/U) interaction can never cause a deadlock. If an administrative (B/U) interaction was directed from the service provider to a service requestor, the following deadlock situation would occur. A service requestor invokes an administrative (B/U) interaction and holds and blocks the mutex M1 until the corresponding response arrives. Since the mutex M1 is locked, no other message can leave the service requestor meanwhile. At the same time, the service provider might invoke an administrative (B/U) interaction, locks the mutex M10 and waits until its response arrives. Since it holds the mutex M10, no other message can leave the service provider meanwhile. That, of course, results in a deadlock since neither the service requestor nor the service provider can send the response that is needed to release the locked mutexes.

Figure 19 shows the same administrative (B/U) interaction on top of a *processed* policy. It is important to note that even the nested callback due to the *processed* policy cannot result in a deadlock.

4.3 Details of the Query Pattern

The *query* pattern is the most complex pattern with respect to the internals since it provides a two-way interaction that is based on an asynchronous request/response protocol. That requires the client side to properly assign the received responses to pending requests and the server side to distribute available answers to the proper clients. Furthermore, disconnecting a client requires to clean up not yet fully processed requests at the server. All patterns are based on the same mechanisms illustrated by the *query* pattern, that is monitors and state automatons to handle the activities of concurrent requests.

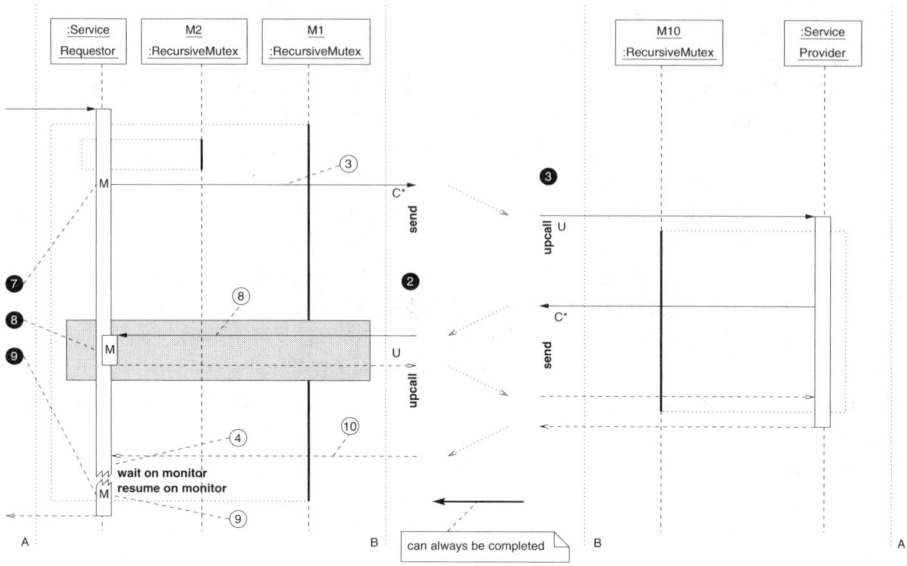

Fig. 19. Administrative (B/U) interaction from the service requestor to the service provider (*processed* policy).

The Client Part

All active queries are represented by their own monitor instance that wraps a state automaton, contains a unique query identifier and provides storage for the received answer until it is retrieved. The client side state automaton is shown in figure 20. The query identifiers manage the basic version of an asynchronous completion token [SSRB00]. Those automatons are the basis for the correct processing of pending calls in case of dynamic rearrangenments of service interconnections by the dynamic wiring pattern.

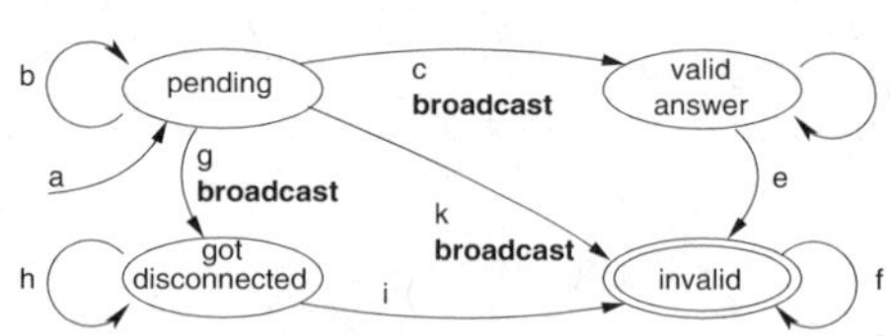

Fig. 20. The client side state automaton to manage the lifecycle of a query.

The sequence diagram of the client side upcall to process an incoming answer is shown in figure 21 and the internals of the client side *receiveWait* method are shown in figure 22.

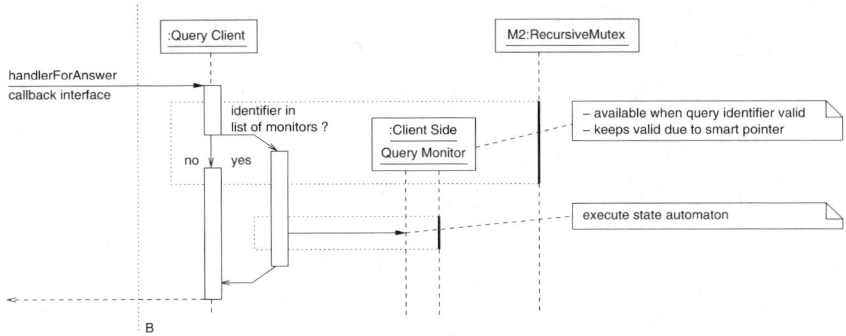

Fig. 21. The internals of the client side handler for the *answer* message.

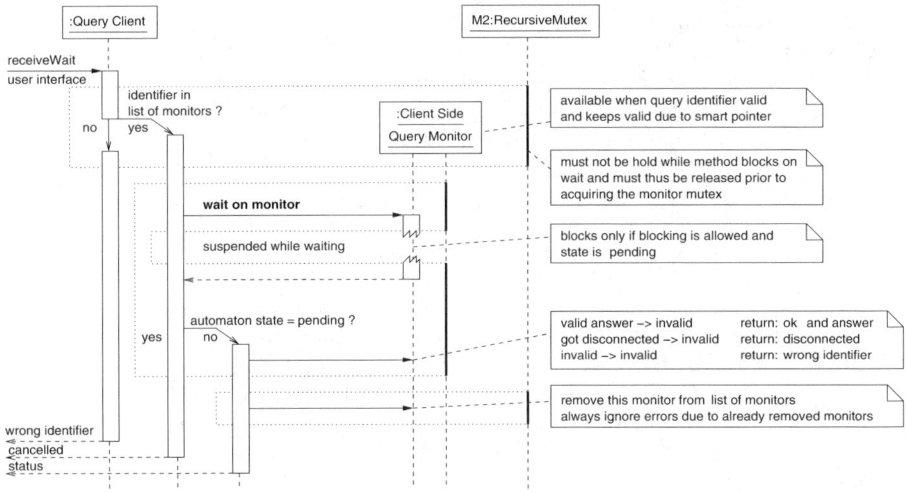

Fig. 22. The internals of the client side *receive wait* method.

5 Summary and Conclusion

Communication patterns are more than just hiding middleware complexity from the robotics expert. They avoid dubious interface behaviors and do not restrict the component internal architecture. The underlying middleware mechanism is fully transparent and can even be exchanged without affecting the component interfaces. Implementations with the very same features and behavior are available on top of *CORBA*, *TCP* sockets, Mailboxes, *RPC* based communication and message based systems. Standardized communication objects for maps, laser range scans and other entities further simplify the interoperability of components. Dynamic wiring is the key towards dynamic and task and context dependent composition of control and data flows.

The approach of communication patterns closes the gap between the general idea of component based software engineering and composability by reducing the diversity of externally visible component interfaces and by prescribing the interface semantics.

References

[AOS00] A. B. Arulanthu, C. O'Ryan, D. C. Schmidt, M. Kircher, and J. Parsons, *The design and performance of a scalable ORB architecture for CORBA asynchronous messaging*, Middleware, 2000, pp. 208–230.

[ED94] J. R. Ellis and D. L. Detlefs, *Efficient Garbage Collection for C++*, Usenix Proceedings, February 1994.

[HV99] M. Henning and S. Vinoski, *Advanced CORBA programming with C++*, Addison-Wesley, 1999.

[MDS01] D. R. Musser, G. J. Derge, and A. Saini, *The STL tutorial and reference guide, second edition*, Addison-Wesley, 2001.

[OSK00] C. O'Ryan, D. C. Schmidt, F. Kuhns, M. Spivak, J. Parsons, I. Pyarali, and D. Levine, *Evaluating policies and mechanisms for supporting embedded, real-time applications with CORBA 3.0*, Sixth IEEE Real Time Technology and Applications Symposium (RTAS), May 2000.

[Scha] D. C. Schmidt, *ACE - Adaptive Communication Environment*, http://www.cs.wustl.edu/~schmidt/ACE.html.

[Schb] D. C. Schmidt, *TAO - Realtime CORBA with TAO*, http://www.cs.wustl.edu/~schmidt/TAO.html.

[Sch04] C. Schlegel, *Navigation and execution for mobile robots in dynamic environments - an integrated approach*, Ph.D. thesis, Faculty of Computer Science, University of Ulm, 2004, http://www.rz.fh-ulm.de/~cschlege.

[Sch06] C. Schlegel, *Communication patterns as key towards component-based robotics*, ARS Special Issue on Software Development and Integration in Robotics, Int. Journal of Advanced Robotic Systems (2006).

[SH02] D. C. Schmidt and S. D. Huston, *C++ network programming, volume 1*, C++ In-Depth Series, Addison-Wesley, 2002.

[SH03] D. C. Schmidt and S. D. Huston, *C++ network programming, volume 2, systematic reuse with ACE and frameworks*, C++ In-Depth Series, Addison-Wesley, 2003.

[Sma05] SmartSoft, *Reference implementation*, 2005, http://www.rz.fh-ulm.de/~cschlege or http://smart-robotics-sourceforge.net/.

[SSRB00] D. C. Schmidt, M. Stal, H. Rohnert, and F. Buschmann, *Pattern-oriented software architecture, volume 2*, John Wiley and Sons, Ltd., 2000.

[SV00] D. C. Schmidt and S. Vinoski, *Object Interconnections - An Overview of the OMG CORBA Messaging Quality of Service (QoS) Framework*, C++ Report, SIGS **12** (2000), no. 3.

[SW99a] C. Schlegel and R. Wörz, *Interfacing different layers of a multilayer architecture for sensorimotor systems using the object-oriented framework* SMARTSOFT, Proceedings 3rd European Workshop on Advanced Mobile Robots (EUROBOT) (Zürich, Schweiz), September 1999.

[SW99b] C. Schlegel and R. Wörz, *The software framework* SMARTSOFT *for implementing sensorimotor systems*, Proceedings IEEE/RSJ International Conference on Intelligent Robots and Systems (IROS) (Kyongju, Korea), October 1999, pp. 1610–1616.

Using MARIE for Mobile Robot Component Development and Integration

Carle Côté, Dominic Létourneau, Clément Raïevsky, Yannick Brosseau, and François Michaud[1]

Université de Sherbrooke, Department of Electrical Engineering and Computer Engineering, Sherbrooke (Québec), CANADA
{Dominic.Letourneau, Carle.Cote, Clement.Raievsky, Yannick.Brosseau, Francois.Michaud}@USherbrooke.ca

1 Introduction

Many existing programming environments, such as those documented in this book, are all proposing different approaches for mobile robotics system development and integration. Most of them are incompatible with each other for different reasons [OC03], such as the use of specific communication protocols and/or mechanisms, different operating systems, robotics platforms, architectural concepts, programming languages, intended purpose, proprietary source codes, etc. This leads to code replication of common functionalities across different programming environments, and to specific functionalities being often restricted to one programming environment.

Reusing existing software components and interconnect them through an integration framework is a possible strategy to benefit from their respective approaches, instead of having to choose only one of them. The main objective of component integration frameworks is to support one or many integration approaches (e.g. communication protocols, central repository, remote procedure call, dynamic and static linkage) to interconnect heterogeneous applications with their own set of concepts and requirements in a larger system.

MARIE (for Mobile and Autonomous Robotics Integration Environment) is a component integration framework oriented towards a rapid prototyping approach to development and integration of new and existing softwares for robotic systems [CLM04] [CBL06]. To achieve the integration challenge, MARIE proposes an extendable collection of blackbox and whitebox components allowing different development techniques to add new functionalities in the system.

The following sections present how MARIE creates a component integration framework, providing tools to create specialized middlewares for dedicated applications. MARIE's efforts have been focused on distributed robotics

D. Brugali (Ed.): Software Eng. for Experimental Robotics. STAR 30, pp. 211–230, 2007.

component-based middleware framework development, enhancing reusability of applications and providing tools and programming environments to build integrated and coherent robotics systems. Section 2 presents MARIE's software architecture ans situates the three principal frameworks used in MARIE: the Component Framework (Section 3), the Communication Abstraction Framework(Section 4) and the Configuration Framework (Section 5). Section 6 presents Spartacus as a study case of how MARIE can be used in a robotic implementation. Conclusions and future work are presented in Section 7.

2 Software Architecture

MARIE's software architecture can be explained from the following three perspectives: component mediation approach, layered architecture and communication protocol abstraction.

2.1 Component Mediation Approach

To implement distributed applications using heterogeneous softwares, MARIE adapted the Mediator Design Pattern [GHJV95] to create a Mediator Interoperability Layer (MIL), as illustrated in Figure 1. The Mediator Design Pattern primarily creates a centralized control unit (named Mediator) which interacts with each class independently, and coordinates global interactions between classes to realize the desired system. In MARIE, the MIL acts just like the Mediator of the original pattern, but is implemented as a virtual communication space where applications can interact together using a common language (similar to Internet's HTML for example). With this approach, each application can have its own communication protocols and mechanisms as long as the MIL supports it. It is a way to exploit the diversity of communication protocols and mechanisms, to benefit from their strengths and maximize their usage, and to overcome the absence of standards in robotic software systems. It also promotes loose coupling between applications by replacing a many-to-many interaction model with a one-to-many interaction model. In addition to simplifying each application communication interface, loose coupling between applications increases reusability, interoperability and extensibility by limiting their mutual dependencies and hiding their internal implementation. By using a virtual communication space approach, the MIL's design reduces the potential complexity of managing a large number of centralized classes, as observed with the original pattern. This is mainly attributed to having limited centralization of communication protocols and mechanisms, leaving most of the functionalities decentralized. Although there is no such thing as an instance of a Mediator in MARIE's implementation, mediation is possible through the Communication Abstraction Framework.

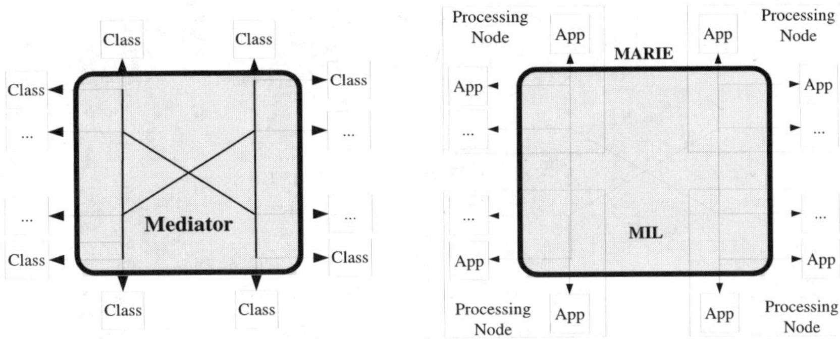

Fig. 1. Original Mediator Pattern (left) and MARIE's Distributed Mediator Adaptation (right)

2.2 Layered Architecture

Supporting multiple sets of concepts and abstractions can be achieved in different ways. MARIE does so by adopting a layered software architecture, defining different levels of abstraction into the global middleware framework. As shown in Figure 2, three abstraction layers are used to reduce the amount of knowledge, expertise and time required to use the overall system. It is up to the developer to select the most appropriate layer for adding elements to the system. At the lower level, the Core Layer consists of tools for communication, data handling, distributed computing and low-level operating system functions (e.g., memory, threads and processes, I/O control). The Component Layer specifies and implements the Component Framework, the Communication Abstraction Framework and the Configuration Framework useful to build new applications using the MIL. The Application Layer contains useful tools to build and manage integrated applications to craft robotic systems.

2.3 Communication Protocol Abstraction

Integrated applications functionalities can often be used without any concerns with the communication protocols, as they are typically designed to apply operations and algorithms on data, independently of how data are received or sent. This eases applications interoperability and reusability by avoiding fixing the communication protocol during the design phase. Ideally, the communication procol choice should be made as late as possible, depending of which applications need to be interconnected together (e.g., at the integration phase or even at runtime). Therefore, a Communication Abstraction Framework, called Port, is provided for communication protocols and applications interconnections.

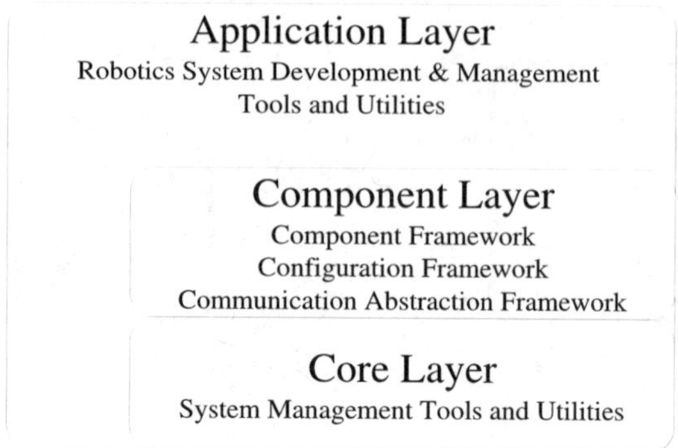

Fig. 2. MARIE's Layered Architecture

3 Component Framework

The development of robotic applications using MARIE is based on reusable software blocks, referred to as components, which implement functionalities by encapsulating existing applications, programming environments or dedicated algorithms. Components are configured and interconnected to implement the desired system, using the software applications and tools available through MARIE. Four types of components are used in the MIL:

1. **Application Adapter (AA)** : component interfacing useful applications within the MIL and to enable them to interact with each other through their standardized interface (i.e. Ports) and using MARIE's shared data types. Interconnections using Port communication abstraction are illustrated in Figure 3 with a small dot between communication links represented by arrows.
2. **Communication Adapter (CA)** : component that ensures communication between other components by adapting incompatible communication mechanisms and protocols, or by implementing traditional routing communication functions. Available Communication Adapters in MARIE are Splitters, Switches, Mailboxes and Shared Maps. A Splitter sends data from one source to multiple destinations without the sender needing to be aware of the receivers. A Switch acts like a multiplexer sending data to the selected output. A Mailbox creates a buffering interface between asynchronous components. A Shared Map is used to share data, in the key-value form, between multiple components.
3. **Application Manager (AM)** : system level component managing, on local or remote processing nodes, Application Adapters and Communication Adapters. Application and Communication Adapters initialization, configuration, start, stop, suspend and resume are handled by the Ap-

plication Manager. When starting the system, the Application Manager initializes the components following the adequate sequence.

4. **Communication Manager (CM)** : system level component dynamically managing, on local or remote processing nodes, the communications mechanisms (socket, port, shared memory, etc.).

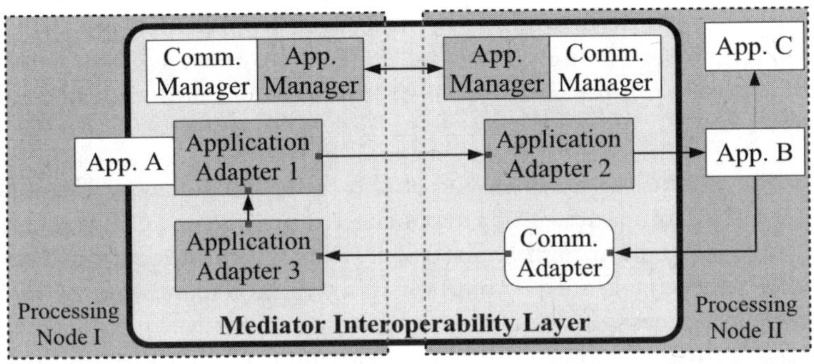

Fig. 3. Component Framework Using the Mediator Interoperability Layer

Although the use of MARIE's frameworks and software tools is highly encouraged to save time and efforts, MARIE is not limited to them. Developers can use the best solution to integrate software applications and interconnect components by having the possibility to extend or adapt existing components and available frameworks. MARIE's underlying philosophy is to complement existing applications, programming environments or software tools, and therefore it is to be used only when required and appropriate.

Figure 3 presents an example of how software applications can be integrated and interconnected in the MIL. Application A represents an integrated application directly linked with the implementation of its AA (e.g., a library or an open source application). When an application is integrated using an AA, it can use the MIL communication mechanisms to exchange data with any other components, as is the case for providing data to Application Adapter 2 and Application Adapter 3. Application B interacts with other applications in two different ways. The first one needs Application Adapter 1 to transmit data to Application Adapter 2, which convert them into a specific communication protocol not supported by the MIL to make them available to Application B. The second one is used to send back data to Application Adapter 3, using a communication protocol supported by the MIL for direct interconnection with any components. However, Application B and Application Adapter 3 do not use compatible communication mechanisms or protocols. Interfacing them requires a CA. Application Adapter 3 implements functionalities directly in the MIL by encapsulating them in a stand-alone AA (e.g., a graphical user interface implemented in the AA directly). Application C can already communicate with Application B, and therefore no interconnection through the MIL is required.

3.1 Main Concepts Supporting the Component Framework

The Component Framework is a whitebox framework enabling developers to extend available functionalities or create new components. Existing functionalities can be reused and extended by inheriting from framework base classes and/or overriding pre-defined hook methods. Five main concepts present in every component are provided by the Component Framework and illustrated in Figure 4 :

- **Handler** : handles the behavior of the component. Every Handler must support the init, start, stop, suspend, resume, reset and quit messages that are received through their Request Interface. Execution flow of the Handler can be customized to adapt its own implementation needs (based on iterations, messages, interrupts, states-machine, etc).
- **VisitorConfig** : holds configuration information for the component. It is extracted from the configuration data structure created by the Configurator. The configuration information will then be used by the Handler in the initialization phase of the component.
- **Director** : handles and manages execution of requests (such as init, start, stop, etc.) and forwards them to the component's Handler to execute specialized functionalities if needed.
- **Configurator** : handles configuration requests, parsing of the configuration file format, and creation of the Configuration Data Structures to be forwarded to the component's Handler.
- **Builder** : builds the component using the specialization of the different elements composing a component : the Director, the Configurator, the Handler, and their respective execution flow mechanisms.

To explain what is required to create a component, Figure 5 illustrates the Splitter Communication Adapter. The Splitter is typically used to route data from one or more source Ports to multiple destination Ports. This mechanism is handled by the Splitter Handler. The Splitter Handler uses an Iterate Execution Mechanism to monitor the Splitter Handler behavior. The Iterate Execution Mechanism refers to an internal loop that runs at a determined cycle inside the Handler. Sending and receiving data from Ports is event-based and is triggered by the main communication loop sending the data as soon as it arrives (not shown in Figure 5). As shown in Figure 5, Ports on the left (A0 to An) and Ports on the right (B0 to Bn) are grouped together to form Group A and Group B. The Splitter can support any number of ports in each group. The configuration of the Splitter supports three modes, which selects how the data flows between Group A and Group B:

1. **Mode AB**. Communicates one way from Group A to Group B.

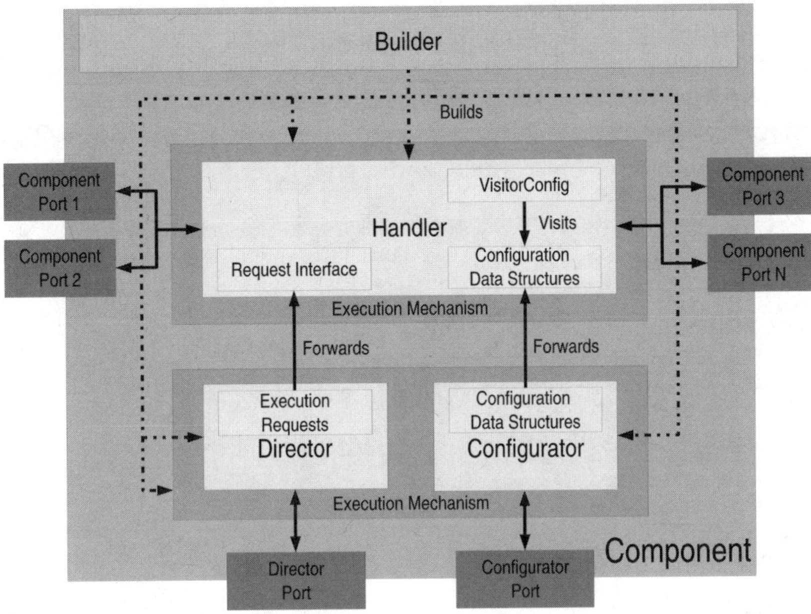

Fig. 4. MARIE's Components Composition

2. **Mode BA**. Communicates one way from Group B to Group A.
3. **Mode ABBA**. Communicates in both ways (bi-directional) between Group A and Group B.

To configure the Splitter Communication Adapter, the XML Configurator is used. It parses the XML configuration files and forwards the Configuration Data Structure to the Splitter Handler. The Splitter VisitorConfig then uses the Configuration Data Structure to get the appropriate configuration at initialization. The Default Director forwards execution requests (start, stop, etc.) to the Splitter Handler.

4 Communication Abstraction Framework

The Communication Abstraction Framework, illustrated in Figure 6, offers an abstraction on how communications are achieved by components, using a Send & Receive Interface to send and receive data. This interface hides implementation details of communication protocols and mechanisms that execute send and receive requests. Those communication protocols and mechanisms are encapsulated in Communication Strategy (CS) classes in order to be more easily reused and extended. The Strategy Design Pattern [GHJV95] is applied to CS to define a set of interchangeable algorithms and to create a loosely coupled relation between CS's clients and implementation details of communication

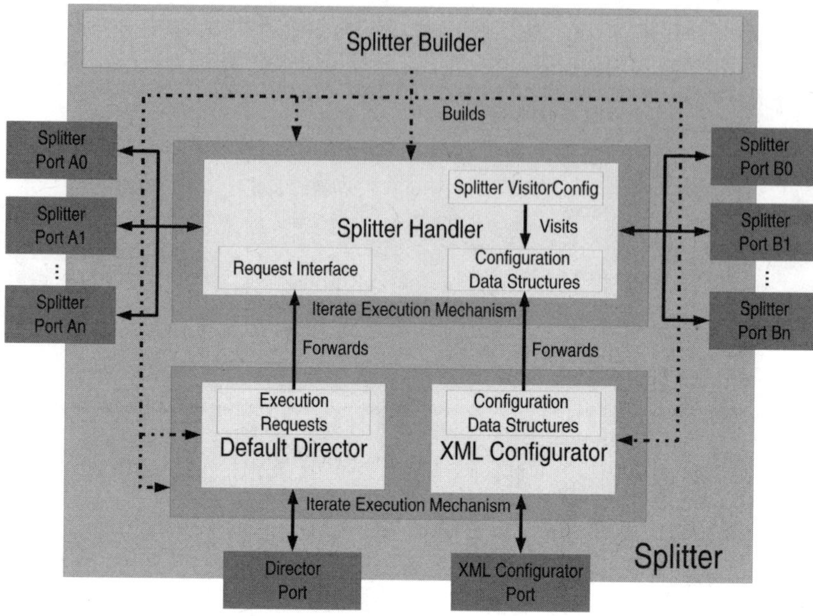

Fig. 5. MARIE's Splitter Composition

protocols and mechanisms. Currently supported Communication Strategies are :

- **SocketAcceptor**. TCP/IP socket-based server accepting one connexion on a specified port.
- **SocketConnector**. TCP/IP socket-based client connecting to the server on a specified port.
- **SharedMemAcceptor**. Memory-based server accepting one local (on the same processing node) connexion.
- **SharedMemConnector**. Memory-based client connecting to the local server.

The Send & Receive interfaces also hides data operations that needs to be applied in order to fulfill communication protocols requirements on data representation. Those operations are encapsulated in Cascading Functional Block (CFB) classes, that can be chained together in a cascaded manner to execute the appropriate sequence of operations on sent and received data. MARIE currently provides four kinds of CFBs :

- **XMLFormatter** : marshalls MARIE's objects (data structures in memory) in an homemade XML representation to be send to a byte stream CS.
- **XMLExtractor** : unmarshalls the XML data representation of MARIE's data objects from the byte stream CS to create new MARIE data objects.

- **ImageFormatter** : marshalls MARIE's Image objects in an optimized serialized data representation to be send to a byte stream CS.
- **ImageExtractor** : unmarshalls the image optimized serialized data representation from the byte stream CS to create a new MARIE Image data.

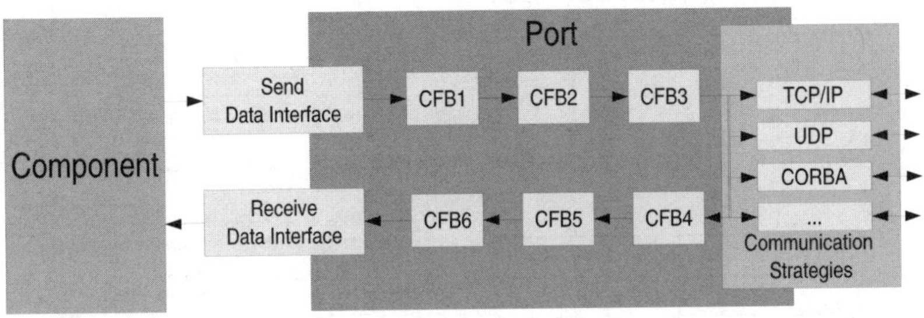

Fig. 6. Communication Abstraction Through the Port Interface

The Communication Abstraction Framework can be extended or specialized knowing that Ports are not tightly coupled to the Component Framework. This means that the Communication Abstraction Framework could be redesigned to handle specialized functionalities (error handling, signals, etc.) without having a major impact on existing implementations. However, it is is highly recommended to use the Send & Receive Interface for standardization of communications with components.

5 Configuration Framework

MARIE's Configuration Framework, shown in Figure 7, offers a generalization of configuration representation in order to use the same data structure for all the components configurations. Four types of configuration elements are available in the Configuration Framework representing the Configuration Data Structures :

1. **Configuration**. Composite configuration element that gives a name to a current configuration structure and contains other configuration elements.
2. **Type**. Composite configuration element that represents a type contained in the component's description domain. A type element can be composed of multiple configuration elements.
3. **Key-Value**. Primitive configuration element that identifies a specific property of an object. The configuration element is represented by a label (Key) and have a value (Value).
4. **Qualifier**. Composite configuration element that represents an attribute for refining configuration element semantics. It can be useful to categorize

or discriminate configuration elements from each other. A qualifier element can be applied on any other configuration element.

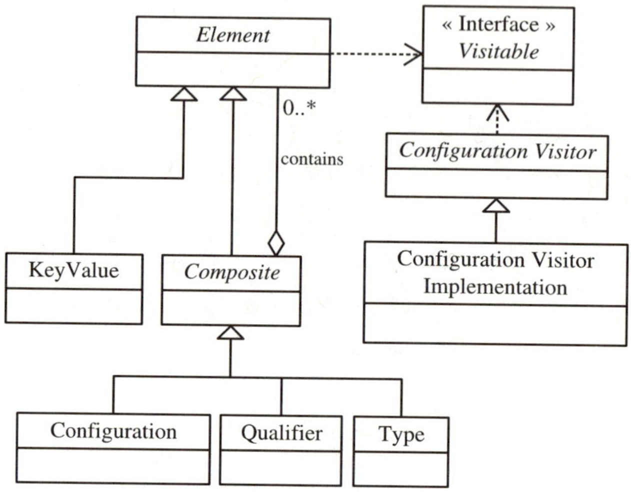

Fig. 7. Configuration Framework Architecture

To support execution of Configuration Data Structures' operations (read, write, modify, etc.), the Visitor and Composite Design Pattern [GHJV95] are used. The of the Visitor Design Pattern is to encapsulate operations to be performed on a data structure in a class, called a visitor, that can traverse the data structure. Having different kinds of configuration elements in the Configuration Framework (primitive and composite), applying the Composite Design Pattern to the Configuration Data Structures' elements permits to the visitor to treat each kind of elements as they were exactly the same, which reduces visitor's implementation complexity.

Generally, the Configuration Framework is used as a blackbox framework by creating required visitors to fetch data configuration in Configuration Data Structures, by extending the Visitor abstract class. Other configuration manipulations, such as parsing configuration files and creating the Configuration Data Structures, are handled automatically. In the current implementation, an XML based generic parser (not shown in Figure 7) is responsible of creating the Configuration Data Structures by parsing the XML configuration file. If required, new parsers supporting other representations and languages can be added in the framework without the need to change existing visitor classes.

The following example shows a sample Splitter configuration file. All four configuration elements are present in this Splitter sample configuration file. Each XML node contains an attribute named "elem" which can be set to four values : *conf* for Configuration element, *type* for Type element, *kv* for Key-Value element and *q* for Qualifier element. This configuration represents a Splitter with one Port in its Group A and two Ports in its group B. Group A and B are represented as qualifiers in the configuration file, providing in which group each Port must be instantiated. Data flows from Group A to Group B as the "mode" Key-Value states. Port A0 is configured to use the SocketAcceptor communication strategy listening on port 30004. Port B0 is configured to also use the SocketAcceptor communication strategy, listening on port 30000. Port B1 is configured to use the Socket Connector communication strategy, connecting to host 192.168.43.67 on port 30030.

```
<?xml version="1.0"?>
<splitter elem="conf">
   <mode elem="kv">AB</mode>
   <groupA elem="q">
      <port elem="type">
         <type elem="kv">Default</type>
         <name elem="kv">A0</name>
         <cs elem="type">
            <type elem="kv">SocketAcceptor</type>
            <portnumber elem="kv">30004</portnumber>
         </cs>
      </port>
   </groupA>
   <groupB elem="q">
      <port elem="type">
         <type elem="kv">Default</type>
         <name elem="kv">B0</name>
         <cs elem="type">
            <type elem="kv">SocketAcceptor</type>
            <portnumber elem="kv">30000</portnumber>
         </cs>
      </port>
      <port elem="type">
         <type elem="kv">Default</type>
         <name elem="kv">B1</name>
         <cs elem="type">
            <type elem="kv">SocketConnector</type>
            <portnumber elem="kv">30030</portnumber>
            <hostname>192.168.43.67</hostname>
         </cs>
      </port>
   </groupB>
</splitter>
```

6 Spartacus' Implementation Using MARIE

Spartacus [MBC05], shown in Figure 8, is a socially interactive mobile robot designed to enter the AAAI Mobile Robot Challenge. Introduced in 1999, the AAAI Challenge consists of having a robot start at the entrance of the conference site, find the registration desk, register, perform volunteer duties (e.g., guard an area) and give a presentation [MSJ04]. The long-term objective is to have robots participate just like humans attending the conference. We became interested by this challenge because of the need to address all design dimensions for such a robot, from the hardware level to the high-level decision-making algorithms.

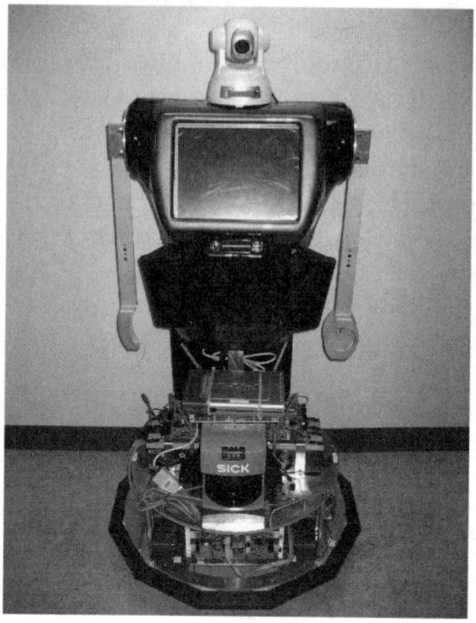

Fig. 8. Spartacus Robot

Spartacus is equipped with a SICK LMS200 laser range finder (for autonomous navigation), a Sony SNC-RZ30N 25X pan-tilt-zoom color camera, an array of eight microphones placed on the robot's body, a touchscreen and a business card dispenser. High-level processing is carried out using an embedded Mini-ITX computer (Pentium M 1.7 GHz). The Mini-ITX computer is connected to the low-level controllers through a CAN bus device, the laser range finder through a serial port, the camera through a 100Mbps Ethernet link and the audio amplifier and speakers using the audio output port. A laptop computer (Pentium M 1.6 GHz) is also installed on the platform and is equipped with a RME Hammerfal DSP Multiface sound card using eight analog inputs to simultaneously sample signals coming from the microphone array. Communication between the two on-board computers is accomplished with a 100Mbps Ethernet link. Communication with external computers can be achieved using the 802.11g wireless technology, giving the ability to easily add remote processing power or capabilities if required. All computers are running Debian GNU Linux.

Numerous algorithms are required to accomplish the Challenge, and here is what we implemented for our 2005 participation to the event:

- **Autonomous Navigation**. When placed at the entrance of the convention center, the robot autonomously find its way to the registration desk by wandering and avoiding obstacles, searching for information regarding the location of the registration desk and potentially following people moving in this direction. Once registered, the robot can use a map of the convention center. The two navigation tools used are CARMEN and pmap. CARMEN, the Carnegie Mellon navigation toolkit [MRT03], is a software package for laser-based autonomous navigation using a map previously generated. The pmap package[1] provides a number of libraries and utilities for laser-based mapping (SLAM) in 2D environments to produce high-quality occupancy grid maps.

- **Vision Processing**. Extracting useful information in real time from images taken by the onboard camera improves interaction with people and the environment. For instance, the robot could benefit from reading various written messages in real life settings, messages that can provide localization information (e.g., room numbers, places) or identity information (e.g., reading name badges). Object recognition and tracking algorithms also makes it possible for the robot to interact with people in the environment. We use two algorithms to implement such capabilities : one that can extract symbols and text from a single color image in real world conditions [LMV04]; and another one for object recognition and tracking to identity and follow regions of interest in the image such as human faces and silhouettes.
- **Sound Processing**. Localization of sound sources provides important clues about the world. However, simply using one or two omnidirectional microphones on a robot is not enough: it proves too difficult to filter out all of the noise generated in public places. Using a microphone array is a better solution for the localization, tracking and separation of sound sources. Our approach is capable of simultaneously localizing and tracking up to four sound sources that are in motion over a 7 meters range, in the presence of noise and reverberation [VMHR1]. We also developed a method to separate in real-time the sound sources [VRM04] in order to simultaneously process vocal messages from interlocutors using software packages such as Nuance[2].
- **Touchscreen Display**. Various information can be communicated through this device, such as: receiving information from people using a menu interface; displaying graphical information such as a PowerPoint presentation or a map of the environment; and expressing emotional states using a virtual face.

MARIE's design and implementation evolved as we worked on Spartacus' implementation. MARIE's current version is in C++ (~10 000 lines of code) and uses the ACE (Adaptive Communication Environment) library [Sch94] for the Core Layer functions (low-level operating system functions). Although ACE met Spartacus' implementation needs, MARIE does not rely on this specific library as the Core Layer and it can be replaced if required. MARIE's Application Manager is partially implemented in MARIE, meaning that AA

[1] http://robotics.usc.edu/~ahoward/pmap
[2] http://www.nuance.com/

and CA must be initialized manually from scripting commands. Also, the CM is not yet implemented, and component configuration must be set manually.

Spartacus' implementation requires 45 components (~50 000 lines of code) composed of 26 AA, 17 CA and two external applications (the Audio Server and NUANCE). Application Adapters are used to interface the different software applications required for decision-making by the robot. Mailboxes, Splitters, Shared Maps and Switches are used as Communication Adapters. Except for the two external applications, component interconnections are all sockets-based using Push, Pull and Events dataflow communication mechanisms [Zha03] with XML encoding for data representation; the Audio Server and Nuance use their own communication protocols.

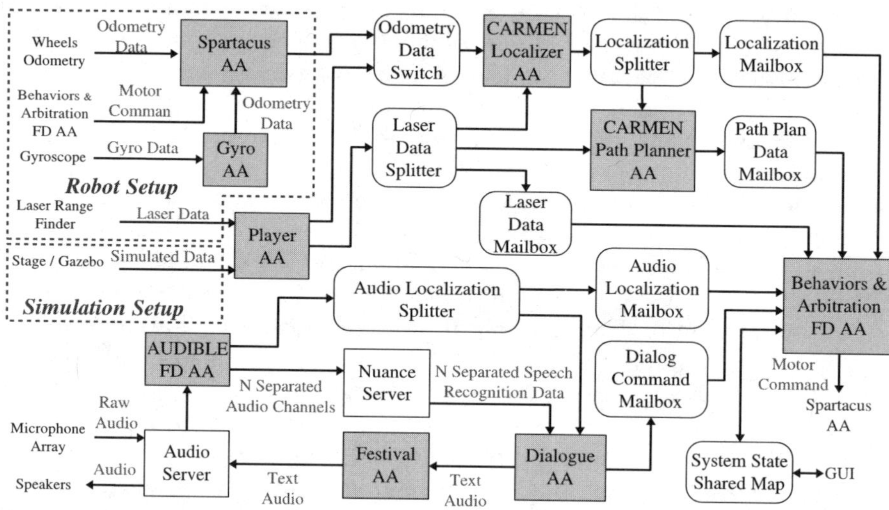

Fig. 9. Partial Representation of Spartacus' Software Architecture

Partial representation of Spartacus' software architecture is illustrated in Figure 9. It covers sensing and acting in simulation and real robot setups, localization, path planning, sound source localization, tracking and separation, speech recognition and generation, and part of the computational architecture [MBC05] responsible for the robot's navigation, reasoning and interactions capabilities. In the real robot setup, SpartacusAA combines wheels odometry and gyroscopic (through GyroAA interfacing a gyroscope installed on Spartacus) data, and pushes the result at a fixed rate (10 Hz) to its interconnected component. Laser data is collected by PlayerAA, interfacing the Player library specialized for sensor and actuator abstraction (see Chapter *The Player/Stage Reusable Robot Software Framework* in this book [VAUG06] and [VGH03]), supporting the SICK LMS200 laser range finder installed on Spartacus. PlayerAA pushes data at a fixed rate (10 Hz) to connected components. In the simulation setup, odometry and laser data are both collected with PlayerAA, as generated using Stage (2D) or Gazebo (3D) simulators [VGH03]. CARMEN Localizer AA and CARMEN Path Planner AA provide path planning and localization capabilities.

RobotFlow and FlowDesigner programs [CLM04] are used to implement Behavior & Arbitration FD AA, handling part of the computational architecture. RobotFlow (RF) [CLM04] and FlowDesigner (FD) are two modular data-flow programming environments that facilitate visualization and understanding of the robots control loops, sensor and actuator processing. They are also appropriate for rapid prototyping since the graphical user interface enables the user to connect reusable software blocks without having to compile the program every time minor changes are made. In the Behavior & Arbitration FD AA component, RF/FD programs implement behavior-producing modules arbitrated using a priority-based approach. It uses data coming from different elements such as localization, path plan, laser, audio localization, dialog command and system states. It uses an asynchronous pull mechanism to get its data, requiring the use of Mailbox CA components, and generates motor commands at a fixed rate (5 Hz).

The Audio Server is interfacing the RME Hammerfal DSP Multiface sound card, and Nuance Server is interfacing Nuance. DialogueAA is a stand-alone AA that manages simultaneous conversations with people. This is made possible with the use of AUDIBLE FD AA, interfacing our sound source localization, tracking and separation algorithms implemented with RF/FD and using Spartacus' microphone array. It generates a number of separated audio channels that are sent to Nuance Server and Behavior & Arbitration FD AA. Recognized speech data is sent to Dialog AA, responsible of the human-robot vocal interface. Speech generated by the robot is handled by Festival [Tay99]. Dialogue AA also provides data to the Behaviors & Arbitration FD AA. The global execution of the system is asynchronous, having most of the applications and AAs pushing their results at a variable rate, based on the computation length of their algorithms when triggered by new input data. Synchronous execution is realized by having fixed rate sensors readings and actuators commands writings.

6.1 Discussion

Using MARIE with Spartacus provided interesting capabilities for software integration and team development. At the peak of Spartacus' software development process, eight software developers, including audio and image processing specialists, AI specialists, robot hardware specialists, Core Layer specialists and the integrator, were working concurrently on the system. Most of them only used Application Adapters (Component Layer) to create their components, conducting unit and blackbox testing with pre-configured system setups (Application Layer) given by the integrator. Communication protocols and operating system tools for component and application developments (CoreLayer) were added by the Core Layer specialists when required. Components were incrementally added to the system as they became available. It took around eight days, spread over a four weeks period, to complete a fully integrated system.

Overall, nine existing specialized applications/libraries were integrated together to build the complete system: Player/Stage/Gazebo, Pmap, CARMEN, Flowdesigner/RobotFlow, AUDIBLE, Nuance, Festival, AUDIBLE, QT3 and OpenCV. Each of these applications required different integration strategies. For instance, Nuance is a proprietary application with a specific and limited interface. Integrating Nuance in an AA was challenging because its execution flow is tightly controlled by Nuance's core application, which is not accessible from the available interface. To solve this problem, we created an independent application that uses a communication protocol already supported by the MIL. CARMEN, on the other hand, is composed of small executables communicating through a central server. CARMEN's integration was realized by creating an AA that starts several of these executables depending on the required functionality and on data conversion from CARMEN's to MIL's format. Having a flexible Component Framework and Ports as the communication protocol abstraction allowed us to adapt application specificities such as external threads execution, dynamic bindings, independent protocols and timing.

Choosing XML data representation for common language communication in the MIL was based on implementation simplicity and ease of debugging. Although it was sufficient for most of the system communication needs, we clearly observed that this solution was not sufficient to support communication-intensive data like audio and vision within MARIE. To avoid using valuable time to support optimized protocols for audio and video, we decided to use FlowDesigner that already provides those protocols.

With regard to component interoperability, the ability to change between simulation and robotic setups with only few system modifications gave us the possibility to do quick simulations and integration tests. Nearly 75% of the system functionalities were validated in simulation and were also used as is in the real world setup. In both simulated and real setups, configurations of components receiving laser and odometry data are exactly the same, abstracting data sources and benefiting from components modularity and the rapid prototyping approach. Moreover, component interoperability can be extended with MARIE to do things like porting a computational architecture on robotic platforms from different manufacturers and with heterogeneous capabilities, or evaluating performances of algorithms implementing the same functionality (e.g., localization, navigation, planning) using the same platform and experimental settings.

Distributing applications across multiple processing nodes was not difficult with MARIE, having chosen network sockets as the transport mechanism. We initially used a shared memory transport mechanism to accelerate communication between components on the same computer. Changing from one transport mechanism to another was transparent using Ports and supporting shared memory interconnection in the MIL. Since no noticeable impact was observed over the global system performances using either of them, we chose to exclusively use socket transport mechanism. It allows us to move components from one processing node to the other easily.

Meeting Spartacus' integration needs using MARIE rapid-prototyping approach highlighted three interesting consequences on robotics system development. First, it revealed the difficulty of tracking decisions made by the system simply by observing its behaviors in the environment, something that was always possible with simpler implementations. The system reached a level of complexity where we needed to develop a graphical application to follow on-line or study off-line the decisions made by the robot. This suggests that creating analysis tools and supporting them in the integration environment can play a key role in working with such a highly-integrated system. The second observation emerged from the number of components involved in the software architecture. Manually configuring and managing the system with many components executed on multiple processors, is an error-prone and tedious task. In this context, MARIE would greatly benefit from having GUI and system management tools to build, configure and manage components automatically. Third, regarding design optimization, being able to quickly interconnect components to create a complete implementation, without focusing on optimization right away, proved beneficial in identifying real optimization needs. Such an exploration strategy gave us the ability to quickly reject necessary applications, software designs or component implementations without investing too much time and effort. For Spartacus, we originally thought that tighter synchronization between components would be necessary to obtain a stable system and support real-time decision-making. For instance, having connected all of Spartacus' components together, we observed that performances were appropriate with the processing power available as long as we did not overload the computers with too many components. Noticing that, we decided to wait before investing time and energy working on component synchronization, to focus on Spartacus' integration challenges.

7 Conclusion

MARIE is a system integration framework oriented towards a rapid-prototyping approach to create robotic systems. To achieve this goal, MARIE is based on the mediation principle and uses a layered framework architecture to facilitate the creation, integration and interconnection of existing applications, programming environments or software tools available in the robotics community. Interconnections of applications are supported by a communication framework that is able to support a wide range of communication protocols, communication mechanisms, and upcoming robotics standards. To ease efforts required to integrate the work of multiple developers, MARIE also supports team development requirements with two design choices : 1) the layered architecture allows each developer to work at the appropriate level of abstraction, related to his contribution to the system, and 2) the component architecture lets developers work independently on each component. This tends to reduce the required knowledge to contribute to the system and accelerates the overall development cycle.

MARIE was experimented during Spartacus' software architecture development, which is the first software architecture implementation using MARIE. From this experience, we have observed that an integrated programming environment such as MARIE helps us focus on the decision-making issues and the high-level capabilities development rather than on low-level software programming considerations and integration issues. MARIE's integration framework was flexible enough to support the integration and interconnection of all the existing and new applications required for Spartacus' software architecture. Using a rapid-prototyping approach is well suited to rapidly identify critical development sections from less-critical ones, just by being able to work with the complete system at the very beginning of the development cycle.

More testing will be performed on Spartacus to validate MARIE's architectural design and implementation. We are currently working on identifying and implementing tools to measure system real-time performances and on software metrics to quantify MARIE's computational overhead. Additional work is also planned on the Application Layer, in which we hope to develop further useful applications and automated tools to manage the overall system and the underlying components.

Acknowledgements

F. Michaud holds the Canada Research Chair (CRC) in Mobile Robotics and Autonomous Intelligent Systems. Support for this work is provided by the Natural Sciences and Engineering Research Council of Canada and the Canada Research Chair program. MARIE is an open-source project available at http://marie.sourceforge.net.

References

[AS04] V. Andronache and M. Scheutz, *Ade - a tool for the development of distributed architectures for virtual and robotic agents*, IEEE Transactions on Systems, Man and Cybernetics, Part B, 2004, pp. 2377– 2395.

[Bru06a] Brugali, D. and Agah, A. and MacDonald, B. and Nesnas, I. and Smart, W. *Trends in Robot Software Domain Engineering*, In D.Brugali (Ed.) *Software Engineering for Experimental Robotics*, Spinger STAR series, 2006

[Bru06b] Brugali, D. and Brooks, A. and Cowley, A. and Côté, C. and Domnguez-Brito, A.C. and Létourneau, D. and Michaud, F. and Schlegel, C. *Trends in Component-Based Robotics*, In D.Brugali (Ed.) *Software Engineering for Experimental Robotics*, Spinger STAR series, 2006

[CBL06] C. Cote, Y. Brosseau, D. Letourneau, C¿ Raievsky, and F. Michaud, *Using marie in software development and integration for autonomous mobile robotics*, International Journal of Advanced Robotic Systems, Special Issue on Software Development and Integration in Robotics (2006).

[CLM04] C. Cote, D. Letourneau, F. Michaud, J.M. Valin, Y. Brosseau, C. Raievsky, M. Lemay, and V. Tran, *Code reusability tools for programming mobile robots*, Proceedings of the IEEE/RSJ International Conference on Intelligent Robots and Systems, 2004.

[GHJV95] E. Gamma, R. Helm, R. Johnson, and J. Vlissides, *Design patterns : Elements of reusable object-oriented software*, Addison-Wesley, 1995.

[GMP01] V. Gazi, M.L. Moore, K.M. Passino, W.P. Shackleford, F.M. Proctor, and J. Albus, *The rcs handbook: Tools for real-time control systems software development*, ch. 1, p. 1, Wiley, 2001.

[LMV04] D. Letourneau, F. Michaud, and J.-M. Valin, *Autonomous robot that can read*, EURASIP Journal on Applied Signal Processing, Special Issue on Advances in Intelligent Vision Systems: Methods and Applications **17** (2004), 1–14.

[MBC05] F. Michaud, Y. Brosseau, C. Côté, D. Létourneau, P. Moisan, A. Ponchon, C. Raievsky, J.-M. Valin, E. Beaudry, and F. Kabanza, *Modularity and integration in the design of a socially interactive robot*, Proceedings IEEE International Workshop on Robot and Human Interactive Communication, 2005, pp. 172–177.

[MRT03] Michael Montemerlo, Nicholas Roy, and Sebastian Thrun, *Perspectives on standardization in mobile robot programming: The carnegie mellon navigation (carmen) toolkit*, Proceedings of the IEEE/RSJ International Conference on Intelligent Robots and Systems (Las Vegas, NV), vol. 3, October 2003, pp. 2436–2441.

[MSJ04] B. Maxwell, W. Smart, A. Jacoff, J. Casper, B. Weiss, J. Scholtz, H. Yanco, M. Micire, A. Stroupe, D. Stormont, and T. Lauwers, *2003 aaai robot competition and exhibition*, AI Magazine **25** (2004), no. 2, 68–80.

[NWB03] I. A. D. Nesnas, A. Wright, M. Bajracharya, R. Simmons, and T. Estlin, *Claraty and challenges of developing interoperable robotic software*, Proceedings IEEE/RSJ International Conference on Intelligent Robots and Systems, 2003, pp. 2428–2435.

[OC03] Anders Orebäck and Henrik I. Christensen, *Evaluation of architectures for mobile robotics.*, Autonomous Robots **14** (2003), no. 1, 33–49.

[Sch94] D.C. Schmidt, *Ace: an object-oriented framework for developing distributed applications*, Proceedings of the 6th USENIX C++ Technical

[Tay99] P. Taylor, *The festival speech architecture*, URL: http://www.cstr.ed.ac.uk/projects/festival/, 1999.

[VGH03] R. T Vaughan, B. P. Gerkey, and A. Howard, *On device abstractions for portable, reusable robot code*, Proceedings IEEE/RSJ International Conference on Intelligent Robots and Systems, 2003, pp. 2421–2427.

[VAUG06] Vaughan, R.T. and Gerkey, B.P., *Reusable Robot Software and the Player/Stage Project*, In D.Brugali (Ed.) *Software Engineering for Experimental Robotics*, Spinger STAR series, 2006

[VMHR1] J.-M. Valin, F. Michaud, B. Hadjou, and J. Rouat, *Localization of simultaneous moving sound sources for mobile robot using a frequency-domaine steered beamformer approach*, Proceedings IEEE International Conference on Robotics and Automation, 1, pp. 1033–1038.

[VRM04] J.-M. Valin, J. Rouat, and F. Michaud, *Enhanced robot audition based on microphone array source separation with post-filter*, Proceedings of the IEEE/RSJ International Conference on Intelligent Robots and Systems, 2004.

[WMT03] E. Woo, B. A. MacDonald, and F. Trépanier, *Distributed mobile robot application infrastructure*, Proceedings IEEE/RSJ International Conference on Intelligent Robots and Systems, 2003, pp. 1475–1480.

[Zha03] Y. Zhao, *A model of computation with push and pull processing*, Master's thesis, University of California at Berkeley, Department of Electrical Engineering and Computer Sciences, December 2003.

Orca: A Component Model and Repository

Alex Brooks[1], Tobias Kaupp[2], Alexei Makarenko[3], Stefan Williams[4], and Anders Orebäck[5]

[1] ACFR, University of Sydney, AUSTRALIA a.brooks@acfr.usyd.edu.au
[2] ACFR, University of Sydney, AUSTRALIA t.kaupp@acfr.usyd.edu.au
[3] ACFR, University of Sydney, AUSTRALIA a.makarenko@acfr.usyd.edu.au
[4] ACFR, University of Sydney, AUSTRALIA s.williams@acfr.usyd.edu.au
[5] Royal Institute of Technology, SWEDEN oreback@nada.kth.se

Summary. This Chapter describes Orca: an open-source project which applies Component-Based Software Engineering principles to robotics. It provides the means for defining and implementing interfaces such that components developed independently are likely to be inter-operable. In addition it provides a repository of free re-useable components. Orca attempts to be widely applicable by imposing minimal design constraints. This Chapter describes lessons learned while using Orca and steps taken to improve the framework based on those lessons. Improvements revolve around middleware issues and the problems encountered while scaling to larger distributed systems. Results are presented from systems that were implemented.

1 Introduction

This Chapter describes Orca: an open-source project which applies Component-Based Software Engineering (CBSE) principles to robotics. Many robotic software projects are presented in this book, however the Orca project stands out by both

- explicitly adopting a component-based approach from the outset, and
- acknowledging the importance of a component market.

While it is not alone in either respect, the Orca project is, to the authors' knowledge, the only framework to combine the two.

The introduction to this Part defined Component-Based Software Engineering, identified its benefits and motivated its application to robotics. The benefits of a component-based approach include

- the ability to build systems quickly by incorporating existing third-party components,
- the ability to build more reliable systems by incorporating components which have been tested across multiple projects,

D. Brugali (Ed.): Software Eng. for Experimental Robotics. STAR 30, pp. 231–251, 2007.

- the software engineering benefits of having a modular system with controlled, explicit dependencies only, and
- the ability to build flexible systems in which individual components can be replaced.

In non-commercial settings, component markets entail the trading of independently-developed components. The recognition of the importance of this has both a practical and a technical impact. Practically, it drives the maintenance of the Orca component repository[6]: an online selection of open-source, interoperable, re-useable robotic components. Significant effort has been devoted to providing sufficient documentation for both the framework itself and for each component, to ensure that components can indeed be deployed by third parties. Technically, the recognition of the importance of component markets drives Orca to make as few assumptions as possible about the systems to which it may be applied, in order to be applicable across as broad a market as possible. Section 2 develops these ideas further, presenting the basic philosophy behind the Orca approach.

The two main principles have remained unchanged since the start of the project in 2001 (under the name of Orocos@KTH). Some implementation details have undergone radical changes, warranting a recent increment of the version number from Orca1 [BKM05] to Orca2. A number of lessons were learned based on experiences with Orca1, resulting in improvements that were made or planned for Orca2. These revolve around middleware-related issues and the problems discovered when scaling to large distributed systems, as discussed in Sections 3 and 4 respectively.

To demonstrate the effectiveness of the framework, Section 5 describes some of the component-based systems that have been built and the re-use that has occurred between those systems. Section 6 compares Orca with the Player project. The comparison is included because Player is the only robotics software project to have established a significant market, and hence has become a de-facto standard in robotics. Finally, Section 7 concludes.

2 Design Philosophy

2.1 Design Minimalism: Impose as Few Constraints as Possible

The design of the Orca framework is conceptually simple. A system consists of a set of components which run asynchronously, communicating with one another over a set of well-defined interfaces. Each component has a set of interfaces it provides and a set of interfaces it requires. The fundamental purpose of the framework is to provide the means for defining and implementing these interfaces. Standardising the definitions and implementations of interfaces ensures that components are likely to be inter-operable, and hence re-useable.

[6] available from http://orca-robotics.sourceforge.net

This statement of purpose is most significant in its omissions. The authors contend that the framework should prescribe as little as possible to achieve the stated aim. This conclusion is reached from the following premises:

1. The benefits of a component-based approach only become apparent when a critical mass of component developers and users arises.
2. There are many different kinds of robotic systems, each with different requirements. For each set of requirements, there are many possible approaches to the system's design.
3. There are many approaches to designing any piece of software (including components) and, correspondingly, many opinions about their relative merits.

The first premise can be motivated by the following simple observation [Szy02]:

> Imperfect technology in a working market is sustainable.
> Perfect technology without any market will vanish.

Indeed, in the absence of a market, the value of a component-based approach is unclear. Producing a general, modular, component-based solution requires significantly more effort and planning than building a specific solution for the specific problem at hand. This extra effort is only worthwhile if there is a reasonable chance of receiving a return on that investment. Szyperski offers the rule of thumb that the break-even point is reached when a component is used across three different projects [Szy02].

In the presence of a component market, the benefits of designing systems to be inter-operable with third-party components become obvious. The alternative to using available components is the re-invention of solutions, which is advantageous only when the home-grown solutions are of significantly better quality. As market size increases this becomes less likely since components act as multipliers in a market, focussing the combined innovation of many individuals.

Examples of the different kinds of robotic systems referenced in the second premise include single-vehicle indoor research systems, large multi-robot outdoor commercial systems and unmanned aerial vehicles. These examples are likely to focus on different requirements, such as a) flexibility, b) scalability, safety and reliability, and c) real-time performance, respectively.

Even for a specific robotic domain there are widely differing approaches. Consider indoor research robots as an example: three-layer architectures [BFG97] involve little direct communication between sensors and high-level deliberative planners. In contrast, different requirements (particularly architectural) are imposed by a reinforcement learning [SB98] approach in which a learner requires direct access to all sensor data.

The third premise is evidenced by the proliferation of libraries that exist for any given purpose. For example, various GUI toolkits exist, all of which perform similar tasks and have large user-bases.

Given that a critical user mass is required, it follows that a framework should strive to be capable of accommodating as broad a range of projects as possible. Furthermore, given that requirements and approaches vary so widely, it follows that a framework should prescribe as little as possible while achieving its stated aims. Since any assumption is likely to be incompatible with some project or approach, unnecessary assumptions should be avoided.

In particular, Orca avoids making assumptions about any of the following:

1. **Architecture:** system developers should be free to compose a system from any set of components, arranged to form an arbitrary architecture, so long as interfaces are connected correctly. There should be no special component which the framework requires all systems to incorporate.
2. **Interfaces:** individual components should be free to provide or require any set of interfaces. There should be no particular interface which all components must provide or require. It should be easy for component developers to define new interfaces.
3. **Internals:** component developers should be free to provide the implementations of their components' interfaces in any way they choose. The implementation details remain opaque to the rest of the system.

To prescribe design rules for any of these aspects is to make assumptions about all robotics projects, present and future. Should those assumptions prove to be unjustified, the re-usability of both the framework and the set of its components is restricted.

2.2 Impose Necessary Design Constraints

While every attempt is made to avoid imposing constraints on component designers, certain decisions do need to be made to ensure component interoperability. In particular, a framework must choose

- a set of pre-defined interfaces, and
- how communication works between interfaces and how components can implement those interfaces.

Orca is distributed with the set of interfaces used by components already in the repository. New interfaces can be defined as needed. The framework is designed such that a new interface can be added without interfering with any existing interface. It is also important to avoid a proliferation of similar but subtly different (and incompatible) interfaces. In this regard no technical solution will replace old-fashioned communication between developers.

The second point involves the selection of middleware. While Section 3 describes the particular choices made by Orca, more generally this decision requires the consideration of the trade-offs listed below:

Efficiency versus Modularity

In many engineering disciplines, a trade-off often exists between building a monolithic, specific implementation versus a general, modular implementation. The former is usually more efficient but less re-useable or maintainable than the latter. Orca (and CBSE in general) favours the latter approach.

Where projects have elements that need to favour efficiency, developers are free to implement the entirety of those elements within a single Orca component. From the point of view of the rest of the system it makes no difference, so long as the same external interfaces are exposed.

Footprint

A framework must make a choice regarding the minimum platform to support. Many robotics projects make use of micro-scale platforms such as MICA motes or Khepera robots. While support for these platforms is desirable, it imposes constraints on those who develop for larger platforms. In particular, it constrains the maximum footprint of the framework and precludes the use of more modern languages such as C++. In this trade-off Orca chooses to favour ease-of-use over portability. This does not prohibit the use of less-capable platforms within an Orca system, however they must be incorporated using non-Orca communication.

Real-Time Capabilities

A trade-off exists in whether a framework chooses to provide real-time guarantees for inter-component communications. Many robotics projects require some element of real-time operation, however ensuring that the entire system is real-time imposes unnecessary constraints on those parts of the system which do not require it. While real-time component-based frameworks do exist [Bru01], Orca chooses not to support real-time inter-component communications.

2.3 Offer Useful Design Guidelines

While the imposition of unnecessary design constraints is unhelpful, a framework can certainly assist component developers by providing design guidelines, for example on the design of components' internals or effective system architectures. Providing working code, in the form of libraries or examples, simplifies the implementation of these guidelines. A framework can thus significantly improve the average reliability and maintainability of its components, especially those written by inexperienced developers, as well as improving the ease with which new components can be developed.

A natural consequence of this approach is that conventions will develop. This is beneficial in general: code is easier to understand when a component's internals follow a convention, and system composers have less to learn if configuration options are consistent. There is an important difference however between these conventions (which are useful and should be allowed to evolve) and interface standards (which are necessary and need to be prescribed). The difference is that the former will arise in those domains in which they make sense, but can be broken, with no disastrous ramifications, in unforeseen cases where they do not apply. Conversely, if interface compatibility is broken the consequences are more severe: if components can no longer communicate they can no longer be replaced between systems, and hence cannot be re-used at all.

The issue of a component's internal behaviour, broached in the Introduction to this Part, provides a useful example. Rather than prescribing the internal behaviour, a framework should define interfaces which expose the necessary functionality (such as *observability* or *controllability*), without the requirement that all components provide those interfaces. If the internal behaviour is suggested rather than prescribed, one can be more aggressive in making those suggestions useful rather than cautiously ensuring that they be general enough to apply to all scenarios.

3 Lessons Learned: Middleware Issues

Middleware refers to the software which acts as an intermediary between different application components [Szy02]. It plays a fundamental role in component-based systems, addressing many issues such as:

1. marshalling and de-marshalling,
2. application-layer communication protocols,
3. versioning,
4. inter-component exception handling, and
5. the means by which interfaces are defined and implemented.

3.1 Middleware for Orca1

There are surprisingly few options for the choice of middleware package. If support for C/C++ on Linux is a requirement (which discounts Microsoft's COM+ [Ewa01] and Sun's Enterprise JavaBeans [Ble01]), the following options exist today:

1. using XML-based technologies such as SOAP,
2. using Ice [Hen04],
3. using CORBA [HV99], or
4. writing custom middleware from scratch.

We rule out XML-based technologies on the grounds that they are too inefficient for low-level robotic control tasks. Ice is a recent development which was insufficiently mature at the time Orca1 was being designed.

Orca1 implemented both of the remaining options. All communication occurred via a set of communication patterns. The user decided, at compile-time, which middleware option would be used to implement the communication patterns. The choice to implement both options was made because neither is without significant drawbacks.

While the CORBA specification is certainly complete in terms of Orca's middleware requirements, it is also large and complicated. This had two main impacts on the project: firstly, from the project maintainers' point of view, the CORBA C++ API is well-known to be complex and difficult to use [Hen04]. The complexity was overcome by wrapping the CORBA API with a simpler Orca1 API, specialised for Orca1's communication patterns. While this sheltered new users who were not CORBA experts, it required considerable time and effort both to write and to maintain. Secondly, from the point of view of component developers, the most apparent impact of the size and complexity of CORBA was the significant increase in compile times, which did have a tangible effect on usability.

The alternative of writing one's own middleware from scratch also requires significant effort. Orca1 implemented simple TCP-based and UDP-based middleware systems from the ground up. While communicating over a socket is simple, implementing middleware sufficiently flexible and reliable to support Orca's requirements involved re-implementing (and maintaining) large parts of CORBA functionality, which is a non-trivial task. Cross-platform support was minimal and cross-language support was never attempted.

3.2 Middleware for Orca2

Providing multiple middleware implementations allowed Orca1 to support projects with different requirements, and avoided the need to make a single choice when the relative merits of the various options were unknown. This flexibility was paid for in two ways. Firstly, it required extra implementation and maintenance effort, which was detrimental to the quality of each individual middleware implementation.

Secondly, rather than being able to take advantage of the specific properties of each middleware option, the abstract communication API supported only the intersection of features provided by the three options[7]. For example, interfaces had to be specified in a middleware-agnostic manner, making it impossible to take advantage of the full capabilities of CORBA's Interface Definition Language (IDL).

The biggest change in Orca2 is the adoption of a single middleware layer. The authors consider Ice to be sufficiently mature at the time of writing and

[7] Interestingly, Player [GVH03] is moving towards offering a choice of middleware options.

vastly superior to the alternatives. It is far more flexible and reliable than any middleware that is likely to be written from scratch by robotics researchers. Ice can be compiled natively under various operating systems including Windows, Linux and Mac OS X and has several language mappings including C++, Java and Python. Ice has far less overhead when compared with CORBA, especially with respect to complexity (see [Hen04] for a comparison). It is released under the GNU Public License (GPL), however commercial licenses are available for closed-source software. Adopting a single middleware layer allows Orca to take full advantage of the features of the middleware layer rather than crippling each type of middleware to fit a common model.

3.3 Interface Definition

Unable to take advantage of an off-the-shelf interface definition language, Orca1 defined interfaces using *communication patterns* as methods for transferring data and *objects* as the data types which were transferred. This approach was sufficient for many applications, however it had the following deficiencies:

1. There was no clear way to relate associated functionality. For example, a laser server provided both scans and the laser's configuration details. With no obvious way to relate the two, the association had to be performed manually.
2. Polymorphism of interfaces was not possible. Therefore a slightly different *object* resulted in an entirely different (and incompatible) interface.
3. On creation of a new *object*, developers were required to write object-specific serialization code for marshalling, demarshalling, logging to files and reading from files. This presented a practical barrier to framework extensibility.

Instead Orca2 uses Ice's interface definition language, called Slice, to define interfaces. To demonstrate Slice's flexibility an example for a laser scanner is presented in Listing 1. Given a reference to a specific Laser interface, one can easily access both the data and the geometric information *for that laser*. The inheritance of Laser from RangeScanner shows how custom fields can be added to new interfaces while maintaining compatibility with existing interfaces. Defining new interfaces is trivial, and Slice automatically generates all serialization code from the interface definition.

4 Lessons Learned: Scaling to Complex Component-Based Systems

An important lesson learned from the Orca1 experience was the difficulty of scaling up. A certain amount of effort was required to implement each component. It eventuated that the amount of extra effort required to actually

```
// RangeScanner interface

sequence<float>        RangeSequence;

exception CannotImplementConfigurationException extends OrcaException {};

class RangeScannerData extends OrcaObject {
    RangeSequence     ranges;
    float             startAngle;
    float             angleIncrement;
};

interface RangeScanner {
    RangeScannerData       getData();
    RangeScannerConfig     getConfig();
    RangeScannerGeometry   getGeometry();

    void setConfig( RangeScannerConfig config )
        throws CannotImplementConfigurationException;
};


// Laser Scanner interface

sequence<byte>         IntensitySequence;

class LaserData extends RangeScannerData {
    IntensitySequence intensities;
};

interface Laser extends RangeScanner {};
```

Listing 1: Definitions of the RangeScanner and Laser interfaces. Some details are omitted for brevity.

configure, deploy and operate a system of those components was significant. When applied to large distributed systems, such as those described in Section 5, this extra effort could even dominate the total effort required.

This eventuality is not uncommon in distributed systems outside of the robotics field [WSO01]. A number of services and utilities have proved useful for mitigating the problems associated with large systems:

1. naming or trading services, to allow location transparency,
2. event services, to de-couple publishers from subscribers,
3. persistence services, to maintain state beyond a component's lifetime,
4. application servers,
5. deployment utilities,
6. system composition and configuration utilities, and
7. remote monitoring utilities.

Availability of these tools is an important consideration when selecting a middleware package or robotics framework. This Section describes the problems that were found to be most problematic and the tools used to solve them.

4.1 Composition and Configuration

Each Orca1 component was configured with a separate file which defined its inter-component connections and modified its behaviour. Defining all the connections by hand in a large system was tedious and error-prone. When system misbehaviour was caused by a misconfigured component, identifying the cause involved inspecting multiple files on multiple filesystems.

The solution attempted in Orca1 was to provide structured information regarding how each component could be connected and configured. This information was used to validate hand-written configurations and to generate configurations using `GOrca`: a graphical system composition and configuration utility similar to Simulink. `GOrca`, shown in Figure 1, parsed a set of component definition files to create a library of available components. Users could drag components from the library onto a project, connect their interfaces graphically and set configuration parameters textually. Finally, `GOrca` would export a set of valid configuration files representing the entire system.

While `GOrca` provided an effective means of generating configuration files, the number of generated files soon became unmanageable. Orca2 simplifies this further using Ice's application server, called `IceBox`. Developing components as dynamically-loadable libraries allows them to be deployed either as stand-alone components or as services within an `IceBox`. Configuration is simplified because an `IceBox` has a single configuration file which specifies the set of components to instantiate and their individual configuration details. In addition, invocations between components within an `IceBox` are optimised by Ice, taking advantage of the fact that they co-exist within the same process[8].

4.2 System Deployment

The deployment of a component-based system entails ensuring binary synchronisation across all hosts (either by copying source code and compiling locally, or by copying binaries), copying configuration files, then starting and stopping components. The amount of time spent tracking down failures resulting from unsynchronised binaries was significant, and having to open one remote shell per component was onerous.

---

[8] A Player server may also be comsidered an application server. The important difference in Orca2 is that communication is identical whether a component is a stand-alone application or part of an `IceBox`. The ramifications of this difference are explained in Section 6.

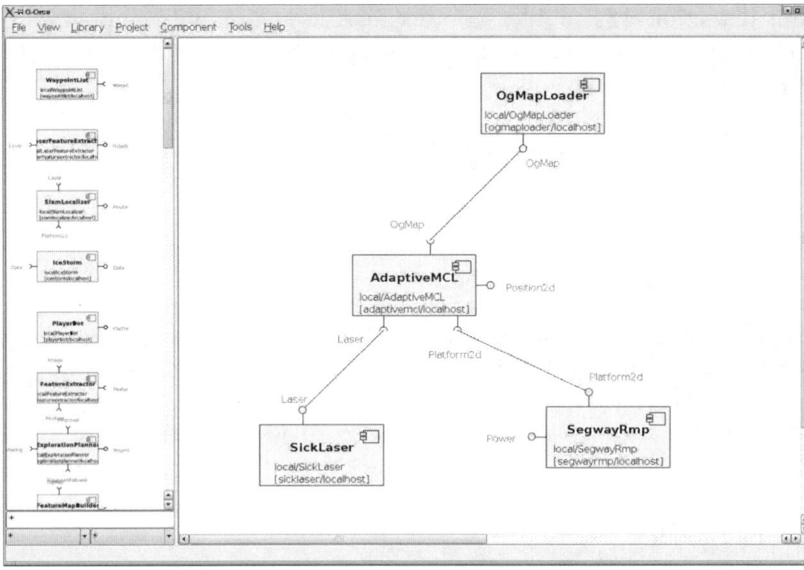

Fig. 1. `GOrca`: a graphical tool to assist component configuration. Components can be dragged from the library on the left to the projcet on the right and connected graphically. `GOrca` ensures that connections are valid. Right-clicking a component allows its configuration parameters to be modified.

To deal with this issue in Orca1, `orcad` was developed. `orcad` is a daemon which deploys all components on a host. It provides an Orca interface over which remote components can provide the names of components to be started or stopped on that host, along with the configuration files with which they should be started. It can then be queried, reporting the set of running components and the exit status of any which have died. This solved the problem of starting and stopping components and allowed all configuration files to be stored centrally.

Ice ships with a deployment service called `IceGrid`, which essentially solves the same problems as `orcad`: allowing a central component to start and stop components remotely. In addition it provides a standard solution for the binary synchronisation problem and reports statistics such as CPU, memory and network usage.

4.3 Remote Monitoring

As a system grew beyond a handful of components or hosts, the mean time between failures in the system became far smaller than the mean time between failures of the least reliable constituent component. In addition to algorithmic failures within components, errors occur due to both communication failures (which are not unexpected when dealing with mobile robots) and the interactions between components. Detecting and localizing faults was difficult.

Fault detection was problematic because, as individual component reliability increased, components would try to detect and recover from faults internally. While this behaviour is desirable, the result was that overall system performance would mysteriously degrade. Even when faults were detected, it could be difficult to determine which component was at fault or even on which host the problem was occurring. These problems were exacerbated by the introduction of `orcad` since trace statements directed to the console became unavailable.

Orca2 provides a remote logging API and a central monitoring component to monitor trace, heartbeat and error messages from any subset of a system (including the entirety). The level of detail required can be filtered on a per-component basis. In-keeping with the Orca approach presented in Section 2 it is not mandatory that components provide this remote logging interface, however the benefits are obvious and Orca2 provides libraries to simplify its implementation.

4.4 Event Service

A common cause of difficult-to-debug faults in larger systems was the coupling between publishers and subscribers. Two problems can occur when communication is unreliable or clients are slow. Firstly, slow clients can delay the publishing component's thread, causing problems that appear to be related to the publisher's algorithm. Secondly, a slow client can starve faster clients, causing problems that appear to be related to the faster clients.

The problem of server-client coupling is solved in Orca2 using Ice's event service, called `IceStorm`. An event service is a stand-alone component which receives data from a single server and disseminates it to multiple clients, as illustrated in Figure 2. It relieves components of the burden of managing subscribers and handling delivery errors. Publishing to a local event service ensures that network delays do not interfere with the operation of the publishing algorithm. `IceStorm` is efficient, since it forwards messages without needing to marshal or demarshal them. It can be configured with multiple threads, to weaken the dependencies between clients.

5 Results: Systems and Re-use

This section describes several systems implemented using Orca1. There is sufficient variation in system types to illustrate the flexibility of the project's philosophy and implementation. The description of the systems highlights the main objective of the project: software reuse. In addition to the highlighted reuse, all projects re-use common infrastructure such as graphical user interface components, `GOrca`, `orcad` and components for logging and replaying data streams and monitoring system status. Robotic platforms used in these systems are shown in Figure 3.

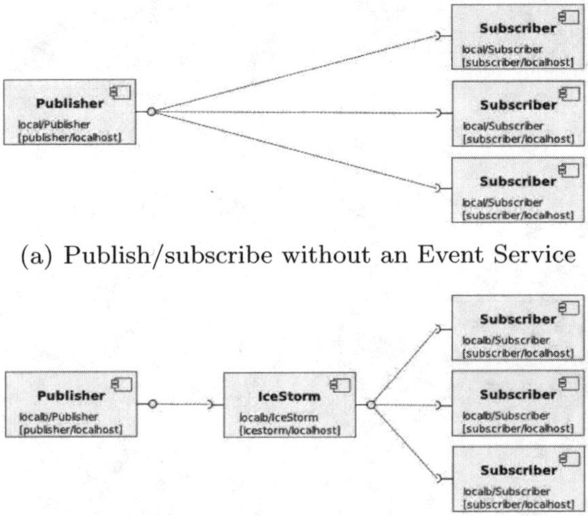

(a) Publish/subscribe without an Event Service

(b) Publish/subscribe with an Event Service

Fig. 2. The use of an event service such as `IceStorm` relieves the publishing component of the burden of managing subscribers and handling delivery failures. A multi-threaded event service weakens dependencies between clients.

5.1 Indoor Multi-vehicle Cooperative Exploration Research

Figure 4 shows the component diagram for a research project on cooperative indoor exploration [MDW06], using three Pioneer robots. Many of the high-level components are specific to the particular exploration approach, however the low-level hardware components, occupancy-grid server, obstacle avoidance algorithm, localisation algorithm and path plan executor are all re-used from other projects.

5.2 Industry-Funded Multi-vehicle Project

This project involves up to eight identical Segway RMP vehicles, where each robot's component diagram is as shown in Figure 5. The project was the source of many of the lessons from Section 4, since system reliability was paramount. The low-level hardware servers, navigation algorithms, map server and joystick interface are re-used between projects. A point of interest is the parallel use of two kinds of localisers: AdaptiveMCL, which uses a particle-based representation of uncertainty [Thr02] and SlamLocaliser, which uses a Gaussian representation [DNDW01]. The two localisers demonstrate the principle of replaceability: the PathPlanExecutor could be switched between the two if problems occurred with either one.

(a) A Pioneer indoor robot equipped with a SICK laser

(b) A Pioneer2AT equipped with two cameras and a SICK laser

(c) A Segway RMP

(d) A Pioneer indoor robot equipped with a camera and a SICK laser

(e) An Argo unmanned all-terrain ground vehicle

(f) A Brumby unmanned air vehicle

Fig. 3. Unmanned robotic platforms used in the systems described in this Section.

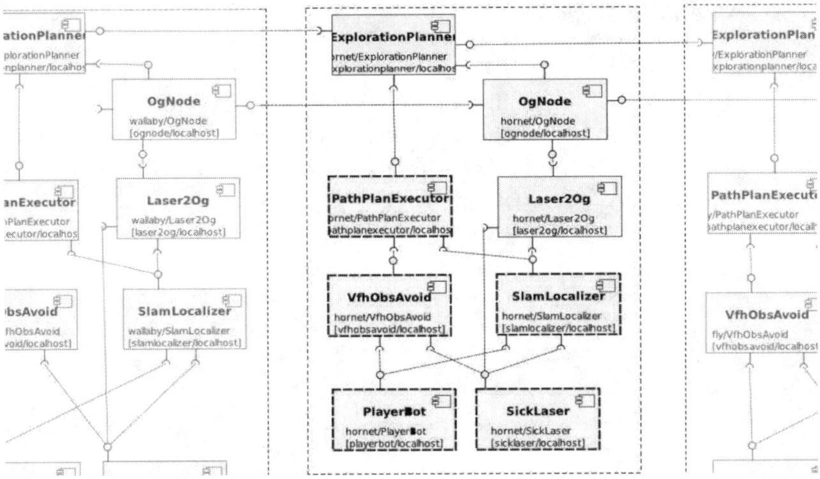

Fig. 4. Component diagram of one of three robots in a multi-vehicle cooperative exploration research project. Links leading left and right connect to other robots. In this and other system diagrams, highlighted components are those which were re-used across more than one project.

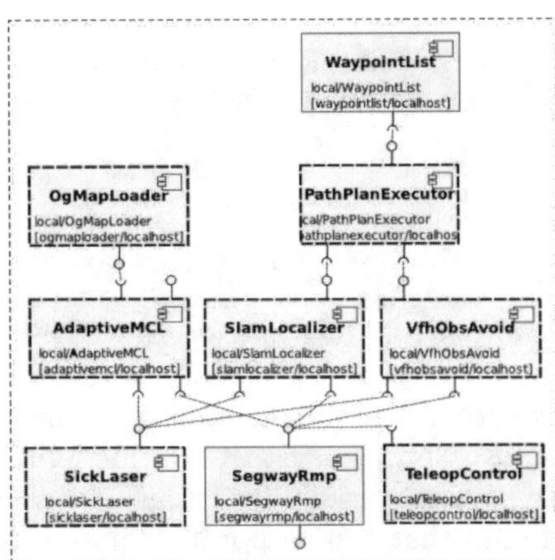

Fig. 5. Component diagram of a single SegwayRMP. The project involves up to eight identical copies of this diagram.

5.3 Single-Vehicle Research

The system in Figure 6 was used for research in sensor-centric localisation on a single Pioneer2AT, both indoors and outdoors [BUM06]. While system management and reliability of the low-level components was not so important for this project, it was helpful to be able to use many of the same components that were hardened by industrial projects.

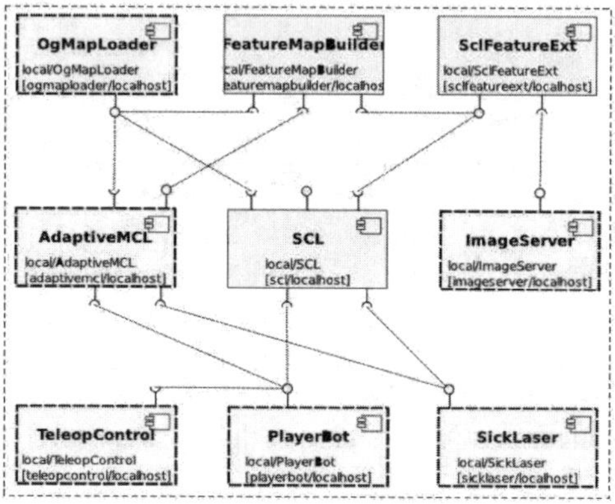

Fig. 6. Component diagram of the system deployed on a Pioneer2AT, used for single-vehicle research in sensor-centric localisation.

5.4 Large-Scale Multi-vehicle Outdoor Research

Figure 7 depicts the component diagram for the ANSER project: a large-scale outdoor decentralised data fusion [URO06] research project involving manned and unmanned ground vehicles, unmanned air vehicles and human operators. This is the most complex system to which Orca has been applied, highlighting many of the system scalability issues discussed in Section 4. The system is also interesting in that it incorporates air vehicles with strong real-time requirements. To accommodate this, the flight control software is written without Orca. To the rest of the system the air vehicles appear as 'black boxes' which provide their positions and observations over Orca interfaces.

6 Comparison with the Player Framework

Player [GVH03], originally designed and developed as a *robot device server*, has become the most successful mobile robotics re-use project to date. Its simplicity, reputation for reliability, good documentation and support from developers all contributed to its success. Player's design is similar to Netscape's

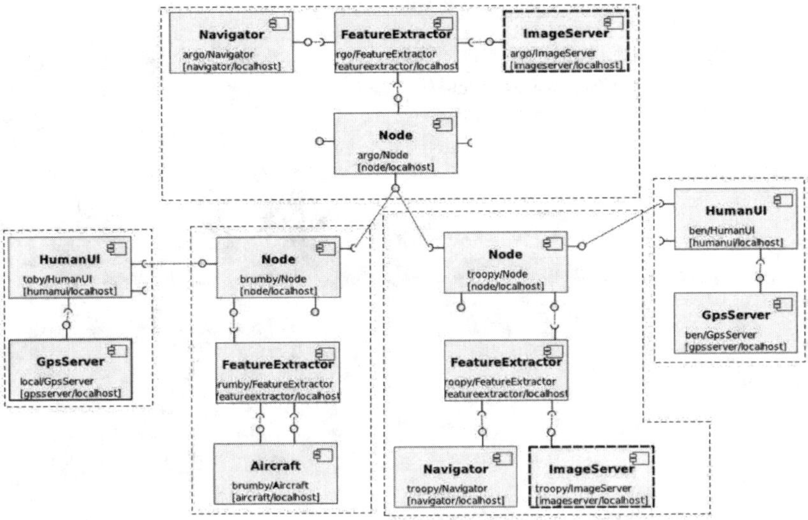

Fig. 7. Component diagram for the ANSER project. Dashed lines indicate the borders between physical platforms. The top-most platform is the Argo unmanned ground vehicle. From left to right, the lower boxes indicate a human operator, a Brumby unmanned air vehicle, a manned four-wheel drive vehicle and a second human operator.

plugin architecture, and is component-based in that sense: a monolithic server houses a set of modular devices. This has contributed to Player's successful re-use: devices conform to well-defined interfaces and correspond to well-understood algorithms or entities, allowing third parties to deploy and configure them without ever seeing the source code. When compared with Orca, Player's most important design decisions are its use of custom middleware and its delineation of server space from client space.

6.1 Middleware

Section 3 discussed both the central role of middleware for component-based frameworks, and the effort involved in implementing and maintaining the required functionality. We believe that it is unrealistic to expect robotics researchers to have the skills or time to develop middleware to the same standards as experienced middleware professionals working full-time on commercial products. By leveraging commercial middleware we expect Orca to provide more flexible and reliable communications infrastructure.

In an attempt to provide a more quantitative analysis, Figure 8 measures the number of lines of source code devoted to implementing middleware compared with the number devoted to actual algorithms and hardware

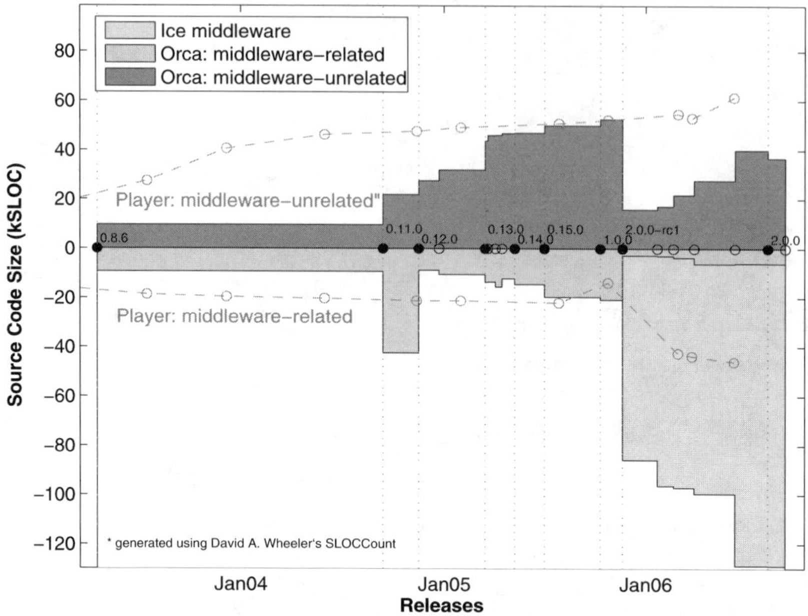

Fig. 8. A measure of the number of Source Lines of Code (SLOCs) devoted to implementing middleware-related (on the positive y-axis) and middleware-unrelated (on the negative y-axis) functionality, for various releases of both the Orca and Player projects. Orca adopted Ice for release 2.0.0-rc1 in late 2005.

drivers, for both Orca and Player[9]. Ideally, the amount of middleware-related code should remain constant and small, while the middleware-unrelated code should grow. Several points are apparent from Figure 8. Firstly, for both projects, the middleware-related effort is clearly significant. Secondly, for the Orca project, the adoption of a single commercial middleware product in late 2005 resulted in a dramatic reduction in the amount of middleware-related code written or maintained by robotics researchers. Finally, the total quantity of middleware-related code used by Orca2 is significantly larger than the quantity used by Player. We believe that this reflects the fact that Ice pro-

[9] Measurements were made using David A. Wheeler's SLOCCount. For recent Player releases, "middleware-related" counts the contents of "libplayercore", "client_libs", and "libplayertcp", while "middleware-unrelated" counts the contents of "server" and "utils". Automatically-generated bindings are not considered. For recent Orca releases, "middleware-related" counts the contents of "libOrcaIce", while "middleware-unrelated" counts the contents of the "utils" and "components" directories. The Ice measurement considers only the Ice features used by Orca, and only the C++ distribution. Full details can be found at http://orca-robotics.sf.net

vides significantly more functionality and flexibility than Player middleware. This also provides some measure of the implementation effort which would be required for Player middleware to provide a similar level of functionality.

6.2 Client-Client Interaction

Player's delineation of server space from client space has several important consequences. Consider a single-robot system including components for hardware abstraction, localisation and control. Figure 9 shows how this might be implemented using either Orca or Player. There are several possible communication mechanisms in the Player model:

1. Inter-device communication, either within or between Player servers
2. Client-Server communication
3. Inter-client communication

Inter-client communication is not defined by Player, and must be supplied by users. As such, software modules that rely on custom inter-client communication mechanisms do not conform to any particular standards, and are not re-useable in the same way as Player devices are. To make the 'MyLocaliser' and 'Controller' modules independently re-useable under the Player model, MyLocaliser must be ported to server space. This requires significant effort,

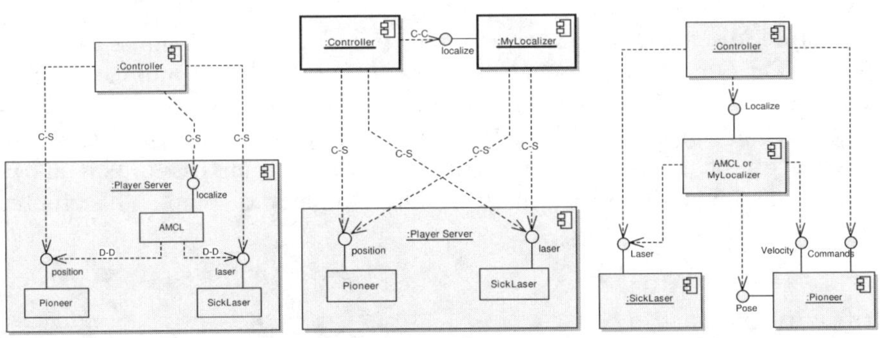

(a) The setup using standard Player devices and a controller in client space.

(b) Using one's own localiser in client space requires the use of client-client communication.

(c) An Orca system: only one type of communication is required.

Fig. 9. Three possible implementations for a system with hardware servers, a localiser (either Player's Adaptive Monte Carlo Localiser (AMCL) or a hand-written localiser) and a controller, using either (a)-(b) Player or (c) Orca. The map-serving component (required for AMCL) is omitted for clarity. For Player, communication links are labelled Client-Server (C-S), Device-Device (D-D) or Client-Client (C-C). The use of non-standard Client-Client communication in (b) affects the re-usability of both the 'MyLocaliser' and 'Controller' implementations.

a possible re-design to fit the Player Device model, and code changes to the Controller.

Rather than developing in client space then porting to server space, one might develop directly in server space. In practice this doesn't often happen due to the extra complexity of conforming to the Player device model, and the problems of developing within a monolithic server where all components must be started and stopped simultaneously.

In contrast, there is only one inter-component communication mechanism in the Orca framework, and no distinction between server space and client space. This improves system flexibility: the localiser in Figure 9 can be replaced without requiring changes to other parts of the system. It also improves the ability of a component repository to grow: both the 'MyLocaliser' and 'Controller' modules can be added to an Orca-based component repository with no modifications. We expect this contrast to be most significant in systems containing modules which act both as clients and servers, as the localiser does in Figure 9.

In summary we hope that, by designing the framework to be general and flexible and by leveraging commercial-grade middleware, Orca will be more able to scale to large complex systems and to provide the reliability required for industrial systems.

7 Conclusion

This chapter discussed the practical application of CBSE techniques to mobile robotics. The Orca project provides the implementation of a component model and promotes the development of re-useable components for the robotics community. While simpler models may be sufficient for small projects, Orca addresses many of the issues that occur as system size increases. This allows the same components to be re-used in both simple and complex distributed systems.

References

[BFG97] R. Bonasso, R. Firby, E. Gat, D. Kortenkamp, D. Miller, and M. Slack, *Experiences with an architecture for intelligent, reactive agents*, Journal of Experimental and Theoretical AI (1997).

[BKM05] A. Brooks, T. Kaupp, A. Makarenko, S. Williams, and A. Oreback, *Towards component-based robotics*, Proc. IEEE/RSJ Intl. Conference on Intelligent Robots and Systems, 2005.

[Ble01] D. Blevins, *Overview of the enterprise JavaBeans component model*, Component-based software engineering : putting the pieces together, Addison-Wesley, 2001.

[BMU05] A. Brooks, A. Makarenko, and B. Upcroft, *Gaussian process models for indoor and outdoor sensor-centric localisation*, submitted to International Journal of Computer Vision (2005).

[Bru01] H. Bruyninckx, *Open robot control software: the OROCOS project*, IEEE International Conference on Robotics and Automation (ICRA'01), vol. 3, 2001, pp. 2523–2528.

[Brug06] Brugali, D. and Brooks, A. and Cowley, A. and Côté, C. and Domnguez-Brito, A.C. and Létourneau, D. and Michaud, F. and Schlegel, C. *Trends in Component-Based Robotics*, In D.Brugali (Ed.) *Software Engineering for Experimental Robotics*, Spinger STAR series, 2006

[DNDW01] M. W. M. G. Dissanayake, P. M. Newman, H. F. Durrant-Whyte, S. Clark, and M. Csobra, *A solution to the simultaneous localization and map building (SLAM) problem*, IEEE Transactions on Robotic and Automation (2001), 229–241.

[Ewa01] T. Ewald, *Overview of COM+*, Component-based software engineering : putting the pieces together, Addison-Wesley, 2001.

[GVH03] B. Gerkey, R. Vaughan, and A. Howard, *The player/stage project: Tools for multi-robot and distributed sensor systems*, Proc. Intl. Conf. on Advanced Robotics, 2003, pp. 317–323.

[Hen04] M. Henning, *A new approach to object-oriented middleware*, IEEE Internet Computing **8** (2004), no. 1, 66–75.

[HV99] M. Henning and S. Vinoski, *Advanced CORBA programming with c++*, Addison-Wesley, 1999.

[MMDW06] G. Mathews, A. Makarenko, and H. Durrant-Whyte, *Information-based decentralised exploration for multiple robots*, Submitted to Proc. IEEE Intl. Conf. on Robotics and Automation, 2006.

[SB98] R. Sutton and A. Barto, *Reinforcement learning: An introduction*, MIT Press, Cambridge MA, 1998.

[Szy02] C. Szyperski, *Component software - beyond object-oriented programming*, Addison-Wesley / ACM Press, 2002.

[Thr02] S. Thrun, *Particle filters in robotics*, Proceedings of the 17th Annual Conference on Uncertainty in AI (UAI), 2002.

[UOK05] B. Upcroft, L. Ong, S. Kumar, M. Ridley, T. Bailey, S. Sukkarieh, and H. Durrant-Whyte, *Rich probabilistic representations for bearing only decentralised data fusion*, Proc. IEEE Intl. Conf. on Information Fusion, 2005.

[VAUG06] Vaughan, R.T. and Gerkey, B.P., *Reusable Robot Software and the Player/Stage Project*, In D.Brugali (Ed.) *Software Engineering for Experimental Robotics*, Spinger STAR series, 2006

[WSO01] N. Wang, D. Schmidt, and C. O'Ryan, *Overview of the CORBA component model*, Component-based software engineering : putting the pieces together, Addison-Wesley, 2001.

[MDW06] G. Mathews and H. Durrant-Whyte, *Scalable Decentralised Control for Multi-Platform Reconnaissance and Information Gathering Tasks*, Proc. Intl. Conf. on Information Fusion, 2006.

[BUM06] A. Brooks, B. Upcroft, and A. Makarenko, *Gaussian Process Models for Sensor-Centric Localisation*, Proc. IEEE Intl. Conf. on Robotics and Automation, 2006.

[URO06] B. Upcroft, M. Ridley, L.L. Ong, B. Douillard, T. Kaupp, S. Kumar, T. Bailey, F. Ramos, A. Makarenko, A. Brooks, S. Sukkarieh, and H.F. Durrant-Whyte, *Multilevel State Estimation in an Outdoor Decentralised Sensor Network*, 10th International Symposium on Experimental Robotics, 2006.

Sidebar – Software Architectures

Patricia Lago and Hans van Vliet

Vrije Universiteit, Amsterdam, The Netherlands {patricia, hans}@cs.vu.nl

Software architecture is becoming one of the central topics in software engineering. In early publications, such as [Sha88], software architecture was by and large synonymous with global design. In [SG96] we read "the architecture of a software system defines that system in terms of computational components and interactions among those components". In a traditional software engineering process, during the software design phase the system is decomposed into a number of interacting components (or modules). The top-level decomposition of a system into major components together with a characterization of how these components interact, was considered as the software architecture of the system under development. In this respect requirements engineering is an activity focussing very much on the problem space, while the subsequent design phase focuses on the solution space. We call this a *pre-architecture development approach*. Here, few persons (also called stakeholders) are typically involved (like for example the project manager, robotic expert, the software analyst, few developers). Iteration involves functional requirements only: once agreed upon, these are supplemented with non-functional requirements to form the requirements specification used as input for design. In particular, there is no balancing between functional and non-functional requirements

Conversely, modern software engineering is moving to an *architecture-centric development approach*. Here, the discussions involve multiple stakeholders: the project manager, different classes of robotic and software experts, future maintainers of the system, owners of other inter-operating systems and services. Iteration concerns both functional and non-functional requirements. In particular, architecting involves finding a balance between these type of requirements. Only when this balance is reached, next steps can be taken.

In the latter approach, software architecture has to bridge the gap between the world of a variety of, often non-software-technical stakeholders on one hand – the problem space –, and the technical world of software developers and designers on the other hand – the solution space.

Software architecture hence describes much more than just the components and the interactions among them. It serves three main purposes [BCK03]:

D. Brugali (Ed.): Software Eng. for Experimental Robotics. STAR 30, pp. 253–256, 2007.

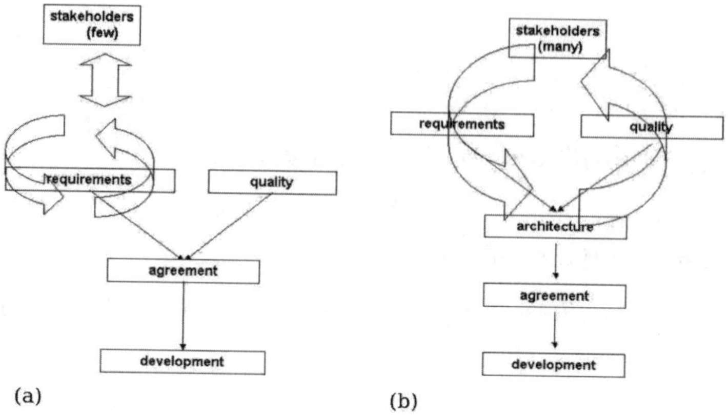

Fig. 1. The software life cycle.

- It is a vehicle for communication among stakeholders. A software architecture is a global, often graphic, description that can be communicated with the customers, end users, designers, and so on. By developing scenarios of anticipated use, relevant quality aspects can be analyzed and discussed with various stakeholders.
- It captures early design decisions. In a software architecture, the global structure of the system has been decided upon, through the explicit assignment of functionality to components of the architecture. These early design decisions are important since their ramifications are felt in all subsequent phases. It is therefore paramount to assess their quality at the earliest possible moment. By evaluating the architecture, a first and global insight into important quality aspects can be obtained. The global structure decided upon at this stage also structures development: the work-breakdown structure may be based on the decomposition chosen at this stage, testing may be organized around this same decomposition, and so on.
- It is a transferable abstraction of a system. The architecture is a basis for reuse. Design decisions are often ordered, from essential to nice features. The essential decisions are captured in the architecture, while the nice features can be decided upon at a later stage. The software architecture thus provides a basis for a family of similar systems, a so-called *product line*. The global description captured in the architecture may also serve as a basis for training, e.g. to introduce new team members.

We do not want here to give yet another definition of what software architecture is or should be. For the interested reader [SEI06] provides a quite complete overview of what is meant by software architecture around the world.

Recently, we observed an interesting evolution of this widely discussed concept. In a more traditional connotation Software Architecture (SA) is considered as the pure, technical result of a software engineering activity: in this sense, it is limited to the modeling of the technical and technological aspects of a computerized information system. SA however is broader than that. Modeling the 'technical SA' is of course needed [Med05]. But to make sense, this 'technical SA' must be integrated with its environment, and this includes non-technical aspects too, like organisation, geographical location, economical, legal aspects, etc. Some efforts have been done in the literature to illustrate these aspects. Architectural standards [IEE00] and reference frameworks [Kru04a, HNS99, Zac87, Oa00] all propose some views showing e.g. business or process aspects next to more technical ones. Still, they all remain focused at the technical level. It remains difficult if not impossible to capture technical and non-technical aspects in an *integrated* way [Kru04b, TA05]. Interesting research questions are, for instance: How does a change in the market influence the architecture of an organization? If the company closes an agreement with a different technology vendor, which parts of the technical architecture are potentially influenced? How much will this cost me? To answer these questions, SA should evolve to cover all aspects in an integrated way. Integration of various technical aspects has been an issue, too. For example, the relation between SA and quality requirements has been investigated for a long time. Relevant results in this respect are represented by the use of tactics in the Architecture Tradeoff Analysis method [BCK03]. This allows for example to analyze how performance is influenced by certain architecture decisions.

A further issue is to keep SA alive. We believe that SA should capture what is likely not to change, or better what is difficult to change. This means that SA gives the relatively long-lasting picture of a software system. Nonetheless, when changes occur (and they will! [LvV05, Fow03]) we must be able to keep the architecture aligned. This does not happen by itself. In practice, organizations have difficulties to align software architecture with the *real situation.* There are various reasons for that. Architects do not work together with those making the changes, hence remain isolated in their ivory tower; sometimes the way SA is documented is not understood, or it differs from the way the various actors are used to document the results of their production processes. Updating SA would require the allocation of additional effort, and such effort is expensive and does not bring an immediate advantage. Hence it is perceived as a loss of time.

References

[BCK03] L. Bass, P. Clements, and R. Kazman, *Software architecture in practice*, SEI Series in Software Engineering, Addison-Wesley, second edition, 2003.

[Fow03] M. Fowler, *Who needs an architect?*, IEEE Software **5** (2003), no. 20, 11–13.

[HNS99] C. Hofmeister, R. Nord, and D. Soni, *Applied software architecture*, Addison-Wesley, 1999.

[IEE00] IEEE, *Ieee recommended practice for architectural description of software-intensive systems*, IEEE Standard 1471-2000, 2000.

[Kru04a] P. Kruchten, *Architectural blueprints - the 4+1 view model of software architecture*, IEEE Software **21** (2004), no. 6, 42–50.

[Kru04b] P. Kruchten, *An ontology of architectural design decisions in software-intensive systems*, Intern. Workshop on Software Variability Management (2004).

[LvV05] P. Lago and H. van Vliet, *Explicit assumptions enrich architectural models*, IEEE 27th International Conference on Software Engineering (ICSE) (2005), 206–214.

[Med05] N. Medvidovic, *Software architectures and embedded systems: a match made in heaven?*, IEEE Software **22** (2005), no. 5.

[Oa00] H. Obbink and al., *Copa: a component-oriented platform architecting method for families of software-intensive electronic products*, Tech. report, http://www.extra.research.philips.com/SAE/COPA/, 2000.

[SEI06] CMU SEI, *How do you define software architecture?*, Tech. report, http://www.sei.cmu.edu/architecture/definitions.html, 2006.

[SG96] M. Shaw and D. Garlan, *Software architecture, perspectives on an emerging discipline*, Prentice-Hall, 1996.

[Sha88] M. Shaw, *Toward higher level abstractions for software systems*, Proc. Tercer Simposio Internacional del Conocimiento y su Ingegneria (1988).

[TA05] J. Tyree and A. Akerman, *Architecture decisions: Demystifying architecture*, IEEE Software **22** (2005), no. 2, 19–27.

[Zac87] J.A. Zachman, *A framework for information systems architecture*, IBM SYSTEMS JOURNAL **26** (1987), no. 3.

Part III

Robotic Software Frameworks

Trends in Robotic Software Frameworks

Davide Brugali[1], Gregory S. Broten[2] Antonio Cisternino[3], Diego Colombo[4], Jannik Fritsch[5], Brian Gerkey[6], Gerhard Kraetzschmar[7], Richard Vaughan[8], and Hans Utz[9]

[1] Universitá degli Studi di Bergamo, Italy brugali@unibg.it
[2] Defence R&D Canada Greg.Broten@drdc-rddc.gc.ca
[3] University of Pisa, Italy cisterni@di.unipi.it
[4] IMT Alti Studi Lucca, Italy diego.colombo@imtlucca.it
[5] Bielefeld University, Germany jannik@techfak.uni-bielefeld.de
[6] SRI International, USA gerkey@ai.sri.com
[7] Fraunhofer Institute for Autonomous Intelligent Systems, Germany gerhard.kraetzschmar@ais.fraunhofer.de
[8] Simon Fraser University, Canada vaughan@sfu.ca
[9] Ames Research Center, USA hutz@mail.arc.nasa.gov

1 Introduction

In the software community, a framework indicates an integrated set of domain-specific software components [CS95] which can be reused to create applications. A framework is more than a library of software components: It defines the common architecture underlying the particular applications built on the framework. Frameworks are a powerful development approach as they consist of both reusable code (the component library) and reusable design (the architecture).

Frameworks have acquired popularity in the object-oriented (OO) community. Here, the interpretation of "framework" ranges from structures of classes of cooperating objects, which through extension provide reusable basic designs for a family of similar applications [JF88], to the more restrictive view [Sch95] of complete, high level modules, which through customization directly result in specific applications for a certain domain. Customization is done through parameterization or by writing functional specifications.

Frequently, the two views of framework, referred to as white box and black box approaches to reuse, may be simultaneously present in one framework. In fact, features which are likely to be common to most applications can be offered and therefore reused as black boxes. Black-box frameworks support extensibility by defining interfaces for components that can be plugged into the framework via composition. Existing functionalities are reused by defining components that conform to a particular interface and integrating these components into the framework. On the other hand, the class library accom-

D. Brugali (Ed.): Software Eng. for Experimental Robotics. STAR 30, pp. 259–266, 2007.

panying the framework usually provides base classes (seen as white boxes) that can be specialized, by adding subclasses as needed, and easily integrated with the rest of the framework. White-box frameworks rely heavily on OO language features like inheritance and dynamic binding in order to achieve extensibility. Existing functionality is reused and extended by (1) inheriting from framework base classes and (2) overriding predefined hook methods using design patterns like the Template Method [GHJV95].

Framework based development has two main processes: development of the framework and development of an application adapting the framework. Every application framework evolves over time; this is called the framework life span [BMA97]. Within this life span, the basic architectural elements which are independent from any specific application are implemented first. When new applications are built using the framework, new components are developed which eventually are generalized and integrated as new black box elements. This means that in its early stages a software framework is mainly conceived as a white box framework. However, through its adoption in an increasing number of applications, the framework matures: More concrete components, which provide black box solutions for common difficult problems, as well as high level objects, which represent the major abstractions found in the problem domain, become available within the framework.

Since developing the framework requires a lot of effort, it is justified only if several applications will be built based on the framework. On the other hand, when the framework is available, developing an application from it offers considerable saving in time and effort compared to the development from scratch. The application is built by slightly modifying the framework, but by accepting the framework's high level design in general. In other words, most of the framework is reused, and reuse happens both at the high level design and at the code level.

Most framework-based software development is organized along the value-added principle [MFB02]. This divides any software system into a set of horizontal layers, each one built on top of another. This approach promotes separation of concerns and enforces modularity. Typically, a software framework is subdivided into a system infrastructure layer, a functional (also called horizontal or business domain) layer, and an application (also called vertical) layer.

The *system infrastructure layer* (usually called *middleware*) consists of a collection of software components that offer basic services like communication between distributed applications, uniform access to heterogeneous resources, independence from processing platforms and distributed location, distributed component naming, service brokering and treading, and remote execution. Examples of distributed middleware systems are Microsoft .NET, the implementations of OMG CORBA, and the Sun Enterprise Java Beans (see Sidebar *Middlewares for Distributed Computing* by Davide Brugali).

The customization of the system infrastructure layer is usually black-box. It consists in the aggregation of the frameworks' elemental components (such as mechanisms for logical communication, concurrency, and device abstraction) in which the whole application has to be written.

The *functional* (also called *business domain*) *layer* is the model of a robotic application and consists of software components, which directly map to real entities like a vision system or to common robot functionalities such as sensor processing, mapping, localization, path planning, and task planning, or even to typical robot capabilities such as reactive behaviors.

Typically, they are implemented on top of the system infrastructure layer. For example, reactive behaviors delegate device abstractions for motion actuation, sensor processing, path planning, and task planning rely on the framework concurrency model in order to execute on different time scale, multi-robot mapping and localization build on data exchanged through a common distributed environment.

The customization of the functional layer is usually white-box. The basic components are intermediary classes, which by their very nature are fairly application independent, although they have been conceived bearing in mind a specific application domain (e.g. robot mobility or visual servoing). They are written using the elemental components of the framework, and they have to be specialized for each concrete application. It is a critical layer in frameworks for robotics applications because often functionalities depends, at least in part, on the underlying robotics architecture.

The *application layer* connects the functional components according to the information and control flows of any robot task. Examples of robot tasks are soccer playing, vacuum cleaning, museum tour guiding, mars surface exploration.

The customization of the application layer is usually grey-box. It consists of the interconnection of pluggable application-specific components through the middleware system. These components are the result of the adoption of the framework for the development of more and more applications and in some cases they are open source components-off-the-shelf (COTS). COTS products are designed to support a standard virtual interface, which consists of a set of rules for accessing the component's functionality in a homogeneous way, regardless of the execution platform, the programming language, and other specifics.

Software frameworks offer a unified view to model and develop robotic software systems at every level of the vertical decomposition from the system infrastructure to the final application through the functional model.

The chapters in the third Part of this book describe different approaches that address to some extent most or all of the aspects of robot framework development (such as black box or white box reuse). Individually, they propose innovative contributions on how to apply framework development concepts in the robotics domain. These approaches are illustrated in the next section.

Thereafter, the Sidebar *Middlewares for Distributed Computing* by Davide Brugali analyzes some of the most widely used distributed middleware frameworks, i.e. OMG CORBA, Sun Java distributed technology, and Microsoft .NET.

2 Opportunities to Exploit Framework-Based Development in Robotics

Robot software developers have often experienced a sense of frustration when developing an application that is (nearly) the same as several other releases for different projects, but they have not been able to capture and exploit the commonalties. These problems frequently and typically occur in four circumstances.

1. When the application migrates to a different robot platform. Despite the differences in the mechanical structure, many commonalties can be identified in the control architecture, such as how to acquire sensory data or transmit commands to the actuators.
2. When the application scales up from a single robot to a multi-robot system or when it is restructured from centralized to decentralized and distributed control paradigm. Despite the differences in the communication mechanisms (shared memory or networked environment) the interactions among the control modules of the robot control architecture remain unchanged or show many commonalties.
3. When the same robotic system is used in different environments or for different tasks. Despite the differences in the control logic, most of the robot capabilities and functionalities are common in every application.
4. When the robotic system is build from the composition of heterogeneous or legacy subsystems. Despite the differences in the programming languages, concurrency models, or data interchange formats, commonalties can be found in the semantics of the components' behavior.

Software frameworks are a possible solution to the above problems, as illustrated in the following four sections.

2.1 Frameworks for Device Access and Control

One vital task for every robot system is to interface with hardware sensors and actuators. Familiar mass-market I/O devices such as mice and joysticks were once specialist experimental devices, comparable to the robot devices we build and buy today. Each device had a unique interface, and software had to be specially written to take advantage of it. Over time, as these devices became common, their external interfaces have converged to well-known logical standards. The hardware details are hidden behind an interface that encapsulates the logical functionality of the device. Other hardware, such as keyboards, disks and printers are treated similarly. A significant fraction of the source code of a operating system such as Windows or Linux is composed of such "device driver" code.

Robotics devices are far more rare than mice and printers, but recently we have seen a limited amount of commoditization in research robots. The Pioneer 2 and 3 robots from Mobile Robots Inc, and the SICK LMS laser scanner are good examples of devices that are sold and used in the hundreds, all over the world. The Pioneers ship with the software framework "ARIA" that contains drivers for the Pioneer's hardware and the LMS. ARIA provides a high-level, structured (Object Oriented) environment for writing robot controllers, and provides some powerful modules for doing common tasks such as mapping and localization. The "Mobility" APIs provide similar functionality for RWI Robots, and also support the SICK LMS. ARIA and Mobility are proprietary systems that work only with their companies' robots. Indeed, they provide a key part of the value of the robot to their customers: they make programming robots much easier. But they target different robots and their APIs are very different. If you move your LMS from a Pioneer to an ATRV Jr. robot, you have to use completely different code to access your range data.

Another approach is taken by the "Player" robot interface, described in the Chapter *Reusable Robot Software and the Player/Stage Project* by Richar Vaughan and Brian Gerkey. The design philosophy of Player is to extend the computer's operating system up to meet the robot controller. Like an operating system, Player is designed to provide convenient access to hardware and software resources, but otherwise keep out of the way. Robot controllers using Player can be written in any programming language, and devices can be accessed via a network socket or via a linked C library. Thus Player places very few restrictions on the user's robot control code.

Player is a popular system, but its lack of constraints means that it imposes no structure or high-level abstractions that can usefully guide and facilitate controller design. Several projects have built on top of Player, wrapping the device drivers and network transparency with a more structured development environment, providing powerful object hierarchies or visual programming interfaces.

2.2 High-Level Communication Frameworks

A very challenging robotic field that demands effective integration techniques is that of personal robot companions. Looking at the more abstract cognitive abilities of robot companions, the integration of multiple modalities for achieving human-like interaction capabilities requires flexible communication frameworks that can be easily used by interdisciplinary researchers. The interdisciplinary researchers involved in integrated projects are neither integration nor middleware experts, so ease of use is crucial for the successful application of such a framework. Another aspect of research prototypes is that specification changes occur frequently during the development of these systems. Thus, the impact of interface changes on the system architecture should be minimal to avoid versioning problems. This, in turn, allows rapid prototyping and iterative development that must be supported by a framework not only for single modules but also for the integrated system.

A framework that addresses these requirements is the XCF-SDK described in Chapter *An Integration Framework for Developing Interactive Robots* by J. Fritsch and S. Wrede. XCF provides a solution for flexible data-driven component integration as well as event-driven coordination by focusing on concepts from different middleware architectures that are relevant for the domain of intelligent systems research. This yields a lightweight API and provides features from both black-box and white-box frameworks. Within XCF, components are mainly described by the XML data they deliver to other system modules either directly or mediated through instances of an active XML database.

Another robotic communication framework is Robotics4.NET a message-oriented framework that supports the implementation of control software for robots designed to operate in human-based environments. It will be presented in Chapter *Increasing decoupling in the Robotics4.NET framework* by Antonio Cisternino et al. Its peculiarity relies in its structure centered around the notion of body of a robot. Body organs, called roblets provide an abstract view of the underlying hardware subsystem they are responsible for. An XML-based, message-oriented communication infrastructure acts like the spine, connecting organs to the brain.

2.3 Frameworks for Functionality Packaging

Supporting specific target applications, such as common tasks in robotics scenarios, is mandatory to allow robotic experts to focus on the problem domain, instead of dealing with the entire problem set of the application domain.

An application framework supports the development of applications for common robotic tasks (such as video image processing or behavior based reactive control), which manage the control flow and organize the data flow for the task at hand, shielding the application programmer from much of the intrinsic difficulties of software development within a concurrent, distributed software environment. Providing a framework for the tasks control and data flow also facilitates to add centralized support for addressing timeliness and scalability problems within the application.

Chapter *VIP: The Video Image Processing Framework based on the MIRO Middleware* by Utz et al. presents a video image processing framework (VIP) that targets the problem of applying computer vision in the field of real-time constraint autonomous mobile robotics. It aims to provide an environment that allows the developer to concentrate on the robot vision task without risking to jeopardize the reactivity and performance of the system as a whole.

The VIP framework features the above discussed different aspects of robotics framework design. The white-box framework manages the control flow and organizes the data flow, shielding the developer from the subtle locking issues and memory management problems. In order to meet the requirements of high-performance real-time-constraint robotics applications, it provides a powerful sensor-triggered, on-demand processing model for image operations. The framework also offers a set of tools that assist in the debugging, assessment and tuning of the vision application. Furthermore, the middleware integration provides a sophisticated communication infrastructure for distributed robotics applications as well as powerful facilities for the configuration and parameterization that are exploited by the framework components to enhance and their flexibility and reusability. A fast growing library of black-box components for the framework confirms the applicability of this application framework in very different robotics scenarios.

Chapter *MRT: Robotics Off-the-Shelf with the Modular Robotic Toolkit* by Bonarini et al. illustrates a modular approach to develop and integrate functionalities of mobile robots that are involved to perform tasks autonomously. The approach is based on the exploitation of a distributed middleware and the customization of an application framework. The middleware makes the interaction among functional modules transparent with respect to their physical distribution. The application framework allows to reuse the same functional units in different applications, thus reducing the development time and increasing the software reliability. Moreover, modules managing the interaction with the environment are designed to make independent the actual interface with physical devices from the needed data processing.

2.4 Frameworks for Legacy Systems Evolution

A well established practice in computing science, frameworks have only recently entered common use in robotics. For industry and government, the complexity, cost, and time to re-implement stable systems often exceeds the perceived benefits of adopting a modern software infrastructure. Thus, most persevere with legacy software, adapting and modifying software when and wherever possible or necessary – adopting strategic software frameworks only when no justifiable legacy exists. Conversely, academic programs with short one or two year projects frequently exploit strategic software frameworks but with little enduring impact. The open-source movement radically changes this picture. Academic frameworks, open to public scrutiny and modification, now rival commercial frameworks in both quality and economic impact. Further, industry is now realizes that open source frameworks can reduce cost and risk of systems engineering.

Chapter *Towards Framework-based UxV Software Systems, An Applied Research Perspective* by G. Broten et al. describes Defence R&D Canada (DRDC)'s experience as they migrated from a legacy, tele-operated system, to a modern frameworks based robotics program. Encountering limits to teleoperation, DRDC reoriented its focus to autonomous, intelligent systems. DRDC reviewed its software development methods and concluded that a new, extensible, flexible, modular and scalable approach was needed and that older work would need to be rewritten or jettisoned. Combatting a "must be invented here" tradition, DRDC readily examined the current trends in robotics and quickly realized a framework approach could mitigate many risks and relieve many constraints incurred by pure in-house software development. DRDC's experience confirms the widely held belief that framework use simplifies application development and maintenance. DRDC adapted the Miro framework, initially designed for the robosoccer competition, to implement an autonomous, outdoor unmanned ground vehicle (UGV). This effort exploited Miro's White-box framework attributes to extend existing classes for DRDC's unmanned systems requirements. Using the Miro framework DRDC was quickly able to create a modern, fully autonomous UGV by integrating legacy code and other open source code sources to a workable application.

3 Conclusions

Although Framework-based software development is an emergent area in Robotics, it reflects the natural evolution in the development of robotic software applications.

As the notion is in its infancy, a proper context and understanding of the concepts, characteristics and challenges are needed for robotic systems developers.

In this chapter we have provided a taxonomy of characteristics for classifying robotic software frameworks and have described these characteristics with a focus on the challenges robotic software engineers commonly struggle with today. In this sense, we believe we have identified a rich set of opportunities for the robotic community to enhance software reusability in Robotics.

References

[BMA97] D. Brugali, G. Menga, and A. Aarsten, *The framework life span.*, Communication of the ACM **40** (1997), no. 10, 65–68.

[CS95] J.O. Coplien and D.C. Schmidt, *Pattern languages of program design*, ch. Frameworks and Components, pp. 1–5, Addison-Wesley, 1995.

[GHJV95] E. Gamma, R. Helm, R. Johnson, and J. Villisides, *Design patterns: Elements of reusable object oriented software*, Addison-Wesley, 1995.

[JF88] R.E. Johnson and B. Foote, *Designing reusable classes.*, Journal of Object-Oriented Programming (1988).

[MFB02] Hafedh Mili, Mohamed Fayad, Davide Brugali, David Hamu, and Dov Dori, *Enterprise frameworks: issues and research directions*, Software Practice and Experience **32** (2002), 801–831.

[Sch95] H.A. Schmid, *Creating the architecture of a manufacturing framework by design patterns.*, Proceedings of OOPSLA'95 (1995).

[Vel05] T.L. Veldhuizen, *Software libraries and their reuse: Entropy, Kolmogorov complexity, and Zipfs law*, Proceedings of the first Library Centric Development (LCSD) workshop (2005).

Reusable Robot Software and the Player/Stage Project

Richard T. Vaughan[1] and Brian P. Gerkey[2]

[1] School of Computing Science, Simon Fraser University, Canada vaughan@sfu.ca
[2] Artificial Intelligence Center, SRI International, USA gerkey@ai.sri.com

1 Introduction

The authors of several well-known robot software systems met at the ICRA 2005 workshop on the *Principle and Practice of Software Development in Robotics*. The meeting was held to examine the role of software engineering concepts and methods in experimental robotics applications. Everyone at the workshop agreed that extensive reuse of robot software should help to make robot development faster, easier and more efficient, and that this was highly desirable. There exist many robot programming tools and frameworks designed to promote this idea, some of which have been actively developed for several years using very fine software engineering techniques. However, very few supposedly reusable systems are extensively used outside their home institution or their immediate collaborators. Many well-engineered systems are never used at all. This suggests that there is more to getting code widely reused than nice code design, however principled.

In terms of enabling experimental robotics, the most elegant, powerful, reusable code objects are of no use until someone actually uses them. After the (excellent but simple and inflexible) Lego Mindstorms system, probably the most reused robot code in the world comes from the Player/Stage Project. We think this is because it solves some common problems and has relatively low barriers to reuse, as well as having a sensible code design.

Based on our experience as authors of Player/Stage, we suggest that to be effectively enable and promote efficiency in robotics software, the practice of "Software Engineering for Experimental Robotics" needs to include a broad consideration of all the barriers to code reuse. Some of these are technical issues of modularity or interoperability, but some, such as licensing, cost, distribution, documentation and support, are not.

This chapter reviews the Player/Stage robot development system (P/S), describing its key models and abstractions, and identifying the opportunities for code reuse presented by its several layers of interfaces. We also discuss some

D. Brugali (Ed.): Software Eng. for Experimental Robotics. STAR 30, pp. 267–289, 2007.

barriers that make code reuse difficult, and describe how P/S has avoided or solved some of these problems to become widely reused.

Before proceeding, it should be made clear that there is already a great deal of reused code running on laboratory robots and workstations every day, but almost all of it is general-purpose computing infrastructure such as Linux, GCC, glibc, GSL, MATLAB, Java, Python and Windows[1]. Some specialized code for computer vision is also commonly used, for example OpenCV[2]. None of this code is specific to robot applications, but we can learn from these successful code reuse examples.

2 Why Reuse Robot Software?

The problem of how to program intelligent robots includes nearly all the sub-problems of AI, from perception, control, and reasoning under uncertainty, to planning, scheduling, and coordination. In addition, robot programming entails solving systems problems, including physical dynamics and real-time constraints, computation, memory and bandwidth limitations, and unreliable communications. Many students and researchers are attracted to the particular challenge presented by this complex combination of AI and systems. These properties, combined with the community's emphasis on demonstrating working systems mean that the products of research can often be quickly and usefully applied in the real world.

However, there are certain systems aspects of robotics work that are simply tedious. Consider the wide variety of wheeled mobile robots used in research labs. Despite the obvious similarity in their functionality and capabilities, the differences in size, shape, kinematics, and communication protocol mean that code developed for one robot (usually) must be ported or otherwise adapted to work on another robot. Simple tasks, like teleoperating a robot or visualizing its sensor state, require a significant amount of work. When they are available, interface libraries for robots often restrict the choice of programming language and/or style. Well-understood algorithms are re-implemented over and over again in laboratories around the world.

In addition to lost time as researchers duplicate engineering tasks, there is a negative impact on the quality of the resulting work. We have little shared equipment, few shared data[3], no shared environments, few shared tasks, and little shared code. As a result, robotic systems are not directly comparable,

[1] Respectively, the GNU/Linux operating system, the GNU Compiler Collection, the GNU C Library, the GNU Science Library, the MATLAB computing environment from The MathWorks, the Java programming environment, the Python programming environment, and Microsoft Windows operating system

[2] `http://sourceforge.net/projects/opencvlibrary/`

[3] An exception is the Radish project [`http://radish.sourceforge.net`], a public repository for localization and mapping data sets.

and competing approaches are often evaluated only in a "trial by video" at workshops and conferences.

Recently, challenges such as RoboCup and the DARPA Grand Challenge have focused research efforts and provided clear metrics for evaluating performance. These challenges are not a panacea, however; the overhead involved in entering such a competition is considerable, as is the danger of overfitting solutions to the peculiarities of the competition environment.

The goal of the Player/Stage Project is to develop Free Software infrastructure that improves research practice and accelerates development by handling tedious tasks and providing a standard development platform. More than simply distributing this infrastructure as a set of tools to the research community, we invite members of the community to contribute to the project. By collaborating on a common system, we share the engineering burden and create a means for objectively evaluating published work. If you and I use a common development platform, then you can send me your code and I can replicate your experiments in my lab. The ability to perform methodical peer evaluation is expected in the natural sciences but lamentably absent from robotics today.

3 Code Reuse from the Player/Stage Project

Our claim that code from the Player/Stage Project is frequently reused needs some supporting evidence. Player's distribution terms, the GNU General Public License[4] allow anyone to use and distribute the sourcecode. This makes it impossible to obtain precise numbers of Player users. There is no reliable mechanism for user registration or reporting. We have four sources of documentary evidence that Player code is being used:

1. Download statistics;
2. Support forum messages and bug tracker items;
3. Submissions to our user laboratory list;
4. Acknowledgements in published robotics articles.

The Player/Stage Project has been hosted by Sourceforge[5] since December 2001. As of 24 July 2006, Sourceforge reported 56,757 downloads of P/S software, and a download rate of around 2,000 packages (2.2GB) per month.

Figure 1 is a graph of the number of downloads and bandwidth supplied by Sourceforge each month of the period December 2001 to December 2005. Downloads show faster-than-linear growth during 2002-4, and were roughly constant in 2005; a year that saw no major new releases. The subsequent release of Player-2.0 in March 2006 produced a noticeable spike of 3,270 downloads, around 150% of the typical number.

[4] `http://www.gnu.org/copyleft/gpl.html`
[5] `http://sourceforge.net`

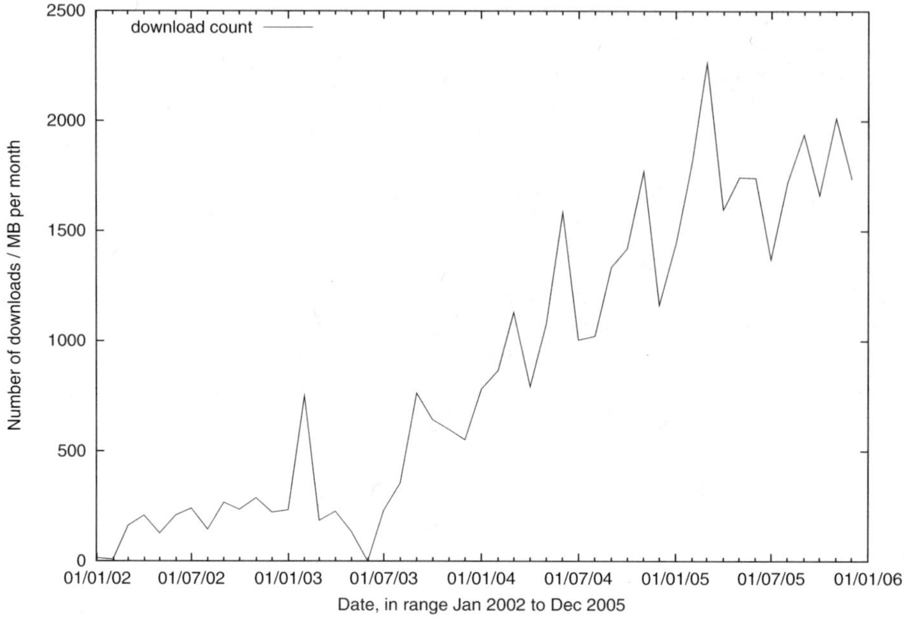

Fig. 1. Four-year download history for the Player/Stage Project

Fig. 2. Geographical distribution of downloads in December 2005

Table 1. Mailing list archive in July 2006

Mailing list	Description	Subscribers	Messages
playerstage-users	Player and Stage user support	293	4136
playerstage-developers	developer discussion	143	1864
playerstage-gazebo	Gazebo user support	121	1345
total messages			7345

Table 2. Issue-tracking database items (including closed items) in July 2006

Tracker	Items
Bugs	191
Patches	121
Feature requests	96
total items	408

Google's web traffic analytics tracking service[6] provides (among other things) a breakdown of the geographical location of visitors to the P/S web site. In the month of December 2005, a total of 70,920 web page requests were served. Figure 2 shows the locations of the visitors, demonstrating the global interest in Player. Unfortunately geographical data for actual software downloads (rather than web page requests) is not available, but we can assume that the distribution of software downloads is similar to that of the web requests.

3.1 Support Forum and Trackers

P/S uses Sourceforge's mailing list and issue-tracker services to communicate with users and developers. Activity on these services suggests active users. Table 1 shows the number of subscribers and messages on each list in July 2006. There were a total of 7,759 messages on the lists since their creation in January 2002; an overal average of 4.7 messages every day, with 12.5 messages per day in the first half of 2006. The total number of subscribers is not meaningful because many people subscribe to more than one list.

Another indicator of user activity is the issue-tracking database, a mechanism designed to manage issues that can not be immediately resolved on the mailing lists. Figure 2 shows the number of items on each list from January 2001 to July 2006: a total of 408. Notable here are the 121 code patches submitted by users to fix bugs or extend P/S software. Many of these patches have been applied to the tree to become 'official' code.

3.2 User Site List

Users are invited to submit the name of their laboratory for listing on the P/S Wiki site[7]. As of July 2006, the list contains 52 entries in 19 countries,

[6] `http://google.com/analytics`
[7] `http://playerstage.sourceforge.net/wiki/`

including the Boeing Company, the Australian Centre for Field Robotics, the Chinese National University of Defense Technology, the Technical University of Munich, the NARA Intitute of Science and Technology, Georgia Institute of Technology, the Mexican National Institute of Astrophysics, Optics and Electronics, and Rochester Institute of Technology's DARPA Grand Challenge Autonomous Race Team.

3.3 References in Scientific Articles

Many authors not connected with P/S development or each other have published papers that acknowledge the use of Player/Stage Project code in their experiments. Examples include the papers [AALB05, HHFS05, TC05, iSW05] all from the IEEE International Conference on Intelligent Robots and Systems (IROS) in 2005. Others are cited elsewhere in this chapter.

Based on these figures, we conclude that Player/Stage is both a well-known and well-*used* source of robot code.

4 Design of the Player Robot Device Interface

All work, including robotics research, is impacted by the tools that are used. Good tools simplify common tasks, while bad tools complicate them. The assumptions that are built into a set of tools bias the researcher who uses them toward particular kinds of solutions.

Our design philosophy is heavily influenced by the operating systems (OS) community, which has already solved many of the same problems that we face in robotics research. For example, the principle function of an operating system is to hide the details of the underlying hardware, which may vary from machine to machine. Similarly, we want to hide the details of the underlying robot. Just as I expect my web browser to work with any mouse, I want my navigation system to work with any robot. Where OS programmers have POSIX, we want a common development environment for robotic applications. Operating systems are equipped with standard tools for using and inspecting the system, such as (in UNIX variants) **top**, **bash**, **ls**, and **X11**. We desire a similar variety of high-quality tools to support experimental robotics.

Operating systems also support virtually any programming language and style. They do this by allowing the low-level OS interface (usually written in C) to be easily wrapped in other languages, and by providing language-neutral interfaces (e.g., sockets, files) when possible. Importantly, no constraints or normative judgments are made on how best to structure a program that uses the OS. We take the same approach in building robotics infrastructure. Though not strictly part of the OS, another key feature of modern development environments is the availability of standard algorithms and related data structures, such as **qsort()**, TCP, and the C++ Standard Template Library. We follow this practice of incorporating polished versions of established algorithms into the common code repository, so that each researcher need not re-implement, for example, Monte Carlo localization. Finally, an important but often overlooked aspect of OS design is that access is provided at all levels. While most C programmers will manage memory allocation with the library functions **malloc()** and **free()**, when necessary they can dig deeper and invoke the system call **brk()** directly. We need the same multi-level access for robots; while one researcher may be content to command a robot with high-level "goto" commands, another will want to directly control wheel velocities.

In summary, our approach to building tools for robotics research is to extend useful abstractions from the OS up *just enough* to enable robotic devices to be used as easily as normal computer devices such as mice and printers. Like an OS, we aim to provide resources as transparently as possible, extending and managing the hardware but otherwise keeping out of the user's way.

5 Abstractions

Player comprises four key abstractions: The Player Abstract Device Interface (PADI), the message protocol, the transport mechanism, and the implementation. Each abstraction represents a reusable and separable layer. For example, the TCP client/server transport could be replaced by a CORBA transport. Alternatively, an entirely different system could be built atop the PADI. These four abstractions, described in the following sections, are the result of a considerable design effort refined over several years of broad use and are in themselves opportunities for reuse, independent of their P/S software implementations.

5.1 The Player Abstract Device Interface

The central abstraction that enables portability and code re-use in Player is the PADI specification. The PADI defines the syntax and semantics of the data that is exchanged between the robot control code and the robot hardware. For ease of use, the PADI is currently specified as a set of C message structures; the same information could instead be written in an Interface Definition Language (IDL), such as the one used in CORBA systems.

The PADI's set of abstract robot control interfaces constitutes a virtual machine, a target platform for robot controllers that is instantiated at run time by particular devices. The goal of the PADI is to provide a virtual machine that is rich enough to support any foreseeable robot control system, but simple enough to allow for an efficient implementation on a wide array of robot hardware. The key concepts used in the PADI, both borrowed from the OS community, are the character device model and the driver/interface model.

The Character Device Model

The "device-as-file" model, which originated in Multics [FO71] and was popularized by UNIX [RT74], states that all I/O devices can be thought of as data files. A distinction is made between sequential and random access devices. Sequential devices such as terminals and tapes produce and consume streams of data one byte after another, and are called *character devices*, while random access devices such as disk drives can manipulate chunks of data in arbitrary orders, usually through a cache, and are known as *block devices.* The nature of sensors and actuators is to produce and consume data in time-extended streams: they are character devices.

The standard interface to character devices is through five well-defined operations. Access to devices is controlled by *open* and *close* operations. Data is collected from the device by a *read* operation, and sent to the device by a *write* operation. The asynchronous read and write are sufficient on their own for many devices, but a third transfer mechanism, the *ioctl* (input/output control) provides a synchronous request/reply channel, typically to access data that is persistent rather than sequential, such as setting and querying the configuration of the device. All five operations will indicate error conditions if they fail.

Player uses the character device interface to access its hardware devices. For example, to begin receiving sensor readings, the appropriate device is first opened, after which data can be read from it. Likewise to control an actuator, the appropriate device is opened, after which commands can be written to it. An ioctl mechanism is used for device configuration and for atomic test/set and read/clear operations required by some devices. For reasons explained elsewhere [GVS01, GVH03], Player does not currently implement exclusive access to devices, so multiple modules can simultaneously control the same device. We are considering adding exclusive access modes, which would be analogous to file-locking mechanisms in operating systems.

The character device model has some drawbacks. In particular, since there is no interrupt mechanism, clients must poll devices to receive new data. This is not the best approach for low-latency I/O with high-speed devices, which are usually interrupt-driven. Player was designed to support update rates of the order of 5-100Hz, covering the majority of research robots. This model is unlikely to suffice for devices that operate on the order of 1MHz. Also, the ioctl channel is often used in a way that breaks device independence and reduces portability, as discussed below.

Apart from the assumption of sequential access (supplemented with the ioctl), the character device abstraction is neutral with respect to programming language and style. Almost every programming language supports this model, and almost any robot control architecture can be (and likely has been) implemented atop the generic read/write/ioctl interface. This model has successfully supported UNIX-like operating systems for decades, and Player for years. We suggest that the character device model is a suitable foundation for a robot device control standard.

The Interface/Driver Model

The character device model defines only the broadest semantics of its three channels (roughly: input, output and configuration), but imposes no other structure on the data streams. Each device could have its own unique data format, requiring controller code to be written specifically for each device. Another powerful abstraction, the *interface/driver* model determines the content of these streams and provides the device independence that is the key to portable code.

The interface/driver model groups devices by logical functionality, so that devices which do approximately the same job appear identical from the user's point of view. An *interface* is a specification for the contents of the data stream, so an interface for a robotic character device maps the input stream into sensor readings, output stream into actuator commands, and ioctls into device configurations. The code that implements the interface, converting between a device's native formats and the interface's required formats is called a *driver*. Drivers are usually specific to a particular device, or a family of devices from the same vendor.

Code that is written to target the interface rather than any specific device is said to be *device independent*. When multiple devices have drivers that implement the same interface, the controlling code is portable among those devices.

Many hardware devices have unique features that do not appear in the standard interface. These features are accessed by device-specific ioctls, while the read and write streams are generally device independent. Interfaces should be designed to be sufficiently complete so as to not require use of device-specific ioctls in normal operation, in order to maintain device independence and portability.

There is not a one-to-one mapping between interface definitions and physical hardware components. For example, the Pioneer's native P2OS interface bundles odometry and sonar data into the same packet, but a Player controller that only wants to log the robot's position does not need the range data. For portability, Player separates the data into two logical devices, decoupling the logical functionality from the details of the Pioneer's implementation. The pioneer driver controls one physical piece of hardware, the Pioneer microcontroller, but implements two different devices: `position2d` and `sonar`. These two devices can be opened, closed, and controlled independently, relieving the user of the burden of remembering details about the internals of the robot.

Since Player was initially designed as an interface to our Pioneer 2-DX mobile robots, early versions of the server provided almost transparent access to specific components and peripherals of the Pioneer as it was used in the USC Robotics Lab. For example, each data packet from the sonars comprised

16 range readings, because the Pioneer has 16 sonar transducers. Likewise, command packets to the wheel motors comprised two velocities, because the Pioneer is a non-holonomic, differentially-driven robot.

This Pioneer-specific device model was extensible, but did not encourage code reuse or portability. When code was added to provide access to a new device, that device presented a unique, device-specific interface that required device-specific support in control programs. As a result, programs that controlled the second Player-supported mobile robot, the RWI B21r, used an API that was completely different from that used to control the Pioneer, despite the fact the two robots are functionally similar.

In order to more conveniently support different devices, we introduced the interface/driver distinction to Player. An interface, such as `sonar`, is a generic specification of the format for data, command, and configuration interactions that a device allows. A driver, such as `pioneer-sonar`, specifies how the low-level device control will be carried out. In general, more than one driver may support a given interface; conversely, a given driver may support multiple interfaces. Thus we have extended to robot control the device model that is used in most operating systems, where, for example, a wide variety of joysticks all present the same "joystick" interface to the programmer.

As an example, consider the two drivers `pioneer-positionand` `rwi-position`, which control Pioneer mobile robots and RWI mobile robots, respectively. They both support the `position2d` interface and thus they both accept commands and generate data in the same format, allowing a client program to treat them identically, ignoring the details of the underlying hardware. This model also allows us to implement more sophisticated drivers that do not simply return sensor data but rather filter or process it in some way. Consider the `lasercspace` driver, which supports the `laser` range-finder interface. Instead of returning the raw range values, this driver modulates them according to the dimensions of the robot, creating the configuration-space representation of free space in the environment.

The primary cost of adherence to a generic interface for an entire class of devices is that the features and functionality that are unique to each device are ignored. Imagine a `fiducial-finder` interface whose data format includes only the bearing and distance to each fiducial. In order to support that interface, a driver that can also determine a fiducial's identity will be under-utilized, some of its functionality having been sacrificed for the sake of portability. This issue is usually addressed by either adding configuration requests to the existing interface or defining a new interface that exposes the desired features of the device. Consider Player's Monte-Carlo localization driver **amcl**; it can support both the sophisticated `localization` interface that includes multiple pose hypotheses, and the simple `position2d` interface that includes one pose and is also used by robot odometry systems.

5.2 The Player Protocol

The **Player Protocol** implements the PADI along with some additional structures and rules for multiplexing, ordering, sending and receiving collections of PADI messages, and commands for inspecting and controlling the behavior of Player itself. For example, part of the protocol specification states that data and messages are asynchronous and not acknowledged, while configurations are synchronous: every request is guaranteed an acknowledgment (positive or negative) in response. The protocol also defines a generic header structure that precedes every message and contains the meta-data necessary to unambiguously interpret the message. Note the the protocol does not specify the manner in which packets are serialized, addressed, or transmitted; those details are handled by the transport layer.

5.3 Transport Mechanisms

The PADI and Player Protocol together are sufficient for building a single-process robot control system. Drivers can be instantiated and bound to interfaces, and the resulting devices can exchange PADI-defined messages (e.g., by function calls) according to the rules of the protocol. However, if we want the ability to move messages among devices or other modules that are in different processes or on different machines, then we need a **transport** mechanism.[8] The job of a transport mechanism is two-fold: handle the addressing and routing of packets, and perform packet serialization and deserialization (also called data marshaling). Some transports may also provide higher-level functionality, such as resource discovery.

TCP Client/Server Transport

Historically Player has relied on a TCP client/server transport, in which devices reside in a *server* and a control program is a *client* to the server. To control a robot, the user first starts the **player** server, which listens on a particular TCP port (by default 6665), on the robot. Then a client program, such as a joystick controller or data visualization GUI, is started and establishes a TCP socket connection to the server. The client can run on-board the robot or on any other machine that has network connectivity to the robot. One client can connect to many servers and many clients can connect to one server. Importantly, clients can be written in any programming language with support for TCP sockets.

In order to safely send messages from one machine to another over a TCP socket, a data marshaling scheme must be defined. The marshaling rules specify the bit-level details of how, for example, numbers are represented on the

[8] With respect to the OSI Network Model [Tan96], we refer collectively to the Network Layers (Transport and below) and the packet-shuffling machinery in the Session Layer as the *transport*.

wire. Earlier versions of Player used a custom encoding of messages as packed C structs that contained integers in network byte-order. Player now uses an open standard called eXternal Data Representation, or XDR [Net87]. The XDR specifies an efficient, platform-independent encoding for commonly-used data types, including integers and floating point values. To reduce the occurrence of marshaling bugs, the library that performs the XDR data marshaling is generated in an automatic fashion directly from the header file that specifies the PADI.

Other Transports

The client/server model has served Player (and many other distributed systems) well for many years, but it is not the ideal transport mechanism for every robot system. In order to allow for the use of other transports, Player was redesigned to be transport-independent. The TCP client/server transport is still available (and will likely continue to be the mostly widely-used), but other transport mechanisms can be substituted in its place. The drivers, message structures, and other underlying details remain unchanged.

For example, we have developed a JINI-based transport for Player. JINI is a Java-based architecture for building distributed systems in a network-centric manner [Wal99]. JINI has been used in many distributed systems, including the largest multi-robot system deployed to date [KOV04]. JINI offers a number of advantages over the TCP client/server approach, including: robustness to network delays and dropouts, automatic resource discovery, and effortless data marshaling using Java's built-in object serialization mechanism. On the other hand, it is non-trivial to install, configure, and run the JINI infrastructure; and of course control programs must be written in Java.

Other trade-offs exist for other popular transport mechanisms, such as CORBA [Obj02], IPC [SW97], and ACE/TAO [SLM98]. It is highly unlikely that the robotics community will agree on a single transport for all robotics software, nor should they; the requirements and constraints exhibited by any given application area will make particular transports more or less appropriate, and it is important to allow the system designer to choose the best one. Because the manual construction of a new Player transport layer is a tedious and bug-prone process, we have taken care with the the PADI and core Player libraries so as to facilitate the automatic generation of transport code. For example, the JINI transport layer uses automatically-generated Java bindings for Player. We expect to see other transports developed similarly.

The choice of transport is critical for a robotics application as it is the transport that determines the real-time peformance of the system. Assuming that the robot software has been designed to avoid any logically unecessary time delays, the transport and the underlying OS that implements it are the only source of timing delays due to buffering, message transmissionn, etc. For control of most slow-moving, statically-stable wheeled robots, TCP over

Ethernet (802.3) or WiFi (802.11b/g) is found to have acceptable timing performance and its ubiquity makes it a good choice of general purpose transport.

5.4 Implementations

We have implemented the PADI, the Player Protocol, TCP client/server transport (with XDR data marshaling), and many device drivers as a set of reusable C/C++ libraries. The most common use of these libraries is in an executable server, called **player**. This server is used to parse configuration files, instantiate drivers, and service client connections to devices. It is customary when using Player to assign to each robot a server that contains all the drivers used in that robot's control system. The user's control program is written as a client that executes outside of the server (the situation is essentially the same, with different terminology, when using JINI instead of TCP).

The server represents a privileged space in which modules have better, faster access to hardware and to each other. Control code on the client side experiences greater latency and a somewhat lessened ability to interrogate devices. On the other hand, the client code has fewer constraints with respect to programming language and structure, and it is relieved of the drivers' burden of behaving properly to avoid crashing the rest of the system.

In this way, Player is analogous to a monolithic kernel operating system, in which a privileged kernel space is separated from a non-privileged user space [SGG05]. Most operating systems, including Linux, employ monolithic kernels. An alternative approach, used by operating systems such as QnX, is the microkernel, in which there is no privileged kernel space, but rather a collection of system processes and a mechanism for efficiently passing messages between them. There are advantages and drawbacks to each approach and the topic has been debated, without resolution, in the OS community for decades. An example of a microkernel-like robot control system is CARMEN [MRT03]. It is worth noting that although Player more naturally operates as a monolithic kernel, a microkernel system can be constructed by connecting multiple servers together, each with a single driver (server-server communication operates exactly like client-server communication).

As an alternative to our C/C++ system, an equivalent implementation of Player could be done in, for example, Java. A Java-only Player would be useful for the many embedded computing platforms that execute Java bytecode natively. Of course the existing C/C++ drivers could not be used on a native Java system. Other possible re-implementations include using strictly C (for systems without C++ runtime support) and removing the reliance on POSIX threads (which Player uses extensively).

6 Higher-Level Drivers

While Player's primary purpose is to provide portable and nearly transparent access to robot hardware, an increasing number of drivers encapsulate sophisticated algorithms that are removed by one or more steps from the physical hardware. These *higher-level drivers* use other drivers, instead of hardware, as sources of data and sinks for commands. The **amcl** driver, for example, is an adaptive Monte Carlo localization system [TFBD00] that takes data from a **position2d** device, a **laser** device, and a **map** device, and in turn provides robot pose estimates via the **localize** interface (as mentioned above, **amcl** also supports the simpler **position2d** interface, through which only the most likely pose estimate is provided). Other Player drivers perform functionality such as path-planning, obstacle avoidance, and various image-processing tasks.

The development of such higher-level drivers and corresponding interfaces yields three key benefits. First, we save time and effort by implementing well-known and useful algorithms in such a way that they are immediately reusable by the entire community. Just as C programmers can call **qsort()** instead of reimplementing quicksort, robotics students and researchers students should be able to use Player's **vfh** driver instead of reimplementing the Vector Field Histogram navigation algorithm [UB98]. The author of the driver benefits by having her code tested by other scientists in environments and with robots to which she may not have access, which can only improve the quality of the algorithm and its implementation. Second, we create a common development environment for implementing such algorithms. Player's C++ **Driver** API clearly defines the input/output and startup/shutdown functionality that a driver must have. Code that is written against this API can enter a community repository where it is easily understood and can be reused, either in whole or in part. Finally, we create an environment in which alternative algorithms can be easily substituted. If a new localization driver implements the familiar **localize** interface, then it is a drop-in replacement for Player's **amcl**. The two algorithms can be run in parallel on the same data and the results objectively compared.

7 APIs

We have described above the various interfaces to Player's components. In practice, the majority of user code will interact with Player through a *client library*; a language-specific interface that the user compiles (or loads, depending on the language) into their client program. Each client library presents an 'Application Programming Interface' (API). The most commonly used client libraries (and hence APIs) are the libplayerc and libplayerc++ libraries, for C and C++ respectively. Several other libraries are thin wrappers around one of these. For example the Python client library is automatically generated from libplayerc using the SWIG (Simplifiedf Wrapper and Interface Generator) tool [Bea96]. Using SWIG, changes to the libplayerc interface are correctly propogated through to APIs in several languages without having to tediously edit each by hand. As well as saving time, this removes a common source of bugs.

There are several advantages for users in using the client libraries instead of talking to the Player server directly; first, client libraries hide the details of the client/server communications almost completely, so the user can largely ignore the socket-level Player protocol, marshalling and serializing data, etc. Secondly, each API is designed to in a way that is natural way for its language; for example libplayerc++ presents client-side proxy objects that correspond to server-side devices. Thus the user can manipulate the proxies as fully-fledged objects, inherit from them, etc. As required by its language, the C client library has a similar proxy-based design, but structures and function calls are used instead.

Client library external APIs must change only when the PADI changes: they need not change with the client/server protocol, though of course the libraries must change internally to communicate with the server. The APIs can often be backwards compatible, for example when the PADI is extended but existing specifications are not changed the client API will still work, it will simply not provide access to the newly-defined structures.

As a 'Presentation Layer' interface in OSI terms, the client library APIs are almost independent of the transport layer in that they hide most details of the transport. However, some high-level transport-related details may leak through this abstraction. For example, all the client libraries that target the Player TCP server must be supplied with the hostname and port number of a running Player server on initialization. After initialization, the TCP connection is completely transparent as user code interacts with Player through client-side proxy objects (C++, Java, Python) or structures and function calls (C).

The popularity of the client libraries, particularly libplayerc++ and libplayerc, means that their design is very important. As the most-used interface anywhere in the Player/Stage system, the quality of their design plays a large part in the utility of the whole system. If the APIs are too complex, or inconsistent, or poorly documented, users will be quickly frustrated. Libplayerc was carefully designed to be as consistent and transparent as the authors could reasonably make it. It was intended to replace libplayerc++ which was written originally to test the Player TCP server and not intended for end-users. The fact that almost all early users used the C++ library instead of talking to the server took us by surprise.

For most users, the internal design of the server is of no interest at all, so long as she is protected from it by a client library. However, code correctness is equally important throughout the system: a bug anywhere upstream will eventually appear to the user through this interface. The existence of clean, maintainable server code is justfied by the increased probability of code correctness alone.

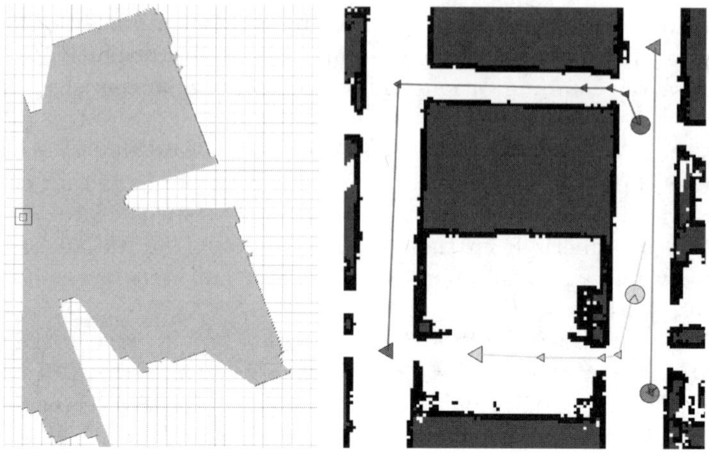

(a) (b)

Fig. 3. Screenshots of: (a) *playerv* displaying range data from a laser-equipped robot; and (b) *playernav* showing poses and planned paths for a team of robots.

8 Tools

As an OS provides the basic services needed to control a computer, Player provides the basic services necessary to control a robot. An OS is also bundled with useful tools for common tasks, like listing files and displaying the process table. Similarly, Player is bundled with several tools:

- **playerprint** : Fetches and prints sensor data to the console.
- **playerv** : Fetches and graphically displays sensor data; also provides teleoperation by mouse movement (Figure 3(a)).
- **playerjoy** : Provides joystick teleoperation.
- **playervcr** : Provides remote control of data logging and playback.
- **playernav** : A graphical operator control unit that provides control over localization and path-planning for multiple robots (Figure 3(b)).
- **playerwritemap** : Fetches grid and vector maps (e.g., from a SLAM driver) and writes them to disk.
- **playercam** : Remotely displays video imagery from robot-mounted cameras.

Third-party tools have also been developed, including a tool similar to **playercam** and an OpenGL-based 3-D application that combines some of the functionality of **playerv** with some of **playernav**. The development of such tools by users outside of the P/S/G project is a testament to the reusability of the system and the extensibility of the architecture.

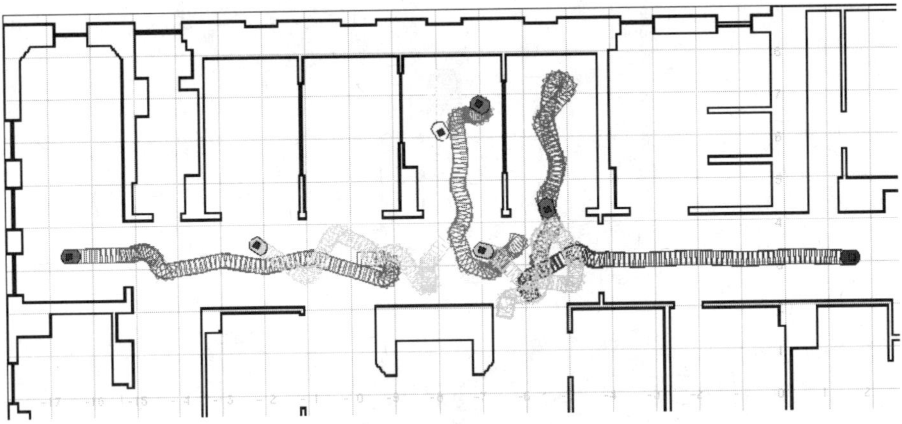

Fig. 4. A screenshot from the Stage multiple-robot simulation, showing several robots leaving trails as they explore a small section of a hospital floorplan.

9 Simulation

A significant contribution of the Player/Stage project is to provide robot simulators. The main benefits to the user of using a simulation over a real robot are convenience and cost: simulated robots are usually easy to use, their batteries need not run out, and they are very much cheaper than real robots.

9.1 Stage

After Player, the next most reused component of the Player/Stage project is the Stage robot simulation engine. Stage provides a virtual world populated by mobile robots and sensors, along with various objects for the robots to sense and manipulate. Designed with multi-agent systems in mind, it provides fairly simple, computationally cheap models of lots of devices rather than attempting to emulate any device with great fidelity. This design is intended to be useful compromise between conventional high-fidelity robot simulations, the minimal simulations described by Jakobi [Jak97], and the grid-world simulations common in artificial life research [Wil85]. We intend Stage to be just realistic enough to enable users to move controllers between Stage robots and real robots, while still being fast enough to simulate large populations. We also intend Stage to be comprehensible to undergraduate students, yet sophisticated enough for professional reseachers.

Stage provides several sensor and actuator models, including sonar or infrared rangers, scanning laser rangefinder, color-blob tracking, fiducial tracking and mobile robot bases with odometric or global localization. Figures 4 and 5 show screenshots from a running simulation.

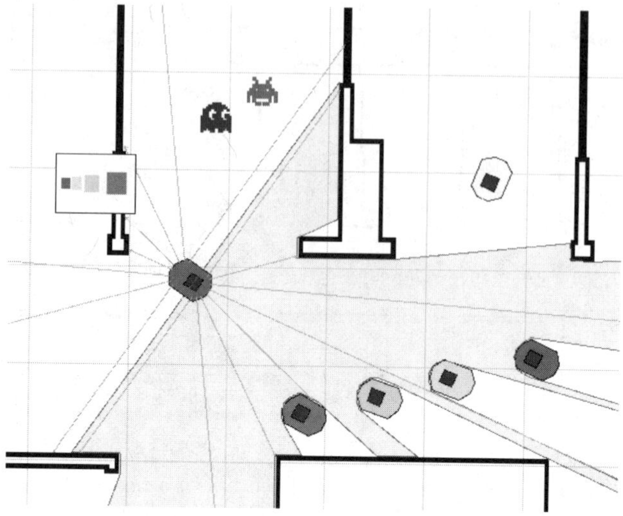

Fig. 5. A close-up screenshot from the Stage multiple-robot simulation, showing rendered laser and sonar data and several robots.

Stage is most commonly used with Player to form the Player/Stage system. Robot controller client programs interact only with Player, so simulated Stage devices appear identical to real devices. Client programs do not need to be rewritten or even recompiled to switch from simulated to real devices: a very convenient feature. Stage is implemented as a Player plugin driver (`libstageplugin`), loaded as Player starts up. This allows Stage to be developed and released on a schedule independent of Player.

```
#include "stage.h"

int main( int argc, char* argv[] )
{
  stg_init( argc, argv );
  stg_world_t* world = stg_world_create_from_file( argv[1] );

  while( (stg_world_update( world,TRUE )==0) )
    {
      /* use the simulation */
    }

  stg_world_destroy( world );
  return 0;
}
```

Fig. 6. Creating a complete multiple-robot simulation in C with libstage.

The Stage simulation engine is implemented as the standalone C library `libstage`; libstageplugin is a wrapper around libstage that connects it to Player's driver architecture. Libstage can be used to very easily create a robot simulation in user code. This is useful for users who wish to have more control over the internals of a simulation than they can get through Player's `simulation` interface. It also allows for perfectly repeatable experiments, without the unpredictable timing inevitably introduced by Player's TCP transport.

Using Stage directly without Player also saves a little computational overhead for those with very high performance requirements. The main downside is the loss of Player's high-level drivers (VFH, AMCL, etc.). Using the libstage in this way is very simple, as illustrated Figure 6 showing the C code required to instantiate a complete multiple robot simulator, similar to that shown in the screenshot above.

Stage is probably the most-used robot simulator, with research papers acknowledging the use of Player/Stage appearing in most major journals and conferences. The first published paper to use libstage directly was [ZV06], but libstage has also the basis of a the commercial product "MobileSim" from MobileRobots Inc. since 2005. This is noteworthy because none of the Stage maintainers have any relationship with the company or MobileSim, but MobileRobots have been active in contributing patches back to libstage.

9.2 Gazebo

Gazebo is also a robot simulator that works with Player. Unlike Stage, it provides realistic kinematics and dynamics in three-dimensional environments. Figure 7 shows a screenshot from Gazebo. Client programs written using one simulator can usually be run on the other with little or no modification. Gazebo is more realistic than Stage, but much more computationally intensive. Stage is designed to simulate a large robot population with low fidelity, and Gazebo is designed to simulated a fairly small population with high fidelity. Thus the two simulators are complimentary, and users may switch between them without cost due to the common Player interface.

Like Stage, Gazebo is implemented as a standalone library and can be used without Player. Due to its complexity, Gazebo is currently more difficult to install and run than Stage, and is less frequently used.

10 Barriers to Reuse

Why are most robot code tools used by no one but their authors? Robot infrastructure code will not be widely used if it is

1. very buggy, or otherwise of poor quality;
2. undocumented;

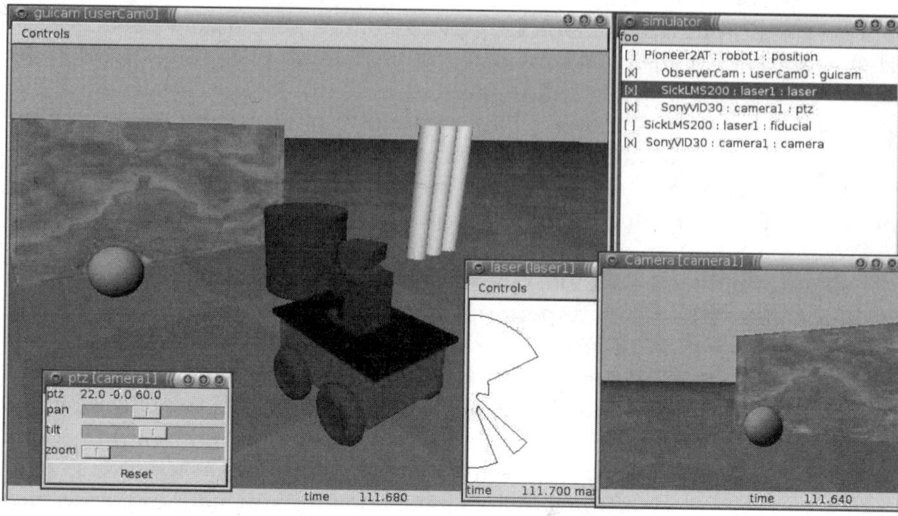

Fig. 7. A screenshot from the Gazebo 3D robot simulator, showing a model of a Pioneer robot carrying a SICK laser scanner and pan/tilt/zoom camera, with some environmental objects.

3. too expensive;
4. tied to a specific robot platform.
5. not solving the problems that lots of people really have;
6. difficult for the user to express their intentions;
7. overwhelmingly complex or cumbersome to use;
8. not distributed.

The first three reasons are self-explanatory, but the others may require some elaboration.

Several well-known pieces of software are produced and/or distributed by robot manufacturers to add value to their products. They deliberately only work with that company's robots. The market for robots is small and diverse, and many robots are still custom-built. Robot companies often (perhaps usually?) go out of business. When evaluating the software from a vendor, the researcher has to decide if the software is worth the investment in time and money (if the software is not free) to invest in software that restricts her choice of robot, and for which the support from someone with access to sourcecode could disappear at any time. These problems are the same with all commmercial software, but the small size and volatility of the research robot business makes this particularly risky.

The next two reasons are closely related. Producing software that does not solve anyone's real problems is a trap that is often fallen in to. For example, one feature that is repeatedly proposed to make robot programming easier is graphical programming, i.e. building systems by connecting boxes with lines using some spatial layout tool. This seems to ignore the fact that the population of robot programmers are overwhelmingly graduates of computer science or engineering, most of whom have no difficulty expressing themselves in code. Being artificially constrained to thinking about a problem *only* graphically, can be frustrating. Some common code structures, such as loops and recursion, are difficult to represent graphically. As anyone who has used Simulink knows, complex programs quickly lead to cluttered screens and so lots of time is spent arranging objects spatially: an arrangement that has no meaning at all for the code. Great complexity is also a problem: users are quick to abandon a system if they feel overwhelmed or can not get their robot equivalent of "Hello World" working in a few hours.

Another problem is of overwhelmingly complex or cumbersome software. Many of the proposed solutions to robot programming problems require infrastructure that may take more time to install and debug than they can save. The application of middleware such as CORBA, TAO and JINI, attractive though it is to software engineers, must be carefully justified to a robot programmer on a tight time budget, with no one in the lab with past experience in these systems. Even world-famous laboratories have "rapid" robot development systems that are so complex that the in-bouse engineers will not use them for real projects.

The final, and very basic reason that many systems are not widely used is that they are not distributed. This is usually due to institutional rules that prevent code from being given away free. The robot software market is very small and few people will pay to license software, so it stays within its development team only.

Notice that of these eight reasons, only the first two would automatically be addressed by conventional software engineering techniques. The rest are strategic, marketing and political problems that can not be solved by applying object-oriented techniques or client-server, publish-subscribe design patterns.

11 Conclusion

The most obvious way is to use code from the Player/Stage Project is to use the software packages as distributed. We have presented evidence above that this is happening frequently. However, the resources developed in and around the Player/Stage Project can be re-used in several different ways:

1. use the software packages as provided;
2. extend Player with new drivers and/or interfaces;
3. extend Stage or Gazebo with new simulation models;
4. use the simulation, driver or server libraries as part of custom software;
5. use the interface specification (PADI) as is, or as a guideline for creating custom interfaces
6. use the non-code resources, such as the Stage environment bitmaps that have become well-known;
7. use Player as middleware

This last option will become important in the near future. The drivers, PADI and Player Protocol specifications in Player are very valuable resources. Player's strength is the transparency it provides by trying to avoid placing constraints on the robot programmer. A Player-equipped robot is a blank slate, with a loose collection of easily accessible, commonly-used devices. A very suitable use for Player is as a substrate, or middleware layer, for more structured robot programming frameworks. If such a framework targetted Player instead of robot hardware directly, it would automatically inherit the large base of supported robots and implemented algorithms, saving the authors a great deal of development time.

Code has a good chance of being widely reused if it is

1. solving a user's problems
2. supported by their robots, or easy to port;
3. easy enough to use;
4. easy to obtain;
5. good enough quality;
6. well documented;
7. affordable;
8. supported by a knowledgable group of people;

We believe that Player/Stage is popular and widely used because it meets these criteria better than any other current system. It is not perfect, it is not finished, and it is not the right choice for every application. It is the product of a community of robot programmers, and it works for us.

References

[AALB05] G. Alankus, N. Atay, C. Lu, and B. Bayazit, *Spatiotemporal query strategies for navigation in dynamic sensor network environments*, Proc. of the IEEE/RSJ Intl. Conf. on Intelligent Robots and Systems (IROS), 2005.

[Bea96] David M. Beazley, *Swig : An easy to use tool for integrating scripting languages with c and c++*, Fourth Annual USENIX Tcl/Tk Workshop (Livermore, California), USENIX, July 1996.

[FO71] R.J. Feiertag and E.I. Organick, *The Multics input/output system*, Proc. of the Symposium on Operating Systems Principles (New York), October 1971, pp. 35–41.

[GVH03] Brian P. Gerkey, Richard T. Vaughan, and Andrew Howard, *The Player/Stage Project: Tools for Multi-Robot and Distributed Sensor Systems*, Proc. of the Intl. Conf. on Advanced Robotics (ICAR) (Coimbra, Portugal), June 2003, pp. 317–323.

[GVS01] Brian P. Gerkey, Richard T. Vaughan, Kasper Støy, Andrew Howard, Gaurav S Sukhtame, and Maja J Matarić, *Most Valuable Player: A Robot Device Server for Distributed Control*, Proc. of the IEEE/RSJ Intl. Conf. on Intelligent Robots and Systems (IROS) (Wailea, Hawaii), October 2001, pp. 1226–1231.

[HHFS05] A. Hassch, N. Hofemann, J. Fritsch, and G. Sagerer, *A multi-modal object attention system for a mobile robot*, Proc. of the IEEE/RSJ Intl. Conf. on Intelligent Robots and Systems (IROS), 2005.

[iSW05] Junzhi Yu inyan Shao, Guangming Xie and Long Wang, *A tracking controller for motion coordination of multiple mobile robots*, Proc. of the IEEE/RSJ Intl. Conf. on Intelligent Robots and Systems (IROS), 2005.

[Jak97] Nick Jakobi, *Evolutionary robotics and the radical envelope of noise hypothesis*, Adaptive Behavior **6** (1997), no. 2, 325–368.

[KOV04] Kurt Konolige, Charlie Ortiz, Regis Vincent, Benoit Morisset, Andrew Agno, Michael Eriksen, Dieter Fox, Benson Limketkai, Jonathan Ko, Benjamin Stewart, and Dirk Schulz, *Centibots: Very large scale distributed robotic teams*, Proc. of the International Symp. on Experimental Robotics (ISER) (Singapore), June 2004.

[MRT03] Michael Montemerlo, Nicholas Roy, and Sebastian Thrun, *Perspectives on standardization in mobile robot programming: The Carnegie Mellon Navigation (CARMEN) toolkit*, Proc. of the IEEE/RSJ Intl. Conf. on Intelligent Robots and Systems (IROS) (Las Vegas, Nevada), October 2003, pp. 2436–2441.

[Net87] Network Working Group, Sun Microsystems, Inc., *RFC 1014 – XDR: External data representation standard*, June 1987.

[Obj02] Object Management Group, Inc., *The Common Object Request Broker: Architecture and Specification, Version 3.0*, July 2002.

[RT74] Dennis M. Ritchie and Ken Thompson, *The UNIX Time-Sharing System*, Communications of the ACM **17** (1974), no. 7, 365–375.

[SGG05] Avi Silberschatz, Peter Baer Galvin, and Greg Gagne, *Operating system concepts*, Seventh ed., J. Wiley & Sons, Inc., New York, 2005.

[SLM98] Douglas C. Schmidt, David L. Levine, and Sumedh Mungee, *The design of the TAO real-time object request broker*, Computer Communications **21** (1998), no. 4, 291–403.

[SW97] Reid Simmons and Gregory Whelan, *Visualization tools for validating software of autonomous spacecraft*, Proc. of the Intl. Symp. on Artificial Intelligence, Robotics, and Automation in Space, July 1997.

[Tan96] Andrew S. Tannenbaum, *Computer networks*, Third ed., Prentice Hall PTR, Upper Saddle River, New Jersey, 1996.

[TC05] E. Topp and H. Christensen, *Tracking for following and passing persons*, Proc. of the IEEE/RSJ Intl. Conf. on Intelligent Robots and Systems (IROS), 2005.

[TFBD00] S. Thrun, D. Fox, W. Burgard, and F. Dellaert, *Robust monte carlo localization for mobile robots*, Artificial Intelligence **128** (2000), no. 1-2, 99–141.

[UB98] Iwan Ulrich and Johann Borenstein, *VFH+: Reliable Obstacle for Fast Mobile Robots*, Proc. of the IEEE Intl. Conf. on Robotics and Automation (ICRA) (Leuven, Belgium), May 1998, pp. 1572–1577.

[Wal99] Jim Waldo, *The Jini Architecture for Network-Centric Computing*, Communications of the ACM **42** (1999), no. 7, 76–82.

[Wil85] Stuart W. Wilson, *Knowledge growth in an artificial animal*, Proc. Int. Conf. Genetic Algorithms and their applications (ICGA85), Pittsburgh PA. (Hillsdale NJ.) (J. J. Grefenstette, ed.), Lawrence Erlbaum Associates, 1985.

[ZV06] Yinan Zhang and Richard Vaughan, *Ganging up: Team-based aggression expands the population/performance envelope in a multi-robot system*, Proc. of the IEEE Intl. Conf. on Robotics and Automation (ICRA), 2006.

An Integration Framework for Developing Interactive Robots

Jannik Fritsch and Sebastian Wrede

Applied Computer Science, Faculty of Technology, Bielefeld University, Germany
{jannik,swrede}@techfak.uni-bielefeld.de

1 Challenges in Interactive Robotics Research

In recent years there is an increasing interest in building personal robots that are capable of a human-like interaction. In addition to multi-modal interaction skills, such a robot must also be able to adapt itself to unknown environments and, therefore, it has to be capable of knowledge acquisition in a lifelong learning process. Moreover, as humans are around, reactive control of the robot's hardware is important, too. Consequently, researchers aiming to realize a personal robot have to integrate a variety of features. Due to the very different nature of the necessary capabilities, interactive robotics research is thus a truly interdisciplinary challenge.

A natural design decision is to take a modular approach to build such a complex system [Ros95]. This allows the different researchers to focus on their respective task and makes integration easier. Because integration is an iterative process in large-scale research systems, it must be possible to incorporate new modules and functionalities in an interactive robot as they become available. Such a continuous extension of the robot's functionality not only results in new or extended data structures provided by the added components but usually also requires modifications of the control flow in the integrated system. This poses questions of how data and control flow can be expressed in a way that the effort and the complexity for continuous integration remains controllable during the integration process [CS00].

As described in Chapter *Trends in Robotic Software Frameworks*, these requirements make the framework-based approach well suited to the development of software applications for Interactive Robots. In this context, a fundamental task for integration frameworks is the ability to distribute modules across different computing nodes in order to achieve the reactivity needed for human-machine-interaction [KAU04]. This applies especially to large-scale systems like personal robots. However, most researchers are no middleware experts, prohibiting the native use of, e.g., CORBA-based solutions. Consequently, the communication framework needs to provide a restricted but suf-

D. Brugali (Ed.): Software Eng. for Experimental Robotics. STAR 30, pp. 291–305, 2007.

ficient set of functionality that enables interdisciplinary researchers to easily integrate their components into a distributed robot system.

Knowledge acquisition and the ability to adapt to unknown situations imposes two additional functional requirements that need to be addressed. An application may need some kind of memory so data management services must be provided by the integration framework. Since adaptation and processes like e.g. attention control induce dynamics into the system, dynamic (re-)configuration of components running in the integrated system is essential, too.

In order to support the envisioned incremental development of a personal robot, not only the functional requirements *modularity*, *communication*, *module coordination*, as well as *knowledge representation and acquisition* and *dynamic (re-)configuration* must be supported by the system infrastructure. Additionally, several non-functional requirements play an important role that are discussed in the following.

Taking into account that the people carrying out research on interactive robots are usually concentrating on single topics and not the overall integration, the developed framework must be very easy to use in order to gain a wide acceptance. The approach we will outline in this chapter tries to tackle this by focusing on *simplicity* and exploiting *standards compliance*.

Another important non-functional requirement we try to address is to provide a framework that enables *rapid prototyping*. Consequently, iterative development should not only be supported for single modules but also for the integrated system. Erroneous directions in system evolution can more easily be identified if integration is performed on a regular basis starting at an early stage even when components are still missing and need to be simulated. For large-scale systems, software engineering research has shown that decoupling of modules is very important. Thus, the framework should support *low coupling* of modules. This facilitates not only independent operation of components but also minimal impact of local changes on the whole system. With a framework that adheres to low coupling, *debugging and evaluation* of a running system architecture can be supported more easily.

The contribution starts with an introduction into the concepts of the XCF framework we developed along these requirements to enable high-level integration and coordination of interactive robots. The process of robot development with this approach and the lessons learned thereby are described in Sect.s 3 and 4, respectively. A conclusion on the presented approach is drawn in Sect. 5.

2 The XCF SDK

Taking into account the requirements outlined in the previous section, we developed a software development kit termed XCF consisting of a set of object-oriented class libraries and the required framework tools to develop, debug and run a distributed robotic system. The concrete implementation of the XCF concepts has been carried out in the context of interactive robotics [Cog05] and cognitive vision [Vam05] research. The complete XCF SDK consists of

- a library named XMLTIO that supports users with an API based on XPath [CD99] for simple XML processing and native data type encoding,
- the XML enabled Communication Framework (XCF) [WFBS04] itself that allows to distribute components over several computing nodes,
- the Active Memory XML server [WHBS04] for event-based coordination and data management, as well as
- the Active Memory Petri-Net Engine which allows for a declarative specification of control flow and easy development of coordination components.

The complete software is GPL-licensed and available for download at Sourceforge [Wr05]. Currently, it is available for Linux only but as all libraries used are available for other operating systems as well, porting to other platforms is possible at moderate costs. Native language bindings are written in C++, even so additional bindings are available for Java and basic XCF functionality is provided in Matlab facilitating rapid prototyping. In the following we will highlight important concepts and the resulting features of the framework that allow for efficient integration of interactive robots developed interdisciplinary.

2.1 Encoding information in XML

Since it is well-known and easy to learn as well as flexible and suited for abstract concept descriptions, the XML language was chosen as a basis to describe content transmitted, stored, and processed by the various robot modules.

Exemplary XML-RPC encoding

```
<member>
  <name>CENTER</name>
  <value>
    <struct>
      <member>
        <name>y</name>
        <value><int>44</int></value>
      </member>
      <member>
        <name>x</name>
        <value><int>32</int></value>
      </member>
    </struct>
  </value>
</member>
```

Information-oriented XML encoding

```
<OBJECT>
  <REGION>
    <RECT x="13" y="27"
               w="80" h="80"/>
  </REGION>
  <CENTER x="32" y="44"/>
  <CLASS>CUP</CLASS>
</OBJECT>
```

Fig. 1. Contrasting XML-RPC with document/literal information encoding.

When XML is used as data exchange protocol as it is done in XML-RPC-based solutions [KAU04] it usually results in a lot of overhead through text-based representation of binary data and parameter encoding rules. Looking at the object recognition example shown in Fig. 1 you see two alternative encodings of a "`CENTER`" element. The example shows an information-oriented encoding of the center item (embedded in an object recognition result) on the right as well as a serialization of the same item in XML-RPC encoding on the left. It is not only this obvious overhead induced by a naive serialization of data to XML that is not desirable, but even worse is the loss of comprehensibility at all processing levels that is imposed by this type of encoding.[1]

Following the concept to encode information and not just data, we developed XML vocabularies for information typically exchanged in interactive robots (e.g., objects, states, events etc.). The contained semantical information is then directly selected and accessed utilizing the XMLTIO library. Declarative, name-based selection of XML nodes with XPath expressions helps in building data types that are extensible and in building systems that will not break as soon as modifications of the exchanged data structures occur. Even though this sounds trivial, carefully designed XML information structures are an important key to facilitate the overall integration progress of large scale systems developed by multiple developers.

An example that explains the benefit of the name-based selection of XML information items for system integration is depicted in Fig. 2. Four modules processing partially equivalent XML data structures are shown. This allows a component (e.g. "Hardware Control" for adjusting the robot base and the pan-tilt camera) to process information from different other system modules. Only the part in the XML document that contains information about center points ("`<CENTER>`") has to be present in an exchanged data item. Starting with a simple partial path specialization to access the context node, e.g. in this example as simple as "`/*/CENTER`", the extraction of contained information is easily possible although the context node itself might appear in varying places of different XML structures. Thus, this path expression works on both documents shown in Fig. 2. As long as no necessary information is removed, this strategy yields loose coupling and facilitates interoperability between separately developed modules.

The functionality of XMLTIO is made up of a combination of XML parsing functionality (Xerces-C) and XPath processing functionality (Pathan). Additionally, XMLTIO helps in encoding and decoding native C++ data structures into the designated XML documents and translating them back to C++ objects by exploiting template-based type lookup.

---

[1] The idea to directly encode information and not data in XML has recently gained more interest even in the Web Services community where SOAP-RPC encoding (which is somewhat similar to XML-RPC encoding) is being more and more replaced by the document/literal encoding we are using.

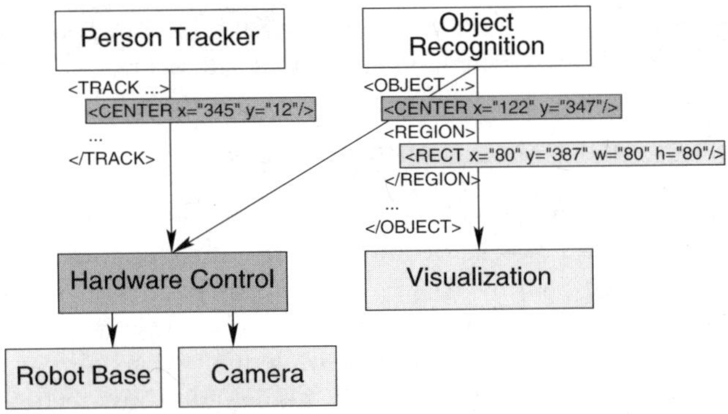

Fig. 2. Accessing common information at arbitrary locations with XPath.

2.2 Specification and Validation of Information

Meta-information, e.g., about data types, is kept separate in corresponding XML schema files and is not encoded in the instance documents themselves. Specifying data types with XML schema has several advantages in contrast to traditional programming language constructs.

First of all, the data types are independent from specific programming languages. Even so, tools for using them are available on almost every platform. Furthermore, XML schemas are able to specify content models and ranges of allowed values in great detail. Providing fine grained sets of semantically grouped declarations in separate schemas with associated XML namespaces makes them reusable throughout different systems. Complex schemas for individual modules can then easily be composed out of these basic type libraries, only adding specific complex types. If taken into account, extensibility of data types is possible with schema evolution. Even complex grammars for components capable of interpreting and validating XML documents originating from different robot modules are easy to compose and well understandable with a sophisticated schema hierarchy.

Information-oriented encoding of XML messages and the use of XML schemas for validation of the exchanged information are both very useful for system integration in interdisciplinary research projects. The focus on simple XML messages to describe exchanged information helps during project inception as almost every developer will be able to contribute to the discussion about the data flow in the system. Later on, XML grammars like XML schema allow for a rigid specification and validation of the datatypes a project consortium has agreed upon. For example, schemas have been defined in the European project COGNIRON [Cog05] to ease the integration of the partner's contributions in the realized robot prototypes. To ease this development task, XCF provides a simple command line tool that encapsulates the schema checking algorithms used in the framework for testing XML messages against the developed schemas without having to start the overall system.

2.3 XML-Enabled Module Communication

For the task of exchanging data in distributed systems a large number of different communication frameworks or middleware solutions have been proposed, ranging from message passing to RPC-style frameworks to implementations of the CORBA standard like ACE/TAO, or novel middleware frameworks like ICE. Note that, while all of the CORBA implementations aim at fulfilling the CORBA standard, most of them differ in many small but technically important aspects. Recently, much interest was awarded to XML-based communication where XML is used foremost for serialization purposes like in XML-RPC, XMPP, or SOAP (see also Sect. 2.1). To build up our integration framework we chose the ICE communications engine [Zer05] as it has a much smaller footprint than many comparable frameworks, allows high-speed data-transfer while being available as open source for different operating systems.

On top of ICE we developed a lightweight communication framework that provides only a limited but useful set of functionality that is necessary for building distributed system architectures. This XCF core library is able to efficiently exchange XML and referenced binary data (e.g., images) structures. The referenced binary data is transmitted natively in a composite object together with the corresponding XML message similar to the recently proposed XML-binary Optimized Packaging (XOP) recommendation [Wor05]. This combines the flexibility of information encoding in plain old XML documents (as explained in Sect. 2.1) with the efficiency of low-level communication semantics for large amounts of binary data.

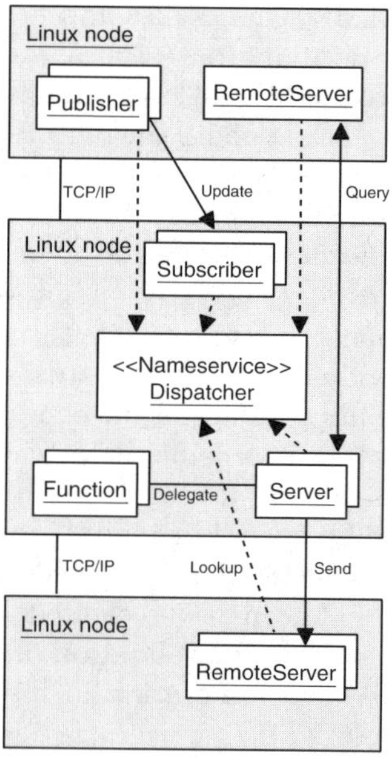

Fig. 3. Core functionality of XCF in an exemplary distributed system architecture

This XCF core features a pattern-based design and offers communication semantics like publisher/subscriber and (a)synchronous remote procedure calls/method invocations (see Fig. 3). All XCF objects and exposed methods can be dynamically registered at runtime, thus no meta-compiler is necessary. In the different communication patterns schema checking can be activated on sending and also on receiving XML data, allowing the system integrator to decide where to verify the correctness of exchanged data. As schema checking can result in a much larger computational load, it is usually switched off in normal operation after the system integration is finished.

Concurrent access of multiple clients on XCF server processes is transparently handled by message queues and exploitation of the dynamic worker thread pools provided by ICE. To achieve location transparency, we developed a name service daemon where processes register the XCF services they provide to other robot modules. This central framework component is shown in the center of Fig. 3. Thus, only the name of a component needs to be known in contrast to direct addressing using the host name of the machine it runs on. Note, that the interaction with the nameserver follows the client-dispatcher-server [BMRS96] pattern which means that after the initial name lookup, all data communication is carried out solely between the involved components. Additionally, we provide an abstract component class that exposes a generic interface for starting and stopping algorithmic processing, shutting down the whole XCF process as well as for reconfiguration and querying components during runtime of a system.

Last but not least, a central logging mechanism is supported through an XCFappenders for log4cxx,log4cpp and log4j which enable component developers to provide their log messages over the network to a central logger for debugging on a system integration level.

2.4 Event-Driven Integration and Coordination

The development of interactive applications and especially robots with sophisticated interaction capabilites imposes several constraints on a software framework for system integration and coordination. Two of the most important requirements are firstly that the system should interact with soft real-time performance and secondly the ability of the system to track the interaction context in terms of perceived episodes, events, and scenes. While the previous sections covered mainly the information representation and process coupling aspects that facilitate the building of distributed system architectures in order to ensure the reactivity needed for interaction, we will now focus on the Active Memory XML Server that forms the basis for coordination and shared data management in our integration approach.

The Active Memory concept and implementation [WWHB05] was integrated into the XCF SDK for use in our robotic framework. Since we focus on high flexibility, relational databases as backend to integrate information are infeasible. Instead, the native Berkeley DB XML database [DBX03] provides the core component of this part of the architecture shown in Fig. 4. DB XML is packaged as an embedded C++ library and provides support for transactions and multi-threaded environments. In contrast to classical database servers it offers only core features and is not a complete server application. Its therefore small footprint allows for an efficient implementation of the XML Server. Through the use of XPath, developers of robot modules can easily specify queries in a standardized and declarative manner. Additionally, runtime reindexing is supported for fast access to database contents. DB XML also allows for mixed XML and traditional data storage which is useful for binary large

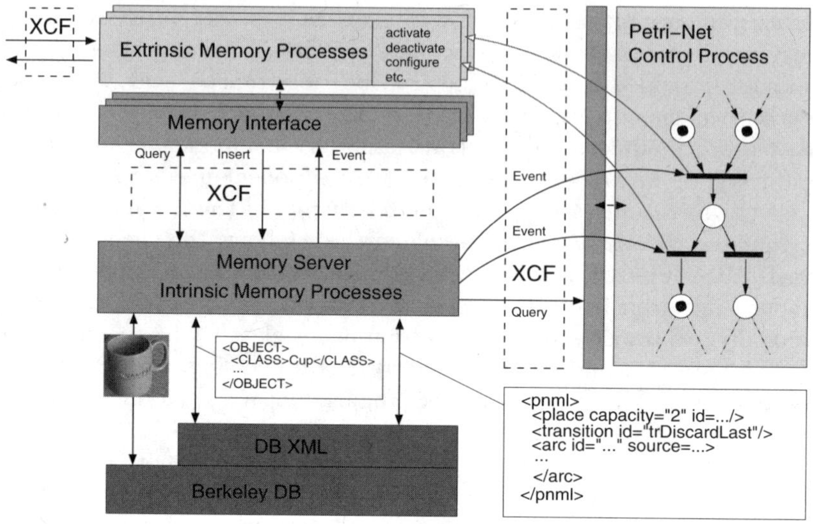

Fig. 4. Architecture for Event-Driven Integration and Coordination

objects like cropped image patches. Utilizing the reference management system we developed, binary data can be linked from and to any XML document in the database. Based on the DB XML API, the client/server architecture shown in Fig. 4 was implemented that encapsulates complexity and connects the basic insert, select and update methods to external XCF processes.

On top of this data integration approach we added an event concept which is the basis for system coordination. The general idea of the event system built on top of the native XML backend is to use XPath expressions as subscription language. For system integrators this yields a very generic and powerful registration method to couple modules of a robot architecture to specific parts of the information flow in the system. For that reason the policy of information modeling described in Sect. 2.1 is highly relevant. With the document/literal encoding of information every developer can easily express his or her interest in specific parts of system information by means of an XPath-based event subscription. To express the semantic action that is associated with a data item processed in the memory server, usually an event specification is enhanced by the specific action that is executed within the memory server on that data item, e.g. insert, query, update or delete. Whenever a basic method is called, the server instance notifies all processes that have been registered for that kind of action and the type of involved memory element. The event message includes the event source which is the matching XML document. If the notification fails, the event notification is discarded and the subscription gets removed.

Coordination is thus implicitly data-driven and *not* bound to explicit links between a fixed set of components present in the system. A simple example

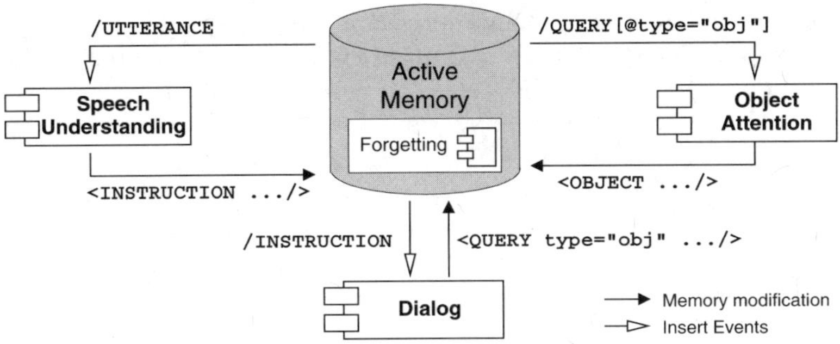

Fig. 5. An exemplary use case for event-driven integration.

of an application of this concept is shown in Fig. 5 where three components of a robot architecture are coordinated by event listeners registered on the corresponding XPaths and memory actions.

Additionally, the Active Memory XML Server features a runtime environment based on python where internal processes[2] like forgetting or information fusion can be realized that need to work on huge parts of the information mediated through an active memory instance.

To provide more complex coordination methods for multiple components operating concurrently in an interactive robot, we developed an application of petri-nets for system integration purposes. Petri-nets, in general [Pet81], extend classic state machines by the ability to represent concurrency. Thus, they are well suited for modeling structure and behavior of parallel distributed systems. The current marking of a petri-net corresponds to a specific system configuration. Dynamic changes in system behavior are controlled by activated transitions. In our approach we extended the classical petri-net concept by so-called *Active Memory Guards* (AMG) which utilize event listeners to connect the model to instances of Active Memory XML Servers. An input arc can only be satisfied after, firstly, the attached place contains a sufficient amount of tokens thus enabling the memory event listener, and secondly, an occurrence of the specified memory event in an active memory instance. If both steps are completed, the input arc as a whole is satisfied and the attached transition is enabled. This concept couples the execution of the specified high-level petri-net model to the overall state. AMGs are context dependent in a sense that they are activated as soon as the place condition of its arc is satisfied.

The realization of the active memory petri-net engine allows a formal and declarative specification of net structure and active memory guards as well as the attached actions in an application of the PNML document format [WK03]. Thus, petri-net coordination models can be extended by new places and tran-

[2] The so-called Intrinsic Memory Processes are explained in greater detail in [WHBS04].

sitions online. As soon as a PNML model is updated in an active memory server, the instantiated petri-net execution engine is reconfigured.

Fig. 4 shows on a conceptual level how a petri-net control process is coupled to the overall system. As soon as a transition fires, a sequence of actions is executed. The set of possible actions which can be attached to a transition can be any number of XCF calls, basic actions on a memory server or local calls to methods of classes that are derived from a basic action interface. Instances of those actions are specified in the PNML model and can be configured with XML parameters.

2.5 Introspection and Compacting

Introspection in general is a very helpful feature for software integration because it helps a lot in debugging and monitoring a running distributed system. In XCF, the basic communication mechanisms as well as the Active Memory XML Server are reflexive in that all kinds of data and process configurations are represented in XML data that can be analyzed and changed on-the-fly. Furthermore, the type of actions and their context can be inspected during runtime. The framework architecture allows to intercept all messages within a running system or to intercept individual communication between specific components. In XCF, the features for introspection are realized with an implementation of the interceptor pattern [SSRB00].

An exemplary application that benefits from the reflection XML itself provides, is the so-called *StreamSimulator* which is able to simulate XCF Publisher components by replaying XML data from previously recorded files or from the contents of an Active Memory XML Server. In order to *peek* inside an active memory server, a generic visualization toolkit, the Memory Content Introspector (MCI), has been developed (see Fig. 6). The MCI is realized as a plugin-framework written in Java where each of the plugins realizes a specific visualization method.

A more complex application of the introspective features is the generic compacting filter for XML streams [LWW05] which we developed to reduce waste of system resources. An example for this is object recognition, where usually parts of the transmitted data may be false when the robot looks around, other data may be redundant, e.g. when the robot gazes at a static scene. In this situation object recognition will work well but (at least the implementation we had at hand) sends nearly identical object recognition results at a high frequency. Both, redundant and obviously false recognition results consume resources of the system in terms of network bandwidth and overall system performance. Therefore, a generic framework component based on reflection was developed that is able to compact processed information and detect relevant changes in a stream of XML information. Thus, only the relevant data is forwarded by the compacting filter to subsequent processing modules. This allows us to integrate existing modules even if their temporal

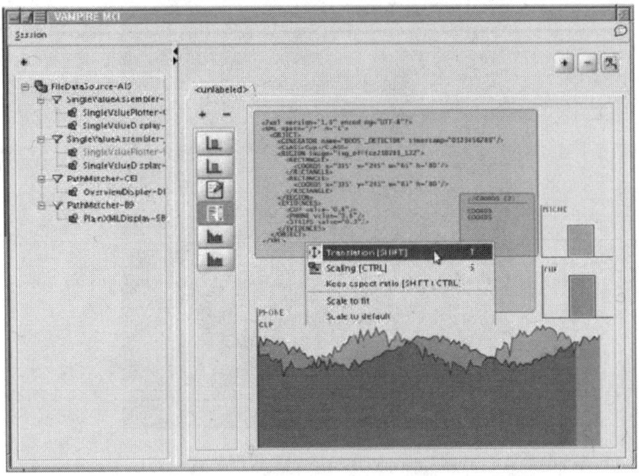

Fig. 6. Screenshot of the Memory Content Introspector.

behavior does not fit the input demands of other modules in the integrated system processing their data.

In the following we show how these foundations are used to realize the integration of different modules for human-robot interaction and which lessons we learned so far by applying our approach during the development of a complex interactive robot.

3 Developing an Interactive Robot

Through integrating a large number of components using the tools presented above we realized a personal robot that has already quite impressive interaction capabilities [LHW05]. The mobile robot BIRON depicted in Fig. 7 is able to pay attention to different persons and engage in a one-to-one interaction with one user if this user greets the robot by saying "Hello Robot". From this point on, the robot focuses on this communication partner and engages in a dialog with him. The user can control the robot's movement behavior by giving commands. For example, the command "Follow me" results in the robot following the human around. More importantly, the user can teach new objects to the robot by pointing at them while giving additional information. For example, giving an instruction like "This <gesture> is my blue cup" enables the robot to focus its attention on the referenced object and acquire an image of it for later recognition. Components for localization and navigation enable the user to teach the robot places and locations as well as to enable the robot to autonomously go to verbally specified locations (e.g., 'Go to the kitchen').

The sketched functionality is achieved by integrating modules for robot control, person tracking, person attention, speech recognition, speech under-

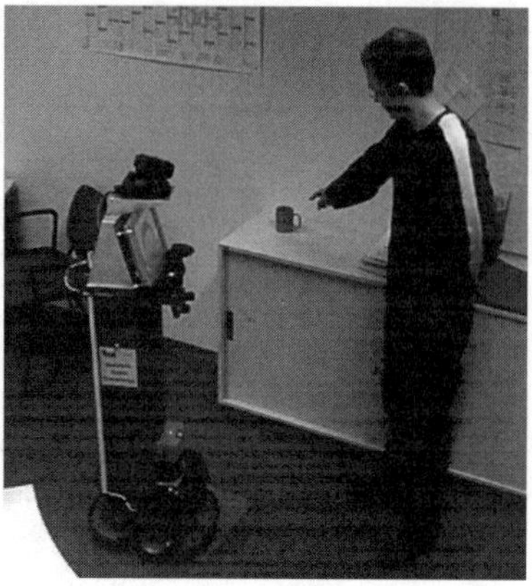

Fig. 7. A user teaching an object to the mobile robot BIRON.

standing, dialog, gesture recognition, object recognition, object attention and an execution supervisor [FKH05]. The module integration task directly benefits from the fact that XCF is very easy to use and the applied concepts are highly standards-based, giving interdisciplinary researchers a quick start. Furthermore, a central system view of the active robot modules and the exchanged data utilizing the active system introspection supports debugging and evaluation of the running system. Loose coupling of modules and the declarative style of accessing system data paid off in ease of modification and the ability to let the system architecture evolve over time as we extended it with new modules and data types. With several researchers in the European project COGNIRON [Cog05] – The Cognitive Robot Companion – contributing to this system and extending the functionality of their respective modules frequently, the use of XML schemas for data type verification proved to be useful for identifying modules providing erroneous data. The active memory enables sharing of knowledge acquired during the robot's interaction with a human instructor. Through event subscription at run-time, different types of newly acquired knowledge result in different modules being notified and, consequently, only processing relevant to the new data is performed.

With respect to the performance of the overall system, the distribution and coordination of modules has resulted in a high reactivity allowing for more natural human-robot interaction. After the end of a user instruction the system produces a response within 400-700ms [FKH05]. Most of this delay is caused by the speech recognition process (200-500ms), whereas the processing by all other modules and the time needed for communication is only around 200 ms. Transferring a message from one module to another module takes only a few milliseconds and shows that the advantage of increased processing power in a distributed system setup for a personal robot exceeds the cost for module communication.

It should be noted here, however, that the proposed communication framework is not intended for integration of low-level robot control functionalities like, e.g., obstacle avoidance. For this task other frameworks described in this Chapter like, e.g., Player or Marie are possibly better suited. While these are especially focused on the needs of robotics researchers, the XCF SDK is especially designed for supporting the development and integration of high-level functionalities like speech processing and gesture recognition.

4 Lessons Learned

Using XCF most of the data flow between all modules in the architecture is event-based and every message is coded in XML. As both, data structures and communication channels, are *name-based* the different modules can easily be replaced by others. It also eases modification concerning the data structures used for communication, as XML allows database-like exchange of information. Experience in different European research projects showed that the use of XML helped in defining data types which are suitable for every involved project partner. XML objects as native data types allow for a flexible schema evolution and separation of concerns within a system architecture. Additionally, generic software modules can much easier be realized within this concept.

In the practical use of schemas during prototyping, exceptions were encountered frequently. This was often due to misunderstandings between the involved developers about data type definitions, but occasionally also serious implementation errors were discovered. For example, a module that provided laser range data threw an exception immediately after the start stating that the XML laser data was not valid. Without schema checking being active, the module started and seemed to operate correctly, the transmitted XML data looked correct, and even the communication worked fine. Only after a more detailed analysis it was found that *at startup* the very first data packet obtained from the laser range finder was corrupt and this caused the exception.

Concerning the goals mentioned in the beginning of this Chapter, we think that *usability* of the XCF toolkit is high. This emerges from two design decisions: On the one hand the component developer only has to deal with a small number of classes in a simple API and on the other hand many technologies used are standards-based. Using XML technologies throughout the whole framework, we meet the *flexibility* requirement. This results in changeability, easier adaptation and integration of modules. Additionally, openness of distributed memory instances allows developers to retrieve information in a standard fashion using declarative queries.

Low coupling is reached through combination of active memory instances and XCF. The memory instances itself decouple the memory processes and serve as an information mediator while the XCF framework provides location and access transparency for components. This leads to easy exchange of components and high robustness against component failure. Both techniques enable *rapid prototyping*.

Debugging and evaluation is supported by simulation of components using XCF for replaying recorded memory data with a so called module simulator. When the memory content has time information associated whole architectural layers can be replaced by simulation tools as if they were online available. Development and evaluation of different algorithms or system configurations on comparable data has been much easier with this feature. For the task of system integration, parts of the system could be tested by simulating the communication with other components. The introspection facilities turned out to be highly valuable for monitoring the overall system state and the correct processing of data at the system integration level. The central logging of information made the task of system integration much easier as it enabled the monitoring of process chains typical for complex interactive robots.

Finally it should be mentioned that all the lessons learned could only be learned by actually having succeeded in motivating different researchers to combine their research systems for isolated interaction aspects in a single integrated robot prototype. We believe that XCF has played a crucial role in this outcome, as it has made it comparatively easy for the involved people to extend their modules with communication facilities.

5 Conclusion

While this Chapter outlined the concepts of the XCF SDK, the interested reader is referred to the webpage [Wr05] for examples and implementation level documentation. Concluding our experiences, the successful realization of a personal robot prototype clearly demonstrates that the tools described in Sect. 2 are usable by interdisciplinary researchers. The proposed framework has shown to adequately support an evolutionary development process and the tools are well suited to be used in such large-scale research projects. Nevertheless, integration is still a challenging issue as it is often underrepresented in research projects but is unavoidable especially if multi-modal interactive intelligent systems are to be developed. The presented data- and event-driven integration method enables the re-use of modules in different application scenarios and we, therefore, consider this coordination scheme to pave the way for incremental development of more complex robot prototypes in the future.

References

[BMRS96] F. Buschmann, R. Meunier, H. Rohnert, and P. Sommerlad, *Pattern-oriented software architecture*, vol. 1: A System of Patterns, John Wiley & Sons Ltd., 1996.

[CD99] J. Clark and S. DeRose, *XML Path Language*, Tech. Report REC-xpath-19991116, W3C, 1999.

[Cog05] Cogniron Consortium, *COGNIRON – The Cognitive Robot Companion*, 2005, http://www.cogniron.org.

[CS00] È. Coste-Manière and R. G. Simmons, *Architecture, the Backbone of Robotic Systems*, Proc. IEEE Int. Conf. on Robotics and Automation (San Franciso, CA), vol. 1, 2000, pp. 67–72.

[DBX03] *Berkely DB XML, Sleepycat Software*, 2003, http://www.sleepycat.com/products/xml.shtml.

[FKH05] J. Fritsch, M. Kleinehagenbrock, A. Haasch, S. Wrede, and G. Sagerer, *A flexible infrastructure for the development of a robot companion with extensible HRI-capabilities*, Proc. IEEE Int. Conf. on Robotics and Automation (Barcelona, Spain), April 2005, pp. 3419–3425.

[KAU04] P. Kiatisevi, V. Ampornaramveth, and H. Ueno, *A Distributed Architecture for Knowledge-Based Interactive Robots*, Proc. Int. Conf. on Information Technology for Application (ICITA) (Harbin, China), 2004, pp. 256–261.

[LHW05] S. Li, A. Haasch, B. Wrede, J. Fritsch, and G. Sagerer, *Human-style interaction with a robot for cooperative learning of scene objects*, Proc. Int. Conf. on Multimodal Interfaces (Trento, Italy), ACM Press, 2005, pp. 151–158.

[LWW05] I. Lütkebohle, S. Wrede, and S. Wachsmuth, *Unsupervised Filtering of XML Streams for System Integration*, International Workshop on Pattern Recognition in Information Systems, 2005, Poster.

[Pet81] J. L. Peterson, *Petri net theory and the modeling of systems*, Prentice Hall, Inc., Englewood Cliffs, Massachusetts,, 1981.

[Ros95] J. K. Rosenblatt, *DAMN: A Distributed Architecture for Mobile Navigation*, Proc. AAAI Spring Symp. on Lessons Learned from Implemented Software Architectures for Physical Agents (Stanford, CA), AAAI/MIT Press, 1995, pp. 167–178.

[Wr05] S. Wrede, *The XML enabled Communication Framework SDK*, 2005, Software and documentation available at http://xcf.sf.net/.

[SSRB00] D. C. Schmidt, M. Stal, H. Rohnert, and F. Buschmann, *Pattern-oriented software architecture*, vol. 2: Patterns for Concurrent and Networked Objects, John Wiley & Sons Ltd., 2000.

[Vam05] Vampire Consortium, *VAMPIRE – Visual Active Memory Processes for Interactive Retrieval*, 2005, http://www.vampire-project.org.

[WFBS04] S. Wrede, J. Fritsch, C. Bauckhage, and G. Sagerer, *An XML Based Framework for Cognitive Vision Architectures*, Proc. Int. Conf. on Pattern Recognition, vol. 1, 2004, pp. 757–760.

[WHBS04] S. Wrede, M. Hanheide, C. Bauckhage, and G. Sagerer, *An active memory as a model for information fusion*, Proc. 7th Int. Conf. on Information Fusion, 2004, pp. 198–205.

[WK03] M. Weber and E. Kindler, *The petri net markup language*, Petri Net Technology for Communication Based Systems., LNCS 2472, Springer-Verlag, 2003.

[Wor05] World Wide Web Consortium, *XML-binary Optimized Packaging, W3C Recommendation 25 January 2005*, 2005, http://www.w3.org/TR/2005/REC-xop10-20050125/.

[WWHB05] S. Wachsmuth, S. Wrede, M. Hanheide, and C. Bauckhage, *An Active Memory Model for Cognitive Computer Vision Systems*, Künstliche Intelligenz **19** (2005), no. 2, 25–31.

[Zer05] ZeroC Inc., *The Internet Communications Engine, ZeroC Inc.*, 2005, http://www.zeroc.com/ice.html.

Increasing Decoupling in the Robotics4.NET Framework

Antonio Cisternino[1], Diego Colombo[2], Vincenzo Ambriola[1], and Marco Combetto[3]

[1] University of Pisa, Computer Science Department, Pisa, Italy {ambriola, cisterni}@di.unipi.it
[2] IMT Alti Studi Lucca, Lucca, Italy diego.colombo@imtlucca.it
[3] Microsoft Research ltd., Cambridge, United Kingdom, marcomb@microsoft.com

1 Introduction

The advent of social robots increases significantly the number and the kind of robotics systems to be controlled. Since CSRS software depends on the particular hardware architecture of a robot, software reuse becomes a significant issue. Reuse is usually achieved by packaging software in modules, by abstracting a well defined set of functionalities.

In the computer domain the variance of hardware has been addressed by abstracting common functionalities with drivers, software modules responsible for interacting with the hardware hiding the details to the rest of the system. In control software for robots [4] this approach presents several problems: first of all, it is very hard to provide an abstract description of robot functionalities independent from its structure. Moreover, the complexity of drivers increases because most of the activity performed by a CSRS is hardware control, located at the lower levels of the system architecture. Last, but not least, driver architectures focus on the notion of interrupt: when an interrupt notification gets lost the whole system may fail because of a lack of robustness in the communication schema.

Recently there have been attempts to define programming frameworks for developing robotics control software like those presented in this book, as well as several others [Bru01] [LBDS03]. Worth of notice is the new Robotics Studio, a new framework announced by Microsoft at the end of June 2006 [RS06] and based on .NET.

In this chapter we introduce the Robotics4.NET framework, its design, and how it contributes to achieve a better organization of control software because of the structure it imposes to CSRS. It is important to keep in mind

[4] In the rest of the chapter we will refer to this kind of systems as CSRS (Control Software for Robotics Systems).

D. Brugali (Ed.): Software Eng. for Experimental Robotics. STAR 30, pp. 307–324, 2007.

that an overall design principle followed throughout the development of this framework is that the software is subdued to a well defined methodology and model. The framework is thus flexible enough within the constraints imposed by the model, avoiding helpful hacks that may break it; as a consequence of this design choice a programmer interested in its use is guided by the framework structure during the development, without having too many options for doing the same thing. We look positively at this design decision because an exceeding freedom in a framework often means that it does not capture the fundamental elements of an application domain, and the programmer will get little benefit by using it since the knowledge of the domain required to use it is significant. Moreover, the software developed under the proposed model (inspired by biology and not by software engineering) has interesting properties about possible reuse of software modules in different systems, even if based on completely different hardware.

At the end of the chapter we also briefly introduce the Microsoft Robotics Studio, and we compare it to our framework, looking for commonalties and differences, since both frameworks target the .NET execution environment.

2 Robotics4.NET

There are many different perspectives to look at the Robotics4.NET framework:

- the architecture it is inspired from a biological model in the hope that in the end the system will have similar properties;
- it provides a concurrent programming model based on message passing;
- it is a multi-agent system;
- it uses XML-based messages to allow heterogeneous architectures, allowing even micro-controlled systems to participate directly in the communication schema;
- it relies on customizable reflection support in order to declare communications aspects in the system;

All these aspects partecipate into defining a methodology that drives a programmer in a set of well defined steps in the software design process. In this section we discuss these perspectives, and finally we depict the overall methodology supporting framework's users in the development process.

2.1 Body Is First Class

The model behind the Robotics4.NET framework revolves around the notion of *body*: it encourages at organizing the software by defining a software counterpart of the physical body responsible to mediate the communication between the "brain" (i.e. the software responsible for control) and the body.

The choice of relying on the body is rather peculiar, if compared with other frameworks that focus on a separation of software in terms of components and a more classical functional decomposition of the problem like MARIE discussed in this book and [Bru01]. Nevertheless this choice has been driven by recent studies on the role of the body in the cognitive process of human beings [Hol04].

In the past two decades the notion of body has become relevant in the research on AI. Brooks [Bro85] [Bro91] [BS93] was one of the first promoters of embodied intelligence, with his hypothesis on complex behavior induced by a complex world. More recently neurophysiologists investigated the relation between the mind and the body and the environment [Dam94] [Dam99] [Dam03] and found that the body is tightly coupled to emotions and rational behavior. Nowadays embodied artificial intelligence [IPSK03] is a multidisciplinary research field whose aim is to study intelligence from a different perspective than the computational approach of traditional AI.

The body plays two fundamental roles in the human architecture:

- It is a controlled environment for the brain which perceives the world through it
- It provides a communication infrastructure to the brain such that a continuous stream of information about its state flows towards the brain

Organs are responsible for processing information acquired through senses and send the result of this activity to the brain for further processing (though there is also an information stream in the opposite direction [MR81]). Thus data coming from sensors is preprocessed before reaching the sensorial cortex.

The body seems to play also an important role in the self-consciousness of humans: as noted in [Dam99] and [Chu02] consciousness depends on the ability of the brain of distinguish the inner world from the outer world representations. This ability seems to be closely related to the continuous communication between the brain the body; besides, computers tend to prefer the notion of interrupt.

2.2 Execution Model

CSRS software based on the Robotics4.NET [CCEP05] framework has the following elements (see Figure 1):

- the brain: a software module responsible for the cognitive reasoning of the robot;
- the bodymap: a sort of black-board used to allow the brain to communicate with the rest of the system;
- a set of roblets: a roblet is a software agent communicating with the brain through the bodymap.

The brain and the roblets execute in parallel and communicate using messages. Roblets write messages into the bodymap; these messages are read by

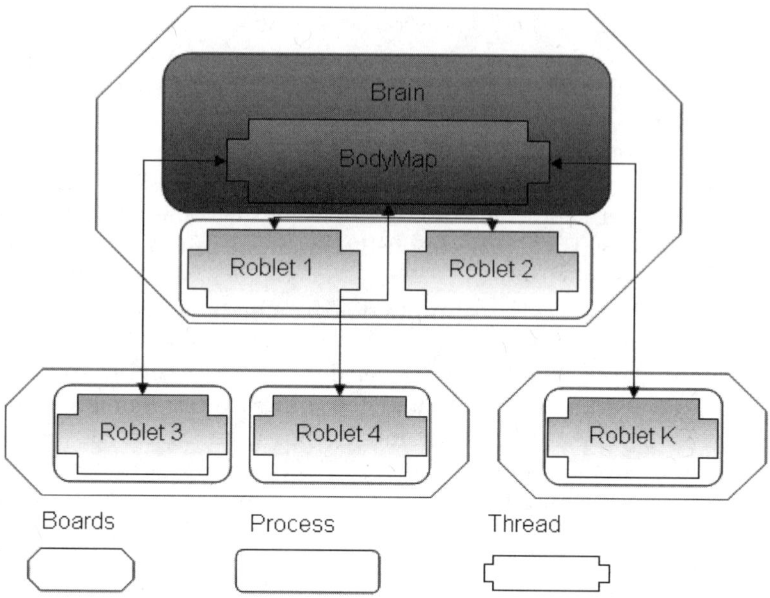

Fig. 1. The Robotics4.NET framework architecture.

the brain that in turn can send messages to them. Although the brain usually mediates communication among roblets, sometimes reactive behaviors require a direct communication between roblets. For this reason it is allowed a declaration of friendship between a pair of roblets that can communicate directly. This relation is asymmetric: a roblet interested in communicating with another one should declare friendship, though the destination of the communication is unaware of this relation.

Roblet execution is temporized: during the initialization phase, a roblet declares the frequency at which its behavior should be executed. A roblet gets notification of incoming messages from bodymap or roblet in friendship relation[5]. The brain has a more traditional organization and is responsible for handling its own control flow; for hosting and controlling the bodymap. The bodymap is also responsible for notifying arrival of incoming messages.

Systems based on the driver model optimize communications between drivers and the rest of the system: interrupts signal when something in the peripheral controlled by a driver changes. However, this communication schema is not robust: if for any reason an interrupt gets lost the entire system may fail. Think for instance to what happens if a software module fails in communicating that some end-run has been reached and a motor should be stopped.

[5] From a roblet standpoint the source of a message, brain or another roblet, is not relevant. However, it is possible to establish the source of the message.

In Robotics4.NET we are interested in robustness; therefore messages are sent asynchronously and the sender does not get any notification of message delivery. In this way, when a module fails the rest of the system continues to run; it is also possible to recover from the failure of a module simply by restarting it. Lost messages are not an issue in this context: roblets continuously communicates with the brain, thus if a message gets lost, there will be a next one bearing the information needed. In fact, roblets communicate continuously with the brain, whether there are changes or not, by pushing the status, in the very similar way as body organs always report to the brain; thus information exchanged is redundant ensuring that the loss of any message is not critical to the system.

The adoption of a message-oriented communication schema implies that there is no state of the connection to be preserved, and there is no acknowledge for received messages. Therefore, roblets may send messages without the bodymap or vice-versa. If a component goes down it can be restarted without affecting the rest of the systems.

The Robotics4.NET framework supports the development of CSRS software by providing suitable programming abstractions for the constituting parts. The communication infrastructure is declared in the program and automatically provided by the framework that gathers the required information from meta-data using the reflection API. Meta-data information is used to annotate types to let the framework know what relation exists among classes[6]. Thus any implementation of this execution model needs a runtime capable of providing reflection support.

2.3 Coordination

Roblets run in parallel and interact with the brain through the *bodymap*. They can also interact one with another either directly (using the *friendship* relation) or indirectly (when the a message from a roblet causes the brain to send a message to another one). It should be noted that, though the direct communication among roblets is possible, we highly discourage the use of this kind of communications since it affects the possible reuse of a roblet in different contextes. Thus the main communication structure of the framework is based on the roblets interacting through the bodymap.

It is natural to wonder how the communication schema proposed by our framework relates to classic coordination models. A roblet can send messages using two different *send* primitives:

- *Send(msg)*: used to send a message to the bodymap;

[6] Annotations in meta-data required by the execution model are limited to expressing relations among types. These can be expressed either implicitly, by defining types containing other types, or explicitly if it is allowed to store custom-annotations together with meta-data (custom annotations are supported both in .NET and Java since version 1.5).

- *Send(roblet, msg)*: send *msg* to the given *roblet*;

the framework is in charge for implementing the effective communication, avoiding explicit location references in the communication primitives. If we assume that the most relevant part of the communications is based on the first primitive, the communication schema closely recall that of blackboard systems of AI [Cor03] or the Linda coordination language [Gel85]. In our case, however, the communication through the bodymap is mediated by the brain software introducing a computation element not present in these models. Robotics4.NET can be placed in the middle between multi-agent systems and blackboard systems, retaining the nature of the first and the communication schema of the latter.

If the communication schema recalls blackboard systems, the multi-agent nature of the framework arises when we come to message delivery. There is no *receive* operation: roblets get notified of incoming messages through a callback. If a roblet is interested in waiting for a message, the main thread of the roblet should be suspended and resumed by the notification callback. This, however, is not a common pattern in roblets, and should be avoided if possible: the robustness of the framework is based on continuous communication that would be paused in this case.

Although it may seem that direct communication among roblets bypass the bodymap, this is true only in part. The bodymap is still responsbile for communicating delivery information about roblets to all the roblets that have a *friendship* relation with them. Yet, once this information has been exchanged, roblet can communicate directly. With this kind of interaction the framework gets closer to a multi-agent system in which roblets communicate through messages. Although this can be considered a more expressive and flexible framework, for the considerations about the role of the body, we believe that this aspect of the framework should be restricted to model of fully reactive behaviors in the robot.

In its first version the communication infrastructure provided by the framework was very simple. As it is discussed in section 2.4 we used the extensible meta-data provided by .NET to allow developers to declare the communication schema as part of the roblet definition. In the natural evolution of the framework we have found that the ability of controlling messages is very valuable and we are currently working to introduce simple message routing in the declarative communication model. We devised the notion of communication state of a roblet, and message delivery can be conditioned by it.

2.4 Implementation

If a contribution of Robotics4.NET is a methodology for designing CSRS software, its implementation is also worth of notice. It relies on the services provided by state-of-the-art virtual machines: dynamic loading of types and the ability of reflect the code structure at runtime. Our prototype runs on the .NET Framework[.NEb], as well as on Mono[Mon], and .NET Compact Framework[.NEa] and it has been developed using the C# programming language[HWG03].

The reasons for choosing the .NET framework are:

- Interoperability: Platform Invoke allows to easily map standard DLLs functions into class methods allowing to develop the lower levels of the architecture in languages such as C or C++;
- Networking: often the software is distributed and facilities help software components to communicate across processes or even computers is useful;
- Dynamic loading: the ability of dynamically load software modules is really useful in robotics, it can be used to load behaviors at need;
- Memory management: although the lower levels of the system are usually implemented in C++, algorithms developed for planning, unification, theorem proving, and similar tasks, greatly benefit from automatic memory management (and traditionally are implemented in programming languages with such a feature);
- Multiple languages: the Common Language Runtime is targeted by compilers of several languages (other than C#, Visual Basic and C++), ranging from Python, SML, Scheme, OCaml; this may help reusing existing software developed in the vast AI field.

In our experience, CSRSs are complex enough to require many of the services provided by virtual machines such as CLR and JVM. Often C++ framework have to implement (or link libraries) providing them, as it happens in the OROCOS system [Bru01]. Moreover the computational power is making possible to run shrunken versions of these runtimes on micro-controllers, making viable to program a robot as a highly distributed system.

In our implementation messages are represented by XML tree structures, obtained through XML serialization, and collected by the bodymap in a larger tree containing all the messages received from the body. In this way the entire status of the body is represented by a single tree, to the advantage of the brain that has a homogeneous data structure to deal with.

The adoption of XML as a wire format for messages has the advantage that even micro-controlled systems can partecipate actively as roblets, though they have to be programmed raw, without any support provided by the framework. We are currently using a micro-controlled board in the a robot that implements a roblet interface for sensors. A meta-program generates the C code hndling newtorking communications reading a .NET annotated set of types describing the messages.

The first time the bodymap receives a message from a roblet, it generates a special message requesting the roblet interface. When the message is received by the roblet the framework automatically generates a description of its interface based on the annotations present on the class definition. An interface consists of:

- an XSD schema (W3C Specification - XML Schemas) describing the input and output message types;
- for each message kind an instance provided as an example to be used by the brain;
- a description of the friendship relations.

The bodymap is responsible for receiving XML messages from running roblets and storing them into the message tree.

Fig. 2. A monitor of a robot implemented as a roblet.

We use UDP as a transport protocol, since we do not rely on an interrupt based model it is not a problem if some message gets lost. Besides, the stateless nature of communication helps to recover from modules or communications failures. UDP is also faster than TCP: we used standard XML messages to generate a video stream from a camera as shown in Figure 2. A drawback of using UDP based communications is that the maximum message size is bound by the particular MTU adopted by the local netework, as well as the network buffer used by the communication stack. In our experience we have found that messages are small enough to be sent over UDP, including those carrying the images in the monitor roblet in a standard ethernet based network.

2.5 A Simple Example

Let us consider the simple example of the *Heartbeat* roblet, which notifies the bodymap that it is alive:

```
public class HeartBeatMessage : RobletMessage {
 public long beat;
 public HeartBeatMessage() {
    beat = DateTime.Now.Ticks;
 }
}

public class SpeedUpBeatMsg : RobletMessage {
   public double freqChange;
}

[OutputMessage(typeof(HeartBeatMessage)),
 InputMessage(typeof(SpeedUpBeatMessage))]
public class HeartBeat : RobletBase {
  public override void Initialize()
  {
    OnSignal +=
      new OnSignalHandler(HeartBeat_Signal);
    Frequency = 0.5;
  }
  public override void Run() {
    SendMessage(new HeartBeatMessage());
  }
  void HeartBeat_Signal(RobletMessage msg) {
    if (msg is SpeedUpBeatMsg) {
      SpeedUpBeatMsg m = (SpeedUpBeatMsg)msg;
      Frequency += m.freqChange;
    }
  }
}
```

The *HeartBeat* roblet sends a heartbeat message every 30 seconds; its type is defined as a standard .NET Class, which inherits from *RobletMessage*, and it will be serialized into XML by the framework inside the method *SendMessage*. We use the UDP/IP protocol for sending serialized messages.

The *HeartBeat* class inherits from *RobletBase* all the behavior required to run. It is worth to note that we use custom attributes (the annotation on the *HeartBeat* class) as a mean to declare input and output messages of a given roblet. In this example we state that *HeartBeatMessage* is an outgoing message of the *HeartBeat* roblet and that the *SpeedUpBeatMsg* is an incoming one. Friendship is expressed through the *Friend* annotation that works in the same way of *InputMessage* and *OutputMessage* annotations.

XML serialization provided by the .NET framework is used to automatically generate XML messages from objects. The serialization process generating the messages is driven by the information stored in the program meta-data and accessed through reflection API.

Incoming messages are received by the framework, deserialized and notified to the class by calling the *HeartBeat_Signal* delegate. In our example the frequency of the heartbeat signal is changed.

2.6 Designing Robotics4.NET Applications

Now that we have introduced the framework, and discussed its design and implementation, we are ready to take a look to its typical use. A CSRS based on Robotics4.NET can be developed following these steps:

- body design;
- message definition;
- communication definition;
- roblet and brain implementation.

Body Design

In the first step the body is defined as a set of organs, each of them defined by a roblet. In this phase the granularity of the system is decided, since we have to establish the functionality of each roblet. A typical system has a vision roblet, one or more motion roblets, and one or more sensors roblets. The principal dilemma here is the way we group sensors and actuators.

In a robot, for instance, we defined a single roblet for all the sensors other than the camera. We also defined a motion roblet responsible for controlling robot movement; a head roblet, responsible for controlling head movements; and a body roblet which controls actuators used to change shape and internal sensors like end-run and batteries voltage detectors.

This process of grouping sensors and actuators of a robot is very important, and has influence over the system and on the ability of reuse software across different systems.

Message Definition

After laying out the overall architecture of the body, we proceed to define the nature of messages exchanged within the system. Although we consider it a separate step in the process, usually messages are defined incrementally, as the system gets refined during the development process.

The definition of messages is a typical problem of data definition, and several considerations can be inspired from those used in systems or database design.

Communication Definition

Most of the communication inside a typical body happen between the bodymap and the roblets. Friendship among roblets, however, are often required in real robots: they are often required to obtain timely delivery among organs, without giving the brain the opportunity of mediate the communication. These reflexes are also present in the human body and are critical to guarantee the most basic istincts such as collision avoidance.

Although this phase does not requires a significative amount of time, it is very important, since introducing friendship among roblets reduces the possibility of reusing roblets in different systems.

Roblet and Brain Implementation

This is the phase where we have to fill all the details in the system. Here we have to take several critical decisions for each elemtent of the architecture.

It is important to note that we can now focuses on the single aspects of the real system, having already addressed several difficult problems related to the overall architecture of it.

In this phase we also establish the timing of roblets: the frequency at which each roblet run (and typically send messages to the bodymap). These values highly depend on the nature of roblets and of the expected reactivity of the system. We believe that there is still room to better define this process, though for the moment we are still observing case studies in the hope of finding common patterns worth to abstract.

Final Considerations About the Design Process

As it can be noted the design process leave small room to the programmer for changing the basic assumptions. Although this may be perceived as a lack of flexibility of the framework, this is by design. We believe that frameworks are good when its clients can leverage on the expertise of the application domain experts that have designed them. If the framework provides only a set of unstructured features there will be more flexiblity in mixing them at the price of a greater knowledge from the client in their use.

3 Reusing by Means of Decoupling

Software reuse refers to the ability of a software unit to be used, unchanged, in different systems. Programming languages have developed many techniques and abstractions to support the creation of these units. Libraries are a popular way to isolate set of self-contained functionalities. Programming constructs like interfaces and classes are used to do the same at a smaller scale.

Software *coupling* refers to the amount of dependencies that a software unit has with other units. Some of these are unavoidable, like those to the standard libraries provided by the runtime support of a programming language. These dependencies, however, are not critical because language runtimes are required to run the program and widely available. Besides, remaining dependencies tend to influence the ability of reusing the software unit. As a rule of thumb the less a software unit is coupled the more is the chance to reuse it.

Robotics4.NET imposes a structure to CSRS software with a low coupling between roblets. A roblet is usually self-contained and has no other dependencies to other components; therefore it can be taken and employed as it is on another system. Whenever a roblet is friend to another one, we have a dependency that may reduce its reuse. However, as it has been pointed out in the previous section, the friendship relation is not symmetric, thus the target of any friendship has no dependencies from other roblets. Moreover the framework recommends avoiding the friendship relation unless strongly needed; this is because the brain is unaware of this kind of communication, losing control over the state of the body.

There is no dependency induced by the compilation of the framework: roblets and brain are not required to share classes (for instance those used to define messages). Message-based communication helps the development of heterogeneous systems.

Decoupling of roblets is what we call *horizontal decoupling* of the architecture: the brain orchestrates communication among modules that do not communicate directly. Robotics4.NET provides also *vertical decoupling*: roblets are a sort of drivers hiding hardware from the brain and defining the amount of abstraction of the body.

Let us consider, for instance, a roblet responsible for controlling a camera; in this case we have the option of sending to the brain the raw data read by the camera (this would be what a driver would do), or we can think of a roblet performing vision tasks and sending the results of the analysis to the brain in the form of a set of objects describing faces, objects, and whatever elements has been detected in the input stream.

Vertical decoupling is important because robotics systems tend to adopt similar hardware, though in different configurations. Therefore it is not infrequent that a roblet can be reused as it is on different robotics systems.

We have positively experienced the ease of reuse, by successfully reusing roblets in different robotics systems.

For instance, we developed a system to support sailors during regattas. The first version of the system has been developed as a distributed application with a Personal Computer reading the boat bus, and some PDAs providing information to people on-board. Afterwards we ported the application to our framework, getting a significantly better implementation. Although the boat is not considered a robot, it can be easily thought as if it is (this is possible for a large category of systems and devices), thus we had to define the body, so we decided how many roblets should have been developed and with which characteristics. During the re-engineering process the data flow of the framework helped us to spot a better software organization than that used in the first place obtained through a standard software design process.

We have been also able to successfully reuse two software modules on ER1, a robot based on the robotic kit by Evolution Robotics (Evolution Robotics Web Site), and on R2D2, an experimental robotic platform developed at our department[CCEP05] - whose appearance and functionality has been inspired from the popular Star Wars saga's droid.

In this case we have used ER1 as a smaller-scale model of the bigger and more complicated R2D2 robot. Roblets concerning interaction with the wireless communication, and camera processing, have been reused without modification. Most surprisingly we have been able to provide similar abstractions for motion and sensors roblets, though the two robots are significantly different in their structure.

An important aspect of the framework is the opportunity to group together different activities and code under the Roblet abstraction. The underlying metaphor of defining the body helps us to find, as it happens for organs, that a roblet may combine different tasks together because they are related to a specific organ of the robot. During the design phase it is reasonably natural to group the functions related to a specific hardware component together: this reduces communication and provides a suitable unit to be exposed toward the rest of the system. Thus we group close and meaningful functionalities together in a coherent module, which can be potentially replaced by another one exposing the same set of messages. With roblets we tend to reduce outbound communication because it is natural to keep information passing inside roblets.

4 Microsoft Robotics Studio

Robotics Studio [RS06] is a new framework for programming robots targeting the .NET execution environment. It is an interesting exercise to compare this framework with Robotics4.NET, both to compare choices made by different software architects and try to understand if there is a common ground shared by different frameworks.

Figure 3 shows the core architecture of Robotics Studio. The framework provides fundamentally three main services to application developers:

- Concurrent programming
- Type-oriented, message based communications in a service world
- Direct access to services through the *Decentralized System Services* (DSS)

Concurrency is an essential element of robotics applications, thus Robotics Studio provides for a concurrency model based on a library called *Concurrency and Coordination Runtime* (CCR). Goal of the CCR is to provide a concurrency layer that is more abstract than the standard concurrent programming

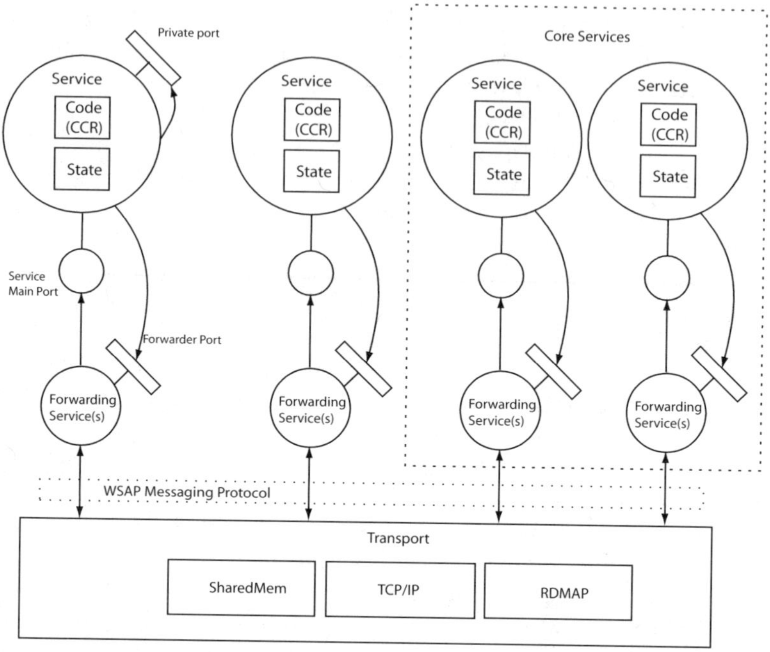

Fig. 3. Robotics Studio execution structure.

paradigm based on threads and synchronization objects. Concurrent activities (tasks) are managed by a dispatcher that is responsible for their execution. Communication is message based, and messages are enqueued into ports (see the figure) waiting for further processing. Coordination is achieved through *Arbiters*, classes responsible for implementing synchronization schemas based on ports. An example of arbiter is the *JoinedReceive* class that monitors two ports waiting for two input in order to execute a task. Tasks are provided in the form of delegates (objects that allow to invoke a method of an object through them), while the *Dispatcher* is responsible to schedule these tasks, possibly managing a thread pool. The following is an example of CCR based program [MSDN06], and shows how to perform an asynchronous read from a file:

```
private static void AsyncStreamDemo(DispatcherQueue dq) {
    FileStream fs = new FileStream(@"C:\Boot.ini",
        FileMode.Open, FileAccess.Read, FileShare.ReadWrite,
        8 * 1024, FileOptions.Asynchronous);

    Byte[] data = new Byte[10000];

    Port<Int32> bytesReadPort = null;
```

```
    Port<Exception> failurePort = null;
    ApmToCcrAdapters.Read(fs, data, 0, data.Length,
        ref bytesReadPort, ref failurePort);

    Arbiter.Activate(dq,
        Arbiter.Choice(
            Arbiter.Receive(false, bytesReadPort,
                delegate(Int32 bytesRead) {
                    Array.Resize(ref data, bytesRead);
                    Msg("Read completed, bytes read={0},
                        data follows:{1}{2}",
                        data.Length, Environment.NewLine,
                        Encoding.ASCII.GetString(data));
                }),
            Arbiter.Receive(false, failurePort,
               delegate(Exception e) {
                   Msg("Read failed, {0}", e.Message); })));

    HitEnter();
}
```

Typeful communication is another element of Robotics Studio; all messages have a type, and ports have a type (the parameter of the generic type *Port*). Support for types is provided by the .NET runtime, as well as reflection and serialization required to serialize messages as XML messages.

Services are based on CCR ports and tasks; each service thus is a set of concurrent activities with a state and a port to communicate. The state of a service is declared using custom attributes to annotate fields of the service class that are to be considered part of the state. The *DSS* infrastructure allows to access, modify and inspect the state of services from a Web browser.

Robotics Studio provides a model that enforces programming patterns that often arise in the development of software for controlling robots. The lightweight concurrent model provided by CCR is such that it is conceivable to define activities with a very fine grain, up to associate a concurrent activity with a sensor to be read. Types and reflection allow providing the communication infrastructure, as well as the Web interface allowing state inspection of a running system. It is important to notice, however, that the framework does not impose a structure to the control software: it is up to the programmer to decide how the concurrent activities are to be used to represent concepts. Besides, the framework is very flexible, allowing very different architectures to be represented in it.

It is interesting to note that Robotics4.NET and Robotics Studio share several elements in common:

- communication infrastructure automatically provided

- custom annotations used to let the programmer declare facts about the program
- concurrency is a fundamental building block of these kind of systems

Besides, the two frameworks also show several differences. The most important is that while Robotics4.NET aims at providing a rather constrained model and defines a methodology for programming robots, Robotics Studio gives a set of mechanisms that the programmer can use to implement the software. Another design choice that differs in the two cases is the granularity of the concurrency model: in Robotics Studio the model is very fine grained while in Robotics4.NET roblets provide a coarse-grain element for concurrency. Finally communication geometry is essentially flat in the first case while it is fixed in the latter.

From these consideration we can draw the following conclusions:

- Concurrency, reflection, and typed, message oriented communication are elements useful for programming robots.
- Virtual machines like .NET CLR are a good start to implement these services. Frameworks not reliying on this kind of execution environments tend to implement several services otherwise already present
- abstractions based on these core services can be very different, and there is no well established practice

As a final consideration, we believe that Robotics4.NET can be implemented on top of Robotics Studio. In this case we can retain the methodology aspects of the framework and the constraints that help the programmer in designing the software without having to maintain two sets of core services.

5 Conclusion

In this paper we presented Robotics4.NET, a framework for developing CSRS. The framework contribution is twofold: on one hand it provides automatically a communication infrastructure among the modules forming a CSRS system in a declarative way based on standard language independent mechanisms; on the other it provides a metaphor to guide programmers into software development. The project is still under development, though the first results are quite promising.

We have successfully reused roblets on very different systems, ranging from embedded systems based on Windows CE to full PCs, as it is. Because of the text-based format of messages exchanged by roblets, we have been able to integrate programs running on microcontrollers to gracefully integrate with the rest of the architecture even without running the framework due to hardware restrictions.

We believe that a major contribution of the framework is the choice of providing to the programmer a metaphor and a programming model that devises a well defined set of abstractions and an execution model. This is rather different from the approach taken by other frameworks tackling the same problem: they provide a set of functionalities together with full freedom in the organization of the modules and the communications among them.

We have also introduced the new framework for robotics announced by Microsoft. After a short presentation we compared the approaches of the two frameworks, comparison made interesting by the fact that both frameworks share the same execution environment; moreover, they tend to build on top of the same core mechanisms provided by it.

References

[Bro85] R.A. Brooks, *A robust control system for a mobile robot*, A.I. Memo 864, Massachusetts Institute of Technology Artificial Intelligence Laboratory (1985).

[Bro91] R.A. Brooks, *Intelligence without reason*, Proceedings of the 12th International Joint Conference on Artificial Intelligence (1991), 569–595.

[Bru01] H. Bruyninckx, *Open robot control software: The OROCOS project*, Proceedings of IEEE Int. Conf. Robotics and Automation (2001), 2523–2528.

[BS93] R.A. Brooks and L.A. Stein, *Building brains for bodies*, A.I. Memo 1439, Massachusetts Institute of Technology Artificial Intelligence Laboratory (1993).

[CCEP05] A. Cisternino, D. Colombo, G. Ennas, and D. Picciaia, *Robotics4.NET: Software body for controlling robots*, IEE Proceedings Software **152:5** (2005), 215–222.

[Chu02] P.S. Churchland, *Self-representation in nervous systems*, Science **296** (2002), 308–310.

[Cor03] Daniel D Corkill, *Collaborating Software: Blackboard and Multi-Agent Systems & the Future*, Proceedings of the International Lisp Conference (New York, New York), October 2003.

[Dam94] A.R. Damasio, *Descartes' error: Emotion, reason, and the human brain*, Avon Books, 1994.

[Dam99] A.R. Damasio, *The feeling of what happens: Body and emotion in the making of consciousness*, Harcourt Brace & Co., 1999.

[Dam03] A.R. Damasio, *Looking for spinoza: Joy, sorrow, and the feeling brain*, Harcourt Brace & Co., 2003.

[Gel85] D. Gelernter, *Generative communication in linda*, ACM Transactions on Programming Languages and Systems (TOPLAS) **7:1** (1985), 80–112.

[Hol04] O. Holland, *The future of embodied artificial intelligence: Machine consciousness?*, Lecture Notes in Artificial Intelligence, vol. 3139, pp. 37–53, 2004.

[HWG03] A. Hejlsberg, S. Wiltamuth, and P. Golde, *The C# programming language*, Addison Wesley, 2003.

[IPSK03] F. Iida, R. Pfeifer, L. Steels, and Y. Kuniyoshi, *Embodied artificial intelligence international seminar, Dagstuhl castle, Germany, july 7-11, 2003, revised papers*, Springer Verlag, 2003.

[LBDS03] T. Lefebvre, H. Bruyninckx, and J. De Schutter, *Polyhedral contact formation modeling and identification for autonomous compliant motion*, IEEE Transactions on Robotics and Automation **19:1** (2003), 26–41.

[Mon] *Mono project web site*, available at http://www.mono-project.com/.

[MR81] J.L. McClelland and D.E. Rumelhart, *An interactive activation model of con-text effects in letter perception: Part 1. an account of basic findings.*, Psychological Review **188** (1981), 375–407.

[.NEa] *.NET Compact Framework web site*, available at http://msdn.microsoft.com/smartclient/understanding/netcf/.

[.NEb] *.NET Framework web site*, available at http://msdn.microsoft.com/netframework/.

[RS06] *Microsoft Robotics Studio*, available at http://msdn.microsoft.com/robotics/, Accessed: 4/9/2006.

[MSDN06] Richter, J., *Concurrent Affairs*, available at http://msdn.microsoft.com/msdnmag/issues/06/09/ConcurrentAffairs, Accessed: 4/9/2006.

VIP: The Video Image Processing Framework Based on the MIRO Middleware

Hans Utz[1,2], Gerd Mayer[1], Ulrich Kaufmann[1], and Gerhard Kraetzschmar[3,4]

[1] University of Ulm, Neuroinformatics, Ulm, Germany
{gerd.mayer,ulrich.kaufmann}@uni-ulm.de
[2] now at Ames Research Center, Mountain View, CA, USA
hutz@mail.arc.nasa.gov
[3] Fraunhofer Institute for Autonomous Intelligent Systems, Sankt Augustin, Germany gerhard.kraetzschmar@ais.fraunhofer.de
[4] University of Applied Sciences Bonn-Rhein-Sieg, Sankt Augustin, Germany
gerhard.kraetzschmar@fh-bonn-rhein-sieg.de

1 Introduction

Application frameworks play a major role in fostering reusability of robotic software solutions. Frameworks foster the reuse of code and design and offer a convenient model of object-oriented extensibility. They provide a powerful concept for providing generic components of well-understood, modular solutions for robotics. Nevertheless, few such frameworks exist in robotics, especially at the application level.

Autonomous mobile robotics is a rapidly evolving field of research, and so far widely agreed upon standard solutions for subproblems have just begun to emerge. Furthermore, the substantial development effort required for developing and maintaining such an application framework has been avoided by the vast majority of research-driven projects, which tend to focus on the development of new ideas and concepts and the implementation of new algorithms.

When analyzing the situation in software development for autonomous mobile robots and the lack of reusable application-level infrastructure, one can identify two additional major obstacles on top of the issues already mentioned: Firstly, the hardware platforms used in autonomous mobile robots integrate a wide variety of different hardware, sensor, and actuator components, resulting in extreme heterogeneity. Secondly, mainly for efficiency reasons one can usually observe a tight coupling of software solutions to the low-level properties of sensory and actuatory devices. This coupling is likely to jeopardize, if not destroy, all high-level portability of robotic software solutions. Thus, as described in Chapter *Trends in Robotic Software Frameworks* it is much more difficult to make the higher development effort for a framework-based

D. Brugali (Ed.): Software Eng. for Experimental Robotics. STAR 30, pp. 325–344, 2007.

solution pay off in the long run by reusing the framework and its components for other scenarios or other robot platforms.

2 Middleware for Robots

Our research on software development in robotics addresses the intrinsic problems of this application domain and resulted in the design and implementation of MIRO, the "Middleware for Robots" [USEK02]. The basic idea of MIRO is to provide generic support for the management of the domain-specific difficulties by addressing them in form of a middleware-oriented architecture. For this purpose, MIRO provides several layers of functionality that reside between the bare operating systems and the robotics applications. The highest layer provides application frameworks for common tasks in robotics. The video image processing framework (VIP), which is the main subject of the discussion in this chapter, is part of this framework layer.

The frameworks make extensive use of the infrastructure provided by the middleware to ensure their portability, reusability, and scalability in long-term robotics projects. For these purposes MIRO provides a standardized distributed systems infrastructure, which was carefully configured and transparently extended in order to meet the requirements of multirobot teams, an advanced toolkit for configuration and parameter management, and generalized abstractions for sensor and actuator devices, which are modeled as network transparent services. So, before we discuss with VIP a particular framework in more detail, we will first survey some aspects of the lower layers of MIRO, in order to provide a clearer picture of their impact on framework-based development in robotics.

2.1 Communications Infrastructure

Robotics applications are inherently distributed. While this property is obvious in multirobot scenarios, single robots are also distributed systems. Many robot platforms are even equipped with multiple computational devices, encompassing embedded microcontrollers and PCs. They are often operated and monitored from remote workstations, or parts of the processing is offloaded to a powerful compute server. Therefore, we consider a distributed communication environment as an essential part of any robotics middleware.

The communication environment provided by MIRO is based on the CORBA standard. It provides a carefully configured, high performance communication environment [SGHP97] to the upper layers. System entities are therefore remotely accessible through object-oriented interfaces which are specified in the CORBA interface definition language (IDL). The design goal of the communications infrastructure provided by MIRO was to encapsulate the complexity of the distributed system domain by a carefully designed default configuration, which is applicable to most robotics scenarios. Nevertheless, the underlying bare communications technology is still easily accessible to developers for special cases of adaptation.

Apart from the method call-based interfaces MIRO also provides event-based communication facilities, using the standard CORBA Notification Service [USM05]. This real-time oriented version of a publisher-subscriber protocol is transparently extended to meet the communication requirements of multirobot teams in wireless networks with heavily varying link quality [USM05]. Thus, MIRO not only offers a scalable, high performance communication infrastructure, but also allows to provide advanced features like generic logging facilities as discussed in [UMK04].

The CORBA technology offers mandatory features such as compile time type safety and programming language interoperability. Its high scalability and sophisticated co-location optimizations, automatically applied for objects that reside in the same address space, make up for most of the introduced computational overhead. Therefore, these interfaces are consistently used for local communication as well.

2.2 Configuration and Parameter Management

Software applications in robotics are full of parameters and magic numbers, such as the *maximum admissible velocity (with respect to the robot's current obstacle avoidance capabilities)* or the *encoder ticks conversion factor*, which allows to translate sensor measurements from the wheel encoders to a distance travelled by the wheels. Whenever better software revisions become available or outdated hardware gets replaced over the lifetime of an autonomous platform, these parameters tend to alter significantly. They also need to be adapted due to changes in the application scenario.

Hardcoding such parameters as constants might possibly allow for some optimizations by the compiler, but limits the flexibility of the design. Furthermore, these kinds of parameters exist throughout the entire robotic application software. Distributing these various parameters over dozens of files ruins quickly the maintainability of the system. Providing a central configuration header does little good either, as every change to this single file will result in an almost complete rebuild of the entire software package.

Simply switching to configuration files that are parsed at run-time results in other disadvantages. Firstly, it eliminates the static type safety provided by the compiler. Secondly, this approach requires every module to add parameter parsing facilities to its implementation for every single parameter that it provides. Also, configuration files introduce a new category of often subtle and hard to find bugs, such as mismatches of parameter names between the file and the parsing functionalities. Furthermore, configuration files usually just provide entries for parameters that differ from their defaults. So a separate documentation has to be maintained, about which parameters are actually processed by the different subsystems.

To support robotics applications in this respect, MIRO provides a sophisticated design for configuration and parameter management. Parameters are specified in their own XML-based parameter description language. These specifications are then transformed by a compiler in classes of the target programming language (currently, C++ is supported). This code holds methods for reading and writing the specified parameters from and to configuration files. Furthermore, the specifications are also used by an interpreter to provide generic GUI-based editing facilities for the configurations files, that ensure type safety and syntactical correctness.

The design of this parameter toolkit not only keeps MIRO highly configurable and maintainable, employing this infrastructure on the framework layer actually propagates this features to the application layer.

2.3 Sensor and Actuator Services

Decoupling robotic algorithms from the physical capabilities of the robot requires to find suitable abstractions for its sensor and actuator devices. This task is the goal of the MIRO service layer. The basic idea is to allow programmers to write their software solutions against an abstract service specification instead of an individual physical device. The challenge in such generalizations is that many robotics applications need to address some of the unique individual properties of such devices in order to perform their task more effectively. Furthermore, a common interface will not abstract away physical differences between robot platforms such as the maximum admissible velocities.

The MIRO service layer offers strictly typed, network transparent, object oriented interfaces for sensor and actuator devices. Sensor devices not only provide method call-based interfaces, but also publish their data using the notification service. Interface hierarchies are used to solve the conflict between generalization of sensor/actuator modalities and access to the special features of an individual device. Meta-information can be queried from the services to make the properties of an individual device, such as e.g. the speed limits of motors, accessible to client programs.

The abstractions provided by the service layer provide an essential ingredient for framework-based development in robotics, as they enable the implementation of components reusable on multiple robot platforms. Furthermore, the frameworks provided by MIRO define network transparent high-level interfaces, introducing themselves as service to other subsystems of the application.

3 Application Frameworks

A good example of an application framework is VIP. VIP covers real-time oriented video image processing and is part of the highest layer of the robotics middleware MIRO. It heavily draws upon the infrastructure for robotics application development discussed above. While a framework mainly provides a design scheme and components that help to meet the challenges of the problem domain (such as image processing), it also deploys infrastructure provided by middleware to support the developer in mastering the immanent challenges of autonomous mobile robotics.

First, the requirements for robotic vision systems are discussed, with a focus on highly dynamic and real-time constrained robotic scenarios, such as robot soccer, followed by a description of how the framework addresses these challenges. A hands-on example illustrates the concepts.

3.1 Video Image Processing

Vision systems for autonomous mobile robots must unify the requirements and demands of two very challenging disciplines: *i)* computer vision and image processing, and *ii)* robotics and embedded systems. While the state-of-the-art in computer vision algorithms is quite advanced, many computer vision methods are intrinsically computationally expensive. Even an efficient implementation of such methods cannot fix this problem. Therefore, the resource demands of computer vision methods are in conflict with the requirements posed by robotics and embedded systems. The latter demand very short execution cycles of the control loops that read out and process sensor data, interpret and fuse them, and determine appropriate actions for the actuators. Particularly, the real-time requirements of robotics seem to rule out most sophisticated computer vision methods, which is one of the reasons why some computer vision experts get discouraged to work on robot vision. As a consequence, robot vision software systems are often inflexible and hard to maintain, because they tend to contain hardcoded quick hacks, which for efficiency reasons try to exploit micro-optimizations like performing multiple operations simultaneously, or because they are heavily model-based or incorporate application-specific heuristics.

In order to mediate between the partially contradictory requirements of advanced vision processing in a real-time constraint environment, proper conceptual support for vision processing architectures is required. Such conceptual support should encapsulate the vision application within its application domain. For a better understanding of the different requirements that need to be supported, we briefly review some characteristics of the two problem domains.

Computer Vision and Image Processing

The basic concept of computer vision is the application of operators to image data, such as logical and arithmetical operations (conjunctions, multiplications), color conversions, morphological functions (erosion, dilation), filtering functions (Gaussian filters, convolutions), or linear transforms (Fourier or wavelet transforms). Often operations transform more than one input image into a new output image. For example, the Canny edge detector [Can86] needs two images which are obtained by convolving a horizontal and a vertical Sobel operator respectively. Other operators, such as optical flow require as input a sequence of images obtained at different instances of time [HS80].In principle, an image operation can be viewed as a mapping of one or more input images into a new one.

More sophisticated operations cover more general input/output mappings, i.e. the result of an image processing operation does not have to be another image, but may be any other kind of data structure, such as a color histogram, a similarity measure between two images, or any other statistical measure on the image. In this case, the definition of a filter is extended to a mapping of one or multiple input images into a new image, or one or multiple classification values associated with the image.

Sequences of such image operators reveal features within the image that can be used to identify regions of interest (ROIs). Some filters do not work on the whole image, but only on parts of it. ROIs are used either to speed up the processing loop, or to make sure that the result of successive operations is not influenced or noisified by image areas which are already known to be irrelevant (e.g. in object classification). Various kinds of feedback loops, such as integration over time, can speed up processing and improve classification results [MMU05]. Because regions of interest can change between individual processing steps, they are associated to images just like the above mentioned classification values.

Robot Vision

Performing operations such as those described in the previous section on the video image stream supplied by one or more cameras on an autonomous mobile robot imposes further constraints on the processing model:

Timeliness Constraints: Robots are situated in a physical world. For tasks like obstacle detection and object tracking the image processing operations must be calculated many times per second and preferably at full frame rate, which is typically 30Hz. The system designer needs to repeatedly assess the performance of the vision system and to ensure its efficiency. Whenever possible, image processing operations should be executed in parallel in order to fully exploit the available resources, such as dual-CPU boards, hyperthreading, and multicore processor technologies. More complex image operations, which need not be applied at full frame rate, should be executed asynchronously in order to ensure that the performance of other image evaluations is not negatively affected. Adequate processing models are required to support such designs.

Fixed Frame Rate Image Streams: New images usually arrive at a fixed frame rate. As the value of the obtained information rapidly decays in a dynamic environment, a sensor-triggered evaluation model is required.

Communication: Vision applications in robotics not only need to communicate their results to other subsystems of the application. For advanced image operations, information from other sensors needs to be accessed: Integration over time on a mobile robot requires access to the dead reckoning sensors of the robot. Visual object detection might be verified with range sensor readings, or even be matched with observations from other robots in an early fusion processing model.

Parameterization: Most image operators need to be parameterized in order to tune the quality of their results. Examples are the width of a Gaussian filter or the number of buckets for an color histogram. The calibration and optimization of parameters is an important part of the development process. Also, the configuration of the filter graph has to be altered frequently during development, which can be significantly facilitated by a flexible and configurable development environment for robot vision systems.

Development Model: Robot vision is performed on live image streams, which are recorded by a moving robot platform. This must be adequately addressed in the development model. For many vision applications, development starts on sets of test images and recorded test image streams. If the application domain implies nondeterminism, or if the robot's actions affect the quality of sensor data by inducing effects like motion blur, the vision system needs to be tested extensively in the real-world environment. This requires effective and stringent support for the test-debug cycle, for inspection, and for adaptation of parameters in the running application.

Related Work

The widely known vision-related architectures can be roughly divided into three categories: subroutine libraries, command languages, and visual programming languages.

Subroutine libraries are by far the most widespread. They concentrate on the efficient implementation of image operators. Therefore, they typically provide a set of functions, each of which implements a different image processing operation. Well-known examples are the SPIDER system [TSTY83] and NAG's IPAL package [CCG89], which are written in C or Fortran. More recent libraries include LTI-LIB [sou05c] or VXL [sou05d], which are both open-source, written in C++, and provide a wide range of image operations covering image processing methods, visualization tools, and I/O functions. The commercial Intel Performance Primitives (IPP) [sou05b] are an example for highly optimized processing routines with a C API. What all these libraries lack is adequate support for flow control. Aside of yet another collection of mutex or semaphore helper classes, and possibly some kind of thread abstraction, there is no special flow control support available.

Command languages for image processing are commonly implemented as scriptable command line tools. In case of the IMLIB3D package [sou05a] the image processing operators can be called from the Unix command line. The

CVIPTOOLS [Umb98] are delivered with an extended TCL command language. Both packages provide programming constructs for conditionals and iteration. While the programmer has complete control over the system in a very flexible way, she also carries full liability over the dynamics of the image processing cycle. Additionally, the scripting-based approach does not make it any easier to meet the required performance constraints of typical robotics applications.

Visual programming languages are currently often viewed as the most sophisticated solution approach. They allow the user to connect a flowchart of the intended processing pipeline using the mouse. They combine the expressiveness and the flexibility of both libraries and command languages. Often, they provide not only a wide spectrum of image processing functions and statistical tools, but also a complete integrated development environment. Many of the available systems are commercial products, with VISIQUEST (formerly known as Khoros/Cantata) being one of the most advanced ones. According to the information available from their web site, it supports distributed computing capabilities for deploying applications across a heterogeneous network, data transport abstractions (file, mmap, stream, shared memory) for efficient data movement, and some basic utilities for memory allocation and data structure I/O.

To the best of our knowledge there is no image processing framework that combines all of the features described above, like processing parts of the filter tree on demand and in a flexible yet powerful way. Such a capability would make the system suitable for a wider range of image processing tasks, like active vision problems on autonomous mobile robots.

3.2 Solution Approach

The challenges robotics imposes on computer vision mostly affect the non-functional aspects of image processing, such as resource management, time constraints and prioritization. Abstractions for the control flow of the programming logic, supplied by a framework-oriented approach, allow for a clean separation of the two aspects. The vision programmer only implements the individual image operations (if not already available in form of a library) and specifies the data flow for the target application. The VIP framework then executes the implemented code as specified by the execution logic, ensuring proper prioritization and synchronization of image processing cascades. The infrastructure provided by the lower layers of MIRO is employed to help mastering the general difficulties of the robotics domain, such as robot platform independence, networked communication, and configuration.

The central base class of the VIP framework is the `Filter`. It denotes not only a (non-)linear image transformation function like a Sobel operator, but any kind of input-output mapping including advanced operations such as a neural classifier on image features. The input of a `Filter` is a set of images and their associated meta-information. It produces an output image as well as associated, user-definable meta-information which is handed to the successor filters. Buffering of image data is provided by a `BufferManager` class, which allows asynchronous image access as necessary e.g. for the processing of consecutive images.

While the control flow is specified in a tree and evaluated in depth-first order, the data flow is much more flexibly organized as a directed acyclic filter graph (DAG). The VIP framework ensures the correct evaluation order.

Robotics Support

The framework performs sensor triggered evaluation of filters to prevent excessive polling or context switching between waiting threads. In order to maximize performance in this highly time constrained environment, VIP keeps track of which filters are actually queried by client modules. Based on this connection management, dynamic graph pruning is performed to process only those filters currently required by the processing tree. If a client module connects to a new filter, the filter is guaranteed to be part of the processing tree as soon as the next image becomes available. The VIP framework also provides support for network transparent as well as co-location optimized access to images of the filter DAG. Additionally, multiple filters can be queried simultaneously for ensuring their synchronizity. For co-located image queries a shared memory-based approach is used with zero-copying, by using the aforementioned specialized buffer manager.

Image Sources

Within the framework, a video `Device` is modeled as a specialized `Filter`. It forms the root node of a processing tree and a source node in the data flow graph. The framework supports various camera connections, such as BTTV, IEEE 1394 and USB cameras (and also a file-based device used for development and debugging as explained below), which exist as black box components for the VIP framework.

Support is also available for processing the image streams of multiple cameras in parallel. Each processing tree is executed within its own thread and is processed in parallel with other source nodes, while the data flow can stay connected, as required for stereo imaging. The framework ensures appropriate synchronization between the image streams in such cases. Note that, as the framework takes care of synchronization, developers do not need to worry about locking issues and the right usage of synchronization primitives.

Additional processing trees can be utilized to decouple time-consuming image operations (a slow path), which cannot be executed at full frame rate, from fast image evaluations, which must be performed at full frame rate in order to ensure reactivity to unexpected events [UKM05].

Real-Time Constraint Image Processing

As one of the dominant features of robot vision is the timeliness constraint, VIP integrates multiple concepts for real-time processing. Each processing tree can be executed with its own thread priority and scheduler choice, which is directly mapped on the OS-native process scheduler by the framework. This is necessary to minimize jitter and ensure correct prioritization, especially under high load situations. Additionally, detailed timing statistics are provided for each individual filter and each filter subtree. Different models for synchronization of filters between different processing trees can be used to either optimize synchronization of image sources (stereo vision) or minimize locking overhead and context switching between threads (slow/fast path processing).

3.3 Example Configuration

The previously described set of features provided by the VIP framework is best illustrated by a small example. Figure 1 describes the derivation of an edge image for a standard test image of computer vision. The original image is blurred using a Gaussian and then transformed into a gray image. A horizontal and vertical Sobel operator is applied next. In the last step, the Canny operator is applied. The screenshots are taken from the generic inspection tool. Meta-information is not provided by these simple filters. Data and control flow are illustrated in Figure 1. The thick, solid lines denote the data flow, while the dashed lines illustrate the control flow.

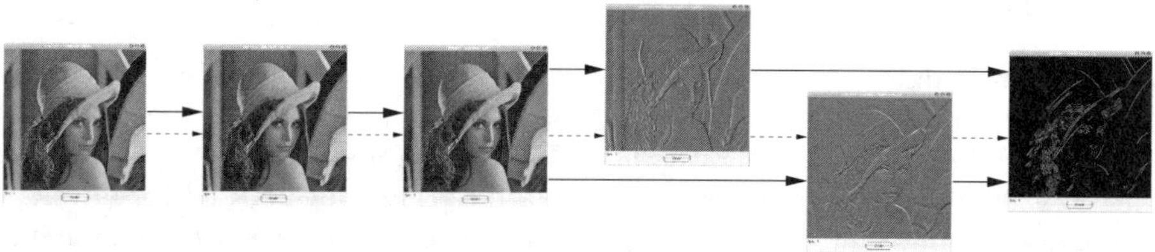

Fig. 1. Original image, intermediate processing steps (blurred, grayed and convolved images), and resulting edge detection. The thick, solid lines denote data flow, the thinner, dashed lines indicate control flow.

4 Middleware-Based Framework Development

The interaction of framework-based application development with infrastructure provided by middleware for robotics is best discussed using a small example. For this purpose, the task of writing a very primitive ball detector for RoboCup soccer robots is used. As the ball is red in this scenario, it is simply assumed that finding the center of all red pixel in the image will suffice.

```
<config_item name="BallDetection" parent="Video::Filter" final="true">
  <config_parameter name="redLow" type="int" default="200" />
  <config_parameter name="redHigh" type="int" default="255" />
  <config_parameter name="greenLow" type="int" default="0" />
  ...
  <config_parameter name="redRatio" type="double" default="0.001" />
</config_item>
```

Fig. 2. Definition of configuration parameters for the thresholds w.r.t. the colorspace dimensions and the overall threshold to decide if there is a ball or not.

This approach will admittedly not work on much more than the provided test image, but illustrates the concepts quite well. Object detection in highly dynamic scenarios with only limited control over illumination conditions requires far more elaborate processing as discussed in more detail in [MKKP05].

4.1 Parameter Definition

The ball detector will be implemented as a filter of the VIP framework. We expect the color of the ball to vary with illumination conditions. Also, even an image without a ball might have some red pixels by circumstance (shadows, reflections). Therefore, a threshold is necessary to decide when there is actually a ball in the image. To achieve high flexibility in the application of image operations, the VIP framework employs the parameter management facilities of MIRO. Therefore, we model these thresholds as MIRO configuration parameters and define them in the XML-based parameter description language provided. The parameter definition is shown in Figure 2. It specifies the different parameters with its respective data type and, potentially, a default value. In this example, the used parameter set contains only simple and plain values, but in general the parameter framework supports also much more complex, nested data structures and even single inheritance (seen here already by use of the `parent="Video::Filter"` attribute). Note that this Figure is a bit simplified for brevity and therefore omits some statements such as include directives.[5]

For debugging purposes the filter will provide an output image that shows all pixels classified as red (aka, belonging to the ball). The actual output of the filter is the center of the ball in image coordinates. In real-life robot soccer the coordinates of the ball would still have to be projected on the floor using the camera model in order to become meaningful. So, to make this data available for successive filters, the output is stored as meta-data of the output image. Meta-data is also defined in parameter definition language, as illustrated in

[5] The full source code of this example is part of the MIRO source code, which is available at `http://smart.informatik.uni-ulm.de/Miro`. For this example, see directory `Miro/examples/videoFilters/`

```
<config_item name="BallFeatureImage" parent="Video::FilterImage">
  <config_parameter name="x" type="int" default="-1" />
  <config_parameter name="y" type="int" default="-1" />
</config_item>
```

Fig. 3. Definition of meta-data using the MIRO parameter description language.

```
class BallDetection : public Video::Filter {
public:
  BallDetection(Miro::ImageFormatIDL const& _inputFormat);
  virtual void process();
};
```

Fig. 4. Filter class specification for a smple ball detection filter.

Figure 3. Again, this data structure is inherited from the appropriate base class `Video::FilterImage`).

The employment of parameter management facilities by the framework saves the developer coding effort, ensures the configurability of the implementation, and enhances its reusability. Furthermore, the infrastructure ensures the maintainability of the application by providing powerful facilities for parameter editing and run-time adaption.

4.2 Simple Filter Implementation

The VIP framework provides the `Video::Filter` base class for image operations. An actual filter implementation derives from this parent and only needs to specify the base parameters of the resulting output image and implement the actual image operation. The specification of the `BallDetection` is shown in Figure 4. In general, only the constructor for filter initialization and the method `process`, which is called to perform the actual image operation, needs to be implemented (another method called `init()` needs to be overwritten in more complex configurations as explained below). Furthermore, generic factory methods need to be provided for the parameter setting functionality discussed above. These are omitted in the example code of the figures.

Only the parameters of the output image need to be specified on construction of the filter, verifying that the input image has the appropriate data format. Our filter implementation expects the image to be coded in 24 bit RGB format and be of any size. The output image will have the same size as the input image (as assumed by default) but using gray scale format (see Figure 5).

The actual filter implementation is shown in Figure 6 for completeness. First a pointer to the image parameters (first three lines) as well as to the

```
BallDetection::BallDetection(Miro::ImageFormatIDL const& _inputFormat) :
  Video::Filter(_inputFormat) {
  if (inputFormat_.palette != Miro::RGB_24)
    throw Miro::Exception("Unsupported input format (RGB required).");
  outputFormat_.palette = Miro::GREY_8;
}
```

Fig. 5. Constructor implementation for the ball detection example filter. It simply checks the format of the incoming image and sets the appropriate output parameters like e.g. the color depth in this example.

```
void BallDetection::process() {
  FilterImageParameters * imageParams = outputBufferParameters();
  ImageParameters * output =
    dynamic_cast<ImageParameters *>(imageParams);
  unsigned char const * src = inputBuffer();
  unsigned char * dest = outputBuffer();

  int sum = 0, sumX = 0, sumY = 0;
  for (unsigned int x = 0; x < inputFormat_.width; ++x) {
    for (unsigned int y = 0; y < inputFormat_.height; ++y) {
      if ((src[y*inputFormat_.width*3+x*3] <= params_->redHigh) &&
          (src[y*inputFormat_.width*3+x*3] >= params_->redLow) &&
          (src[y*inputFormat_.width*3+x*3+1] <= params_->greenHigh) &&
          (src[y*inputFormat_.width*3+x*3+1] >= params_->greenLow) &&
          (src[y*inputFormat_.width*3+x*3+2] <= params_->blueHigh) &&
          (src[y*inputFormat_.width*3+x*3+2] >= params_->blueLow)) {
        sum++;
        sumX += x;
        sumY += y;
        dest[y*inputFormat_.width+x] = 255;
      } else {
        dest[y*inputFormat_.width+x] = 0;
      }
    }
  }
  if (sum > params_->redRatio * inputFormat_.width * inputFormat_.height) {
    output->x = sumX / sum;
    output->y = sumY / sum;
  }
}
```

Fig. 6. Process method implementation for the ball detection example. It includes image parameter query, input and output image requests, ball finding and output parameter creation.

input and output images (next two lines) are obtained. Then it iterates over all pixels of the image and calculates the average position of all red pixels on the x- and y-axis and counts the number of red pixels. If the number of red pixels lies above the parameterized threshold, the filter concludes, that there is a ball in the image and assumes it to be at the center of the red pixels. These coordinates are stored in the output image metaq-data (last two lines).

```xml
<instance type="Video::Parameters" name="BallDetection" >
  <parameter value="BallDetectionBroker" name="VideoBroker" />
  <parameter value="288" name="Width" />
  <parameter value="384" name="Height" />
  <parameter value="rgb" name="Palette" />
  <parameter name="Filter" >
    <parameter value="DeviceDummy" name="Name" />
    <parameter value="DeviceDummy" name="Type" />
    <parameter name="Successor" >
      <parameter name="Filter" >
        <parameter value="BallDetection" name="Name" />
        <parameter value="BallDetection" name="Type" />
      </parameter>
    </parameter>
  </parameter>
</instance>
```

Fig. 7. Filter tree configuration.

VIP provides standard filters for image acquisition from various types of cameras, as well as for reading stored images from disk. In order to put this simple ball detector to work all left to do is to combine it with an existing filter for image acquisition. Therefore it is necessary to introduce it to the VIP framework configuration machinery. For this purpose, a factory instance for this filter type needs to be added to the repository of existing filters of the VIP framework (Details on this step can be found in the complete source of this example).

4.3 Configuration

The configuration of the filter tree is provided to the VIP framework in form of an XML-based specification. A simple configuration is shown in Figure 7. Note, that values for the parameters specified in Section 2.2 can also be specified at this point. In this case we assume the defaults to be sufficient.

It is not necessary to write a single line of code for reading and parsing such an XML file, as this code is generated automatically by the use of the MIRO configuration parameter framework. The filter tree root node in this example is a file device filter, which reads images from disk to test the implemented algorithm offline. By simply exchanging the root node by a real device filter (such as e.g. a IEEE 1394 device filter), exactly the same XML file can be used on a real robot.

5 Development and Tool Support

The example developed so far demonstrated, how the VIP framework *minimizes the coding efforts* by providing generic functionality in the form of super

```
<instance type="Video::BallDetection" name="BallDetection" >
  <parameter value="true" name="InterfaceInstance" />
  <parameter name="Interface" >
    <parameter value="Ball" name="Name" />
  </parameter>
</instance>
```

Fig. 8. Interface definition for access to the output image.

classes and standard filters which can be easily customized. The infrastructure provided by MIRO allows for flexible configuration and parameterization. This enhances the portability of applications built upon VIP, as it e.g. allows to easily replace image acquisition filters, if the application would have to run with another type of camera on another robot.

This section shows how the MIRO infrastructure *supports the development process* and briefly sketches additional functionality which is targeted to ensure the scalability and maintainability of large scale image processing applications for robotics.

5.1 Interactive Inspection

Testing, debugging, and refinement of applications for autonomous mobile robotics requires specialized support, as monitoring the internal state of a robot in operation proves to be very difficult without altering the system under observation. For example, single-stepping through the code is not an option for a moving robot that needs to do reactive obstacle avoidance. For this purpose MIRO provides multiple concepts that support the development process. Firstly, the event-based communication can be generically written to persistent storage without further coding effort. Thereby streams of sensory and cognitive data from a robotics experiment can be logged to disc and replayed offline for inspection or statistical analysis. Secondly, the communication and configuration infrastructure allows to remotely query system information and alter parameter settings at run-time.

Each filter can also instantiate a service interface for network transparent access to its output image. This is exemplified in the specification of the filter parameters for the `BallDetection` filter in Figure 8. The interface is assigned an individual name to avoid ambiguities.

Figure 9 shows the result of the filter operation on a test image. It is visualized with `QtVideo`, a generic remote inspection tool provided by MIRO.

Additionally, the values of the parameters of each filter can also be inspected and altered at run-time, by the use of the configuration editor. Figure 10 shows the dialog for the parameters of the ball filter. The parameters of the filter can be adapted interactively at run-time, based on real-world data acquired by the robot in operation. Again, no code has to be written for all

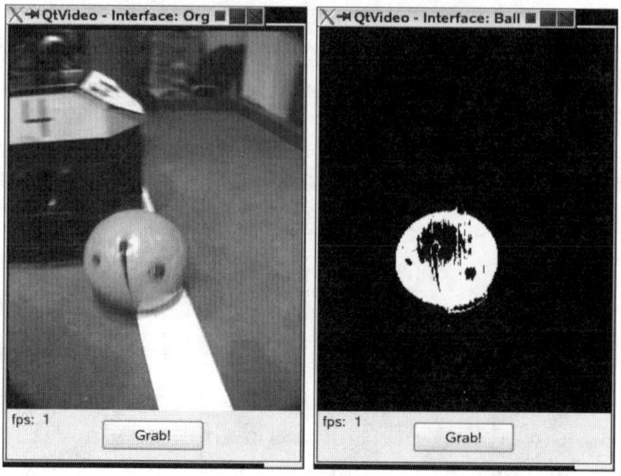

Fig. 9. Screenshots of the generic inspection tool QtVideo, showing an original image and the result of the BallDetection filter.

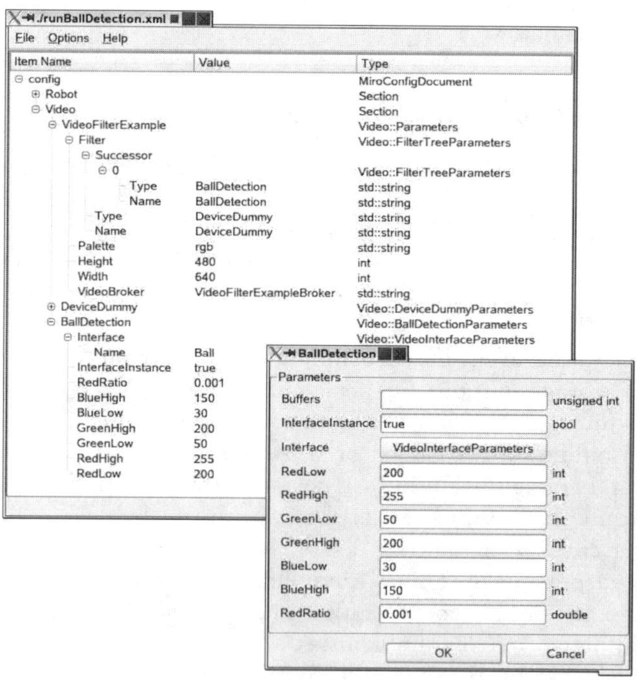

Fig. 10. Screenshot of the configuration editor, including the generically generated configuration dialog.

these tools. The parameter configuration is parsed at runtime to generate the appropriate configuration dialogs.

```
struct BallResult {
  long x;
  long y;
};
```

Fig. 11. Definition of the data structure in CORBA IDL, used for communication of the ball's position within the image.

```
EventChannel_var ec =
  _server.resolveName<EventChannel>("EventChannel");
supplier_ =
  new StructuredPushSupplier(ec.in(), _server.namingContextName.c_str());

StructuredPushSupplier::initStructuredEvent(event_,supplier_->domainName(),"BallResult");
supplier_->addOffer(event_);

...

BallResult result;
result.x = output->x;
result.y = output->y;
event_.remainder_of_body <<= result;
supplier_->sendEvent(event_);
```

Fig. 12. Example code to set up event communication via an event channel (upper part) and to finally fill and send the event itself (lower part).

5.2 Communication

So far, the result of the filter operation, i.e. the ball position in the image, is stored as part of the meta-data of the output image. With this method it is readily accessible for all successor filters. However, it would also be desirable to make this data available to other modules of the system. In our example, we choose to publish this data as an event, so that everybody interested can subscribe to this event and is automatically notified each time the ball filter processed an image.

For this purpose, a ball event is defined in CORBA IDL, as shown in Figure 11. The code required in the ball filter itself is also minimal, as illustrated in figure 12. The init() method of the filter is extended (upper part of the example). An event supplier is instantiated (first line), linked to the event channel of the robot (second line) and the event is initialized with the robots name and the chosen event name, for identification through the consumers (next two lines). At the end of the process method, the filter result is copied into the `BallEvent` data structure, and the data is copied into the payload of the event. Finally the event is sent to the event channel (lower part of Figure 12).

6 Application

The VIP framework is successfully applied in different robotic scenarios, such as biologically-motivated neural network learning, neural object classification in an office environment [FKSP04], and reliable high speed image processing in the RoboCup middle-size league [MKKP05]. The application of VIP in these very different scenarios resulted in the implementation of a large library of over 100 filters, which are shared between these projects.

Some statistics illustrate the complexity of these robot vision applications which are managed with the help of the VIP framework: The application in RoboCup consists of a dual camera setup, combining a directed camera for object detection, classification, and tracking with an omnidirectional camera for obstacle avoidance and near-distance ball tracking. The full RoboCup application currently consists of more than 120 filters with over 200 connections. One of the fastest path, a simple color-based football goal detection, takes only around 4 msecs to execute, while one of the slowest paths, a full, neural network-based classification of robots, requires around 20 msecs on average when seeing one robot per image (measured on 1.4GHz Pentium M).

7 Conclusions

Middleware-based application frameworks facilitate the reusability and scalability of robotics applications and help the development process.

In the first part of this chapter, the infrastructure provided by the Middleware for Robotics MIRO was introduced. Its lower layers, the communication and configuration infrastructure and the service based abstractions for robotics sensors and actuators, are nowadays also found in other robotics middleware architectures. Their implementation based upon open, mature standards such as CORBA and XML ensures highly robust and stable code and ensures a high amount of scalability.

A particular application framework of MIRO, VIP – the framework for video image processing, was introduced in the second part of the chapter. The design of the framework was discussed, based upon the requirements of advanced robotics vision applications in dynamic environments.

The third part of the chapter exemplified how the middleware provided infrastructure ensures reusability and maintainability of VIP-based vision applications and how the development process is supported by the infrastructure provided by MIRO.

The concepts of the application framework VIP discussed in the example are deliberately mostly not vision-specific, but are mandatory for robotics applications in general. Replicating them in every robotics application framework would not only result in considerable code bloat but furthermore introduce a significant development overhead. Furthermore, without middleware infrastructure like generalized sensor and actuator services, a robotics vision framework could not provide advanced components such as information fusion with other sensory sources of the robot without loosing its portability. So the discussion of VIP in this section not only provides an introduction to application framework design in the context of autonomous mobile robotics, but exemplifies the necessity as well as the benefits of middleware-based application frameworks.

References

[Can86] John F. Canny, *A computational approach to edge detection*, IEEE Transactions on Pattern Analysis and Machine Intelligence (1986), 679–698.

[CCG89] M. K. Carter, K. M. Crennell, E. Golton, R. Maybury, A. Bartlett, S. Hammarling, and R. Oldfield, *The design and implementation of a portable image processing algorithms library in fortran and c*, Proceedings of the 3rd IEE International Conference on Image Processing and its Applications, 1989, pp. 516–520.

[FKSP04] Rebecca Fay, Ulrich Kaufmann, Friedhelm Schwenker, and Gnther Palm, *Learning Object Recognition in a NeuroBotic System.*, 3rd Workshop on Self-Organization of AdaptiVE Behavior SOAVE 2004 (Horst-Michael Gro, Klaus Debes, and Hans-Joachim Bhme, eds.), VDI Verlag, Dsseldorf, 2004, pp. 198–209.

[HS80] B.K.P. Horn and B.G. Schunck, *Determining optical flow*, Tech. report, Massachusetts Institue of Technology, 1980.

[MKKP05] Gerd Mayer, Ulrich Kaufmann, Gerhard Kraetzschmar, and Gnther Palm, *Biomimetic neural learning for intelligent robots*, ch. Neural Robot Detection in RoboCup, Springer, Heidelberg, 2005.

[MMU05] Gerd Mayer, Jonas Melchert, Hans Utz, Gerhard Kraetzschmar, and Gnther Palm, *Neural object classification and tracking*, 4th Chapter Conference on Applied Cybernetics, IEEE Systems, Man and Cybernetics Society, 2005.

[SGHP97] Douglas C. Schmidt, Andy Gokhale, Tim Harrison, and Guru Parulkar, *A high-performance endsystem architecture for real-time CORBA*, IEEE Comm. Magazine **14** (1997), no. 2.

[sou05a] sourceforge, *imlib3d*, Available via http://imlib3d.sourceforge.net/, last visited 2005.

[sou05b] sourceforge, *Intel performance primitives (ipp)*, More information on http://www.intel.com/software/products/perflib/, last visited 2005.

[sou05c] sourceforge, *Lti-lib*, Available via http://ltilib.sourceforge.net/doc/, last visited 2005.

[sou05d] sourceforge, *Vxl*, Available via http://vxl.sourceforge.net/, last visited 2005.

[TSTY83] H. Tamura, S. Sakane, F. Tomita, and N. Yokoya, *Design and implementation of spider-a transportable image processing package*, Computer Vision, Graphics and Image Processing **23** (1983), no. 3, 273–294.

[UKM05] Hans Utz, Ulrich Kaufmann, and Gerd Mayer, *Advanced video image processing on autonomous mobile robots*, 19th International Joint Conference on Artificial Intelligence (IJCAI), August 2005, Workshop on Agents in Real-time and Dynamic Environments.

[Umb98] Scott E. Umbaugh, *Computer vision and image processing: A practical approach using cviptools*, Prentice Hall, June 1998.

[UMK04] Hans Utz, Gerd Mayer, and Gerhard Kraetzschmar, *Middleware logging facilities for experimentation and evaluation in robotics*, 27th German Conference on Artificial Intelligence (KI2004), Ulm, Germany, September 2004, Workshop on Methods and Technology for Empirical Evaluation of Multiagent Systems and Multirobot Teams.

[USEK02] Hans Utz, Stefan Sablatng, Stefan Enderle, and Gerhard K. Kraetzschmar, *Miro – middleware for mobile robot applications*, IEEE Transactions on Robotics and Automation, Special Issue on Object-Oriented Distributed Control Architectures **18** (2002), no. 4, 493–497.

[USM05] Hans Utz, Freek Stulp, and Arndt Mhlenfeld, *Sharing belief in teams of heterogeneous robots*, RoboCup 2004: Robot Soccer World Cup VIII (Berlin, Heidelberg, Germany) (D. Nardi, Riedmiller, Sammut M., and J C., Santos-Victor, eds.), Lecture Notes in Artificial Intelligence, vol. 3276, Springer-Verlag, 2005.

MRT: Robotics Off-the-Shelf with the Modular Robotic Toolkit

Andrea Bonarini, Matteo Matteucci, and Marcello Restelli

Politecnico di Milano, Department of Electronics and Information, Milan, Italy
{bonarini,matteucci,restelli}@elet.polimi.it

1 Introduction

Since robotic applications are becoming more and more sophisticated, the design, development, and maintenance of the related software may benefit from a modular approach both in terms of flexibility and reusability. By decomposing a complex system into several simpler functional modules, it is possible to separate responsibilities and parallelize efforts. Many robotic systems, even operating in really different application domains, share several functionalities. The use of a modular approach allows to reuse the same functional units in different applications, thus reducing the development time and increasing the software reliability.

In our bi–decennial experience in making mobile robots, we have identified a certain amount of functionalities that are involved to perform tasks autonomously. We have devised a modular approach to their implementation, based on the principle that each specific module can interact with others by simply exchanging messages in a distributed setting. Within this framework, different modules may even run on different machines, and data may be integrated by modules providing aggregated information to the others. Moreover, modules managing the interaction with the environment are designed to make independent the actual interface with physical devices from the needed data processing. A middleware is also provided to make the interaction among modules transparent with respect to their physical distribution. This makes also possible to implement multi–agent and multi–sensor systems integrated in a unique network, which may change its structure in time.

Section 2 describes the main functionalities we have identified for autonomous robots applications, and how they interact. We have implemented modules for all these functionalities in the Modular Robotic Toolkit (MRT). In Section 3, we summarize the main features of each of our modules, and we present DCDT, our middleware to support interaction among them. MRT has proved effective to provide the functionalities needed for different successful applications. In Section 4, we present how some MRT modules have been

D. Brugali (Ed.): Software Eng. for Experimental Robotics. STAR 30, pp. 345–364, 2007.

integrated in these applications: Roby, a robot able to explore hostile environments, FollowMe, a guide for indoor environments, and the Milan RoboCup Team, a robotic soccer team participating to the RoboCup [KAK95] competition.

2 Functional Modules and Their Interaction in Autonomous Robots

Recently, the interest in modular architectures is growing in the robotic field [BKM05], with the final aim of producing frameworks where a set of off-the-shelf modules can be easily combined and customized to implement robotic applications with minimal effort and time. All the proposed approaches consist of the definition of basic modules and an integration layer, which often relies on commercial standards (e.g. CORBA, JINI, etc.). In the following, we present a general functional decomposition of autonomous robot applications and we motivate our choice of adopting the publish/subscribe paradigm to support interaction.

2.1 Functional Modules

In a robotic application, the typical functional modules can be classified into three main categories: sensing modules, reasoning modules, and acting modules. Sensing modules are directly interfaced with physical sensors (e.g., sonars, laser range finders, vision systems, encoders, bumpers, gyroscopes), with the aim of acquiring raw data, processing them, and producing higher level information for the reasoning modules.

Acting modules are responsible for controlling actuators, implementing the decisions of reasoning modules. Thus, the reasoning modules need to know neither which sensors, nor which actuators are actually mounted on the robot; this allows their reuse in different contexts. We decompose each sensing and each acting module into two sub-modules: the *driver*, which directly interacts with the physical device, and the *processing* sub-module, which is interfaced on one side with a driver, and on the other with some reasoning modules or the world model. So, thanks to its driver, a processing sub-module may abstract from the physical characteristics of the specific sensor/actuator, and it can be reused with different devices of the same kind. For instance, let us consider a mobile robot equipped with a firewire camera. The driver should implement some functionalities such as the interface for changing the camera settings and the possibility to capture the most recent frame. The processing sub-module should extract from the acquired image the most relevant features, which can be used by reasoning modules like the ones devoted to localization, world modeling, and planning. If we decide to change the firewire camera with an USB or an analog camera, or if we have several robots equipped with different cameras, we may have to re-implement the driver sub–module, but we can reuse the same processing sub–module.

Reasoning modules represent the core of each robotic software, since they code the decision–making processes that determine the robot behavior. In MRT, reasoning modules are in principle independent from sensors that have been used in the specific application. Thus, all the information gathered through different sensors can be integrated into a world representation, on which the other modules may perform their inferential processes. The outcomes of reasoning modules are high level commands to be processed by acting modules, and then executed by actuators. Most real–world robot applications can be built using some of the following modules:

- *localization module*: it has to estimate from sensor data the robot pose with respect to a global reference frame, using a map of the environment;
- *world modeling module*: it builds and maintains a representation of the external world, integrating the information gathered through sensors and, eventually, the interaction with other robots;
- *planning module*: based on knowledge about the environment and the robot itself, this module selects the most appropriate actions to reach the given goals, while respecting task constraints;
- *sequencing module*: this module is responsible for the execution of a given plan, monitoring its progress, and handling exceptions as they arise;
- *controlling module*: it contains all the control laws, typically implemented as reactive behaviors [Bro], to decide which actions the robot has to execute;
- *coordination module*: multi-robot applications often require a module that allows robots to share perceptions and to communicate intentions in order to perform effective task allocations.

Figure 1 depicts the robotic functional modules described in this section, arranged in a typical hybrid control architecture [Gat98], in which the deliberative and reactive layers are combined through a middle layer that implements a sequencing mechanism. At this point it is worth noticing that a key factor in software reuse is the configurability of these modules. In fact, it is really hard to build general modules that can be usefully employed in different applications, unless it is possible to configure them to fit specific needs. For each module, it is important to identify the application specific parameters, so that they can be easily specified on a case by case basis. According to this view, the development process of each robotic application should consist of: implementation of the driver sub-modules for each sensor/actuator device, selection and parametrization of the reasoning modules required by the application.

2.2 Publish/Subscribe Middleware for Interaction

Having different modules that cooperate to obtain a complex control architecture shifts the focus from the classical monolithic approach to a more flexible

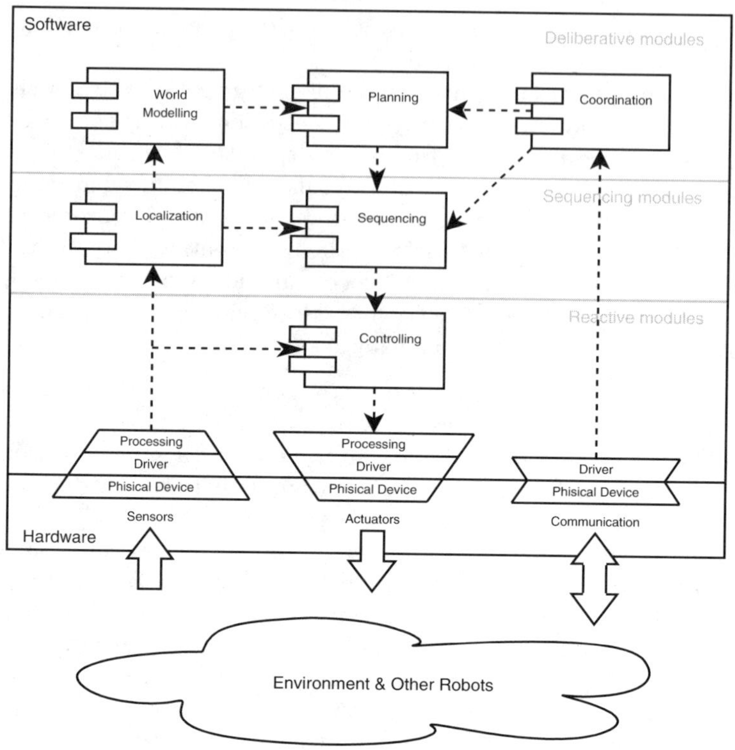

Fig. 1. The MRT general architecture.

distributed approach. This shift in perspective introduced by programming in a distributed, concurrent, and modular environment has been clearly described in [BF02], and the adoption of an integration middleware becomes a key for the effectiveness of this approach. At the same time, autonomous robotic systems, particularly those involving multiple robots and environmental sensors, are becoming increasingly distributed and networked; thus, people involved in robotics are getting conscious that developing multirobot systems might require tools supporting integration and communication [GS02].

No definitive answer has been given about which is the proper middleware for integration, but a key issue to be analyzed in making this choice is certainly how data are exchanged. In the past we have proposed the publish/subscribe model as the most interesting communication model in this new development scenario providing a compared analysis of the publish/subscribe middleware tools currently used in robotic applications [Mat03]. While the publish/subscribe approach is not the only approach that can be used for integration in robotics (among others are: client/server, distributed blackboards [HRPL95] or request/response models [USEK02]), it has been widely used for this purpose as testified by [PSZ00].

In the publish/subscribe model, data providers publish data and consumers subscribe to the information they require, receiving updates from the providers as soon as this information has been published. This paradigm is based on the notion of event or message. Components interested in some class of events subscribe expressing their interests. Components providing information publish it to the rest of the system as event notification. This model introduces a decoupling between producers and consumers since publishers and subscribers do not know each other. This model offers significant advantages in situations where data transferred correspond essentially to time-changing values of an otherwise continuous signal (e.g., sensed data, control signals, etc.). A single subscription replaces a continuous stream of requests and the data is transferred with minimum delay since the exchange is one way and asynchronous. The publish/subscribe model is also notification-based and this is very beneficial when a system needs to monitor a great number of perhaps infrequent events, while in a client/server model (or shared memory model), the monitoring process must continuously poll for possible changes. One-to-many communication is supported by the publish/subscribe model in a natural way, and its implementation can often take advantage of multicast and broadcast mechanisms at the network level to improve the efficiency of event notification.

3 The Modular Robotic Toolkit

Our research effort in modular software for robotics has its roots in the development of autonomous systems in an academic environment, where generations of students have to face necessarily limited parts of the whole robot development, and they should transfer their knowledge about the evolving robots to the next generations. The modular approach began to become effective with the experience made developing the Milan RoboCup Team (MRT) [BMRS05], a team of soccer robots for the RoboCup competition. This effort has resulted in a more general Modular Robotic Toolkit (MRT again) that we are presenting in this chapter and that has been used to develop other robotic applications in different domains, thus experiencing the benefits of a modular approach.

3.1 MRT Modules

We have implemented all the reasoning modules described in Section 2 and, according to the specific application, we have selected and customized the required modules, as discussed in Section 4. In the following, we describe each module and the publish/subscribe integration middleware developed. In this section, we describe each module and in the following one the publish/subscribe integration middleware developed.

MUREA: The Localization Module

We have adopted *MUREA* (MUlti-Resolution Evidence Accumulation) [RSM02], a mobile robot localization algorithm for known 2D environments. The localization space is divided into subregions (cells) and then an evidence accumulation method is applied, where each perception votes those cells that are compatible with the map of the environment. In order to reduce the complexity of working with a fine grid, we have adopted a multi–resolution scheme. We start applying evidence accumulation on a coarse grid, with few large cells. Then, we select and refine (i.e., divide into smaller cells) those cells that have collected highest votes. This module has several configurable parameters that allow its reuse in different contexts. The map of the environment is specified through a configuration file in which are listed the elements that can be perceived by the robot. Each element is defined by a name, its geometrical shape (point, segment, circle, etc.), and a list of attributes that describe the element. Other parameters that can be set in the configuration file are the range of feasible positions, the required accuracy, and a timeout. In order to grant the reusability in different robotic applications, MUREA completely abstracts from the sensors used for acquiring localization information, since its interface relies on the concept of *perception*, which is shared with the processing sub-module of each sensor.

MAP: The World Modelling Module

The world modelling module is a nodal point for many robotic applications. We have studied and proposed *MAP* (Map Anchors Percepts) [BMR01], a module that builds and maintains in time, through an anchoring process, the environment model on the basis of the data acquired by sensors. MAP contains a hierarchical conceptual model, which must be specified for each application, where the classes of objects that can be perceived, and their attributes, are defined. The MAP module is divided into three sub-modules. The *Classifier*, that receives from sensors descriptions of perceived objects in terms of percepts (symbolic features), and identifies, for each perceived object, the concept that has the best matching degree. The classification process generates conceptual instances with an associated degree of reliability. These are passed to the *Fusion* sub-module. Since any physical object may be perceived at the same time by distinct sensors, the output of the Classifier may contain several conceptual instances that are related to the same physical object. The goal of the Fusion sub-module is to merge the conceptual instances that are supposed to be originated from the perception of the same object. Finally, Map performs the tracking phase, which consists of maintaining in time a coherent state of the model and a correct classification of the instances. The *Tracker* tries to match the conceptual instances perceived in the past with those generated by the latest perception. Then, the dynamic properties of the conceptual instances are updated through Extended Kalman filtering.

This approach is well suited also for multi–robot domains, where each robot may take advantage of the data acquired by its teammates. Each agent can be seen as an intelligent sensor that, instead of producing symbolic features, generates conceptual instances. These can be processed directly by the Fusion sub-module of any other agent that shares the same hierarchical conceptual model.

SPIKE: The Planning Module

Trajectory planning is obtained by *SPIKE* (Spike Plans in Known Environments) a fast planner based on a geometrical representation of static and dynamic objects in the environment. The environment to navigate, in SPIKE, is modeled as a 2D space, the robot is considered as a point with no orientation, and static obstacles (e.g., walls, bins, furniture) are described using basic geometric primitives such as points, segments and circles. SPIKE exploits a multi-resolution grid over the environment representation to seek a proper path from a starting position to the requested goal; this path is finally represented as a polyline that does not intersect obstacles. Moving objects in the environment can be easily introduced in SPIKE representation of the environment as soon as they are detected and they can be considered while planning. Finally doors or small (w.r.t. the grid resolution) passages can be manged by the specification of *links* in the static description of the environment. The resolution of the plan is customizable and computation simply requires the description of the environment, the starting point and the goal point; the output is given as a sequence of positions the robot has to reach from the starting point to the goal. This trajectory is computed by using an adapted A^* algorithm on the multi-resolution grid, and then optimized by using geometrical considerations. The grid representation, inspired by the work from Sven Behnke [Beh04], uses a coarse resolution for the (mostly free) area to be navigate, and a finer resolution near static objects and around the robot to manage, respectively, small passages and dynamic obstacles. Path computation as well as the resolution of the grids used can be easily customized by considering robot size, required accuracy, and a safety distance from obstacles.

SCARE: The Sequencing and Coordination Module

In our applications, the sequencer and the multi-robot coordination functionalities are implemented as two sub-modules of a module called *SCARE* (Scare Coordinates Agents in Robotic Environments) [BR02]. Through a two-phase distributed process, SCARE assigns activities to each team member. Activities may belong to one of two categories: *jobs* that are single robot activities, and *schemes*, multi-robot activities ruling the performance of a synchronized sequence of jobs by a team of robots. In the *decision making* phase, SCARE gathers up (through sensors and, in case, communication with other robots) all the available information about the current situation, and evaluates for each activity several parameters such as the physical attitude (e.g., how much the robot is *physically suited* for the job), the *opportunity* (e.g., how much, in the current situation, the robot is suited for the job), and the *need* (e.g., how much, in the current situation, the job is useful for the team). By employing multi-objective analysis techniques, SCARE produces for each robot an ordered list of activities called *agenda*. In the *coordination* phase, SCARE searches for the best job allocation to agents by observing coordination constraints such as *cardinality* (the minimum and maximum number of robots that can be assigned to the activity at the same time) and *scheme coverage* (a scheme can be assigned only if there is an appropriate robot for each functionality). Once the job allocation has been performed, the sequencing sub-module decomposes the assigned job into simpler subjobs, schedules their execution, and monitors the environment to react to unexpected events.

MrBRIAN: The Control Module

Our control module is *MrBRIAN* (Multilevel Ruling Brian Reacts by Inferential ActioNs) [BMR04], a fuzzy behavior management system. Behaviors are implemented as set of fuzzy rules whose antecedents match context predicates, and consequents define actions to be done. Behavioral modules are activated according to the CANDO conditions defined for each of them as fuzzy predicates. Actions are proposed by each behavioral module with a weight depending on the degree of matching. The proposals are then composed with weights coming from the evaluation of WANT fuzzy predicates, defined for each behavioral module to take into account the desirability to apply a given module in a given situation. Among the WANT conditions we have context conditions, which are evaluated on the information coming about the environment and define the opportunity to apply a behavior in a given situation, and coordination/planning conditions, eventually set by other modules of the architecture, to model the opportunity to apply a behavior given intra-agent plans and coordination needs.

In MrBRIAN, it is also possible to organize behaviors in a hierarchy. It this case, behaviors at the higher levels match also the action proposed by others with lower priority, and may decide to inhibit them, either partially or completely. MrBRIAN is completely configurable for the specific application, by specifying which behaviors are required, their implementation, their priority, their activation conditions (CANDO), and desirability conditions (WANT), and the fuzzy sets involved in the definition of predicates.

3.2 DCDT (Device Communities Development Toolkit)

To integrate modules in MRT we use *DCDT* (Device Community Development Toolkit) [CMBM01], a publish/subscribe middleware able to exploit different physical communication means in a transparent and easy way. DCDT is a multi-threaded architecture that consists of a main active object, called

Agora, hosting and managing various software modules, called *Members*. Members are basically concurrent programs/threads executed periodically or on the notification of an event. It is possible to realize distributed applications running different Agoras on different devices/machines, each of them hosting many Members. It is also possible to have on the same machine more than one Agora (hosting its own Members), to emulate the presence of different robots without the need of actually having them connected and running. Agoras on different machines are able to communicate with each other through particular members, called *Finder*, which are responsible for finding each other dynamically with short messages via multicast. The peculiar attitude of DCDT toward different physical communication channels (e.g., RS-232 serial connections, USB, Ethernet or IEEE 802.11b , etc.) is one of the main characteristics of this publish/subscribe middleware.

DCDT messages are characterized by header and payload fields. In the header we have: the unique identifier of the message type (i.e., the content/-subject of the message), the size of the data contained in the payload and some information regarding the producer (e.g., producer id, time of construction, etc.). Members use unique identification types to subscribe and unsubscribe messages available throughout the community. Messages can be shared basically according to three modalities: without any guaranty (e.g. UDP), with some retransmissions (e.g. UDP retransmitted), or with absolute receipt guaranty (e.g. TCP).

3.3 MRT Messages

In MRT, the interfaces among different modules are implemented by exchanging messages. According to the publish/subscribe paradigm, each module may produce messages that are received by those modules that have expressed interest in them. In order to grant independence among modules, each module knows neither which modules will receive its messages nor the senders of the messages it has requested. In fact, typically, it does not matter which module has produced the acquired information, since modules are interested in the information itself. For instance, the localization module may benefit from knowing that in front of the robot there is a wall at the distance of 2.4 m, but it is not relevant which sensor has perceived this information or whether this is coming from another robot.

Our modules exchange XML (eXtensible Markup Language) messages whose structure is defined by a shared DTD (Document Type Definition). There are several motivations for our choice of using XML messages, instead of messages in a binary format. Since XML messages are in text format, they can be directly read by humans and can be edited by any text editor. The use of DTDs allow to specify the structure of each message and check that the messages received are well-structured. Furthermore, XML messages can be easily changed and extended. Finally, the input interface for XML messages is greatly simplified by the existence of standard modules for syntactic

```
<message robot="Roby" module="Vision" data="Perceptions"
         timestamp="11982374383" />
  <object name="RedObject0" class="Object" />
    <attribute name="Color"    value="Red" />
    <attribute name="Distance" value="332" std_dev="10" rel="1"  />
    <attribute name="MinAngle" value="175" std_dev="2"  rel="0.8"/>
    <attribute name="MaxAngle" value="185" std_dev="2"  rel="0.9"/>
  </object>
  <object name="WhiteGreenWhite0" class="Transition" />
    <attribute name="Color"    value="WhiteGreenWhite" />
    <attribute name="Distance" value="505" std_dev="24" rel="0.7"/>
    <attribute name="Angle"    value="245" std_dev="1"  rel="0.9"/>
  </object>
  <object name="YellowObject0" class="Object" />
    <attribute name="Color"    value="Yellow" />
    <attribute name="Distance" value="178" std_dev="10" rel="1" />
    <attribute name="MinAngle" value="275" std_dev="2"  rel="1" />
    <attribute name="MaxAngle" value="331" std_dev="2"  rel="1" />
  </object>
</message>
```

Fig. 2. Example of an XML message used in the MRT framework

parsing. These advantages are partially offset by some drawbacks: the amount of transferred data typically increases, parsing may need more time, and binary data can not be included directly in the message. However, none of these drawbacks has been a limitation in our applications.

Each message contains some general information (e.g., the time-stamp) and a list of objects each characterized by a name and its membership class. For each object it is possible to define a number of attributes, that are tuples of name, value, and, optionally, variability and reliability. In order to correctly parse the content of the messages, modules share a common ontology that defines the semantic of the used symbols. In Figure 2 we have reported an example of XML message produced by the vision sensor module.

4 Applications

We have used the MRT framework to develop robotic applications with different robot platforms and in a few contexts. The general structure of the Modular Robotics Toolkit is reported in Figure 3 including all the implemented modules and the typical message passing connections between them.

Since we have used different robotic platforms, from custom bases with differential drive or omnidirectional wheels to commercial all-terrain platforms, the use of a proper abstraction layer in sensing and actuation modules allowed us to focus mainly on the development of robot intelligence.

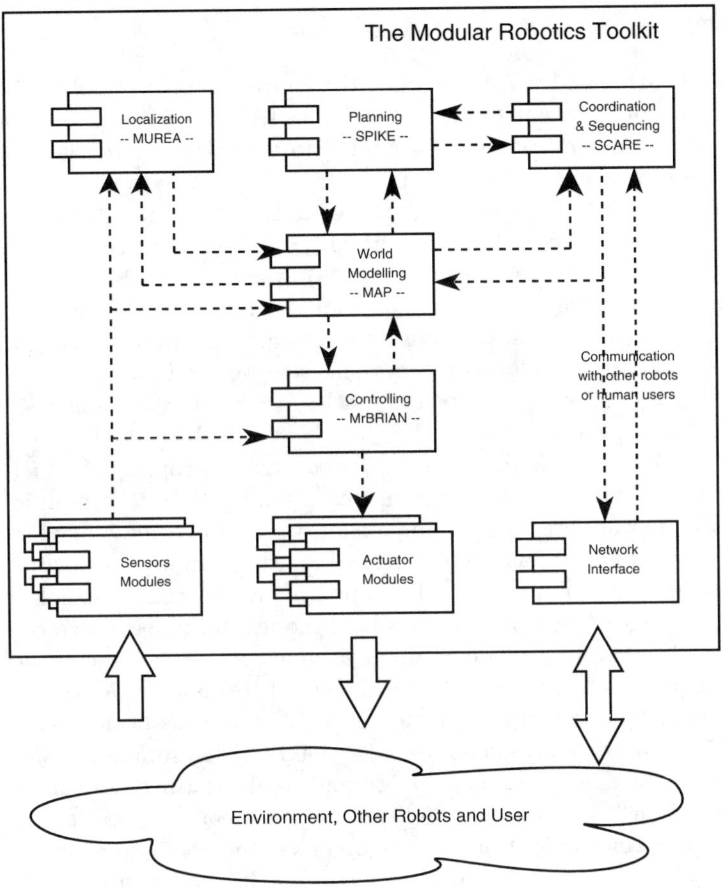

Fig. 3. The general architecture implemented by the Modular Robotics Toolkit; all the reasoning modules are present and typical message passing paths are reported using dashed arrows.

An application might not need all of the modules implemented in MRT, and the modular structure makes it possible to integrate also only a subset of the available modules. In the following, we present three different architectures implemented with MRT to solve specific tasks. We are just interested in presenting three case studies where MRT was successfully used to reduce in a sensible way the "time to market" for the requested application, thanks to its modularity.

4.1 Roby

ROBY is a project (in the framework of the European project GALILEO) that we have developed in collaboration with the Italian company InfoSolution[1]. In this project, an ActiveMedia P3-AT platform with a differential drive kinematics, equipped with a Differential GPS device, must travel between two points, chosen over a pre-defined map of an outdoor environment. A path planner, different from SPIKE, was already available and run on a machine not on the robot. It computed the sequence of points that describe a free path, and sent the set of way points to the robot through a wireless connection.

The map of the planner contains only information about static, known objects (e.g., buildings, bridges, etc.); for this reason, the robot is also equipped with a sonar belt in order to detect obstacles, so that it can manage situations in which the trajectory produced by the planner is obstructed by something, possibly not included in the map (e.g., cars, trees, people, etc). In Figure 4 we show the architecture of the ROBY case study. Only few modules of the general architecture are used to implement a path-following behavior. In this case, MRT modularity has been exploited for the seamless integration in the control architecture of a planner developed by an external contractor.

In this application, no complex world modelling is needed since the differential GPS already provides a good estimation of the robot position and this is sufficient to accomplish the task. Even if this is a single robot domain, the sequential functionality, implemented in SCARE, has been used to enable MrBRIAN to follow the sequence of path–points with a simple reactive behavior. Each job `ReachPathPointJob` terminates when either the robot reaches the related path point (success condition), or a timeout expires (failure condition). In the former case, the current job ends, and the `ReachPathPointJob` job related to the next path point is activated. In the latter case, the whole task is aborted, and the planning module is requested to produce, starting from the current robot position, a new sequence of path points.

During navigation, data collected from the sonar belt are used directly by MrBRIAN to feed the reactive `AvoidObstacle` behavior, since we want our robot to promptly react in avoiding collisions. This behavior has a higher priority w.r.t. `ReachPathPoint`, so that when and obstacle is faced, `AvoidObstacle` tries to avoid it by leaving as less as possible the path that `ReachPathPoint` is suggesting to follow.

Although the implemented structure is very simple, we have obtained satisfactory results in several trials with different conditions, and the robot was always able to reach the goal point dealing with unforeseen obstacles. The wireless connection has been also used by a person to drive the robot in a teleoperated fashion while keeping active sensing modules to implement safety obstacle avoidance.

---

[1] Sample movies on the real robot are available on the at URL `http://www.infosol.it/Movies/offer_roby_movies.htm` on the InfoSolution web site.

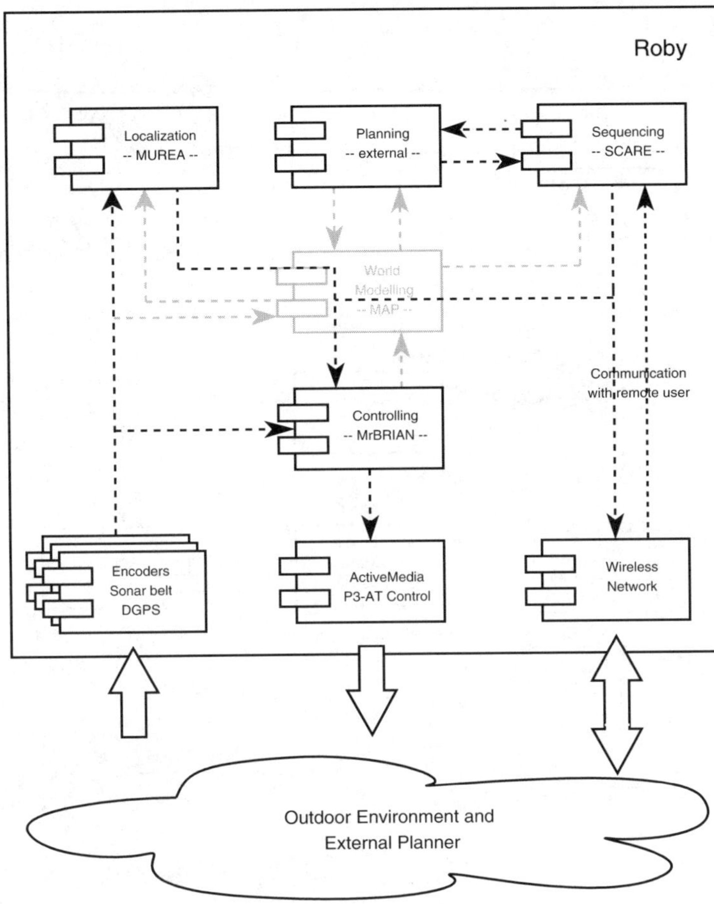

Fig. 4. MRT modules involved in the ROBY case study.

4.2 FollowMe

FOLLOWME is a project to develop a guide robot to be deployed in a museum. In this case, the task to be accomplished is more complex than the one faced by ROBY: the robot has to guide a visitor in the museum; whenever the visitor stops next to an exhibit, it has to wait and move again on the tour as the visitor starts moving again. Sometimes it may happen that the visitor moves away attracted by some interesting exhibit and in this case the robot has to follow her, re–plan the tour and start again the visit as soon as the visitor is ready.

The robot used in this indoor application has a differential drive kinematics with shaft encoders and it is provided with an omnidirectional vision system able to detect obstacles as well as landmarks in the environment. Not having a

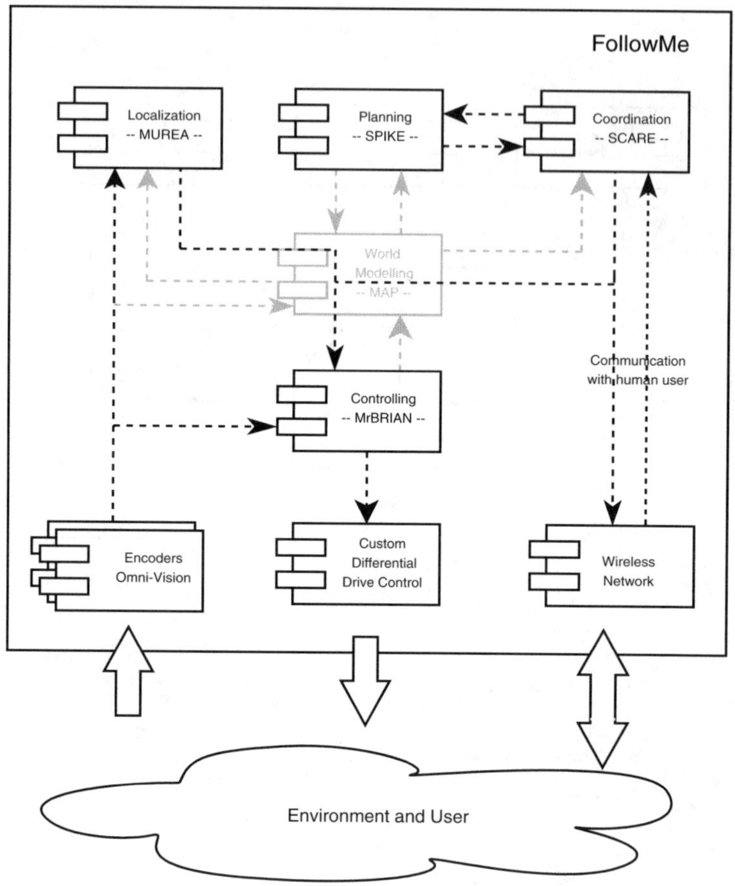

Fig. 5. Modules involved in the FOLLOWME architecture.

reliable positioning sensor like the Differential GPS, this application requires a more complex setup for the localization module that has to fuse sensor information and proposed actuation to get a reliable positioning. MUREA has thus a twofold role in this application: sensor fusion and self localization. Information coming from different sources is fused by using the framework of evidence, and the robot position is estimated as the pose that accumulate the most evidence [RSM02].

Figure 5 describes the MRT modules involved in the FOLLOWME application. In this case the architecture includes also the trajectory planner SPIKE, triggered by the sequencing module SCARE whenever a new tour is required or a path has to be re–planned. Almost any behavior used in MrBRIAN and job in SCARE for the ROBY application have been re–used, as the message containing points generated by SPIKE is equivalent to the message transmitted by the external planner in that application. In addition to the forementioned `ReachPathPoint` and `AvoidObstacle` behaviors, here we have also the `SeekVisitor` and `WaitForVisitor` behaviors respectively in charge of getting close and leading the visitor. In the behavior hierarchy the latter behaviors subsume `ReachPathPoint` in order to properly perform the job and are subsumed by `AvoidObstacle` for obvious safety reasons. Notice that `AvoidObstacle` is exactly the same used in the ROBY application.

This time, the coordination is limited to the interface with the user to acquire the characteristics of the tour (i.e., destination, object of interests, and so on) and SCARE is mostly used as sequencer. Jobs as `SeekVisitorJob` and `ReachPathPointJob` have been implemented, but this time the latter was implemented as the combination of the `ReachPathPoint` and `WaitForVisitor` behaviors.

4.3 Milan RoboCup Team

We participate to the Middle Size League of the RoboCup competition [VBS02] since 1998, at the beginning in the Italian National ART team, and since 2001 as Milan RoboCup Team. RoboCup provides a multi-robot domain with both cooperative and adversarial aspects, and it is perfectly suited to test the effectiveness of the Modular Robotics Toolkit in an environment different from the ones mentioned above.

The development of the Modular Robotics Toolkit followed a parallel evolution with the Robocup Team. At the very beginning there was only a reactive architecture implemented by BRIAN (i.e., the first version of MrBRIAN), after that, SCARE and MUREA followed, to realize an architecture resembling the ROBY case study presented before. While the planning module was developed to fulfill the needs of the FOLLOWME application, MAP has been developed to model the complex task of dealing with sensor fusion in a multi-robot scenario and opponent modelling.

The robots used in this indoor application have different kinematics; we adopt three differential drive custom platforms, two omnidirectional robots exploiting three omnidirectional wheels each, and one omnidirectional robot base with four omnidirectional wheels. Only differential drive platforms are provided with shaft encoders, while odometry for omnidirectional robots is computed using measures acquired by optical mice [BMR05]. Omnidirectional vision is used to detect obstacles as well as the ball and landmarks in the environment.

The software architecture implemented with MRT to be used in these robots is quite different from the previous ones. Since RoboCup is a very dynamic environment the robot behaviors need to be mostly reactive and planning is not used. On the other hand, RoboCup is a multi–robot application and coordination is needed to exploit an effective team play. From the schema reported in Figure 6 it is possible to notice the central role that MAP, the world modelling module, plays in this scenario. Sensor measurements (i.e.,

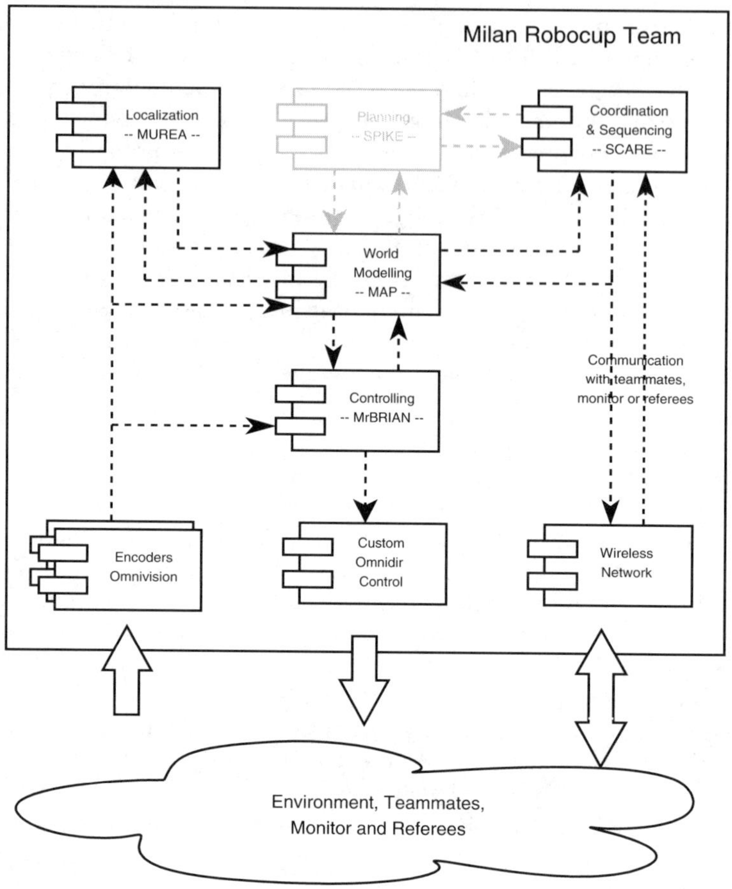

Fig. 6. Modules involved in the MRT architecture for the Milan RoboCup Team.

perceptions about visual landmarks and encoders when available) are fused with information coming from teammates to build a robust representation of the world. This sensor fusion provides a stable, enriched and up-to-date source of information for MrBRIAN and SCARE so that controlling and coordination can take advantage from perceptions and believes of other robots.

Coordination plays an important role in this application domain. We have implemented several jobs (`RecoverWanderingBallJob`, `BringBallJob`, `DefensJob`, etc.) and schemes involving few robots (`DefensiveDoubling`, `PassageToMate`, `Blocking`, etc.). These activities require the interaction of several behavioral modules. Also MrBRIAN has been fully exploited in this application; we have organized our behaviors, in a complex hierarchy, reported in Figure 7. At the first level we have put those behavioral modules whose activation is determined also by the information coming from the coordination module. At this level, we find both purely reactive modules and parametric modules (i.e., modules that use coordination information introduced in MAP by SCARE). The higher levels contain only purely reactive behavioral modules, whose aim is to manage critical situations (e.g., avoiding collisions, or leaving the area after a timeout).

In order to give an idea of how the behavioral system works, let us consider the case of the `AvoidObstacle` behavior. The related behavioral module contains some rules which become active when it is detected any obstacle on the path followed by the robot, and they produce actions that modify the current trajectory in order to avoid collisions. Since avoiding obstacles is an important functionality, we have placed the related behavior at a high level of the hierarchy. The `AvoidObstacle` behavior is composed by a set of 32 rules similar to this one:

(AND (ObstacleNordNear)
        (ProposedTanSpeed FORWARD)) ⇒ (DEL.TanSpeed.*FORWARD)
                                                        (TanSpeed STEADY)

that can be read as: "*if* I am facing any obstacle that is near *and* I would like to move forward *then* discard all the actions that propose a forward movement *and* propose to stay steady". Whenever the robot has no obstacle in its proximity, the `AvoidObstacle` module leaves all the proposed actions unchanged to the following level. Notice that the actions proposed by the behaviors at lower levels may be considered, but there is no need to know which behaviors have made the proposals. This interface among behaviors allows to implement independent behavioral modules that can be reused both in different hierarchies (e.g., we may exchange the levels of `AvoidObstacle` and of `KeepBallInKicker` if we would like a less aggressive behavior when the robot has the ball control) and in different applications (e.g., the same `AvoidObstacle` behavior has been used in all the three applications presented, although the subsumed behaviors are different)

5 Conclusions

In this chapter, we have presented MRT, a modular software architecture that has been successfully adopted in several robotic applications. MRT consists of modules, each implementing one of the functionalities we have identified for the application. The modules interact with each other by exchanging XML messages supported by a dedicated middleware.

The modular implementation has made possible to reuse the functional modules and even the basic control modules implemented as behaviors, thus speeding up the development process. We have proved that with few weeks of work, and off–the–shelf modules integrated in a structured architecture such as MRT, it is possible to implement effective applications for autonomous robots.

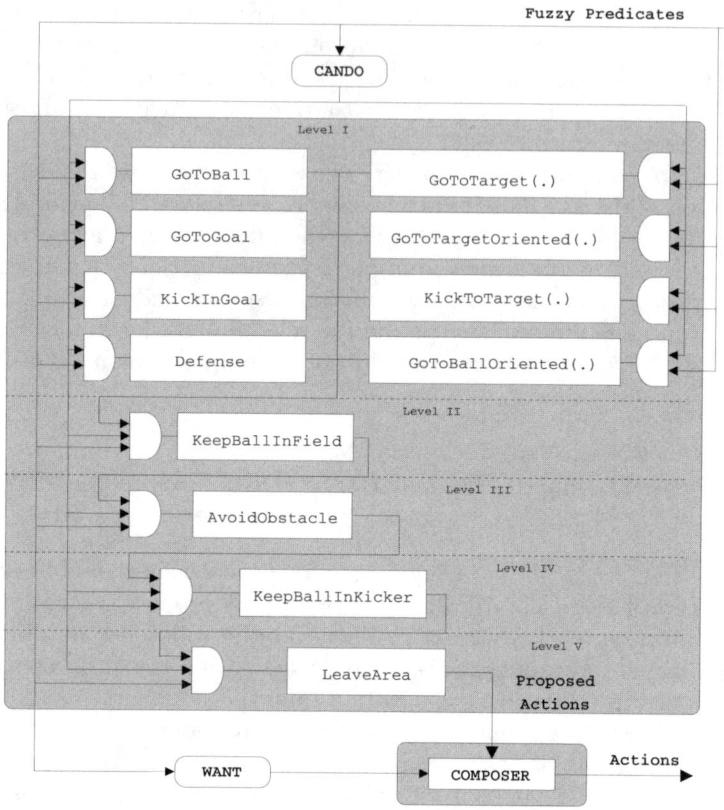

Fig. 7. The behavior architecture used by the Milan RoboCup Team.

Acknowledgements

This project has been partially supported by the PRIN 2002 Project MADSys (Software technologies and knowledge models for design, implementation and testing of multi-agent robotic system in real environments) co-funded by the Italian Ministry for Research, University and Technology (MURST).

References

[Beh04] Sven Behnke, *Local multiresolution path planning*, RoboCup 2003: Robot Soccer World Cup VII (Daniel Polani, Brett Browning, Andrea Bonarini, and Kazuo Yoshida, eds.), Lecture Notes in Computer Science, vol. 3020, Springer, 2004, pp. 332–343.

[BF02] D. Brugali and M. E. Fayad, *Distributed computing in robotics and automation*, IEEE Transactions on Robotics and Automation **18** (2002), no. 4, 409–420.

[BKM05] A. Brooks, T. Kaupp, A. Makarenko, S. Williams, and A. Oreback, *Towards component-based robotics*, Proc. IEEE/RSJ Intl. Conference on Intelligent Robots and Systems, 2005.

[BMR01] Andrea Bonarini, Matteo Matteucci, and Marcello Restelli, *Anchoring: do we need new solutions to an old problem or do we have old solutions for a new problem?*, Proceedings of the AAAI Fall Symposium on Anchoring Symbols to Sensor Data in Single and Multiple Robot Systems (Menlo Park, CA), AAAI Press, 2001, pp. 79–86.

[BMR04] Andrea Bonarini, Matteo Matteucci, and Marcello Restelli, *A novel model to rule behavior interaction*, Proceedings of IAS-8, 2004, pp. 199–206.

[BMR05] Andrea Bonarini, Matteo Matteucci, and Marcello Restelli, *Automatic error detection and reduction for an odometric sensor based on two optical mice*, Proceedings of ICRA 2005, IEEE Press, 2005, pp. 1687–1692.

[BMRS05] A. Bonarini, M. Matteucci, M. Restelli, and D. G. Sorrenti, *Mrt, milan robocup team 2004*, RoboCup 2004: Robot Soccer World Cup VIII (D. Nardi, M. Riedmiller, C. Sammut, and J. Santos-Victor, eds.), Springer, 2005.

[BR02] Andrea Bonarini and Marcello Restelli, *An architecture to implement agents co-operating in dynamic environments*, Proceedings of AAMAS 2002, 2002, pp. 1143–1144.

[Bro] Rodney A. Brooks, *A robust layered control system for a mobile robot*, IEEE Journal of Robotics and Automation **RA-2**, no. 1, 14–23.

[CMBM01] Riccardo Cassinis, Paolo Meriggi, Andrea Bonarini, and Matteo Matteucci, *Device communities development toolkit: An introduction*, Proceedings of EUROBOT 01, 2001.

[Gat98] E. Gat, *Three-layer architectures*, Artificial Intelligence and Mobile Robots: Case Studies of Successful Robot Systems (D. Kortenkamp, R. P. Bonasso, and R. Murphy, eds.), AAAI Press/The MIT Press, Menlo Park, CA, 1998, pp. 195–210.

[GS02] C. D. Gill and W. D. Smart, *Middleware for robots?*, Proceedings of the AAAI Spring Symposium on Intelligent Distributed and Embedded Systems, 2002.

[HRPL95] B. Hayes-Roth, K. Pfleger, P. Lalanda, P. Morignot, and M. Balabanovic, *A domain-specific software architecture for adaptive intelligent systems*, IEEE Transactions on Software Engineering **21** (1995), no. 4, 288–301.

[KAK95] H. Kitano, M. Asada, Y. Kuniyoshi, I. Noda, and E. Osawa, *Robocup: The robot world cup initiative*, Proceedings of IJCAI-95 Workshop on Entertainment and AI/Alife, 1995.

[Mat03] Matteo Matteucci, *Publish/subscribe middleware for robotics: Requirements and state of the art*, Tech. Report 2003.3, Politecnico di Milano, Milan, Italy, 2003.

[PSZ00] M. Piaggio, A. Sgorbissa, and R. Zaccaria, *A programming environment for real-time control of distributed multiple robotic systems*, Advanced Robotic Journal **14** (2000), no. 1, 75–86.

[RSM02] Marcello Restelli, Domenico G. Sorrenti, and Fabio M. Marchese, *Murea: a multi-resolution evidence accumulation method for robot localization in known environments*, Proceedings of IROS 2002, 2002, pp. 415–420.

[USEK02] H. Utz, S. Sablatnog, S. Enderle, and G. K. Kraetzschmar, *Miro - middleware for mobile robot applications*, IEEE Transactions on Robotics and Automation. Special Issue on Object-Oriented Distributed Control Architectures **18** (2002), 493–497.

[VBS02] Manuela M. Veloso, Tucker R. Balch, Peter Stone, Hiroaki Kitano, Fuminori Yamasaki, Ken Endo, Minoru Asada, Mansour Jamzad, B. S. Sadjad, Vahab S. Mirrokni, Moslem Kazemi, Hmid Reza Chitsaz, A. Heydar Noori, Mohammad Taghi Hajiaghayi, and Ehsan Chiniforooshan, *Robocup-2001: The fifth robotic soccer world championships*, AI Magazine **23** (2002), no. 1, 55–68.

Towards Framework-Based UxV Software Systems: An Applied Research Perspective

Gregory S. Broten, Simon P. Monckton, Jared L. Giesbrecht and Jack A. Collier

Defence R&D Canada – Suffield, Canada Firstname.Lastname@drdc-rddc.gc.ca

1 Introduction

Defence R&D Canada changed research direction in 2002 from pure tele-operated land vehicles to general autonomy for land, air, and sea craft (UxV). The unique constraints of the military environment coupled with the complexity of autonomous systems drove DRDC to carefully plan a research and development infrastructure that would provide state of the art tools without restricting research scope.

DRDC's long term objectives for its autonomy program address disparate unmanned ground vehicle (UGV), unattended ground sensor (UGS), air (UAV), and subsea and surface (UUV and USV) vehicles operating together with minimal human oversight. Individually, these systems will range in complexity from simple reconnaissance mini-UAVs streaming video to sophisticated autonomous combat UGVs exploiting embedded and remote sensing. Together, these systems can provide low risk, long endurance, battlefield services assuming they can communicate and cooperate with manned and unmanned systems.

Clearly computational, structural, and control requirements will vary greatly in both complexity and scale between platforms. Yet future forces must share a common infrastructure, command and control to participate in joint operations. This chapter describes DRDC's experiences during the migration from a legacy, tele-operated system, to a modern frameworks based robotics program. To assist this migration DRDC chose a whitebox framework approach that extended the capabilities of the existing, academic based, Miro framework. By adapting an existing framework DRDC leapfrogged directly to the application development phase of porting legacy systems into modern robotic application and waived the need to develop an applicable framework. This whitebox approach saved DRDC a considerable amount of time and effort over the alternative "develop from scratch" option.

D. Brugali (Ed.): Software Eng. for Experimental Robotics. STAR 30, pp. 365–393, 2007.

Fig. 1. Tele-Operated Vehicles

2 Background

Over the past 20 years, Defence R&D Canada developed numerous tele-operated unmanned ground vehicles, many founded on the ANCÆUS [GBV03] command and control system. First fielded in 1990, ANCÆUS solved difficult wireless networking and vehicle control problems five years or more before similar military systems appeared internationally. Further, ANCÆUS demonstrated the viability of network architectures for fieldable robotic systems and multivehicle control. With remarkable foresight, the developers decoupled the vehicle platform from both the commands and communications interface. The ANCÆUS design proved adaptable to other platforms such as the Barracuda sea target, the combat CAT, the ILDP de-mining vehicle and the tele-operated Bobcat with a backhoe, depicted in Figure 1.

Hardware and software development eventually overtook ANCÆUS, however. Substituting today's technology, ANCÆUS becomes a modest command interface similar to the Joint Architecture for Unmanned Systems (JAUS)[JAU04]. Nevertheless, DRDC learned some valuable lessons in this early program, specifically that systems become more expensive, less reliable, and time consuming with:

- the need to port, extend and maintain customized software across platforms and payloads.
- platforms and payloads that must evolve continuously to avoid obsolescence.
- platforms and payloads that are task dependent.

Despite the power and flexibility provided by human operators, tele-operation is fundamentally manpower intensive and communications links can be fragile. Thus, DRDC launched the Autonomous Land Systems (ALS) Project shifting its research focus to *autonomous* UGVs. The project objectives were to build personnel and technical capability in multivehicle robot systems targeted at reconnaissance in medium complexity terrain.

2.1 The ALS Project

To pursue the long term objectives within and beyond the ALS project, DRDC needed an R&D strategy that would permit multiple investigators to collaborate on unmanned systems. Unlike commercial system integrators on an industrial project, DRDC relies on voluntary contributions from a loosely organized team across the agency to a develop and test a research system. Only a set of long term objectives, some specific performance expectations, a sketched hybrid arbitrated control system, a target vehicle, and sensor suite constrained system development. Over a period of 15 months, DRDC-Suffield assembled two robotic systems: one founded on the SegwayRMP and the other on converted Koyker Raptor. The SegwayRMP project familiarized the team with Player/Stage[BGH03], basic sensing, and the perils of systems integration. The Raptor project would represent new development using distributed processing software, sophisticated sensing and mapping.

2.2 Requirements Analysis

Within this general context, the ALS project captured specific requirements, treating the UGVs as 'black boxes' and establishing the outside requirements of vehicle performance. The requirements document outlined, in detail, the team's expectations of a land multivehicle system in two *Autonomous Search* (AS) demonstrations: *Terrain Discovery (AS-TD)* and *Map Guided (AS-MG)*. In AS-TD, the demonstration would show a multivehicle, cooperative, silent reconnaissance task with a minimum of stop-think behaviour. Quoting from the document:

1. a *complement* of vehicles,
2. through an MMI accept a *search mission specification* .
3. *navigate* in *formation* to a search area at least 300 to 500 metres distant, *identifying* and *reporting* targets if any
4. *cooperatively search* the area, *identifying* and *reporting* targets if any
5. return to a base following a different route, *identifying* and *reporting* targets if any

AS-MG would be identical with the change that a mission plan would be developed autonomously based on a-priori elevation maps. These loose objective statements were expanded into more detailed, often numerical, requirements

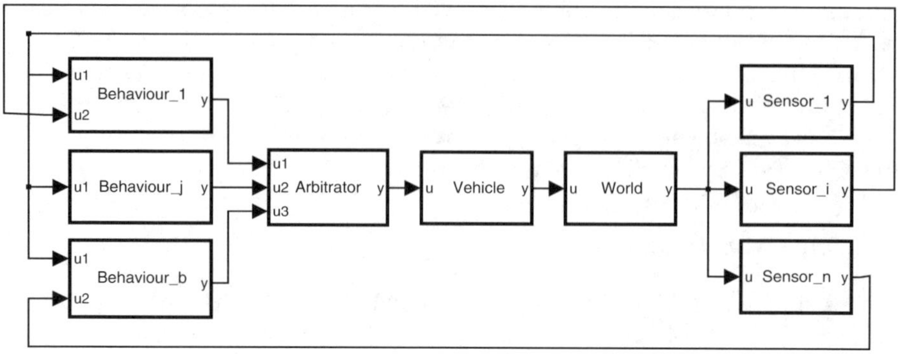

Fig. 2. An abstract control block diagram of arbitrated hybrid control.

such as ground speed and obstacle dimensions. Ultimately, though, the requirements served both as a measure of progress towards an ideal and as a means to constrain software development.

2.3 Conceptual Design

With the 'black box' characteristics of the system prescribed, the team could explore plausible vehicle control architectures. Early in the project the team settled on the increasingly popular hybrid arbitrated control philosophy (e.g. [Ros95]) exemplified in Figure 2. In this structure, an 'arbitrator' combines or selects a course of action based on the input from multiple controllers. Generating a common output format, each behaviour controller regulates a subset of observed variables based on unique inputs and algorithms. For example: though obstacle avoidance and waypoint following both issue steering and speed commands, obstacle avoidance regulates the clearance distance between the vehicle and an obstacle, while waypoint following regulates steering to follow a desired track. The arbitrator combines or switches between these inputs to ensure safe travel to the objective. This platform-neutral control structure structure offers scalability and extensibility through computing parallelism, but requires sophisticated computing middleware. The following sections detail the system's sensing and vehicle components.

Platform

DRDC selected a modified Koyker Raptor as the unmanned ground vehicle (UGV) demonstrator. In each, a 25Hp gasoline engine powered a 4x4 hydrostatic drivetrain and generated 1.5 kW surplus electrical power. The vehicles' payload included power inverters, quad and single Pentium servers, ethernet and USB hubs. A SpeedLan router provided 802.11b class wireless mesh networking.

Proprioceptive Sensing

The sensing system collected raw position and orientation data from a GPS, an IMU, and odometry. The Sokkia GSR2600 GPS, combined with a Pacific Crest PDL RVR radio, supplied differentially corrected GPS positions at 4 Hz. Two Honeywell 1GT101DC hall effect sensors provided on-board software with quadrature pulse trains describing the displacement, velocity and direction of each front wheel. A Microstrain 3DM-GX1 produced orientation and angular rates with respect to gravity and magnetic north.

Exteroceptive Sensing

The Raptor system captured range data through active laser and stereo devices. DRDC and Scientific Instrumentation Ltd. developed a nodding mechanism for a 2-D SICK, creating 3-D data scanning system. Communicating through an ethernet interface, the embedded RTEMS controller nods the laser up to 90°/sec with 0.072 degree resolution and 4cm accuracy over 1 - 30 metres. DRDC adopted Point Grey's Digiclops system to provide high speed range image streams. The Digiclops develops a disparity map between three camera image streams, publishing the resulting 3D range image stream over an IEEE-1394 Firewire digital connection.

3 Managing Software Complexity

The relentless march of ever increasing microprocessor computational capabilities, as given by Moore's Law, has resulted in small, efficient and portable computers that deliver traditional "super computer" capabilities. This growth in computational processing power is an enabling technology that allows programmers to implement evermore elaborate software.

Today's UxVs, with an extensive suite of sensors and capabilities, require sophisticated algorithms that yield complicated software implementations. Such UxV software engineering challenges development teams to efficiently and effectively manage the complexity, size, and diversity of software components. Robtics researchers have seized upon two important software concepts, middleware and frameworks, in their quest to control this complexity.

3.1 Networking Middleware

Given its long term objectives, DRDC identified communications middleware as a key tool for managing software complexity[Gow00b] and evaluated a number of middleware toolkits. The evaluation process centred on mitigating the research life cycle risks found in design, development, maintenance, publication and commercialization of research software that, nevertheless, approaches industrial quality and reliability. The requirements listed below, though subjective, captured DRDC's primary concerns:

- Open Source: Toolkits with Open Source Initiative licensing encourage the rapid publication and growth of new techniques within and between institutions without licensing encumbrances[Ini05].
- Open Tools: Support and use of open source tools and operating systems ensures that developers do not become orphaned by proprietary techniques.
- Component Based: Software based on component services, rather than single object hierarchies, to encourage modular, scalable, and distributed autonomous systems.
- Messaging: Multiple messaging paradigms frees developers to adopt any messaging protocol, process distribution, and structure.
- Comprehensiveness: Additional services, such as logging or exception handling, can often simplify or improve research software.
- Portability: Multiple platforms are a certainty in DRDC's UxV program, making toolkit portability across operating systems a crucial consideration.
- Reconfiguration: Mission and payload variation makes dynamic reconfiguration highly desireable.
- Applicability: The toolkit must not restrict the application domain and must accommodate current and future robotic architectures, payloads, and missions, particularly multi-robot tasks.
- Resource Usage: With drastic variation in possible scale, memory and storage space should be conserved and/or support off-loaded computation.
- Performance: Autonomous vehicles need fast processing to move at human rates on the battlefield, making high control rates with low overhead very important.
- Usage: Popularity and maturity of the toolkit influences adoption and the number of comparative experiments.
- Ease of use: Similarly, ease of use will widen the toolkit's popularity and lower development costs.

Table 1 lists communications middleware toolkits reviewed by DRDC. The middleware attributes were subjectively ranked on a 5 point scale, with 1 denoting a low ranking. Table 2 lists the middleware name acronyms used in Table 1 and provides references.

While no consensus has emerged on the best toolkit, clearly many recognize the significance of UxV middleware. From DRDC's perspective only three middleware toolkits, ACE, TAO and ICE, ranked as possible candidates. The Carnegie Mellion University trio, IPT, RTC and IPC, were too narrow in their application domain. NML, with its close links to the synchronous NASREM/RCS architecture[Alb97], lacked flexibility. The proprietary NDDS and .NET products deny impartial software review while restricting the distribution and sharing of software. Promising the capabilities of CORBA without its complexity, the ICE middleware is such a new product that both the adoption

Table 1. Summary of Communications Middleware Toolkits and evaluation criteria

Criteria	IPT	RTC	NML	NDDS	IPC	ACE	TAO	.NET	ICE
Open Source	5	2	5	1	5	5	5	1	3
Open Tools	5	5	5	1	5	5	5	1	4
Component Based	2	2	3	4	2	5	5	5	5
Messaging	2	2	2	3	2	3	5	5	5
Comprehensive	1	1	3	3	1	3	5	5	5
Portability	3	3	4	3	3	5	5	1	5
Reconfiguration	3	3	2	4	3	4	5	5	5
Applicability	3	3	3	3	3	5	5	5	5
Resources	5	5	5	4	5	3	1	1	3
Performance	5	5	5	5	5	5	3	2	4
Usage	1	2	3	3	1	5	5	5	3
Ease of Use	4	4	3	4	4	3	2	2	3

Table 2. Middleware Acronyms

Acronym	Name
IPT	Interprocess Communications Toolkit[Gow96]
RTC	Real-Time Communications[Ped98]
NML	Neutral Messaging Language[MP00]
NDDS	Network Data Distribution System[GPCH99]
IPC	Inter-Process Communications[Sim91]
ACE	Adaptive Communications Environment[SH02, SHS04]
TAO	The ACE ORB[HV99, Bol02, SK00, DSM98]
.NET	Microsoft Web Services Strategy
ICE	Internet Communications Engine[Hen04]

and long term viability are unknown. Thus DRDC concluded ACE and TAO are the most suitable communications middleware for UxVs.

3.2 Robot Middleware

Changing focus from tele-operation to autonomy, DRDC reviewed the history of robotic architectures and examined current architecture trends. Over the last 20 years, robotics has shifted from philosophically grounded deliberative artificial intelligence and reactive subsumption-esque architectures to pragmatic make-it-work hybrid architectures. This migration to pragmatism, captured by Gowdy[Gow00a], resonates with DRDC's long experience with UGVs. Gowdy groups robotic architectures into those driven by either reference or emergent architecture philosophies. Reference architectures establish idealized guidelines for component design and data flow. The emergent philosophy eschews any single reference design, advocating software frameworks that allow architecture to emerge and evolve. In retrospect, DRDC's ANCÆUS was a reference architecture that, ultimately, could not provide the necessary flexibility, extensibility, and scalability for autonomous military robotics. Additionally, ANCÆUS was a "closed system", supported and used by only DRDC and a few selected contractors, thus making it difficult for DRDC to leverage and utilize other sources of research.

DRDC concluded from its ANCÆUS experience that open source software and collaborative research are a crucial means of capturing and communicating science, the ultimate product of DRDC's UxV research.

DRDC evaluated candidate robotics frameworks using the same criteria as networking middleware, with the addition that each framework should:

- support an emergent architectural design strategy.
- support modular, extensible, flexible and scalable software systems.
- use standard communications middleware toolkits.

Numerous robotics architectures were reviewed[GBD04] including Player/Stage[BGH03], DRCS[Alb97], NASREM[JAH87], TCA[Sim94], Dervish[INB95], CLARAty[RVD01], RAP[Fir89], Xavier[SKS98], Saphira[KM98] and 3T[RBS97], Berra[MLC00], Marie[CCT04], Carmen[MMT03], OCP[LWV01], Miro[HUK02, HUS05] and Orca[ABW05]. Table 3 provides more details on a selected list of architectures. The rating given to each category ranges from a low of 1 to a high of 3 and are subjective values from the view points of the authors.

Table 3. Candidate Architectures

Architecture	Open	Tools	Emergent	Comms	Modular	Usage
Player/Stage	3	3	3	1	3	2
Carmen	3	3	2	2	2	1
Marie	3	3	3	3	2	1
Miro	3	3	3	3	3	1
Orca	3	3	3	3	3	1
CLARAty	1	?	?	?	?	1
OCP	1	3	?	3	?	1
Saphira	1	?	?	?	?	2
DRCS/NASREM	3	3	1	3	2	2

Using the above criteria the list of potential candidates was quickly reduced to Player/Stage, Carmen, Marie, Miro and Orca. Obvious candidates such as CLARAty, OCP and Saphira are either closed source or unavailable in Canada, while DRCS and NASREM define reference architectures.

The five remaining frameworks were examined for their ability to meet the needs of implementing autonomy on vehicles at DRDC. This investigation concluded that frameworks based upon the CORBA communications middleware held the most promise it terms of allowing a modular, extensible, flexible and scalable architecture to emerge. Two of the reviewed frameworks were based upon CORBA and only one of these frameworks is completed and in current use. It was concluded that Miro, while originally designed for soccer playing robots, had both the flexibility and expandability to serve as a foundation for DRDC's UxV program.

4 Miro and CORBA

4.1 Introduction

Though very powerful and robust, CORBA has a steep learning curve. As a measure of this complexity, the CORBA bible, *Advanced CORBA Programming with C++*[HV99], at a 1000 pages is surpassed by the two-volume *TAO Developer's Guide* [TAO03] at approximately 1500 pages. Fortunately, for robotics researchers, the Miro framework encapsulates much of this complexity within a small, manageable toolset[Utz03].

Implementing services under Miro requires understanding six key concepts, depicted in Figure 3:

- Interface Description Language
- Naming Service
- Servers
- Clients
- Event Channels and Polling
- Data Exchange Patterns

Figure 3 shows the relationships between these concepts, while the following sections introduce these concepts in more detail.

Interface Description Language

The Interface Description Language (IDL) describes the properties of a CORBA object in terms of data types and access methods. The IDL compiler transforms this description into client and server side code using a language binding. Thus, via the IDL language, a CORBA object represents a component with an interface that allows independent processes to exchange information in a network transparent manner. While TAO supports numerous language bindings including Java, C++, C and Fortran, Miro currently uses only the C++ language binding.

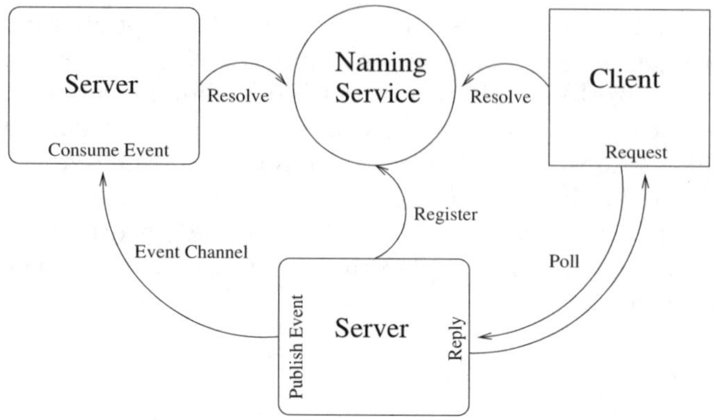

Fig. 3. Key Miro and CORBA Services

Naming Service

Much like a phone book's *white pages* that maps names to phone numbers, the CORBA Naming Service facilitates data exchange between processes by mapping object names and event channel names, both simple text strings, to object references and event objects respectively. In the Miro context, the Naming Service allows a Miro server or client to resolve an object name to an external object reference and instantiate the object reference locally, even though it exists on another Miro server. For event channels the Naming Service enables a Miro server to subscribe to an event channel using a simple text string.

Miro Server

Using the Miro server framework, both traditional client-server transactions and event driven processing can occur simultaneously. *Events* can be broadcast or *published* on an event channel to multiple *subscribers*, while the server can respond to traditional polling requests from a client [1]. The Miro server uses the server side IDL object binding to setup, maintain and allow remote (polling) access to an internal CORBA object.

Miro Client

The Miro client defines a framework, using the client side IDL object binding that allows a client process to poll a Miro server in a familiar client-server transaction.

[1] A CORBA derived hybrid server is both a server and a client.

Event Channels and Polling

Using CORBA capabilities Miro implements both the message and information paradigms, as polling and event channels respectively. Under the polling paradigm a Miro client directly requests and receives data from a Miro server. Events channels, implemented using the CORBA Notification Service, allow a Miro server to anonymously subscribe to events published by another Miro server. Figure 3 shows both of these data exchange implementations.

Data Exchanges Patterns

The Miro framework enables data exchanges that adhere to specific patterns. For a Miro server these patterns include:

- Register an event channel under a text string name.
- Register an object reference using a text string name.
- Subscribe to an event channel and receive published events.
- Resolve an object name into an object reference and polling the object interface for data.

The Miro client is limited to resolving an object name into an object reference and polling the object interface for data.

4.2 Design Patterns

The Miro server and client processes, referred to in Section 4.1, are constructed using one of three basic design patterns. All software implemented using these design patterns have defined CORBA object interfaces, or in software parlance these design patterns yield *components*.

The following sections provide details on the implementation and use of these design patterns.

Subscribe-Publish Server

The *Subscribe-Publish server* design pattern, illustrated by the collaboration diagram shown in Figure 4, allows a Miro server to receive events, process the data, and publish events. Using this design pattern the Miro server becomes an independent component with its interfaces defined by the CORBA objects that facilitate event subscription and publication as well as the CORBA objects that enable polling.

This design pattern results from the collaboration of the following classes:

- The Structured PushConsumer class responds to events published on event channels by one or more Miro servers.
- The Algorithm processes the data provided by the event.
- The Implementation class contains:

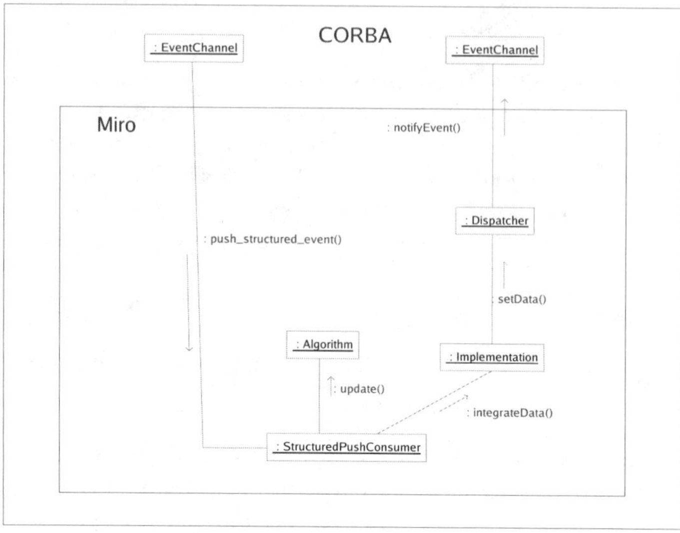

Fig. 4. Publish-Subscribe Server Design Pattern

  - Defines the the CORBA object's polling methods thus allowing the server to respond to polling requests.
  - Invokes the Dispatcher class to publish events.
- Finally the Dispatcher class is responsible for supplying the CORBA objects to the event channel where they can be distributed to other Miro servers.

Publish Server and Reactor

The *Publish Server and Reactor* design pattern serves as the basis for handling external hardware/software entities. Its fundamental objective is to separate the physical device driver from the software that publishes events and enables polling. Figure 5 shows a collaboration diagram detailing structure of this pattern.

The following classes collaborate to implement this design pattern:

- The Connection class is responsible for managing the connection between the physical device and Miro, which includes establishing the connection and writing commands to the device.
- The interrupt driven EventHandler class receives and processes data from the Connection class when data is available. This processing includes packet synchronization and the construction of a message from the low level byte stream.
- The Message class handles the routing of completed messages to the Consumer class.

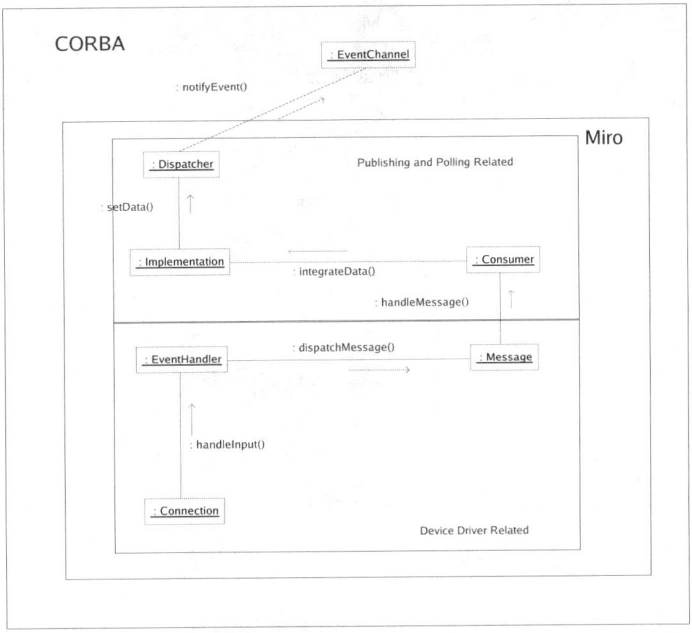

Fig. 5. Publish Server and Reactor Design Pattern

- The Consumer class is responsible for taking the device specific message and converting it into a CORBA object. It is important to note that all classes up to this point are device specific while the output from the Consumer class uses a generic CORBA object and thus different devices of the same type may reuse the same Implementation class.
- The Implementation class:
  - Defines the polling methods of the CORBA object, thus allowing the server to respond to polling requests.
  - Invokes the Dispatcher class to publish events.
- Finally the Dispatcher class is responsible for supplying the CORBA objects to the event channel where they can be distributed to other Miro servers.

Client

The client design pattern, shown in Figure 6, allows a client process to poll data from a Miro server. The client application's Implementation class allows:

- The resolution of a CORBA object reference using an object name lookup in the Naming Service.

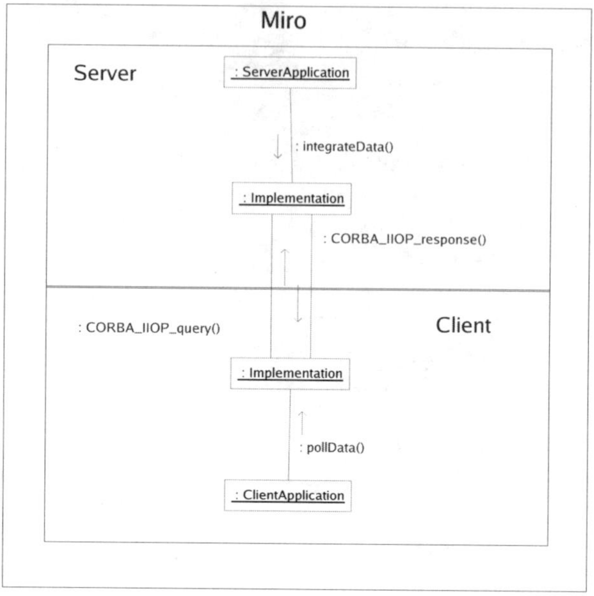

Fig. 6. Client Design Pattern

- Polling uses the CORBA Internet Inter-ORB Protocol to request query and receive data from Miro server application[2].

4.3 A High Level Example: World Representation

Introduction

Section 4.1 introduced basic Miro and CORBA concepts, while Section 4.2 gave an overview of the design patterns used to implement Miro services. This section illustrates how the Miro framework, design patterns, and CORBA services are used to implement world representation capabilities for a UxV.

World Representation

An autonomous UxV requires a representation of its world, which allows it to safely avoid obstacles while navigating. Figure 7 shows a flow diagram representation of the software that implements range data acquisition and the creation a terrain map representation of the world.

The generation of a world representation utilizes three services, the CORBA Naming Service, the Laser component and the Map component. The Naming Service is a standard CORBA service whose purpose and usage was described

[2] The identical Implementation class is instantiated on the client and server processes.

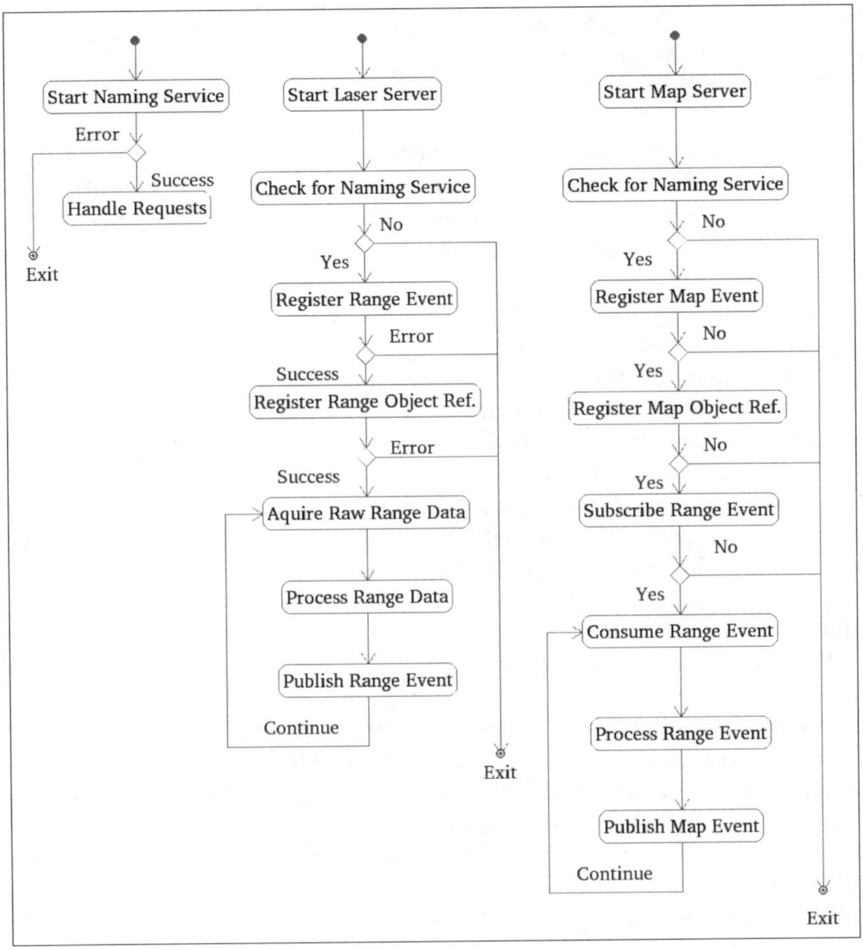

Fig. 7. Range Data and Map Generation Flow Diagram

in Section 4.1. The following sections describe the implementation of the Laser and the Map components using the Miro framework and design patterns.

Laser Component

The Laser component acquires raw range data from a laser range finder, processes this data into a 3D representation and then publishes the data as a *Range* event. It requires a physical device driver for the external nodding SICK laser data source and thus uses the *Publish Server and Reactor* design pattern described in 4.2 as a template.

Upon startup the Laser component first verifies a Naming Service is running. It then registers the *Range* event object, under the event channel named

EventChannel, with the Naming Service. It also registers the *Range* object reference with the Naming Service which, when resolved, allows a client to poll for *Range* object data.

Having successfully completed the event and object registration, this component then initializes the laser and begins to acquire raw range data. This raw range data is converted to a 3D representation and the 3D data is published to the *EventChannel* as a *Range* object event. This process is repeated on the 26.6 ms update period of the SICK laser.

Map Component

The Map component uses 3D range data to create a $2\frac{1}{2}$ D terrain map representation of the world. This component receives and publishes events on an event channel, thus the *Publish-Subscribe Server* design pattern described in 4.2 served as the template for this implementation.

As a first step the Map component verifies the existence of the Naming Service and then registers the *Map* event object against the event channel named *EventChannel*. It also registers the *Map* event object, under the *EventChannel*, with Naming Service and registers the *Map* object reference with the Naming Service.

With the registration completed the Map component then resolves and subscribes to *Range* events. The 3D Range data, provided by the *Range* object events, is delivered under the information based paradigm, as there is no direction connection between the Laser and Map components. The Laser component anonymously produces *Laser* events, while the Map component anonymously consumes these events. Thus as detailed in Section 9, the Laser component can easily be replaced by the Logging component as the producer of *Range* events.

Upon receiving a *Range* object event the Map component processes this 3D range data, creates the $2\frac{1}{2}$ D terrain map and publishes a *Map* object event.

5 Design Process

The requirements analysis and the available hardware placed necessary constraints on the development process. Working down from the requirements and up from the available hardware, an initial system architecture was constructed which detailed software components and established the dataflow between elements. Negotiation between the suppliers and consumers resulted in skeleton CORBA objects, defined via IDL interfaces.

The Miro infrastructure and CORBA object discipline provide a flexible development environment with firm deliverables. Miro decouples the algorithm from the interface implementation, limiting a given module's sensitivity to 'upstream' process failures and reducing the impact of internal failure on other downstream processes. Consuming from and publishing to established CORBA object interfaces, the developer may adopt any algorithm or technique within the process. Thus, the CORBA object based skeleton provides the freedom to implement any algorithm, the security from other poorly behaved processes, and the responsibility to deliver to agreed interfaces.

Miro and the CORBA object skeleton approach effectively seeds system development. Interfaces mark both a start and endpoint of a development effort and Miro's design patterns provide basic communication mechanics. Together, a raw skeletal image of a system can be rapidly assembled. If the algorithm development method is iterative – often the case in research– successive improvements can be incorporated to working skeleton code with the confidence that the mechanical details of interprocess communications are fully functional.

Both the skeleton and design patterns freed DRDC staff to develop algorithms within only partially working systems. Routinely, components were exercised in isolation using the minimum Miro services, while some algorithmic elements were developed in Player/Stage [BGH03] and later 'dropped' into the appropriate skeleton. By adhering faithfully to CORBA object interfaces, investigators engaged in iterative development with minimal impact on others.

Based on a rough hybrid control scheme and the requirements analysis, development began with the negotiation/establishment of the CORBA object interfaces between software components. The early establishment of these CORBA objects benefited development by ensuring that developers understood the information flow through their modules and that holes in the design could be found quickly.

Developers chose to either develop CORBA objects directly using IDL derived data types, or to develop outside of the framework using custom or third party data types. Algorithm efficiency typically drove custom type adoption, while throughput and simplicity drove the use of IDL types.

CORBA object based design guaranteed simple integration of the algorithm into established Miro design patterns using relatively simple data types. However since CORBA does not marshall and unmarshall pointer types, structures containing pointers cannot be transferred between processes. This forced IDL-based code to use simpler data types, often at the cost of algorithm efficiency.

Developers using custom data structures freely adopted the most suitable data types but ultimately had to convert to and from IDL equivalents. So while this technique favoured algorithmic efficiency, performance was sacrificed in type conversion.

Components were developed using the three design patterns described in 4.2. Low-level hardware drivers were developed using the *Publish Server and Reactor* design pattern as seen in Figure 5, while information processing and control services were developed using the *Publish-Subscribe Server* and *Client* design patterns seen in Figures 4 and 6 respectively.

A number of the low level drivers were leveraged off the Player/Stage project. The ease of integrating Player drivers into Miro is a testament to the design of both systems. Since both Player and CORBA decouple the interface from the implementation a majority of the Player source could be used directly in the Miro based driver. Even though the low-level drivers would undergo minor changes throughout the development cycle, these drivers were and utilized early in the development process. This functionality allowed developers to focus their efforts on algorithmic development.

5.1 Testing

Like development, testing benefits from the process isolation afforded by CORBA objects. Since Miro was not a monolithic system, fragments of the Raptor controller could be launched or killed with little impact. In conjunction with LogPlayer, and archive data, such partial testing could occur on the Desktop and across the network.

Testing Tools

In addition to standard testing tools and practises, DRDC used Miro's Logging toolset and Log Level facilities.

Two components comprise the Miro Logging toolset: the LogNotify and LogPlayer duo, Miro Debug Logging and visual interfaces.

LogNotify and LogPlayer

In realtime, the LogNotify component captures desired events published on an event channel. The graphical LogPlayer component, shown in Figure 8, controls the sequenced, variable rate, data playback over an event channel. Event streams could be accelerated, decelerated, paused, and reversed, permitting close inspection of process failure modes. Further, any event stream could be disabled in the player, permitting selective testing of process event handling. While most processes tolerate such input control, any process using an internal clock would only perform accurately with logged data played back at original rates.

Since downstream processes relied on raw sensor data streams, DRDC usually archived only raw sensor data from trials and not downstream process output. This strategy reduced log size while permitting comparative off-line optimization and debugging of downstream processes.

Figure 9, where the Laser Server component from 7 has been replaced by the LogPlayer component, illustrates the power of these logging tools. In this collaboration diagram the Map component consumes logged *Range* object events published by the LogPlayer and this is accomplished without changing the Map component code or configuration.

The CORBA *Range* event object enables this seamless integration between *device* data and *logged* data by providing a generic, uniform interface.

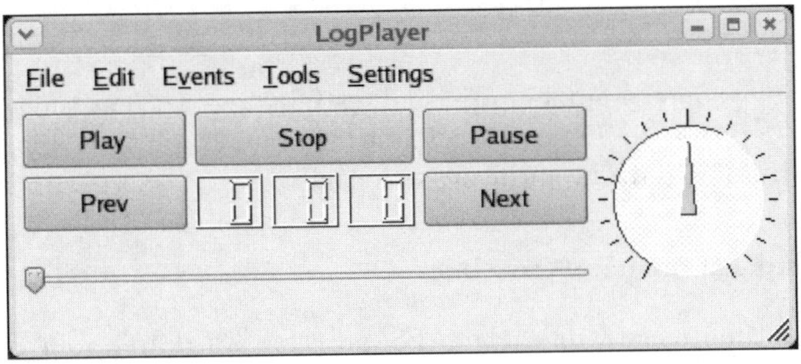

Fig. 8. Miro Log Player

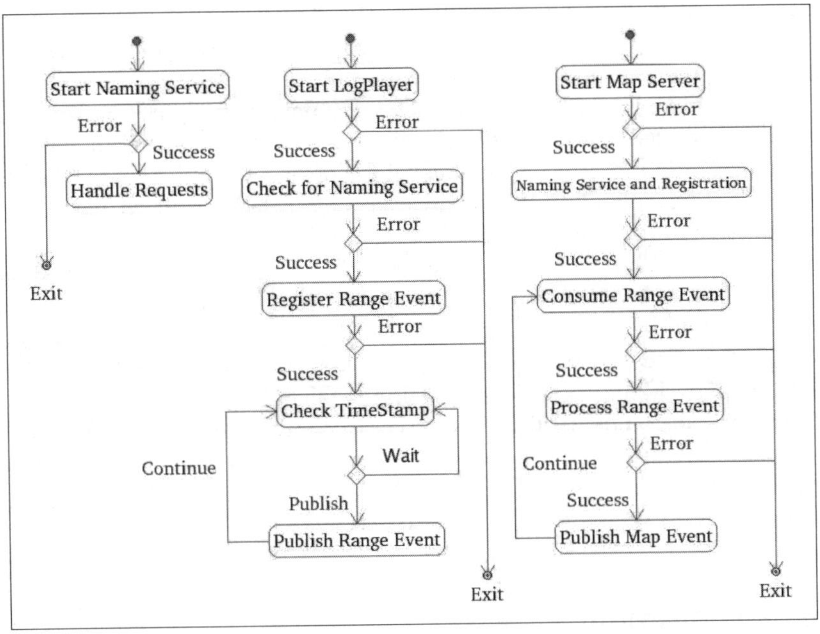

Fig. 9. Map Generation with Logged Range Data Flow Diagram

Miro Debug Logging

Debugging a multi-process, distributed environment frustrates the best development efforts. Invariably, testing methods fall to the simplest mechanism: printed debug statements. Fortunately, Miro's LogLevel utilities permitted testers to output filtered debug statements based on a unique log levels. Processes supporting Loglevels take command line switches that display or suppress debugging information without recompiling.

Visual Interfaces

DRDC also developed a number of visual testing aides, such as the traversability display depicted in Figure 12. These utilities graphically display event channel and polled data in an intuitive format.

Module and Integration Testing

Under tight time constraints, the ALS program tried to avoid integration problems common to large, complex software systems by integrating software onto the Raptor as early as possible. DRDC leveraged the sensor drivers by logging data from field trials and using the LogPlayer to perform desktop testing. Together these strategies maximized the time available for algorithm development and testing.

In the final two weeks before demonstration day, system integration testing incrementally added subsystems to the control network to assess whole-system performance and stability. While unusual for projects of this scope, this short test period sprung from DRDC's arbitrated control design and Miro's run-time properties that, together, permitted extensive preintegration subsystem testing. In general, preintegration testing established stability, correctness, and performance of individual control threads. Final integration testing brought multiple threads together to expose system vulnerabilities and reveal total system performance. In most cases, final system improvements were reduced to parameter tweaking to achieve a desired vehicle behaviour.

This approach lends itself well to a research environment where algorithm content and stability fluctuate constantly. Though desireable, a more rigorous approach would not be possible given the time constraints and scope of the system.

6 Implementation

Using the software facilities discussed in Section 3.1 and design methodology detailed in Section 4 DRDC-Suffield implemented a flexible and modular control system for autonomous UGVs. Trials completed in Fall 2005 demonstrated this software on the Koyker Raptor vehicle and sensor compliment described in Section 2.3, as shown in Figure 10.

The vehicle control system, based upon separate reusable components and reusable interfaces is portable to a variety of autonomous vehicles. These components, derived from CORBA object interfaces and event delivery mechanisms, impart a number of desirable attributes:

- Extendability, allows changes in vehicle platform, sensing or algorithms.
- Flexibility to distribute processes to where computing power is available.
- Reactivity to incoming sensor data.

Fig. 10. One of two modified Koyker Raptors used in DRDC's ALS Project. Each Raptor used one or more roof mounted SICK lasers and stereo cameras; Differential GPS and IMU; and wireless mesh networking routers.

- Scalability to varyings levels of vehicle capability and autonomy.

This section describes the software control system, its components and interfaces, and usage.

6.1 System Overview

The vehicle controller consists of many components, each a process, possibly co-located on one processor, or distributed over a network. They can be organized into four broad categories:

1. Hardware Interfaces - Components that interface directly with hardware, providing data services from GPS, IMU, Laser and Stereo Range sensors, as well as managing the vehicle platform.
2. Information Processing Services - Components providing world modelling and data fusion from the raw sensor data, including Terrain and Traversability Mapping, and pose estimation via the Model Server
3. Vehicle Control Services - Those components that produce vehicle behaviour, such as Tele-operation, Obstacle Avoidance and Waypoint Following, Path Planning and Hazard Detection.
4. Decision Making Services - The Arc Arbiter combines the output of the reactive and deliberative control threads to provide vehicle commands.

These components and their interconnections are shown in Figure 11. The CORBA event channel facilitates the delivery of CORBA event objects between information suppliers and consumers, either local or remote, and is indicated by the data flow connectors in Figure 11. In general, information flows from the Hardware Interfaces, through the Information Processing Services, to the Vehicle Control Services. The use of components, with their defined interfaces, allows for exceptions such as the Hazard Detection algorithm that directly uses laser data.

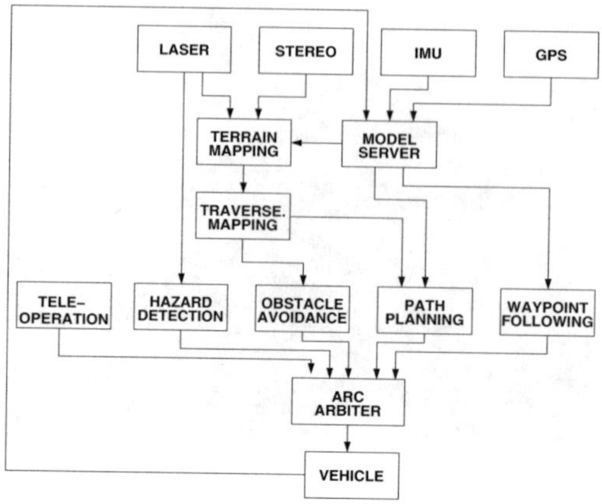

Fig. 11. ALS architecture flow diagram. Each box represents a component that can reside anywhere on a network and each connector represents data flows accessible to any subscriber component.

6.2 Hardware Interfaces

The GPS, IMU, Laser, Stereo and Vehicle components interface with the physical hardware, acquiring data and publishing CORBA event objects. As data abstractions, CORBA objects divorce physical devices from their data representations.

6.3 Information Processing Services

In general, information services such as pose estimation and world representation draw from the hardware interfaces and feed into vehicle control services.

Pose Estimation

Accurate sensing, world modelling and control on large outdoor vehicles requires accurate estimates of both vehicle and sensor locations in local and global coordinates. As shown in Figure 11, Model Server provides these services. With every generated *GPS*, *IMU*, and *WheelOdometry* event object, Model Server updates a direct state Kalman filter of vehicle pose and publishes a CORBA *Pose* event object containing the location of the vehicle in global coordinates.

Model Server maintains an internal geometry database of the system's rigid body assemblies. By polling Model Server through the CORBA *Platform* object reference, clients can interogate Model Server for either a *body* list, a given body's *frame* list, or transformations between any frames in the system. A client process can combine static internal transformations with filtered *pose* events to establish the global coordinates of any vehicle body frame.

World Representation

Intelligent navigation requires a world representation. For the ALS project DRDC represented the world through a Terrain Map and its companion, the Traversability Map.

Terrain Map

The Terrain Map component subscribes to CORBA *Range3d* event object published by the laser and stereo camera exteroceptive sensors. The service fuses range data into a grid map, a rectangular array of regions, to build a 2.5D or digital elevation map[Kel97, HK93, KK92]. The Terrain Map estimates the height of the world at each cell in the 20cm x 20cm grid, as well as the elevation variance. This map is local to the area directly in front of the vehicle, scrolling and wrapping as the vehicle moves in both the X and Y directions. With the new range data fused into the terrain map a CORBA *MapEvent* event object is published.

Traversability Map

The Obstacle Avoidance and Path Planning components of the Vehicle Control Services do not subscribe directly to the Terrain Map, but to a simpler Traversability Map service, depicted in Figure 12. The traversability analysis transforms the terrain map into a traversability map, containing a *Goodness* rating the terrain hazards in a given area, similar to the Gestalt system[SGM02]. For this map the grid cells are typically 50cm x 50cm. An analysis of step and slope hazards in the corresponding Terrain Map grid generates a traversability element's Goodness rating. The Traversability Map component publishes a CORBA *TravMap* event object with each map update.

6.4 Vehicle Control Services

Asynchronous multiple behaviour components produce a sliding scale of autonomy. Each independent behaviour publishes CORBA *ArcVote* event objects. The Arc Arbiter consumes these vote objects and directs the Raptor's movement. The following describes the five behaviour components implemented thus far.

Tele-Operation

This component allows direct joystick control of the vehicle with either a direct or network connection.

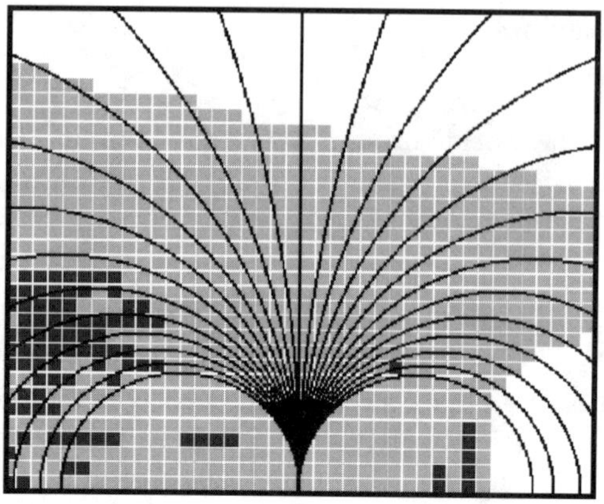

Fig. 12. A traversability map overlaid with 25 candidate arcs. Obstacles are dark cells and traversable areas are light cells.

Hazard Detection

Similar to other obstacle detection systems [HK], this component inspects raw range sensor data from a laser range finder or stereo vision camera. Received as a CORBA *Range3dEvent* event object, this component detects obstacles higher or lower than a user defined safety range. The detection of a qualifying obstacle will halt the vehicle by publishing a veto within a CORBA *ArcVote* event object.

Obstacle Avoidance

This component examines local terrain to establish safe candidate steering trajectories, similar to the Gestalt system [SGM02], and also provides maximum speed recommendations. Subscribing to the CORBA *TravMap* event object, the Obstacle Avoidance component estimates the vehicle's stability and safety over candidate steering arcs using Traversability Map data to construct a CORBA *ArcVote*.

Waypoint Following

The Waypoint Following component use the Pure Pursuit algorithm [Cou92] to follow a path of user defined waypoints. It subscribes to the CORBA *Pose* event objects published by the Model Server, executes the Pure Pursuit algorithm and publishes a CORBA *ArcVote* event object.

Path Planning

This component provides high level path planning and goal directed behaviour based upon a global map of accumulated local Traversability Maps. It accounts for both desirability of terrain as well as goal directed behaviour in making its decisions using the D* Lite algorithm [KL02]. The data for the D* Lite algorithm arrives as CORBA *TravMap* and *Pose* event objects, the algorithm executes, and a CORBA *ArcVote* event object is published.

6.5 Decision Making Services

In order to combine the outputs of all the Vehicle Control Service's behaviours, a vote based arbitration scheme was implemented in the Arc Arbiter component.

All Vehicle Control modules publish their desired behavior using the CORBA *ArcVote* event object. The Arc Arbiter component subscribes to the *ArcVote* event and fuses the individual behaviours into a global action using an arbitration scheme. DRDC expanded a DAMN-like [Ros95] arbitration scheme. Each behaviour votes for each arc in a set of arc candidates, for example as shown in Figure 12. The behaviour gives a desirability, a certainty, a maximum speed and, if unacceptable, a veto to each arc. The IDL code listing shown below illustrates the implementation of CORBA *ArcVote* event object.

```
struct ArcVoteIDL     // Defines a behaviour's vote for a single
arc {
    float curvature;  // Curvature (1/meters)
    float desire;     // Desireability: between 0 (worst) and 1 (best)
    float certainty;  // Confidence: between 0 (least) and 1 (most)
    float max_speed;  // Maximum acceptable speed (meters/second)
    boolean veto;     // "true" vetos this arc
};

// Define the available voting behaviours
enum Voting_Behaviour{HAZARDDETECT, OBSAVOID, PATHPLANNER,
WAYPOINT};

// Define the number of candidate arcs in the system
const long NUM_CAND_ARCS = 25;

struct ArcVoteEventIDL                    // The CORBA ArcVote event
object {
  ArcVoteIDL VoteSet[NUM_CAND_ARCS]; // The array of votes
  TimeIDL time;                      // The time that the vote was generated
  Voting_Behaviour votingBehaviour;  // Identify voting behaviour
};
```

The Arc Arbiter component's event driven design gives the voting components the flexibility to run asynchronously. Further, behaviours may implement as much of the *ArcVote* object as they need. For example, the Hazard Avoidance module uses only the veto field, while the Obstacle Avoidance and Path Planning modules use all fields. Nevertheless, the arbiter combines all available outputs into one coherent decision packaged in a CORBA *Motion* object reference.

The vote based Arbitration scheme and the event based delivery of Vote events creates a sliding scale of autonomy, through a subset of available components. For example, to run teleoperation control, the system needs only three modules: Teleoperation, Arc Arbiter and Vehicle. For autonomous operations in simple environments, Waypoint Following replaces Teleoperation. As the environment increases in complexity, Hazard Detection, Obstacle Avoidance, and Path Planning can be added incrementally to provide greater autonomous capabilities.

7 Conclusions

This chapter summarised DRDC-Suffields's experiences during the migration from a tele-operation focused, to a general autonomy based program.

As part of this migration DRDC-Suffield conducted an in-depth literature review of network middleware, robot middleware, and robot control architectures. This review concluded the most suitable communications middleware is TAO. TAO is a real-time CORBA implementation based on the ACE communications middleware. Fortunately DRDC-Suffield discovered open source, CORBA based Miro framework prior to developing in-house CORBA tools. Though originally developed for Robosoccer robots, Miro possessed a modular, flexible, scalable and extensible design well suited to more complex robot control problems.

DRDC modified Miro for an outdoor UGV, revealing both the advantages and disadvantages of both middleware toolkits and resulting design process. After experience with ANCÆUS tele-operated platforms, DRDC sought a platform-neutral control structure offering scalability and extensibility through computing parallelism. Miro achieved this goal by providing well designed CORBA object interfaces and, therefore, transparent, network-enabled software.

CORBA IDL interfaces decouple the control system both from the sensors and the vehicle. Creating well defined interfaces creates flexibility by removing dependence on specific command sets, for example: the specific range sensor has little impact on terrain mapping. Similarly, vehicle commands (translational and rotational velocity) remain the same regardless of vehicle. With different platform or sensor configurations, integrators need change only the drivers. More importantly for DRDC, the internal control algorithms can be easily changed, allowing direct comparison of their performance.

The CORBA event based delivery mechanisms provided two key benefits. Processes could be distributed over network nodes as needed. Secondly, processing and control modules run on the receipt of new data, using only the most recent data delivered if a recipient is too slow. In this way, no explicit timing is necessary anywhere in the system. Since sensors and algorithms are free to run at any rate, the hybrid design can react to sensory information over any time scale.

CORBA's network transparency allows components to be distributed across processors and networks as required, without the need to change code or configurations. The use of CORBA events or polling has the added benefit of decoupling the execution threads of components, thus allowing components to execute at independent rates.

This modular approach has some shortcomings, however. The complexity and reconfigurability of the system requires significant run-time management. Depending on the sensors and level of autonomy, the user may have to configure, launch, and monitor up to 20 different processes. To date, these tasks have been imperfectly managed through a single XML configuration system for all processes. For monitoring, DRDC has developed a process *watchdog* component that monitors and reports on the status of each component in the system.

CORBA's services and capabilities represented a significant, but worthwhile, learning curve DRDC. The Miro framework served as a crucial aid to understanding and using CORBA by providing templates and examples from which new components could be constructed. Despite CORBA's apparent complexity, robot subsystems remained responsive, often faster than Player-based equivalents. CORBA provides DRDC with industrial strength, platform neutral interprocess communication capabilities, and a plausible military software environment. Miro provides a much needed, accessible toolset focussed on robot systems, greatly easing DRDC's introduction to CORBA.

References

[ABW05] A. Makarenko A. Oreback A. Brooks, T. Kaupp and S. Williams, *Towards component-based robotics*, IEEE/RSJ International Conference on Intelligent Robots and Systems, August 2005.

[Alb97] J. Albus, *Drcs: A reference model architecture for demo iii*, Tech. report, 5994, National Institute of Standards and Technology, Gaithersburg, MD., 1997.

[BGH03] R. T. Vaughan B. Gerkey and A. Howard, *The player/stage project: Tools for multi-robot and distributed sensor systems*, Proceedings of the 11th International Conference on Advanced Robotics, 2003, pp. 317–323.

[Bol02] F. Bolton, *Pure corba: A code intensive premium reference*, Tech. report, SAMS, 2002.

[CCT04] F. M. J.-M. Valin Y. Brosseau C. Raievsky M. Lemay C. Cote, D. Letourneau and V. Tran, *Code reusability tools for programming mobile robots*, IEEE/RSJ International Conference on Intelligent Robots and Systems, 2004.

[Cou92] R. C. Coulter, *Implementation of the pure pursuit path tracking algorithm*, Tech. report, Tech Report CMU-RI-TR-92-01, Carnegie Mellon University, 1992.

[DSM98] D. Levine D. Schmidt and S. Mungee, *The design and performance of real-time object request brokers*, Computer Communications **21** (April 1998), 294–324.

[Fir89] R. Firby, *Adaptive execution in complex dynamic worlds*, Tech. report, 1989.

[GBD04] J. Giesbrecht S. Verret J. Collier G. Broten, S. Monckton and B. Digney, *Towards distributed intelligence*, Tech. report, Technical Report TR 2004-287, Defence Research and Development Canada - Suffield, December 2004.

[GBV03] J. Giesbrecht S. Monckton G. Broten, D. Erickson and S. Verret, *Engineering review of ancaeus/avatar an enabling technology for the autonomous land systems program*, Tech. report, tech. rep., DRDC Suffield, Dec. 2003.

[Gow96] J. Gowdy, *Ipt: An object oriented toolkit for interprocess communication*, Tech. report, Tech. Rep. CMU-RI-TR-96-07, Robotics Institute, Carnegie Mellon University, Pittsburgh, PA, March 1996.

[Gow00a] J. Gowdy, *Emergent architectures: A case study for outdoor mobile robots*, Tech. report, PhD thesis, Carnegie Mellon University, 2000.

[Gow00b] J. Gowdy, *A qualatative comparision of interprocess communications toolkits for robotics*, Tech. report, Tech. Rep. CMU-RI-TR-00-16, Carnegie Mellon University, June 2000.

[GPCH99] S. Schneider G. Pardo-Castellote and M. Hamilton, *Ndds: The real-time publish-subscribe middleware*, Tech. report, Real-Time Innovations, Inc., 1999.

[Hen04] M. Henning, *A new approach to object-oriented middleware*, IEEE Computer Society **8** (January-Febuary 2004), 66–75.

[HK] L. Henriksen and E. Krotkov, *Natural terrain hazard detection with a laser rangefinder*, IEEE Int. Conf. On Robotics and Automation.

[HK93] M. Herbert and E. Krotkov, *Local perception for mobile robot navigation in natural terrain: Two approaches*, Workshop on Computer Vision for Space Applications, Sept. 1993, pp. 24–31.

[HUK02] S. Enderle H. Utz, S. Sablatnog and G. Kraetzschmar, *Miro - middleware for mobile robot applications*, IEEE Transactions on Robotics and Automation (June 2002).

[HUS05] S. Enderle H. Utz and S. Sablatnoeg, *Miro - middleware for robots*, Tech. report, http://smart.informatik.uni-ulm.de/MIRO/content.html Accessed, 2005.

[HV99] M. Henning and S. Vinoski, *Advanced corba programming with c++*, Addison-Wesley, 1999.

[INB95] R. Powers I. Nourbakhsk and S. Birchfield, *Dervish: An office navigation robot*, AI Magazine **16-2** (1995), 53–60.

[Ini05] The Open Source Initiative, *The open source definition.*, Tech. report, http://www.opensource.org/docs/definition.php, 2005.

[JAH87] R. Lumia J. Albus and H.McCain, *Hierarchical control of intelligent machines applied to space station telerobots*, Tech. report, 1987.

[JAU04] JAUS, *The joint architecture for unmanned systems, reference architecture specification ra v3.2 parts 1-3*, Tech. report, http://www.jauswg.org/baseline/refach.html, August 2004.

[Kel97] A. J. Kelly, *An approach to rough terrain autonomous mobility*, International Conference on Mobile Planetary Robots, January 1997.

[KK92] S. Kweon and T. Kanade, *High-resolution terrain map from multiple sensor data*, IEEE Transactions on Pattern Analysis and Machine Vision **14** (Feb. 1992), 278–292.

[KL02] S. Koenig and M. Likhachev, *D* lite*, Proceedings of the National Conference on Artificial Intelligence, 2002, pp. 476–483.

[KM98] K. Konolige and K. Myers, *'the saphira architecture for autonomous mobile robots*, Artificial Intelligence and Mobile Robots: Case Studies of Successful Robot Systems (1998).

[LWV01] S. Sander M. Guler B. Heck J. Prasad D. Schrage L. Wills, S. Kannan and G. Vachtsevanos, *An open platform for reconfigurable control*, IEEE Control Systems Magazine (June 2001).

[MLC00] A. Oreback M. Lindstrom and H. Christensen, *Berra: A research architecture for service robots*, International Conference on Robotics and Automation, 2000.

[MMT03] N. Roy M. Montemerlo and S. Thrun, *Perspectives on standardization in mobile robot programming: The carnegie mellon navigation (CARMEN) toolkit*, Proceedings of the Conference on Intelligent Robots and Systems, 2003.

[MP00] J. Michaloski and W. S. F. Proctor, *The neutral message language: A model and method for message passing in heterogeneous environments*, Proceedings of the World Automation Conference, (Maui, Hawaii), June 2000.

[Ped98] J. . Pedersen, *Robust communications for high bandwidth real-time systems*, Tech. report, Tech. Rep. CMU-RI-TR-98-13, Carnegie Mellon University, 1998.

[RBS97] E. Gat D. Kortenkamp D. Miller R. Bonasso, R. Firby and M. Slack, *Experiences with and architecture for intelligent, reactive agents*, Journal of Experimental and Theoretical Artificial Intelligence **9-2** (1997), 237–256.

[Ros95] J. Rosenblatt, *DAMN: A distributed architecture for mobile navigation.*, Proceedings of the 1995 AAAI Spring Symposium on Lessons Learned from Implemented Software Architectures for Physical Agents , H. Hexmoor and D. Kortenkamp (Eds.) AAAI Press, Menlo Park, CA., March 1995, pp. 317–323.

[RVD01] T. Estlin D. Mutz R. Petras R. Volpe, I. Nesnas and H. Das, *The claraty architecture for robotic autonomy*, IEEE Aerospace Conference, March 2001.

[SGM02] M. Maimone S. Goldberg and L. Matthies, *Stereo vision and rover navigation software for planetary exploration*, IEEE Aerospace Conference Proceedings, 2002.

[SH02] D. Schmidt and S. Huston, *C++ network programming volume 1*, Addison-Wesley, 2002.

[SHS04] J. Johnson S. Huston and U. Syyid, *The ace programmer's guide*, Addison-Wesley, 2004.

[Sim91] R. Simmons, *The inter-process communications (ipc) system*, Tech. report, http://www-2.cs.cmu.edu/afs/cs/project/TCA/www/ipc/ipc.html, 1991.

[Sim94] R. Simmons, *Structured control for autonomous robots*, IEEE Transactions on Robotics and Automation **10** (February 1994).

[SK00] D. Schmidt and F. Kuhns, *An overview of the real-time corba specification*, IEEE Computer special issue on Object-Oriented Real-time Distrubuted Computing (2000).

[SKS98] R. Goodwin S. Koenig and R. Simmons, *Xavier: A robot architecture based on partially observable markov decision process models*, Artificial Intelligence and Mobile Robots: Case Studies of Successful Robot Systems (1998).

[TAO03] TAO, *Developer's guide*, Tech. report, vol. 1 and 2, Object Computing Inc., 12140 Woodcrest Executive Drive, Suite 250, St. Louis, MO, 63141, 2003.

[Utz03] H. Utz, *Miro manual*, Tech. report, University of Ulm, Department of Computer Science, November 2003.

Sidebar – Middlewares for Distributed Computing

Davide Brugali

Universitá degli Studi di Bergamo, Italy brugali@unibg.it

1 Introduction

A framework for distributed computing consists of an integrated set of service components that allow distributed systems to operate together. Typically, distributed component frameworks offer at least the following services:

- Distributed event management supports dynamic notification of events raised by remote objects.
- Location-transparent access to remote objects allows distributed objects to cooperate regardless of their network location, of the operating platforms where they are executed, and of their implementation language.
- Distributed object location allows client objects to determine at run time which server objects offer the functionality they need. Client and server objects can appear and disappear on the network dynamically.
- Persistency and transaction management supports persistent storage, access to distributed data, data replication and data consistency.

Using a distributed component framework consists in developing end-user applications by delegating the execution of common functionalities to middleware services. The following distributed component frameworks are commonly used to build large-scale distributed systems.

2 OMG CORBA

The Common Object Request Broker Architecture (CORBA) (Vinoski 1997) is a standard for distributed middleware defined by the Object Management Group (OMG), a consortium of more than 700 organisations including software industry leaders such as Sun, HP, IBM, Microsoft and Rational. This architecture has reached a good level of maturity and is now implemented in more than ten commercial products.

D. Brugali (Ed.): Software Eng. for Experimental Robotics. STAR 30, pp. 395–398, 2007.

The basic component of the architecture is the Object Request Broker (ORB), which should be installed on each connected host. The ORB uses the Stub/Skeleton mechanism for remote communication between client and server objects. The stub and the skeleton of a server object are generated at compile-time from a declarative specification of the server's interface in the language-neutral Interface Definition Language (IDL). The interface describes which methods the server object supports, which events the object can trigger, and which exceptions the object raises. The server object can be implemented in any programming language (C, C++, Java, etc.). The ORB installed on the server side is in charge of translating the incoming IDL service requests from the remote clients into the server's method invocations. The IDL supports the declaration of only basic type arguments, which have to be passed to a remote server. The connection between the stub and the skeleton is established using the Remote Procedure Call (RPC) mechanism.

The ORB supports location transparency, as it provides the client with a reference to the server object regardless of its network location. The client side ORB dispatches service requests to the server side ORB transparently. Since CORBA can be implemented using different technologies, the Internet Inter-ORB Protocol (IIOP) defines the standard communication protocol for inter-vendor ORB compatibility.

The ORB supports the control of the threading policy used by the servers, such as one-thread-per request and one-thread-per-objects. This threading capability is at the basis of recent CORBA extensions towards a real-time ORB (see [FWDC00] for a comprehensive survey on recent results in developing a standard real-time CORBA).

Server objects can publish the IDL specifications of their interface using the Interface Repository API. The Interface Repository is used to implement two services for object location.

The *Trader Service* is similar to the yellow pages. It allows client objects to find out which distributed server objects support a given interface. The Trader Service returns a list of references to the objects, which have registered that interface and a description of the usage constraints for each object.

The *Naming Service* allows client objects to identify which interface is supported by an object that has been registered with a given name. In some cases, the client object has to request the service of a server object, whose interface was not known at compile-time. This means that the client object does not have a stub of the remote server object, but it can use a generic interface, called Dynamic Invocation Interface that makes it possible to construct method invocations at run-time [Vin97].

3 Sun Java

SUN Java Java is the Object Oriented programming language from Sun Microsystems [Joy00]. The most relevant characteristic (called Write Once/Run

Anywhere) is its portability: Java programs written on one type of hardware or operating system can run unmodified on almost any other type of computer. This is possible because Java programs are compiled in an intermediate format called byte-code that is interpreted by the Java Virtual Machine. The most important packages for distributed computing are:

The Remote Method Invocation (RMI) is the Java mechanism for remote object communication. It implements the Stub/Skeleton approach but, different to CORBA IDL, the server's interface is written in Java. The stub is not linked within the client's address space at compile time, but it is downloaded from the server side when the client needs a connection to the server object.

Applets are small graphical applications that run within a Web browser. When a remote user accesses a web page containing a reference to an applet, the applet's code is downloaded from the web site and executed inside the browser. The applet is not allowed to access the local resources on the client side (e.g. its file system), but it exchange data with the remote server.

Servlets are small applications that run within a Web server. The programmer implements specific servlets on the server side and gives them network names in the same way as is done for standard web pages. A remote user connects to a servlet's URL using a web browser. JavaServer Pages (JSP) are an extension of Servlet technology.

Enterprise Java Beans (EJB) are a component-based technology for developing server-side applications. The basic concept is the EJB Container, a database management application that provides persistence and transaction services to the enterprise beans.

IBM Aglets [LO98]. Aglets are active objects (mobile agents) that can suspend their activity, migrate to a remote host and resume their execution on that host. Before dispatch, the aglet serializes its internal state into a standard form. When the aglet arrives at the receiver side, it is assigned a new thread and executed. An Aglet belongs to an AgletContext, a container class that provides an interface for the runtime environment on the local host.

The *Sun JINI Platform* [Arn99] is an extension of the RMI framework. The stub (called proxy) is no longer a simple interface. Instead, it is a fully-fledged object, which embeds the co-operation protocol between the client and the server and executes (part of) the service logic within the client's address space. Clients download the proxy object from the corresponding device and are ready to use its services. The Jini platform implements the LookUp service in order to identify available services on the network.

4 Microsoft.NET

Microsoft.NET [TL01] is a new technology for developing distributed systems. It borrows many successful concepts from the Java world and from CORBA in order to achieve interoperability of heterogeneous applications. The following four aspects characterize the .NET framework:

The Common Type System (CTS) is an object model that extends the previous COM and DCOM models with the goal of supporting multiple language software development.

The Intermediate Language (IL) is an object-oriented language that conforms to the CTS. Various Microsoft and third-party language compilers (for C++, Java, etc.) generate code in the IL language.

The Common Language Infrastructure (CLI) is a run time environment (similar to the Java Virtual Machine) that executes code compiled in IL.

The .NET Software Development Kit (SDK) is an object-oriented collection of reusable components. It provides run time hosts for the CLI for a variety of execution platforms. Internet Explorer is an example of a run time host. It also supports the development of customer run time hosts.

The essence of the Microsoft proposal is the possibility of compiling existing code and new code written in the programmer's preferred language. The resulting applications can interoperate with class libraries and components written in different languages using the .NET run time environment.

.NET Remoting allows objects to interact over the Internet using binary encoding where performance is critical, or XML encoding where interoperability with other applications is essential. Similar to Java RMI, it is based on the Stub/Scheleton model: the stub (called proxy) is created at run-time using the metadata published by the server.

Web Services are reusable components which are used to develop server-side applications. They combine the distributed computing capability with the Web portal concepts and can be compared to the Java Servlets and JSP. The ASP.NET is the hosting environment for Web services. Clients use the ubiquitous wsdl.exe utility (a Web Service Directory supplied with the .NET SDK) to discover and find Web Services.

Serialization allows a memory object or graph of objects to be converted into a linear sequence of bytes that can be stored on a disk or sent to another networked computer. This mechanism is at the basis of the .NET persistence service that uses metadata information to automatically store and to reconstruct memory object.

References

[Arn99] Ken Arnold et al., *The jini(tm) specification.*, Addison-Wesley, 1999.

[Joy00] Bill Joy et al., *The java language specification.*, Addison-Wesley, 2000.

[FWDC00] V. Faye-Wolfe, et al., *Real-time corba*, IEEE Transactions on Parallel and Distributed Systems **11(10)** (Oct. 2000).

[LO98] D.B. Lange and M. Oshima, *Programming and deploying mobile agents with java and aglets.*, Addison-Wesley, 1998.

[TL01] Thuan Thai and Hoang Lam, *.net framework essentials.*, O'Reilly, 2001.

[Vin97] S. Vinoski, *Corba: Integrating diverse applications within distributed heterogeneous environments.*, IEEE Communications Magazine **14(2)** (Feb. 1997).

Part IV

Software Environments for Networked Robotics

Trends in Software Environments for Networked Robotics

Davide Brugali[1], Moisés Alencastre-Miranda[2], Lourdes Muñoz-Gómez[2], Debora Botturi[3], and Liam Cragg[4]

[1] Universitá degli Studi di Bergamo, Italy brugali@unibg.it
[2] Tecnológico de Monterrey - Campus Estado de México, México {malencastre, lmunoz}@itesm.mx
[3] Department of Computer Science University of Verona, Italy botturi@metropolis.sci.univr.it
[4] University of Essex, Department of Computer Science, Colchester, CO4 3SQ, UK. lmcrag@essex.ac.uk

1 Introduction

Networked Robotics [SG02, SG03] is a kind of Human-Robot Interaction [HRI04]; the human operator collaborates with the robot for the execution of difficult tasks for which the robot cannot achieve a high level of autonomy due to the complexity of the task or the structure of the operating environment.

The peculiarity of Networked Robotics is the physical distance between the robot and the human operator that prevents the operator from seeing the robot during the execution of its tasks. Therefore, Networked Robotics strongly relies on information and communication technology for the efficient transmission of the operator's commands and of the robot sensory feedbacks.

Mainstream application fields related to teleoperation of physical devices are Space Robotics and Remote Surgery, where effective human-robot interaction builds on dedicated and reliable communication networks (e.g. ATM or satellite) and specialized hardware devices (e.g. exoskeleton master linkage). In teleoperation applications of haptics, a loop is closed between the human operator's motion "inputs" and forces applied by the haptic device via a communication link, robot manipulator, and the environment. Key challenges for the advancement of teleoperation technology include the development of high performance control paradigms, advanced mechanism and quantitative measures on the teleoperation system including haptic displays, and new software methodologies that permit a very careful and precise design of high performance architectures at a reasonable effort in terms of cost and development time.

D. Brugali (Ed.): Software Eng. for Experimental Robotics. STAR 30, pp. 401–408, 2007.

The adoption of teleoperation technology in other application areas, such as factory automation, rescue robots (earthquake/terrorism), and nuclear decommissioning is enabled by the use of standard PC-based user interfaces as robot console and of conventional networks (e.g. Internet and wireless LANs) as communication medium between the robot and the human operator.

While PC-based and Internet-based systems provide widely accessible, and cost effective mechanisms for teleoperation technologies, they are subject to indeterminate transmission delay and data loss. Software engineering environments can aid in overcoming the drawbacks of adopting Internet as a communication mechanisms by facilitating the appropriate sharing of autonomy and exchange of information within a distributed system between the human operator and the robotic system. When Networked Robotics is employed in tasks where the complexity of the task or the structure of the operating environment prevents full robot autonomy, such systems may still benefit from the application of intelligence to increase the efficiency of user commands.

Furthermore, distributed software environments allow the construction of Networked Robotics systems that seamlessly scale up from one-to-one interaction (one human operator and one robot) to many-to-many interaction (multiple human operators and multiple robots). Users can collaborate between them in this distributed interaction in order to work together in some robotics application (e.g. teleoperation of several robots in a common task). Users can be located in different places. Commonly, robots are in the same global environment (e.g. a factory). However, several robots can be sharing the same local environment (e.g. production area, offices, outdoor area, etc), or each one can be in a different local environment.

2 Opportunities to Exploit Software Development Techniques in Networked Robotics

The chapters in Part IV of this book describe Networked Robotics systems that have been build exploiting several software development techniques, such as domain analysis, components and frameworks which have been thoroughly discussed in the previous parts of this book. In the following sections we analyze three Networked Robotics paradigms from three points of view:

1. The operator's role. Depending on the task to be executed, the operator can participate to the robot's operations by reacting to its feedbacks and sending new commands immediately, or he can elaborate the robot's strategy and behavior off-line.
2. The robot autonomy. Networked Robots can have autonomy levels that range from full autonomy (the robot is given a task which it then accomplishes safely), to shared autonomy (the operator influences or decides the robot's behaviour and strategy), and to limited autonomy (the operator controls the robot directly).

3. The robot-operator information exchange. From the communication point of view, the human-robot interaction is characterized by the amount and quality of information that has to be exchanged in order to perform the task correctly and to exploit the communication medium efficiently.

The three points of view serve as convenient schema to highlight strengths and limitations of each Networked Robotics paradigm. For each paradigm a suitable software development approach is illustrated which is described in details in a subsequent chapter.

Finally, the Sidebar *Java3D for Web-based Robot Control* by I. Belousov illustrates some examples on how to exploit the Sun Microsystems Java3D library for robot teleoperation.

2.1 Direct Control

The simplest model of interaction between the robot and the human operator is the Master-Slave control. The operator sends to the remote robot (e.g. a manipulator arm) low-level commands (e.g. increment a joint's position) and receives sensory feedbacks (e.g. images from a camera mounted on the arm). This model of interaction is also known as "move-and-wait" strategy: the operator induces a small movement and waits to observe the results of the movement before committing to further action.

The operator plays the role of remote controller in the human-robot feedback control loop. The dynamic of the task is assumed to be quite slow since teleoperation of a robot requires intense mental attention. The advantage of having a direct control is that the operator can feel the effects of his commands, since there is a direct correspondence between action and sensing.

The robot autonomy can be very limited as the robot is under the full control of the remote operator. Usually the robot is at least able to react to unexpected events (e.g. a collision) in a simple and predefined way, such as stopping the command execution and informing the remote operator. The direct control paradigm is highly versatile, since it does not require the robot to build and use a detailed world model, nor it assumes the environment to be static or structured.

The main drawback of direct control is that the communication load is high and that the operator feels the control of the robot under important communication delays very uncomfortable. This is due to two characteristics of Internet communication: *unpredictable time delays* and *low bandwidth.*

Unpredictable time delays means that data are received with variable time interval that irregularly changes. As a consequence, the control loop between the robot and the remote operator may become instable. The operator is instinctively induced to repeat or magnify a command to the robot if he does not feel its effects within a given time interval. At the other side of the communication link, the robot behavior is affected by the duration of a command or the temporal distance between two consecutive commands.

Low bandwidth means that the network traffic affects the flow of data that can be profitably exchanged between the robot and the operator. TV images are substantially delayed, and the size and quality of the images are in many cases insufficient for the operator to perform the required task, e.g. estimating the position and distance between moving objects in the scene. In order to reduce remote control traffic, feedback data can be compressed without loosing relevant information. When video feedback is not essential, information on the robot's operating state can be transmitted to the remote operator in the form of relevant events occurred in the scene (e.g. the object was grasped or dropped, an unexpected object appeared in the working environment, the robot has reached the target position, etc.).

Chapter *Advanced Teleoperation Architecture*, by Andrea Castellani et al. presents an approach to develop Internet-based teleoperation systems that deals with the problems of unpredictable delays and low bandwidth. The approach builds on distributed middleware technology and in particular on the real-time extension of the Common Object Request Broker Architecture. The proposed software framework for bilateral, i.e. force reflecting, teleoperation system takes into account the main features of a haptic system: real-time behavior, distributed resources and general purpose structure. The framework support the interconnection of different devices, the exploitation of different control algorithms, the processing of data from several and different sensors and the use different connection media. RT-CORBA has been selected for its characteristics that make it a suitable middleware for teleoperation systems: standard object interface in support of incremental modularity; hidden object location, implementation and transport protocols independence, suitable for a distributed software infrastructure; platform (programming language, CPU architecture) independent, thus capable of using optimal module implementation; real-time support; optimization of memory management, network protocols, code generation; support of different transport protocol; performance optimization, i.e. the run-time scheduler maps client requests to real-time threads priorities.

2.2 Shared Control

Virtual Reality is often used as a convenient tool for visually reconstructing the robot's environment and behaviour from raw sensory data (e.g. the distance measurements of the robot's sonars), from aggregate data (e.g. the 3D shape of objects detected with a stereo vision system) or from geometric data of a previously known environment (e.g. the blueprint of a known building). Virtual Reality models can be effectively defined to represent industrial environments, which are usually well structured, static or their dynamic is known. Nevertheless, accurate VR models are difficult to define and maintain. Typically, they are build off-line using automatic reconstruction techniques based on stereo vision or active vision. Sensory information are affected by uncertainty and are usually incomplete.

A Virtual Environment (VE) is a 3D visual simulation that represents the computational geometric model of a real environment created with computer graphics techniques. A VE includes several virtual objects that represents the real objects that are found in the real environment. Some of that virtual objects can be virtual robots. Visual simulation can run much faster than the real robot, thus the operator can predict the final state of the robot before the command has been received and executed. The operator can then use this information for preparing the next command and issuing it in advance in order to compensate the transmission delay.

For each task to be performed by one or more robots, one or more operators analyse the problem, plan the sequence of actions that solves it, verify the robot behavior in simulation, and finally, transmit the batch of commands to the remote real robot. The VE simplifies the task analysis and command definition, since each operator is no more constrained by the single point of view of the robot's video camera. Instead, operators can observe the virtual robots, their tools, and the surrounding objects from every position in the space by changing the viewpoint to an arbitrary position, zooming in/out of the scene to an unlimited extent, and using semitransparent images. Once defined, the batch of commands can be executed repeatedly every time the same task has to be performed. During the task execution, the operator plays the role of task supervisor. If the environment characteristics or the task specifications change, the operator must define a new sequence of commands and transfer it to the remote robot.

The robot autonomy becomes now significant for the correct execution of the batch of commands received from the remote operator. The feedback control loop is completely local, thus the robot must be able to react to unexpected events in a more robust way. In order to increase robustness, commands can be transmitted to the robot along with the predicted final state, in order to allow the robot to compare its state with the predicted state and check for consistency. If significant differences are identified, the robot notifies the remote operator that an unexpected event occurred.

The information exchange between the operator and the remote robot is not only limited to strings of textual commands but involves code fragments in a script language and high level data structures. The goal is to avoid or at least minimize the operators' intervention during the robots' operations. The exploitation of VEs to reconstruct the robot's environments allows to the operators to work off-line in tasks where the robots have more autonomy (e.g. testing planning methods on the visual simulation of a known environment), monitoring the right execution and also to save communication bandwidth. The robot can simply transmit information related to variations in the scene (such as the position of a moving object) instead of a video stream.

Chapter *A Multi-Robot-Multi-Operator Collaborative Virtual Environment* by Moises Alencastre-Miranda et al. addresses the issues involved in developing a Collaborative Virtual Environment (CVE) for robotic applications, that is a system that allows multiple users in different geographical locations to interact, share, communicate and collaborate in a given task at the same time in a common virtual environment. In particular, the chapter illustrates specific software development design choices related to:

1. How to model a shared sense of space, presence and time.
2. How to implement this sharing model on a distributed networked environment using a software middleware.
3. How to synchronize distributed applications that interact in the virtual environment and that operate robots in the real environment.
4. How to promote modularity and interoperability of the CVE.

2.3 Mobile Control

The two Networked Robotics paradigms described in the previous sections are characterized by different mixes of robot autonomy and operator's intervention which determine the information exchange between the operator site (the Client) and the robot site (the Server).

In most distributed systems, the information that can be passed between the Client and the Server is limited to a small number of primitive data types (e.g. strings, integers, bytes, etc.). Code Mobility allows clients and servers to exchange full-fledged software components (mobile objects) carrying over both the data and the code that manipulates it. Mobile objects might be passive components or threads that start running when the downloading has been completed. An object can migrate at run-time from the Client's to the Server's execution environment (Mobile Agent) or vice-versa (Code on Demand).

The main advantage of moving the code through the network is to perform the computation where the resources are located (data, physical devices, human operators). The goal is to reduce the interactions between distributed subsystems and thus the network traffic. While Code on Demand offers the opportunity to send knowledge and action to a remote location as a single software component, a Mobile Software Agent encapsulates knowledge, action and past experience in an autonomous, asynchronous, dynamic and potentially intelligent (i.e. able to learn) software entity, able to move multiple times within a distributed but unified execution environment (which supports the execution of individual mobile agents but protects an underlying network of host computers and robots). The flexibility, autonomy and intelligence which mobile agents offer directly translates into increased flexibility, autonomy and intelligence in the host architecture. In addition, the mobility of mobile agents within such an architecture can be used specifically to provide a range of functional benefits, including increased system adaptability and fault tolerance, and the ability to dynamically allocate control.

The adaptability of a distributed architecture can be improved by employing mobile agents to automatically move and integrate updated control code at run time, and in parallel, in multiple networked robots; fault tolerance can be improved by using mobile agent mobility to move important mission level data away from failing system components, or store such data to persistent media such as hard drives for later retrieval; while dynamic allocation of control can be achieved by relocating control in a distributed multiple telerobot architecture in the event of significant changes in network topology (caused by possible robot failures or additions); so as to maximize control and reduce the effects of communication delay between a user and the controlled robots, as found in Internet based telerobot control architectures.

Typically the operator interacts with the robot through the Internet using a desktop PC or even a mobile device, such as a laptop or a Personal Digital Assistant (PDA). If the robot is accessible through a Web portal, code mobility allows the operator to download the front-end to the robot that best fits the computational capabilities of its client device, e.g. an Applet with a VRML environment for a desktop PC or a Middlet with a simple graphical interface for a PDA. For example, the operator can supervise the robot activity using a PDA to visualize the current values of the robot's state variables, such as its current task, location, operating status, the battery level.

Usually, when the operator needs to upgrade the robot's functionality, the robot must be stopped and a new version of the software control application installed. Code mobility can simplify software upgrading as it allows to replace single software components while the application is running. Furthermore, it can be performed remotely. For example, the operator could upgrade the sensory data fusion algorithm by sending the mobile object that encapsulates it to the remote robot. This is useful when a new type of loom inspection requires a different video images processing algorithm.

The information exchange between the operator and the remote robot could be enhanced using code mobility since this technology can reduce network traffic and improve communication reliability. If the needed relevant data (e.g. a binary information) is much smaller than what the robot will answer to the operator request (e.g. a video stream), the algorithm (encapsulated in a mobile object) that elaborates the information can be transmitted to the robot's place obtaining a significant optimization. This is the case of the operator who has to achieve a succession of possible interdependent interactions with the remote robot (for example successive loom inspection, sensor data filtering and interpretation) before getting the relevant data (e.g. the identification of a malfunctioning component). This kind of interaction can be improved by sending a mobile object to the robot's site that handles the composite request at once and gets back the result.

Code mobility enhances the concept of autonomy transfer, as it allows the operator to transfer the code implementing the control algorithm that generates the sequence of batch commands or that carries on abstract plans along with the procedures to decompose them in simple robot actions.

The benefit of code mobility for autonomy transfer can be appreciated in those application scenarios for which the maximum level of flexibility is required, such as when the robot operates in ever changing environmental conditions (e.g. rescue robotics) or when the robot functionality must be customized for remote training of students or technicians.

The robot can play the Client and Server roles interchangeably. It plays the role of Server when it receives mobile objects from the operator's site, and it plays the role of Client when it downloads mobile objects from other sites or resources.

Chapter *Modularity and Mobility of Distributed Control Software for Networked Mobile Robots* by Liam Cragg et al. presents the authors' experience with a number of implementations, namely autonomous, telecontrol and AI robot architectures with which they have explored the properties of mobile agent mobility. Although many advantages of mobile agents are claimed in literature, our investigation has systematically examined whether mobile agents are applicable and beneficial in the multiple robot domain. In particular the chapter addresses the following issues:

1. Which kind of functionality does a mobile agent implement?
2. Which support does this technology provide to the software development process?
3. How to enforce secure and reliable interactions between the mobile agent and the robot?
4. How to test and debug the agent's code that will interact locally with the remote robot?

3 Conclusions

The development of Networked Robotics systems involves the combination of several technologies, such as computer graphics, computer network, distributed computing, and multithreading execution in addition to the most sophisticated techniques to build autonomous, fault-tolerant, and efficient robots. Advanced software environment play a key role in the development of Networked Robotics systems as they can simplify the integration of heterogeneous distributed technologies and enhance the performance of the overall system. The three approaches presented in the following three chapters face similar problems but from different perspectives. The challenge for future research is to explore their synergism in order to exploit them in more complex scenarios.

References

[HRI04] Special Issue on *Human-Robot Interaction*, IEEE Transactions on Systems, Man and Cybernetics, Part C, Volume: 34, Issue: 2, May 2004

[SG03] Siegwart, R. Goldberg, K. Ed., Special Issue on *Internet and Online Robots*, Autonomous Robots, Volume 15, Number 3, November 2003

[SG02] Siegwart, R. Goldberg, K. Ed., *Beyond webcams, An Introduction to Online Robots*, MIT Press, Cambridge, MA, 2002.

Advanced Teleoperation Architecture

Andrea Castellani, Stefano Galvan, Debora Botturi, and Paolo Fiorini

Department of Computer Science University of Verona, Italy
{castellani, galvan, botturi, fiorini}@metropolis.sci.univr.it

1 Introduction

In this Chapter we report on the efforts carried out at the Robotics Laboratory ALTAIR of the University of Verona (Italy) towards the development of a high performance architecture for bilateral, i.e. force reflecting, teleoperation system. Key issues for the advancement of teleoperation technology include high performance control dealing with problems such as unpredictable delays and low bandwidth, advanced mechanisms and quantitative measures on the complete system including haptic displays. In the area of software development, new methodologies have been developed that permit a very careful and precise design, and allow to implement high performance architectures at a reasonable effort in terms of cost and development time. Our architecture, called Penelope, takes into account very important features of a haptics system: real-time behavior, distributed resources and general purpose structure. However, we are also studying features that make it particularly suited for high reliability and safety operations, such as space teleoperation and robot-assisted surgery.

In summary, the main features provided by the architecture proposed here, are:

- Communication between elements implemented with ACE CORBA.
- Applications written in C++ under the RTAI Linux Operating system.
- CORBA class allowing object communication using the C++ CORBA IDL interface.
- Inheritance from CORBA class for each "real" class.
- Communication independent of the real location of the implemented class and of the communication mechanism.
- Methods called by procedures without knowledge of their physical location within the teleoperation system.
- Security and robustness ensured by UML based design, enforcing software specifications.

D. Brugali (Ed.): Software Eng. for Experimental Robotics. STAR 30, pp. 409–430, 2007.

This architecture is under test with the ALTAIR setup in force feedback demonstrations. The teleoperation system consists of a PUMA 200 and a PUMA 560, two NASA-JPL Force Reflecting Hand Controllers (FRHC) with force and torque reflection capability in any arbitrary direction and orientation, and force sensors mounted on the robot wrists.

In the following, we present a our design choices, we give their motivations, and describe the laboratory setup that we are using to test the various features. Emphasis is given on the performance expected from a software architecture developed for a teleoperation purpose.

2 Teleoperation

Teleoperation is defined as the control over a distance of one or more devices by a human operator. In this context, we assume that the controlled devices are robots. Usually teleoperation refers to a system with a master/slave configuration, where the operator works on a master joystick that is kinematically compatible with the slave manipulator (see Figure 1).

It has been shown that operator performance is improved by providing force information to the human operator [DZK92]. Force information can be presented visually to the operator on a monitor, but the most significant performance improvement is achieved by providing force feedback to the operator, i.e. by generating forces directly with the motors of the master device. In this case the operator is said to be kinesthetically coupled to the slave and the teleoperator system is said to have bilateral control or to be force reflecting. Haptic feedback devices were pioneered in teleoperation systems as far back as the 1940's. In teleoperation applications of haptics, a loop is closed between the human operator's motion "input" and forces generated by the haptic device based on data received via a communication link from the robot manipulator and the environment. Teleoperation is used in cases where the environment is not directly accessible by a human, or when it is too dangerous for an operator, or when it is necessary to scale the force exerted on the environment. Almost all of the applications of teleoperation involve contact with the remote environment, e.g. grasping, welding and puncturing. In particular, robotic applications to surgery include puncturing as one of the most usual actions. The trade-off between stability and performance is the main consideration in the design of a teleoperation system. Mainly, the stability problems are due to the drawbacks of the communication over a distance, *unpredicatable time delays* and *low bandwidth*. Therefore, those problems reduce perception of environment, making uncomfortable and unsafe to work with the system. Teleoperation control architectures analyzed in the literature can be classified in terms of their stability and performance trade-off's [Law93]. Control algorithms for ideal kinesthetic coupling [YY94] are at one end of the spectrum, whereas passivity based algorithms [AS92] are at the other end.

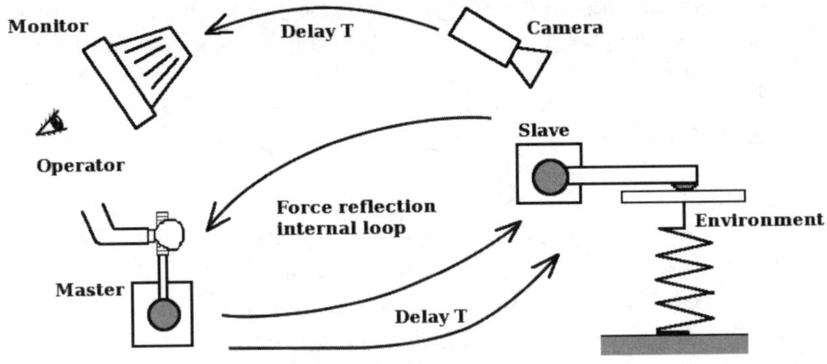

Fig. 1. Teleoperation schema.

Conventional algorithms such as position error based force feedback and kinesthetic force feedback lie in the middle. Both performance and stability are inherently dependent on the task for which the system is designed. Thus the need to design system controllers for the specific applications, including local and remote environments, not only in terms of parameter tuning, but also with respect to the overall control scheme.

In a teloperation environment, and especially for research purposes, we want to have the possibility to easily interconnect different devices, test different control algorithms, process data from several and different sensors and use different connection media. From this assumption, the idea of a middleware integration framework arises. The framework has to be designed to enhance the ability of modularize, extend and reuse the software infrastructure in order to handle a distributed environment. In addition some real-time capability is needed to maintain transparency and stability while data are sent over the communication channel, dealing with unpredictable time delays and low bandwidth. Moreover, the main advantages of distributed real-time middleware for teleoperation are reduced systems cost, remote expertise on demand and dynamic access to the system [She92a]. One possible approach is the so-called whitebox framework [GHJV95]. This framework relies on the object-oriented nature of the programming language used (inheritance, abstraction, dynamic binding) to achieve modularity, reusability and extensibility properties. Whitebox frameworks require a good knowledge of the complete architecture in order to create a suitable inheritance hierarchy. A top-down approach is required, and to reduce complexity and to ensure robustness of the software a UML schema can be provided.

The Unified Modeling Language (UML) [Gro] is a standard language for specifying, visualizing, constructing, and documenting the artifacts of software systems, as well as for business modeling and other non-software systems. There are many different UML diagrams to cover different software engineering problems that are described in papers and/or books [Ric99]. With respect to this project, we have used only the *Class Diagram.* We have used the UML Class Diagram firstly in specification perspective to understand which classes were necessary and what should be their relationships. During implementation, we used the implementation perspective to organize the complex software structure and to test its operation.

In distributed telerobotic applications, systems are extremely heterogeneous, they are often implemented using previously available devices, based on specialized hardware, running on different operating systems, and programmed in many different languages. Therefore, middleware software for communication and for distributed object computing is a necessity to guarantee the overall system functionality. Available solutions to resolve the heterogeneity problem are: Microsoft Distributed Component Object Model (DCOM), Java Remote Method Invocation (Java RMI) and Common Object Request Broker Architecture (CORBA). In order to have general-purpose architecture and language, vendor, and operating system independence, CORBA is the solution adopted in the research described here.

CORBA is an emerging open distributed object computing infrastructure being standardized by OMG [OMG]. The basic CORBA paradigm is that of a request for services of a distributed object. The services that an object provides are given by its interface. Interfaces are defined in OMG's Interface Definition Language (IDL). Distributed objects are identified by object references, which are typed by IDL interfaces. More about CORBA can be found in [OMG] and about TAO Real Time CORBA in [HV99].

Deterministic performance is achieved using standard Linux drivers, applications and functions, in the RTAI (Real Time Application Interface) [DIA] of the standard Linux kernel, providing the ability to make it fully pre-emptable. RTAI [DIA] extends the standard (vanilla) kernel by providing different schedulers that permit to give the highest priority to a specific time-critical process, delaying interrupts from the Linux operating system. The real-time code has to be written in the form of a Linux kernel module. This permits its full integration with the system and the possibility to define the new scheduler and activate/deactivate preemption. Several communication functions are provided to permit exchange of information between the kernel "realm" and standard user space programs. All traditional IPC (Inter-process Communication) functions are provided such as semaphores, mailboxes, FIFO's and shared memory. Dynamic Memory Allocation (DMA) and memory lock in hard real-time tasks are also available.

Another enhancement to the aim of deterministic performance can be achieved moving the bottlenecks present in the control loops of the devices to hardware. This can be done by using specific interfaces to fulfill the middleware framework requirements and preserving aspects such as transparency and software independency. Using CORBA, other important aspects such as security and fault tolerance are ensured by using distributed and redundant resource management. Local control and the network are decoupled guaranteeing system robustness. This approach helps in the design of new modules by providing a common interface structure to match the architecture requirements and, by using this architecture during implementation a more conscious software evolution can be acheived. Another advantage is that developers do not have to deal with the communication layer and thus they can reuse the code more easily. However, the top-down approach is very difficult to follow during the software development phase. In same cases in fact, it is necessary to modify the software specifications and to change some design characteristics because new needs have arisen during implementation. Therefore, another goal of the middleware development should be to make the modification process easier.

3 Penelope ($\Pi\eta\nu\epsilon\lambda o\pi\epsilon\iota\alpha$)

Penelope is a modular and distributed software architeture, aimed specifically at teleoperation systems. With this architecture, it is possible to control, in a distributed framework, several types of robots in a simple way and test different teleoperation configurations, including new devices, sensors and boards (e.g. I/O boards, FPGA boards and Multi-Axes Controller boards). Initially, to develop this type of structural software we have analyzed how to implement the structure and which tools or technologies to use. Software for teleoperation must be modular and easy to expand or modify. In a robotic laboratory there are different robots controlled by different devices or boards that can be used together with different sensors. For these reasons, we have chosen to implement Penelope using Object Oriented Programming (OOP). In this way, every real object in the teleoperation framework is represented by a specific object in the software. With this programming choice, it is also possible to use hierarchy, polimorfism, function overloading and all the other OOP mechanisms to develop a complex and modular software system.

In a general teleoperation system, it is possible (and appreciated) that objects (robots (master or slave), sensors, graphic interfaces) are used, controlled or connected to different PC stations. In some cases, different Operating System are used and therefore a communication protocol or, in general, a communication infrastructure is necessary. A possible approach is to develop a dedicated system to ensure master/slave (or other device) communication [She92b]. However, this solution, which is the natural choice in a critical teleoperation architecture, is economically unacceptable in most cases. To obtain a general and more powerful communication infrastructure, we have chosen to use CORBA. With this communication framework, it is possible to distribute Penelope over different PC stations simply by using the object interconnection service of CORBA (OOP helps in using CORBA in the code). Communication is implemented using the real-time implementation of CORBA, The ACE ORB middleware (TAO), which allows to control the communication timing more precisely, and to use different communication protocols or scheduling priorities, when necessary.

Penelope is an architecture developed with research objectives in a research environment. For these reasons, we have implemented all the infrastructure in a free Operating System using Linux under a GPL (General Public License). To respect strict timing constraints, for example in control algorithms, we have used the RTAI Linux extension presented in the previous Section.

Given the structure of Penelope, the three most important features that must be described in detail are:

- The architecture skeleton (structure of the classes);
- Object communication management when they are located on different PC's;
- The robot control to satisfy real-time constraints.

Another important feature of Penelope is its Graphical Interface independence. Penelope can be used by different types of Graphical User Interfaces (GUI), by means of different user interface builders and different graphical libraries. GUI can run and communicate with Penelope on various OS and also using a web interface. Therefore, a GUI is not a key feature of this architecture and it will not be described here. Each user can implement his own personal interface and use Penelope only knowing the IDL class interfaces.

After describing Penelope main features, we will present an example of a master/slave teleoperation system and we will discuss early performance results in communication and control tasks.

3.1 Class Structure of Penelope

A teleoperation system, and Penelope in our case, must use many different object types, for example robots controlled by different algorithms through a variety of Input-Output boards. Moreover, in many cases master/slave systems use external devices, such as visual and audio displays, or force sensors. Because of the complexity of this structure, we have designed the architecture of Penelope using the UML Class Diagram shown in Figure 2.

As shown in the Figure, Penelope main classes are `Robot`, `Board`, `Sensor`, `Controller`, and `Trajectory` that are briefly described in the following[1].

- **BoardImpl**. Represents all devices used as interface from a real robot to a PC station, e.g. to read encoders or write voltage/torque to the motors. This represents a general class from which the specific boards inherit their interface.
- **RobotImpl**. Manages all types of robots. A manipulator, a joystick, a mobile robot or a multiple robot (for example a mobile robot platform connected physically with a manipulator) are sons of RobotImpl class via

[1] The "Impl" suffix after a class name is used to differentiate real implemented classes from CORBA interface classes.

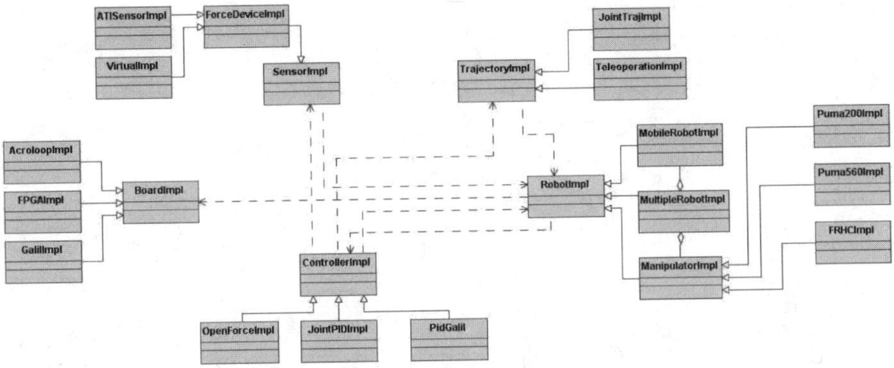

Fig. 2. Penelope UML Class Diagram. It represents the diagram at the specification perspective and does not consider the interface classes for communication.

inheritance. In the same way, a specific manipulator inherits from ManipulatorImpl class.

- **ControllerImpl**. This class contains all the feature necessary to control a robot; for example controller algorithms and controller parameters. As robot and board, this represents only a general class; specific types of controllers and their implementations inherit from it.
- **SensorImpl**. Interface class to different sensor devices. With inheritance it contains the force sensor class, the sonar and laser device class, and the camera device as a visual sensor.
- **TrajectoryImpl**. It produces all references followed by a robot, for this reason this class can be used in two different way:
  – as a trajectory generator (of different types) for a single robot in a simple robotic system (not in a teleoperation system);
  – as a communication channel between master and slave using, if necessary, all the sensors present in the teleoperation system. In this case, for example, TrajectoryImpl can manage the master/slave data transfer invoking the correct methods.

Clearly, there are complex relations between these classes, as shown in Figure 2. For example an instance of ControllerImpl (a ControllerImpl object) can be in relation with some instance of other objects. The control algorithm manages a robot and there must be a relation between this two types of objects. Again, a control algorithm can use a sensor, for example an instance of the force sensor in the case of a force control algorithm.

In all cases, these relations can be translated into code by means of a class attribute (a pointer to the specified class) as shown in the following example.

```
#include "trajectory.h"
#include "sensor.h"
#include "robot.h"
```

```
class controllerImpl
{
protected:
    robotImpl *robot;
    trajectoryImpl *trajectory;
    sensorImpl *sensor;

// Other attribute ...

public:
    controllerImpl();
    ~controllerImpl();
    int setRobot(int robotNumber);
    int setTrajectory(int trajectoryNumber);
    int setSensor(int sensorNumber);

// Other methods ...
};
```

As it will be explained in the next Section on communication, methods to set object relations can not use pointers and then input are codes that represent the specified object type. To understand the complexity of the Penelope objects, we briefly describe the interface of an instance of ManipulatorImpl, Puma200Impl, shown in Figure 3 and in the followinf code fragment.

```
class puma200Impl:public virtual manipulatorImpl
{
// Some Attributes

public:
    puma200Impl();
    ~puma200Impl();

public:
    int getJointPosition(float *joint_position);
    int getJointVelocity(float joint_velocity);
    int getJointAcceleration(float *joint_acceleration);
    int getJointMovements(float *position,float *velocity);
    int WSPosition(const float *angles,float *work_space_position);
    int WSPositionOther(float *work_space_position);
    int WSForce(float *fortor);
    int WSVelocity(float *work_space_velocity);
    int WSAcceleration(float *work_space_acceleration);
    int computeJointPosition(const float *work_space_position,
                             float *angles);
    int dynamic(float *vect);
    int jacobian(const float *angles,float jaco[][]);
    int transposeJacobian(const float *angles,float jaco[][]);
    int inverseJacobian(const float *angles,float jaco[][]);
    int jointTorque(const float *angles,const float *force,
                                        float *torque);
    int directKinematic(const float *angles,float *position,
                        float *versor1,float *versor2,
                                       float *versor3);
    int inverseKinematic(const float *position,const float *versor1,
                         const float *versor2,const float *versor3,
                                              float angles[]);
    int step_rad(const float *step,float *rad);
    int rad_step(const float *rad,float *step);
    int goNested();
    int goStart();
    int goWork();
```

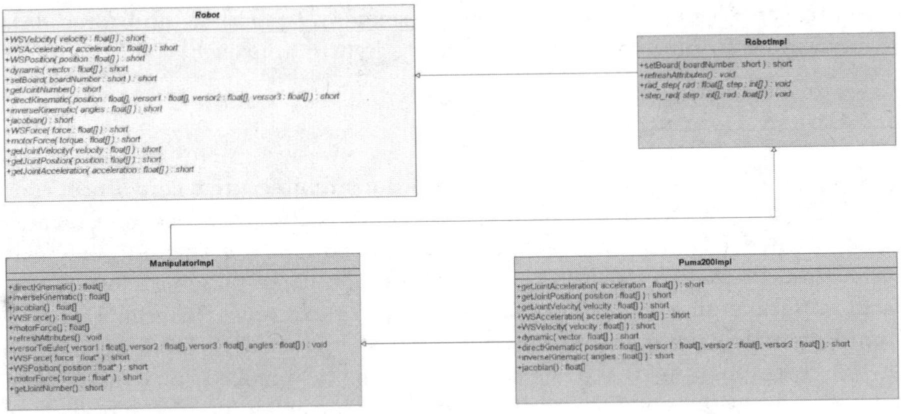

Fig. 3. Particular of Penelope class Diagram with RobotImpl inheritance structure and Puma200Impl interface (attributes and methods).

Some of the methods present in this class interface use the relations with other object. For example when `goWork()` is called, the PUMA 200 manipulator goes to its working position. To do this, methods use a controller object and a board object. The first to start the joint control algorithm and the second to communicate data and receive robot information. It is also possible that a single board object is connected to more than one robot, or a single robot object connected to multiple boards (for example three robot joints connected to one board and three other robot joints connected to another board). Penelope has this versatility also respect to robot control algorithms: a set of joints can be controlled with a joint-based algorithm and the remaining joints can be controlled with a different algorithm.

Another important class in Penelope is represented by TrajectoryImpl. As mentioned above, it can be used in two different ways. In general, when an object calls the TrajectoryImpl method "`getReferiment(float time,float *ref)`" it returns the current reference followed by the control algorithm in that control cycle:

- as a trajectory generator, it generates different types of trajectory in joint-space or in operational space; step, ramp, velocity or position profiles with cubic splines.
- in teleoperation, it sets the parameters to create the reference connection for controller. When the connection is ready, `getReferiment` is able to send references to controllers that can be very different. It can send joint positions if the master controls the slave movements by sending position in configuration space, or $3D$ positions if the control is in operational space, or by sending forces measured by the sensors at the slave or the master sides. Is also possible to send hybrid references as a mixture of different

data types (for example, the master sends $3D$ positions and force data measured to the slave controlled with a hybrid approach).

3.2 Object Communication

Penelope represents a distributed teleoperation architecture where objects can be on different PC's. Thus a communication infrastructure is needed. As mentioned earlier, CORBA TAO middleware as been used for this purpose. Any main class in the architecture inherits from the corresponding class (interface) in IDL, as shown in Figure 4. Therefore, for example, RobotImpl inherits from Robot IDL interface. All interfaces of methods in IDL are implemented in the corresponding real *C++* class and can be invoked in a distributed way. The following code shows this concept as an example of ControllerImpl where `Controller` is the interface name in IDL, ROB is an IDL module corresponding to a *C++ NameSpace* and `CORBA::Short` represents the mapping of CORBA for basic type `int`.

```
class controllerImpl:public virtual POA_ROB::Controller
{
// Attribute ...

public:
   controllerImpl();    // Constructor
   ~controllerImpl();   // Destructor
   CORBA::Short setRobot(CORBA::Short robotNumber)
                  throw(CORBA::SystemException);
   CORBA::Short setTrajectory(CORBA::Short trajectoryNumber)
                          throw(CORBA::SystemException);
   CORBA::Short setSensor(CORBA::Short sensorNumber)
                    throw(CORBA::SystemException);

// Other methods ...
};
```

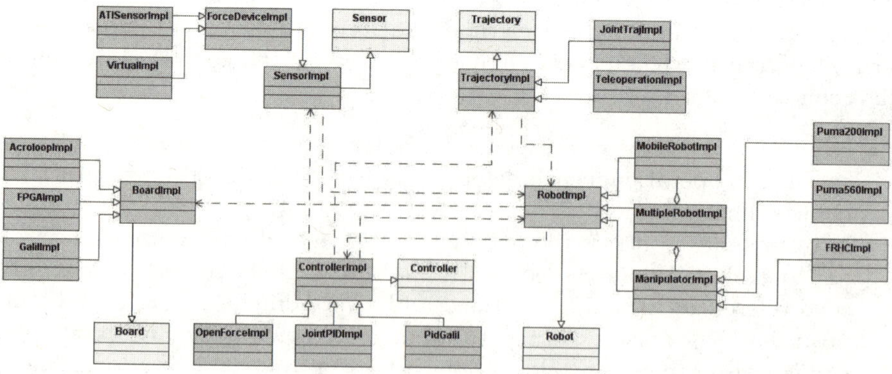

Fig. 4. Penelope UML Class Diagram with IDL interface for object communication.

To invoke the distributed methods of a specific object, the architecture must know their references. To do this, CORBA defines the Naming Service. Penelope uses this service to read and write references from a complex tree managed by the Naming Service. The tree structure is divided into contexts to maintain a logical representation of the architecture hierarchy as shown in Figure 5.

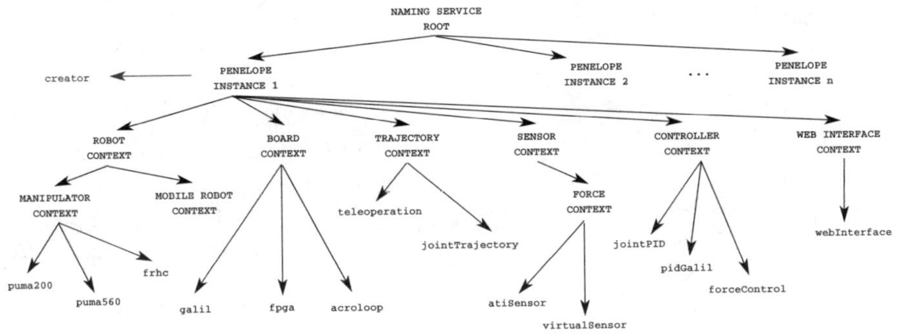

Fig. 5. The logical structure of Penelope CORBA Naming Service.

By means of this service, it is possible to receive any object reference and use this as a distributed object. However, a teleoperation system is more complex and Penelope must be able to create or destroy an object when demanded. When the architecture starts, no object is created and there are not references in the Naming Service. Only after the operator sets the objects the references are created. To allow this feature, Penelope uses an important class (inherited from the corresponding IDL class) named creatorImpl shown in Figure 6. When Penelope starts, only this object is active and its reference is in the Naming Service. Methods of creatorImpl can be invoked on demand, to create or destroy any other architecture object, as shown in the code fragment below, representing the interface of the creator class.

```
using namespace CORBA;
class creatorImpl:public virtual POA_ROB::Creator
{
public:
  char* rootName;                     // Name of Naming Service Root
  vector<std::string> robotNames;     // Vector Names of Robots
  vector<std::string> boardNames;     // Vector Names of Boards
  vector<std::string> controllerNames; // Vector Names of Controllers
  vector<std::string> trajectoryNames; // Vector Names of Trajectories
  vector<std::string> sensorNames;    // Vector Names of Sensors

  vector<int> robotID;       // Vector ID of Robots
  vector<int> boardID;       // Vector ID of Boards
  vector<int> controllerID;  // Vector ID of Controllers
  vector<int> trajectoryID;  // Vector ID of Trajectories
  vector<int> sensorID;      // Vector ID of Sensors
```

```
  vector<robotImpl *> robotVect;                // Robot pointers
  vector<boardImpl *> boardVect;                // Board pointers
  vector<controllerImpl *> controllerVect;// Controller pointers
  vector<trajectoryImpl *> trajectoryVect;// Trajectory pointers
  vector<sensorImpl *> sensorVect;              // Sensor pointers

public:
  Short objectCreator(const Short ref,const char *name)
                               throw(SystemException);
  Short objectDestroyer(const char *name) throw(SystemException);
};
```

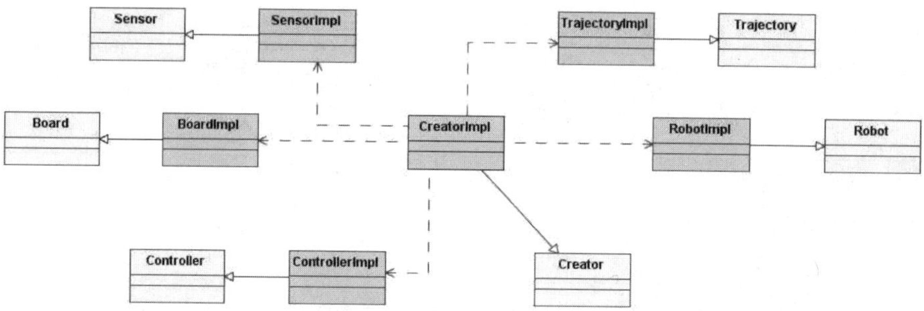

Fig. 6. UML Class Diagram of Penelope with Creator dependencies.

To create an object, `objectCreator()` must be invoked (using the root name and the creator reference in the Naming Service). This stores name, ID and the pointers to the created objects in a structure, and a reference with the specified name is created in the Naming Service under the correct context. When `objectDestroyer()` is invoked, the specified object with its name, ID and pointers is destroyed (deallocating occupied object memory) and the reference in Naming Service is removed. In other words, with the `objectCreator()` it is possible to manage Penelope dynamically creating the necessary distributed object at runtime, or destroying then on demand.

3.3 Robot Control

Some features of a teleoperation system have real-time properties. The more interesting example of this is well represented by the Robot class. A robot must be controlled and typically a control algorithm is executed on a Board and not on a PC station, because of time constraints. However, in a flexible teleoperation system such as Penelope, it must be possible to use different, and often new control algorithms. For this reason, board-based control is not the best choice. In Penelope control algorithms run on a Linux PC equipped with RTAI to satisfy real-time constraints. When the controller is started, two different methods are executed:

- communicationThread(). An RTAI thread that reads the reference by invoking `getReferiment()` method of the corresponding trajectoryImpl object, and writes information in an attribute of the controller class. This sequence is repeated during each control cycle at every T_{com} ms defined by the user.
- controlThread(). An RTAI thread that reads the reference in the controller class attribute, computes the correct signal (for example torque/voltage signal) and sends it to the robot using the specific board object. This sequence is repeated cyclically at every T_{ctrl} ms defined by the user.

The threads above use a variable represented by the controller attribute object that includes the controller reference; this is a shared variable that must be synchronized using a semaphore paradigm (Penelope uses RTAI semaphore services).

3.4 Penelope for Teleoperation

When Penelope is used to manage a teleoperation system the Penelope process must be running. In the case of multiple robots, sensors or general devices, it is possible to use more PC stations, each one running a copy of Penelope. In this case, in any instance of Penelope, only a subset of the system objects is created and the communication is managed by CORBA through the Naming Service and the "creator" object. In other words, any station generates only the objects (for example its robot, board and controller object) relevant to its operation, but it can also use objects of other instances of Penelope, by using the Naming Service. In the GUI, it is necessary to choose which object is created in a specific Penelope instance, and which distributed object is used by the other Penelope instances. To use them, it is necessary to obtain only the reference through object names stored in the Naming Service. Briefly, when Penelope is running, through a GUI it is possible to:

- create new objects on a PC and add new object references in the right Name Service context (for example, create an instance of a robot, of a controller or of a board);
- use objects created to control a robot or for teleoperation experiments (for example, reading joint position or setting torque voltage on a specified motor joint);
- use distributed objects via the Naming Service structure (for example, read joint positions of a robot or read force sensor data acquired by a different PC);
- destroy objects and delete references in the Name Service structure (for example, destroy the current robot instance to use another robot or destroy all objects and delete all Naming Service references to restart a test with Penelope).

In a teleoperation system, Penelope is robot independent since it can control either Master or Slave. The only difference is the configuration procedure when specific objects are created and the communication channel is established, setting the right configuration parameters in the architecture by using a GUI. It is important to understand that when more instances of Penelope are running, only one Naming Service tree is used. Every Penelope instance uses the same tree with the same structure but in different contexts, as shown earlier in Figure 5.

4 Performance Analysis

In [Ark98] Arkin gives a list of desiderata for behavior-based architectures to measure an architecture's utility for a specific problem. This list contains items such as: support for parallelism, hardware targetability, niche targetability, support for modularity, robustness, timeliness in development, run time flexibility, performance effectiveness. These criteria are used in Arkin's book to compare different types of architecture, focusing on mobile robots. In [OC03] a comparative study is carried out of a few successful software architectures for mobile robots, based on features such as portability, ease of use, software characteristics, programming and run-time efficiency. In the evaluation process the authors assess software and programming characteristics of the test system: OS, language support, standard libraries, communication facilities and their performance, hardware abstraction, porting and application building, documentation and programmer efficiency. Then an evaluation of the run-time performance is given, based on application dependent goals, such as localization, mapping and planning. In this approach to system evaluation, it is difficult to measure the coupling between the specific program developed for a task and the software framework that the program is running in.

We can address this problem by proposing the strategy that we use to evaluate our architecture. In order to understand what part of system performance is due to the task-specific algorithm, and what is due to the architecture itself, we measure some of communication times on the critical data paths, i.e. between software modules instead of the more typical "beginning-to-end" time of a complete task. We test the real-time performance degradation of the architecture by making the processor busy with other tasks, and by observing its response to higher priority jobs. In order to develop a repeatable test, we setup a meaningful benchmark that can demonstrate the benefits of an architecture for teleoperation. We will also consider development time as a measure of the environment quality and of the support provided to developers by a structured architecture.

These tests lead to the conclusion that the global evaluation of a software architecture must be user referenced, and should consider performance effectiveness, timeliness in development, support for modularity and hardware targetability. The other criteria indicated in the literature are more appropriate to robotic applications, but not strictly depending on a software architecture. For instance, with respect to the very important criteria of run time flexibility and robusteness the framework has to take into account reconfigurability and fault recovery, but these problems are mainly matter of good programming practices than of the overall architecture.

4.1 Test Results

In order to test Penelope and evaluate the performance of our architecture a teleoperation test is presented. To define a demanding test task, we use the maximum number of classes implemented in Penelope in a time critical operation. We consider a simple teleoperation task with force feedback. We use a NASA-JPL Force Reflecting Hand Controller (FRHC) as the system master. The FRHC, shown in Figure 7, is a six degree of freedom (dof) joystick with force feedback in every joint. It permits to operate with full dexterity in a cubic workspace of 30x30x30 cm developing forces up to 10 N and torques up to 0.5 Nm. The joystick is driven by a custom designed controller implemented mostly on a FPGA board (NI PCI-7831R).

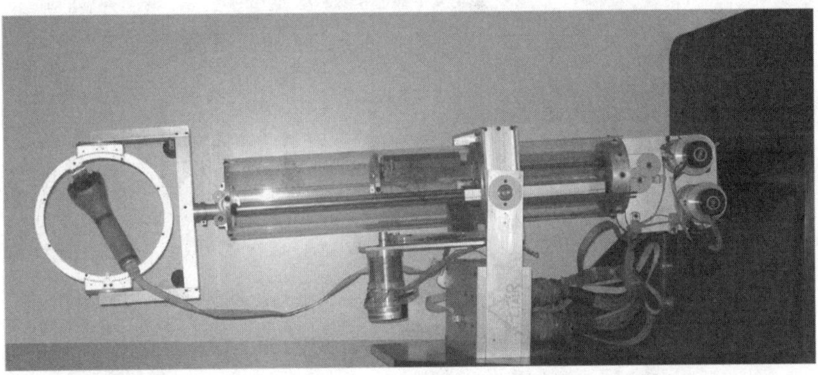

Fig. 7. The NASA-JPL Force Reflecting Hand Controller (FRHC).

The idea behind this appraoch is to improve the computational speed of the master while maintaining accuracy and integrating the controller in the architecture. The FPGA is a good instrument to test hardware task implementations at a reasonable cost, reducing the time between the system design and final implementation. Our approach relies on porting the most time-consuming teleoperation algorithms to hardware. In a force feedback teleoperation task, the study of the structure of the joystick highlights that the main computational bottlenecks are the forward kinematic computation and the force transformation from the sensor frame to the joint frame of the joystick. Our FPGA permits to implement in hardware the forward kinematic algorithm. The focus here is on ease of software and hardware integration: the developer has only to maintain the interface expected by the IDL in order to make the hardware implementation as transparent as possible to the overall software architecture.

The slave robot is a Unimation PUMA 200 arm with six dof, controlled by a Galil board, shown in Figure 8. We use a virtual sensor to simulate different force/torque profiles reflected to the user. Every movement the operator imposes to the joystick is transferred to the robotic arm with the appropriate workspace transformations and scaling. The readings of the robot sensors are fed back to the joystick and, through the activation of the FRHC motors, to the operator. The communication is carried out on¡ a standard 100MB ethernet LAN.

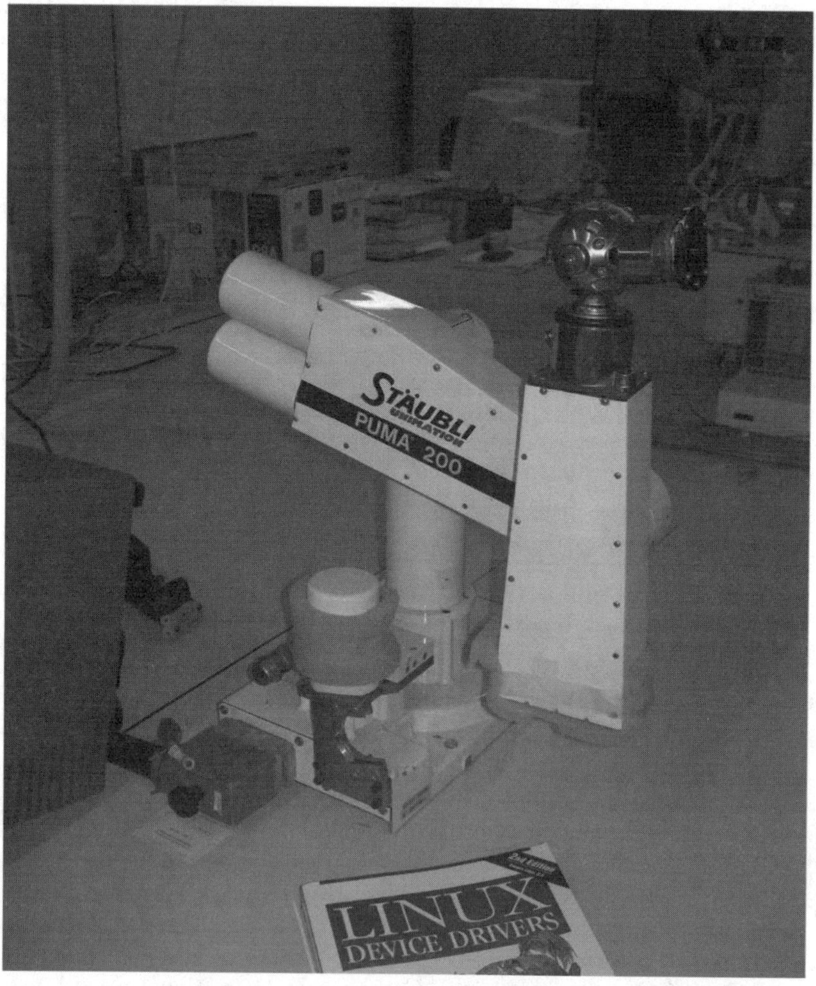

Fig. 8. The Unimation PUMA 200.

We run several tests to "tune" and evaluate our architecture. Most of them study the operator's perception, the stability of the entire system and the correctness of the movements of the teleoperated robot. With the final setup, operators have good perception of precision positioning and force reflecting.

Since it is difficult to quantify the architecture performance from these usability tests, we also set up two other experiment groups. The first one investigates the performance of the communication and the control implementation in Penelope. The second group focuses on the advantages of a modular and transparent design with respect to data exchange.

One of the tests objectives is to evaluate whether the use of CORBA adds a significant communication overhead with respect to more traditional communication strategies such as sockets. A CORBA version of `PING` has been implemented on top of our architecture and results are compared with the traditional `ICMP PING` of the Linux operating system. The comparison however, is not completely straightforward. In fact, CORBA supports only TCP and UDP protocols, so a real CORBA-based PING is not feasible. To compare equivalent methods then, we added a similar functionality on top of Penelope using TCP, paying attention to use IP datagram of the same size of ICMP echo-request packets. Instead of waiting for an echo-reply we use the ACK of TCP flow control. The packet size of the acknowledgment is smaller, but this compensates the overhead used by the more reliable protocol. The test is repeated on our laboratory LAN under different load conditions: with a dedicated point to point link between two computer, with normal intranet and Internet traffic and with an overloaded LAN, as shown in Figure 9.

To analyze data we have computed the mean and variance of all data sent from the two PC's; this is well represented by the normal distribution shown in Figure 10 where black distributions refer to normal PING's while red distributions refer to CORBA-based PING's.

We can see that CORBA-based PING is slower than normal PING and this difference is easy to see specially with the PC called Hector. In this case, in fact, the mean of CORBA-based PING is higher than of the ICMP PING of the other PC: with respect to CORBA-based PING, Hector's mean is about $1.7596ms$ and Vincent's mean is about 2.1443, whereas with Linux PING, Hector's mean is about $1.5726ms$, which is very similar to Vincent's mean ($1.5359ms$). This can be explained by considering that the overhead introduced by CORBA has a larger impact on a slower PC than on a newer and faster hardware.

The results of our PING tests represent a lower bound on the communication velocity achievable by our architecture during teleoperation tasks. The ICMP PING provides a measure of fast, but unreliable, communication.

Another important performance measure refers to the maximum frequency of the control loop achievable under RTAI with our hardware. This information alone is not completely significative, because real-time is not synonymous of high frequency, thus the second step is to verify that this performance is guaranteed by the real time operating system even under heavy system load

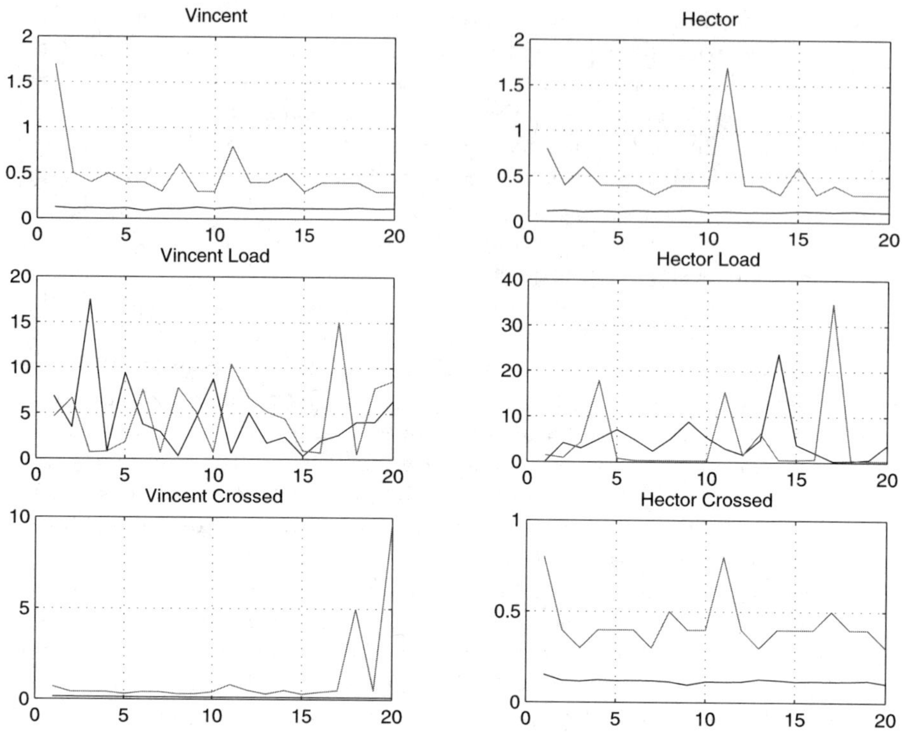

Fig. 9. Test communication performance with different PC (Vincent: Pentium 3, 866Mhz, with 256M RAM, Hector: Pentium 4, 2.4Ghz, with 512M RAM). Figures in the first line are with normal laboratory traffic; figures on the second line are under overloaded traffic and on the third line with point to point dedicated connection (Crossed). Black lines represent traditional ICMP PING in Linux while red lines represent CORBA-based PING. All data are expressed in milliseconds.

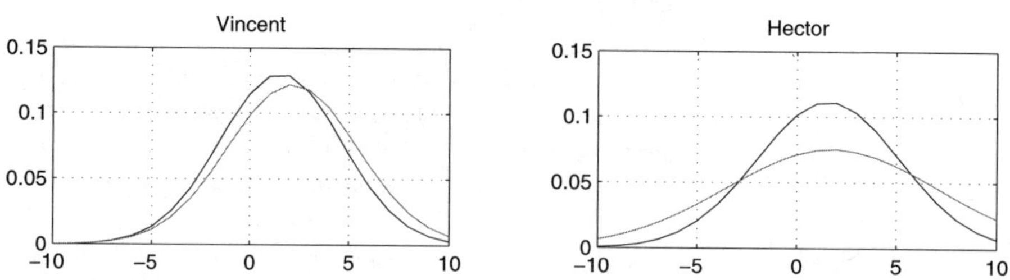

Fig. 10. Mean and Variance expressed by normal distributions. All data are in milliseconds.

or typical high latency I/O operations. Once the maximum control loop frequency is known with respect to the hardware limits of our host, a hard real time task with this time constraint is started. The computation must work without loss of performance while the system is loaded with different applications and data transfers, as shown in Figure 11.

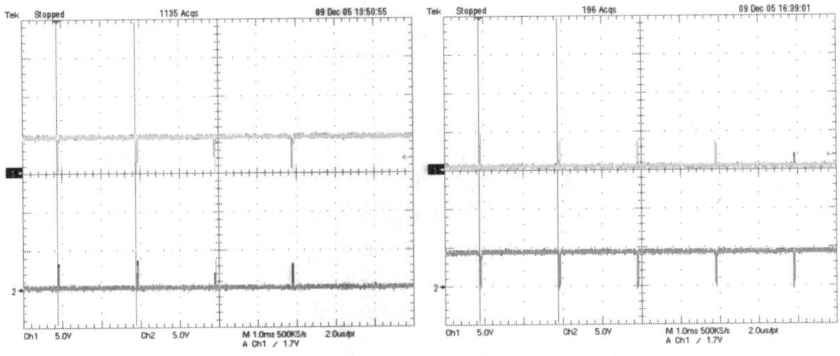

Fig. 11. RTAI performances with Hector PC with normal load (left figure) and with heavy system load (right figure) with a period of $2ms$ measured by Digital Oscilloscope. Every peak is a pulse train (25 pulses). No performance loss is visible.

The second group of tests is designed to demonstrate the modularity and reusability of Penelope. In these tests we want to exploit the capability of CORBA of handling distributed components. Different configurations have been distributed over our system paying attention to how many reconfigurations or modifications, with subsequent recompilations, Penelope required. We first tested the standard teleoperation task in a point to point connection, one PC controlling the joystick, collecting movements and driving the motors for the force feedback, and the other PC controlling the robot arm, simulating the force/torque sensor and running the CORBA naming service. All the matrix transformations needed by the FRHC and the robot are carried out by the respective hosts. Then, a different setup is used. Four computers are involved in the task: one for every device (robot, joystick and sensor) and one for the CORBA naming service. The performance of the two systems are compared.

One important fact arising from these tests is that no recompilation is needed. The selection of the service needed in the task is made at run time. CORBA takes care of identifying the host that can provide the elaboration requested in a very transparent way. If we decide to change the control type of one of our devices, for example transferring the forward kinematic computation of the FRHC to our FPGA to speed up the computation, only the selected classes on the selected host needs to be changed. Intercommunication interfaces are already defined and coded into the IDL.

Another test that involves both performance and modular capabilities of Penelope is to read sensor data from two different computers. Even in this case no change is needed to the code. The GUI's of the two hosts permit to choose the same service and no delay in the communication is introduced by the parallel requests to the sensors. Each device is read once and then transmissions are made in parallel over the net but not high load is detected on the net, according to our monitoring devices.

5 Conclusion and Future Work

In this Section, we have described Penelope, the architecture developed at the Robotics Laboratory ALTAIR of the University of Verona (Italy) and we presented the results of preliminary tests on a real teleoperation system. The tests are very encouraging and suggest that more devices and software algorithms can be added effortlessly to the system. Our goal is to demonstrate the feasibility of the solution proposed for a global system structure in a research laboratory where fast prototyping is a very important development aspect, together with performance improvement, especially with respect to the system real-time behavior. We want to have a common software basis to support the migration of algorithms across the laboratory and to our technology transfer partners. The use of a single common architecture would have a tremendous impact in this specific environment. We do not think that a unique framework can be possible across laboratories because from our experience a performing architecture is strictly tied to specific hardware and software. However, we think that it is possible to exchange high level modules when the architecture is truly modular.

RTAI and CORBA are difficult instruments to learn and to use. A lot of time has to be spent for an in-depth understanding of their mechanisms and functionalities. However, the design of Penelope and the current implementation relieves the programmer from a heavy involvement in communication and timing. For example, if a new type of controller is coded in our architecture it will perfectly fit in the already made IDL definitions, and the structure of communication/control synchronization may be reused.

Hard real-time thread must be free of unreliable instructions such as device driver access and network direct utilization. That's way the communication and the control loop are decoupled. Specific software is available in order to use simple, but widely used, communication channels with deterministic timing capabilities. USB4RT [Kis] and RTNet [Mar] are projects that will permit the integration among USB, ethernet and RTAI. Serial and parallel port support are already part of RTAI. Specific devices, such as axis boards, need kernel modules customization in order to achieve this goal.

One of the most important open issues regards the synchronization of data exchanged through CORBA. Due to the distributed nature of Penelope the possibility that each sensor is driven by a different PC has to be taken into consideration. If the teleoperation task needs the utilization of both cameras and force sensors, a synchronization mechanism must be developed to ensure that streaming video and force feedback reach the operator in a synchronized way, or a loss of realism will occur. The solution to this problem is not simple because of network uncertainty and data acquisition speed. We are working on a heartbeat strategy to maintain a common clock on the distributed environment. To achieve this goal we want to define a high priority channel using RT-CORBA, but we must still develop this technology. Some new real-time features like RT-Net may be useful even if these "drivers" rely on specific chipsets of ethernet board and are mostly for point to point connection. A different approach consists in using an NTP (Network Time Protocol) such as the one described in [Pro].

A more precise and general evaluation testbed for this type of architectures is needed, to make comparisons between different strategies easier and more useful. We are addressing this problem while our architecture becomes more usable and stable, letting us focus on performance and task related problem.

Finally, we are implementing a web interface for Penelope. The objective is to use it from Internet in various types of robot control experiments. User will be able to connect with Penelope, choose the controller type and its parameters, the trajectory type and its parameters, and run the remote experiment sending command through Internet. We have only added an IDL class (webInterface) and implemented the correspondent C++ class (webInterfaceImpl). In this manner, a client must only know this class interface and not the entire Penelope structure. In the same way, additional external tools or systems will be able to communicate with Penelope either using the standard Penelope interface (with all classes, methods and attributes) or using a built-in interface to hide complex structure to the client (for example with Matlab or Simulink toolboxes).

Acknowledgements

The authors would like to thank Davide Moschini for his participation in the design and in the early development of Penelope.

References

[Ark98] Ronald C. Arkin, *Behavior-based robotics*, The MIT Press, 1998.

[AS92] R.J. Anderson and W Spong, *Asymptotic stability for force reflecting teleoperators with time delay*, Journal of Robotics Research **11** (1992), no. 2, 135–148.

[Con] Scilab Consortium, *Scilab - a free scientific software package.*

[Des] Mathieu Desnoyer, *Ltt next generation (lttng).*

[DIA] DIAPM, *Real time application interface (rtai).*

[DZK92] H. Das, H. Zak, W. S. Kim, A. K. Bejczy, and P. S. Schenker, *Operator performance with alternative manual control modes in teleoperation*, Presence **1** (1992), no. 2, 201–218.

[GHJV95] Erich Gamma, Richard Helm, Ralph Johnson, and John Vlissides, *Design patterns: Elements of reusable software architecture*, Addison-Wesley, 1995.

[Gro] Object Management Group, *Uml: Unified modeling language.*

[HV99] M. Henning and S. Vinosky (eds.), *Advanced corba programming in c++*, Addison Wesley, Massachusetts, 1999.

[Kis] Jan Kiszka, *Usb real time support (usb4rt).*

[Law93] D.A. Lawrence, *Stability and transparency in bilateral teleoperation*, IEEE Transaction on Robotics and Automation **9** (1993), no. 5, 624–637.

[Mar] Ulrich Marx, *Rtnet hard real-time networking for real-time linux.*

[Mat] MathWorks, *Matlab - the language of technical computing.*

[OC03] Anders Orebačk and Herik I. Christensen, *Evaluation of architectures for mobile robotics*, Autonomous Robots **14** (2003), 33–49.

[OMG] Object Management Group OMG, *Common object request broker architecture*, Tech. report, http://www.corba.org.

[Ope] Opersys, *Linux trace toolkit (ltt).*

[Pro] Network Time Protocol.

[Ric99] Charles Richter, *Designing flexible object-oriented systems with uml*, Sams, 1999.

[She92a] T. B. Sheridan, *Telerobotics, automation, and human supervisory control.*, Mit Press, Cambridge, MA, 1992.

[She92b] T. B. Sheridan (ed.), *Telerobotics, automation and human supervisory control*, MIT Press, Cambridge, MA, 1992.

[YY94] Y. Yokokohji and T. Yoshikawa, *Bilateral control of master-slave manipulators for ideal kinestetic coupling–formulation and experiment*, IEEE Transactions on Robotics and Automation **10** (1994), no. 5, 605–620.

A Multi-robot-Multi-operator Collaborative Virtual Environment

Moisés Alencastre-Miranda, Lourdes Muñoz-Gómez, Carlos Nieto-Granda, Isaac Rudomin, and Ricardo Swain-Oropeza

Tecnológico de Monterrey - Campus Estado de México (ITESM-CEM), Mechatronics Research Center (CIMe), Atizapan, Estado de México, México
{malencastre, lmunoz, A00461680, rudomin, rswain}@itesm.mx

1 Introduction

In this chapter, we describe the design of an object oriented and distributed architecture and the development of a modular networked system that allow us to have a multi-robot-multi-operator system. On that system, users can collaborate on different robotics applications (e.g. teleoperation, planning, off-line programming, mapping, etc) using robots of different types (e.g. manipulator arms, wheeled mobile robots, legged robots, etc). For this chapter, not only will be addressed the design details in the architecture but also implementation details in the system. Also, is presented an extensive literature and concepts about the multi-user robotics systems.

There are several applications for such systems. In industry, operators must frequently program off-line several manipulators in a workcell. In research, researchers in different cities or countries want to test a planning method with several wheeled mobile robots. In education, students must execute a practice session with several virtual and real robots in a robotics course; etc. For all these cases, the development of software that allows several users to interact in a shared VE with many virtual robots and also communicate with the real robots is very useful.

Multi-robot-multi-operator interaction in a shared VE is a n-to-m relation where there are n users or operators that interact with m virtual robots (n and m are natural numbers). Each virtual robot can control a corresponding real robot of the same type.

A virtual robot is only the visual simulation of a real robot. This means that the virtual robot only has the appearance of the real one. But, the functionality and the features of the real robot must be simulated considering for example the kinematics and dynamics of the robot in order to make a virtual robot move and work inside the VE in a way that is similar to a real robot. However, it is so difficult and computationally expensive to model the exact

D. Brugali (Ed.): Software Eng. for Experimental Robotics. STAR 30, pp. 431–458, 2007.

functionality and features of the real robot in the virtual one. In order to integrate the virtual robots to a VE software is necessary to have the geometric model of each robot in some common format (e.g. OBJ, VRML, etc) and a way to load the models from a program (e.g. in C++, Java, etc). These virtual robots and VE features are the basic elements to develop a Collaborative Virtual Environment for robotics explained in next section.

2 Collaborative Virtual Environments

A Collaborative Virtual Environment (CVE) or Networked Virtual Environment (Net-VE) is a system that allows multiple users in different geographical locations to interact, share, communicate and collaborate in a given task at the same time in a common virtual environment [SZ99] [CPMTT99]. A CVE has the following features:

- A shared sense of space. All participants (users) have the feeling of being located in the same place (inside of a shared VE). The shared space must have the same visual features for all participants; however, each one can see it from a different point of view.
- A shared sense of presence. Each participant can be identified by using a graphical representation in the VE. This representation can be an avatar (digital character animated) or a Graphical User Interface (GUI) showing who is connected to the system at every moment.
- A shared sense of time. Participants must see another participant behaviors or actions in the moment in which these behaviors are taking place.
- A communication mechanism. Because of the participants' remote locations, there must be voice or text communication existing between them (e.g. chat).
- A sharing mechanism. The importance of a CVE lies in the ability of the participants to interact with other participants in the VE, and maybe exchange files.

In other words, a CVE is a multiuser 3D VE system where: all users are running the same software on its computer, everybody are working on the same scenario or environment (VE) that contains the same elements (virtual objects) in the same positions and orientations at any time, one or more group of users can be collaborating in one or more tasks in the VE, but each user can observe a desired area of the VE from a different point of view and zoom configuration. CVEs had been mainly used for games and military training. Four well known CVEs architectures were developed as militar simulation systems during the 80's and 90's decades by the Department of Defense (DoD) and the Naval Postgraduate School (NPS) (e.g.[MZP99], [SZ99], [KWJ99]): Simulator Networking (SIMNET), Distributed Interactive Simulation (DIS), Naval Postgraduate School Network (NPSNET), High Level Architecture (HLA). First three were based on three basic components:

1. The object-event component. Each element in the visual simulation (virtual objects) was called 'object' or 'entity' and their interactions with other objects were called 'events'.
2. Autonomous distributed simulation nodes. Each node (participant) must place the corresponding packets onto the network for the changes-in-state of each object or entity that this participant included to the VE.
3. Dead reckoning methods. A set of prediction and convergence algorithms are used to reduce network packet traffic and synchronize the objects and their copies in other computers called 'ghosts'.

HLA is the glue that allows to integrate separate and remote computer simulations into a large simulation. Each individual simulator application is called 'federate', and a set of federates working together, through a Runtime Infraestructure (RTI) based in objects and components, are called 'federation'.

CVEs are useful Virtual Reality (VR) tools to develop Networked Robots (NR) systems where all users are running the same robotics software on their computers. In this chapter is developed a Multi-Robot-Multi-Operator CVE that needs to deal with following issues:

- GUI. Users need a set of interfaces organized in several levels of windows and menus to interact with the system.
- 3D Graphic Display. A graphics library is needed to create and display the virtual objects and the whole VE.
- Modular software. The design of an object oriented architecture, the use of a database and the implementation based in software components helps to obtain modularity. Final software must allow users to include new virtual robots, to integrate new robotics applications, and to exchange some programs for the new versions modifying as less as possible the source code.
- Distributed simulation. A middleware software must be chosen in order to get a distributed system where doesn't exist a server that performs all visual and non-visual simulation processing. The use of remote objects avoids to send long network packages reducing the low bandwidth.
- Network protocol. In a low level, middleware software must communicate between computers through a network protocol. The network protocol chosen must help to diminish the low bandwidth.
- Delays. A dead-reckoning method helps to deal with unpredictable time delays synchronizing the visual simulation on all the participants in the CVE.
- Real robot controller. Virtual robots must have a way to replicate their movements to the correponding real robots.

3 Related Work

As described in Chapter *Trends in Software Environments for Networked Robotics*, in last years NR systems in experimental robotics research have increased. Robot teleoperation systems and visual simulation systems for other robotics applications are two types of NR systems. Most of those NR systems are not CVE systems, it means that they don't have a n-to-m interaction. Common teleoperation systems are networked robotics systems that have four types of interactions (defined in [GSL03]):

1. Single-operator single-robot (SOSR) teleoperation systems. A human user teleoperates the robot one at a time, it meaning a one-to-one interaction. This is the most common type of teleoperation systems. Most SOSR systems don't have a VE, for instance in Telegarden users control one manipulator [Gol00], in KheponTheWeb users work with one mobile robot [MSM97], in [DQBM98] vision tasks and teleoperation of one mobile robot are performed, and many others SOSR systems are mentioned in [GS02]. Only some SOSR systems include a VE like RoboSiM [SK99], UJI [MSDP02] and the work in [TCB04] where a virtual industrial manipulator is used to teleoperate the real robot, but where that VE is not shared with other users.
2. Multiple-operator single-robot (MOSR) teleoperation systems. Several users send different commands or positions to the same robot in a n-to-one interaction (e.g. [GSL03]).
3. Single-operator multiple-robots (SOMR) teleoperation systems. One user controls multiple robots existing a one-to-m interaction (e.g. [LHHY03]).
4. Multiple-operator multiple-robots (MOMR) teleoperation systems. Each operator controls only one robot and the robots have overlapping workspaces. This is a n-to-m interaction but where $n = m$, it means that exist n one-to-one interactions (e.g. [OKC99]).

Therefore, a CVE system that allows robot teleoperation is a n-operators m-robots ($nOmR$) teleoperation system for any number of operators n and robots m. In fact, the four types of teleoperation systems mentioned are particular cases of a $nOmR$ system depending on the values of n and m. Common teleoperation systems have a client-server scheme. In a CVE is less expensive in processing and bandwidth to have a distributed scheme. This $nOmR$ system applies for all the robotics applications that can be integrated in the CVE (e.g. teleoperation, planning, off-line programming, mapping, etc).

On the other hand, visual simulation systems including VEs are mainly single user stand-alone robotics systems used for specific robotics applications. Some of them are commercial systems like IGRIP from Delmia [Don98] that include several models of manipulators and are useful for programming and manufacturing tasks on industrial workcells, Webots from Cyberbotics [Mic04] that allows modelling and programming several wheeled mobile robots and several legged robots, etc. Other stand-alone systems that are non commercial systems are Operabotics [Say98] that is useful for path planning and teleoperation of up to four wheeled mobile robots and one aquatic robot, and Move3D [SLL01] used in motion planning for manipulators, wheeled mobile robots and legged robots. Most of those systems are used only for one user at the same time, they are for a few number of robotics applications, and allow to work with only one or a few number of robots commonly of the same type (for example only manipulators); two exceptions of this are Move3D and Webots.

All related work mentioned above are not CVEs. We have found only a few CVEs for robotics developed mainly by two NASA laboratories:

- Virtual Environment Teleoperations Interface for Planetary Exploration (VEVI) from NASA Ames Research [HHF95] was the first CVE developed for robotics. VEVI was a software that allows to several users to teleoperate mobile robots for spatial and submarine exploration. This was a distributed system processing different modules on several computers, it means that there were multiple severs, one for the database, one for the render of the VE, one for the kinematics calculus, etc.; however, the visual simulation was not yet distributed. VEVI used the Task Control Architecture (TCA) to communicate tasks on different computers by sockets, and the commercial graphics library called World Toolkit (WTK) from Sense8 for the 3D graphic display of the VE. This system was developed only for SGI workstations and displayed until two images monitoring the remote robot actions.
- Web Interface for Telescience (WITS) from Jet Propulsion Laboratory (JPL) [BTT97] [BTN00] was used in the Mars Polar Lander Mission. WITS was a client-server architecture that allows to several researchers on different countries to work with rovers for Mars exploration. Each client computer had three module applications: teleoperation, path planning and mapping (using an stereo system). Database with all information for the simulation was in one server. WITS used remote objects to communicate with the server, Java for portability of the system on different platforms, graphics library Java3D for the 3D graphic display of the VE with VRML models, and multiple cameras for monitoring the rover.
- Mission Simulation Facility (MSF) [PLC04] from NASA Ames Research offers a simulated testing environment including robotic vehicles with manipulators onboard, terrains, sensors and vehicle subsystems. MSF uses the High Level Architecture (HLA) and is developed for autonomous tasks of rovers, spacecrafts and submarine robots. This CVE system is developed by several components that allows to generate virtual terrain surface, virtual environmental conditions, virtual robots, simulated equipment and graphical display. MSF have a database where are the models for the simulation and the results.
- Science Activity Planner (SAP) [NPV05] from JPL plans daily activities of the Spirit and Opportunity robots in the Mars Exploration Rover project. SAP is also useful as the operation interface for research rovers in develpment at the JPL. SAP renders a collection of images obtained by robots. With these images are generated terrain meshes in a 3D VE. Everybody sees the robot's environment but each user works with a dif-

ferent instance of the software to generate different plans in parallel. New information about the terrain and defined targets are updated in all the instances. SAP is implemented with Java, VRML, MySQL and XML.

The Development of a large CVE for multiple robots and multiple robotics applications, integrating the best features of all related work mentioned, involves the combination of computer graphics techniques, object oriented and components modeling, computer network protocols, distributed software, databases issues, GUIs, multithreading execution, and a lot of different algorithms for several applications. In next sections, the architecture, functionality, software implementation and examples cases of a Multi-Robot-Multi-Operator CVE are explained in detail.

4 Object Oriented Distributed Virtual Reality Architecture

Object Oriented Distributed Virtual Reality Architecture (OODVR#) is an object-oriented architecture designed for running VR applications in a distributed visual simulation system [AMMGR03] [MGAMR03].

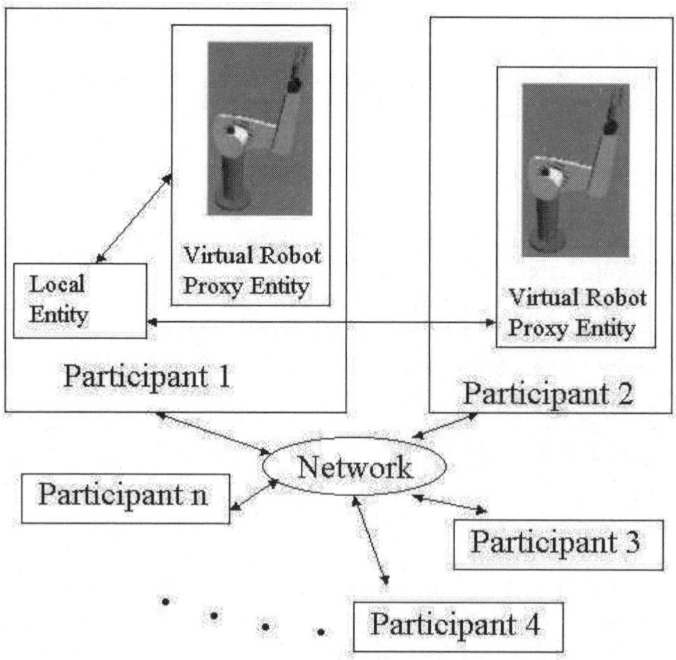

Fig. 1. OODVR# Architecture

OODVR# is composed by 'participants' and 'entities'. A participant is a new instance of the software running on a computer connected to the network, commonly each participant represents one user running the software on a computer. However, depending on computational resources, this architecture allows to one user execute one or more participants on a single computer. Entities are all objects that lie inside of the VE. Entities can be inserted in the VE at any time. The architecture is designed to have multiple entities in a distributed VE where the complete visual simulation can be accomplished by multiple participants running in several machines (see Fig. 1). This means that there is no need of a central server for the visual simulation, the entire simulation doesn't go down if one participant has some trouble. A new participant can enter to the system, at any moment during the simulation, seeing the current state of the VE. Each participant can insert zero, one or more entities in the VE. It doesn't matter how many entities a participant inserted (even zero), it can either start to work alone, to collaborate with other participants in a given task, or only see what is happening during all simulation. Entities are classified in 'dynamic entities' and 'static entities'. Dynamic entities represent objects in the VE that are able to perform movements during the simulation, for example robots and conveyors. Static entities are motionless things like floor and furniture in general.

For each entity there is a 'local entity' and several 'remote entities', called 'proxies'. A proxy entity is in charge of the 3D graphic display (render) of the virtual object that represents that entity (as in the Fig. 1 where a virtual manipulator is displayed by all the corresponding proxies). The local entity is only in the participant that inserts the entity in the simulation. For each participant there is one proxy entity, even in the participant that has the local entity. The participant who inserted a new entity is in charge of the movements (change-in-state) of that entity in all the participants through its local entity (similar to the autonomous distributed simulation nodes mentioned in section 2); it means that, if this participant starts a movement of that entity, its local entity will activate all its proxies to replicate the movement automatically in all the participants. If a participant having only a proxy entity wants to start a movement, then the proxy communicates the movement to its local entity (on the other participant), activating all proxies to execute the same movement. If for some reason, one participant that inserted entities leaves the simulation, then the next participant that moves those entities will have new local entities to control them.

This mechanism is a remote objects scheme that allows all participants in the simulation to observe multiple movements of multiple entities at the same time, doesn't matter which participant moves which entity. With this, an object oriented and distributed visual simulation is achieved, having an architecture useful to implement a CVE system.

5 System Modules

The Multi-Robot-Multi-Operator CVE System implements the OODVR# architecture adding all extra functionality needed to integrate several robotics applications. As in Brugali and Fayad [BF02], a recommendation for distributed computing in robotics is to use an object-oriented software development and a middleware framework for distributed issues. For this CVE implementation, the Java programming language is choosen in order to have a portable and multiplatform system, the Java Remote Method Invocation (RMI) is selected to manage the remote objects scheme explained in the OODVR#, and software components composed by several Java Beans are used to create a modular system. Therefore, CVE system is divided in software modules. All modules that a participant has and their main relations are shown on Fig. 2 [AMMGR03] [MGAMR03] [MGAMR04]. Main modules representing the core architecture and the basic robotics application (teleoperation) are shown inside and to the left of the big box, and will be explained in next subsections. Extra modules to the right of the big box represent other robotics applications that can use the main modules to work and will be described in next section.

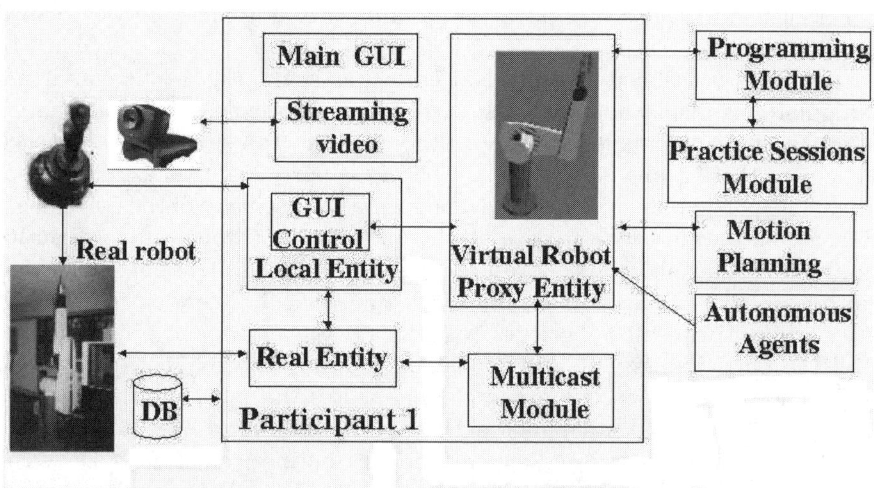

Fig. 2. CVE Modules

5.1 Core Module

This module, represented on Fig. 2 by boxes called 'Participant' (big box), 'Local Entity' and 'Proxy Entity', implements the main issues of the OODVR#

architecture (Java package called 'oodvr'): functionality of the participants, local entities and proxies as is described in section 4.

This was implemented using RMIs [PK01] but in a different way from what is common practice. It means that there isn't a common stub/skeleton approach with a client object and a server object [BF02]. Difference is that a direct communication between the stub and the remote object is performed through a Java socket (in this case, a multicast socket explained in next subsection). Is like there are no server objects, only multiple client objects controlled by the local entity (also there is no need to run the 'rmiregistry' command). Even the class *Participant* extends to the class *Entity*. Each participant has a hash table of all the other participants and other hash table for its local entities. Local entities have a hash table including all their corresponding proxies. Movements of each entity are executed in a different thread.

5.2 Communication Module

This module, represented on Fig. 2 by a box called 'Multicast Module', allows the participants to send and receive the remote objects information using a transport layer networking protocol called 'Internet Protocol Multicasting (IP Multicasting)'. For this, a Java Socket class called *MulticastSocket* is used (see [HSH99]).

IP Multicasting is a faster and more efficient protocol than Transmission Control Protocol/IP (TCP/IP), User Datagram Protocol/IP (UDP/IP) and IP Broadcasting for large-scale distributed CVEs [SZ99] [Die01]. Here are the main reasons:

- TCP/IP is the only reliable protocol of that four, but is the slowest because its packets have the biggest size and those packets may be delayed to ensure an ordering.
- UDP/IP is a not reliable protocol, therefore is faster than TCP/IP because use packets of small size and doesn't need an ordering of packets, however lost and corrupted packets happens only if the network is saturated with a lot of traffic.
- Nevertheless, both TCP/IP and UDP/IP are no so efficient during programming a final application because they are point-to-point protocols; therefore in a communication component would be necessary two things: to have always actualized a table with IP addresses of all possible computers that will connect to the CVE at any moment (this is no always possible because the locations of the computers, or will be very tedious to all users to recollect all IP addresses and give them to the software), and is needed a high network bandwith to send the same packet N times for the N computers.
- IP Broadcasting is based on UDP/IP but is a multi-point protocol that delivers the packets to all computers on a sub-network. Therefore, there is no need to have a table of IP addresses, but IP Broadcasting only can

be used in a LAN environment and still there is expensive in network bandwidth because each computer must receive and process the packet sent, even if the CVE system is not running on that computer.
- IP Multicasting is also a multi-point protocol based on UDP/IP, but solves the problems of IP Broadcasting because only sends one packet that is received only by the computers that need that packet and are joined to the multicast group (computers that are running the CVE system), preventing bottlenecks by load balancing. Information only travels through routers along networks that lead toward an interested destination host.
- Due to the scalability of a Multi-Robot-Multi-Operator CVE is better to use a faster and efficient protocol to diminish delays on the visual simulation.

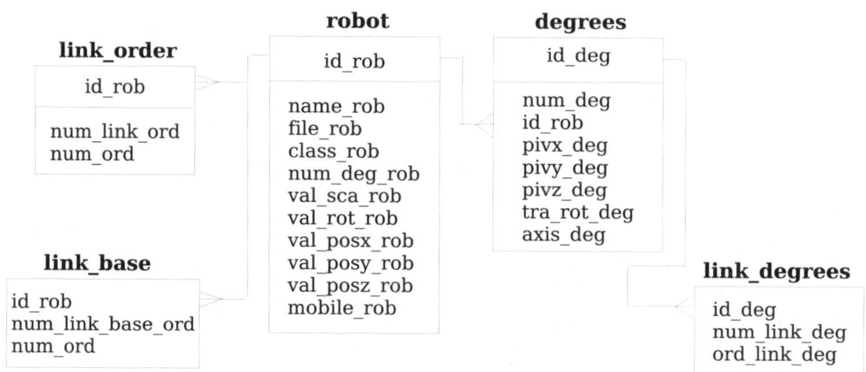

Fig. 3. Entity-relation diagram of some tables on the DB

5.3 Data Module

This module, represented on Fig. 2 by a cylinder called 'DB', is a data layer in charge of store and load information from a DataBase (DB). Robot's information stored in the DB is needed to insert a virtual robot in the VE and to move its different degrees of freedom at any moment. One example of this is represented in an entity-relation diagram shown on Fig. 3.

The tables in the figure, are the ones corresponding to the information about a robot, its geometry and its degrees of freedom. The 'robot' table has the main information about each robot: name (name_rob), number of degrees of freedom (num_deg_rob), file with the geometry (file_rob), type of robot (mobile wheeled, manipulator or legged indicated in field mobile_rob), and so forth. The 'link_order' and 'link_base' tables are used for parsering the files with the geometry of the robot. Finally, the 'degrees' and 'link_degrees' tables have the information about each degree of freedom, the pivot point (pivx_deg, pivy_deg, pivz_deg), the type of joint (translation or rotation indicated in the field tra_rot_deg), the part of the geometry that will be moved with the degree of freedom (table 'link_degrees') and so forth. The system allows two types of geometry files: VRML and OBJ.

The implementation of this module was performed with Java DataBase Connector (JDBC) and the freeware DB MySQL. Each time that a new virtual robot needs to be added to the system it is only necessary to fill the corresponding information in the DB (without recompile).

5.4 Real Robot Module

This module, represented on Fig. 2 by a box called 'Real Entity' and by the real robot image, allows the connection of the system to each type and specific model of real robot through its own library (e.g. Aria libraries for mobile robots Amigobot and Pioneer3DX from ActivMedia, Tekkotsu libraries for Sony Aibo legged robot, etc.). Tekkotsu is already in Java. We have made the connection between Java and the corresponding C++ libraries for some real robots tested using Java Native Interfaces (JNI) [Gor98] and threads. To achieve this connection, four classes are important, three classes implemented in Java and one in C++. For example, if is required to teleoperate two degrees of freedom of an AmigoBot (translation and rotation), therefore:

- First class called *RealRobot* (made in Java) instanciate all corresponding Java classes in charge of teleoperating or controling each real robot through a Java reflexion mechanism (use the Java classes *Class* and *Method*) taking the name of the corresponding robot class from the data module. Only is needed one class of this type for all new robots added on the system.
- Second class called *AmigoBot* (made in Java) is the class that teleoperates or controls the real robot AmigoBot, instanciates the third class called *TeleoperaAmigoBot* (made in Java) and loads the fourth class also called *TeleoperaAmigoBot* (but is made in C++) via the dynamic library called *libAmigoBot.so*. Down is an example of the code structure of this class.
- *TeleoperaAmigoBot* is the class in the Java system that corresponds to the class made in C++. There must be defined as native the method that shares this class and the class in C++ (e.g. private native void IsRobot();).
- *TeleoperaAmigoBot* is the class in C++ that includes the Aria libraries functions to move the real robot. This program needs to include a '.h' file generated by the command javah (e.g. #include "TeleoperaAmigoBot.h"), and the needed methods must have specific names (e.g. JNIEXPORT void JNICALL Java_TeleoperaAmigoBot_IsRobot(JNIEnv* env, jobject thisObj)).

```
... // Here goes some packages needed
public class AmigoBot
{
  ... // here goes some variables and objects definitions needed
  TeleoperaAmigoBot teleop;
  static
  {
   System.loadLibrary("AmigoBot"); // To load the dynamic library
  }
  public AmigoBot()
  {
  }
  public void open() // Method that opens the connection to the real one
  { ...
   teleop.start();
  ... }
  public boolean isRobot()   { ...
   teleop.isRobot(); // Method that check if the real robot is available
  ... }
  public void rotate(int grades) // Sends rotate command to the robot
  { ...
   teleop.rotate();
  ...}
  public void translate(int mm) // Sends translate some distance to the robot
  { ...
   teleop.translate();
  ...}
  }
```

Code structure of class *AmigoBot*

5.5 Virtual Environment Module

VE module, represented on Fig. 2 by a virtual robot inside of the box 'Proxy Entity', is the one that performs the render of the 3D virtual objects in the VE using the Java3D libraries [BP99]. Due the slowness of Java3D when there are a lot of 3D virtual objects, some tests have been done using OpenGL libraries [SWND05] connect them to the system also with JNIs. This module allows each participant to see the same VE from the desired point of view. Left side on Fig. 4 shows an example of a VE including a workcell with four manipulators, furniture, floor and a conveyor.

5.6 User Interface Module

This module, represented on Fig. 2 by the boxes called 'Main GUI' and 'GUI Control', and the image of a joystick, implements the GUIs and the external device interfaces. GUIs display all menus, windows and options that needs

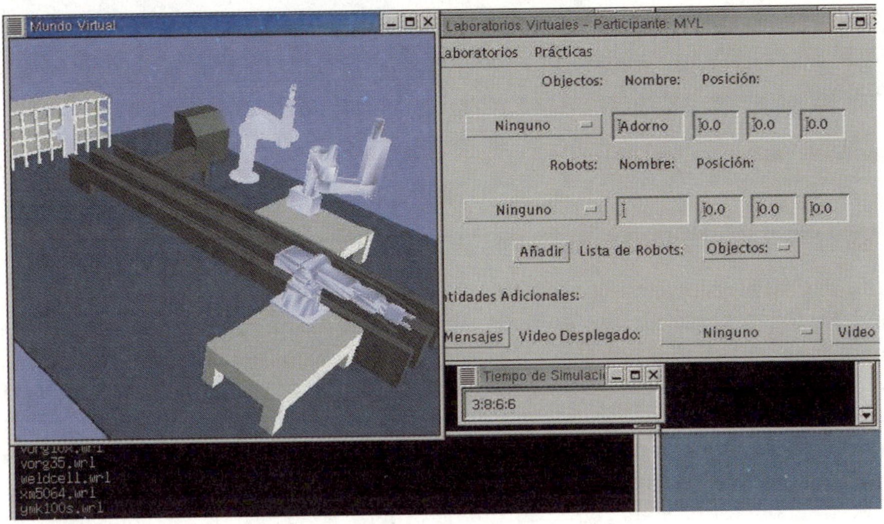

Fig. 4. VE and GUI

each user to control the virtual or real robots (right side of Fig. 4 shows the main GUI in the big window and the simulation clock in the small one). External devices include joysticks or steering wheels. This module was implemented using Java Abstract Window Toolkit (AWT) and threads.

5.7 Telemonitoring Module

This module, represented on Fig. 2 by the box called 'Streaming video' and the image of a webcam, uses videostreaming through a multicast protocol to allow remote monitoring for the real robots. This was implemented with Java Media Framework (JMF) [GT99]. Telemonitoring can be accomplished from multiple video cameras that could be connected to different participants; one camera maybe seeing the robot environment and other mounted on robot. Each participant can decide how many cameras' view will open to telemonitor the scene (even can decide only to see the virtual robot without see any camera's view).

6 Robotics Applications

Until now, we have integrated four robotics applications: teleoperation, programming practice sessions, reactive agents, and path planning. The first one, teleoperation, is achieved by the core module and the real robot module. Last three were easily implemented in our modular CVE because they use mainly previously developed modules. We need to take care of the own features of each application, because the general functionality is done on the system (virtual robots display, control of robot movements, etc); only is needed to add some new menus and bottons to the GUI, add some tables to the DB, and add some classes to the system (as in the case of the real robot module) to connect each application.

6.1 Teleoperation

In our CVE, each participant can teleoperate multiple manipulators, mobile and legged robots (virtual or real). With the core module implementing the main issues of the OODVR# architecture, we have a natural way to perform the 'teleoperation of virtual robots'. Teleoperation of real robots is done with the real robot module. Tests were done on Linux and Windows.

A simple control mechanism can be activated in order to allow only one user to control the one real robot at the same time. In the GUI each user can take the control of that robot and none other user can move it. The user can release also the control of a robot and so take it other users.

A simulation clock allows the movement synchronization of the robots, considering that can exist delay in the network. When a user moves a degree of freedom of a robot, this information is sent to the local entity with the start time of the movement, and the speed. When a proxy receives the information about the movement, if the start time is different of the current time, this proxy will move the robot with a bigger value of speed. With this mechanism all proxies finish the movement at the same time.

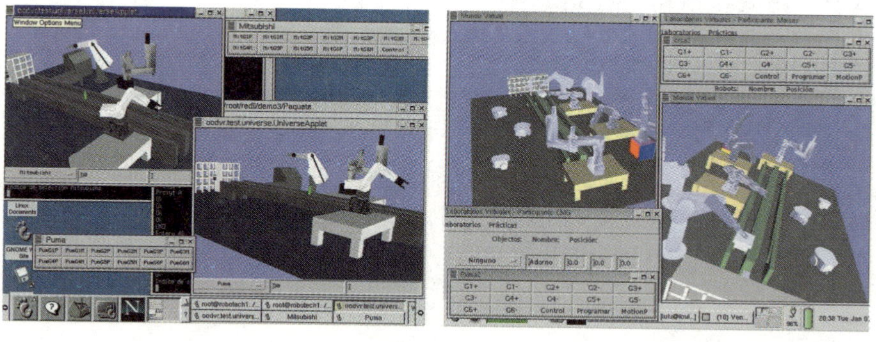

Fig. 5. Two examples of teleoperation of multiple virtual robots

Fig. 5 shows two examples of teleoperation of multiple virtual robots, both examples show two participants on the same computer (but this two participants were interacting with more participants in other machines) and the same workcell with multiple robots but each static and dynamic entity has different colors. Left side only has four manipulators (Mitsubishi Movemaster

EXR, Amatrol Jupitel XL, Amatrol ASRS and Puma560) but right side has seven manipulators (same first three manipulators than in the first example plus two Puma 560 and two CRSA 465) and four mobile robots (two Pioneers 3DX, one AmigoBot and the conveyor that works as a mobile robot). In that figure can be seen the different points of view of the VE on each participant and the GUI windows that are used to control each degree of freedom of each robot.

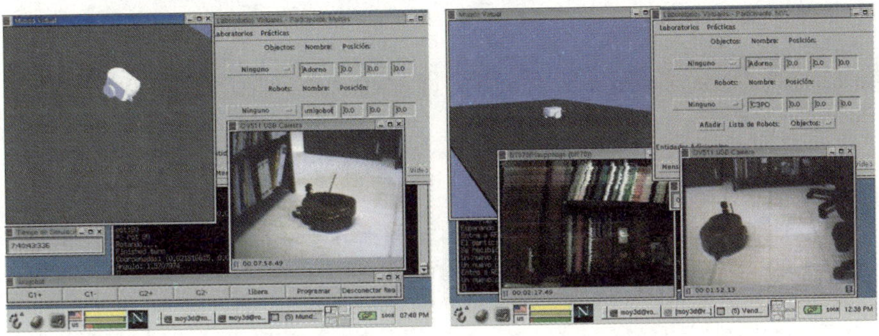

Fig. 6. Two snapshots of two different computers executing teleoperation of a real mobile robot

Teleoperation of a real mobile robot AmigoBot is shown on Fig. 6. In that figure there are two snapshots of two participants in a test with more participants. The real robot can be observed by the video sent by two cameras and there also exists a virtual representation of the robot in the VE. The two cameras are a webcam watching the environment where the robot moves and the camera onboard of the robot. While a participant moves the AmigoBot, the virtual robot will follow the same movements. Left snapshot shows the computer of the participant that is controlling the movements of the robot, this participant is only telemonitoring the robot with the webcam view. Right snapshot shows other participant monitoring the robot with both cameras.

Teleoperation of a real Aibo legged robot is shown on Fig. 7, there appears the VE window with the virtual legged robot and one image taken from the monitoring window.

6.2 Programming Practice Sessions

The idea of having practice sessions as an application of the system, is to provide training and learning tools for students or researchers improving distance learning programs, reducing investment and providing a flexible experimental framework. With this practice sessions the risk of damage caused by accidents can be reduced.

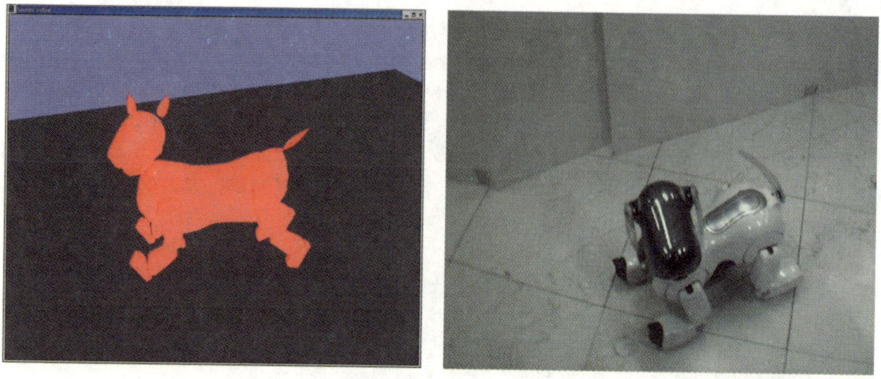

Fig. 7. Teleoperation of a real Sony Aibo legged robot

With this application, an instructor is able to define a practice session, and students (or simply other users) are able to perform programs and tasks to accomplish a practice session. This application can be considered as a Virtual Laboratory in which users have access to virtual equipment, in this particular case there are virtual manipulators and virtual mobile robots.

For the implementation of this virtual laboratory, the following features are considered:

- The DB is used for saving information about the practice sessions, for example robot positions, environments, programs and instruction files.
- The chat included in the system provides a communication channel between users in a practice session.

A programming language that can be used for both manipulators and mobile robots has been developed. There is also an interface for teaching positions to manipulators, as is performed with real robots using a teach pendant. With these tools users can write, compile and run programs for each robot and after that there is another interface that allows to coordinate several programs for different robots to simulate a more complex task [MGAMR03].

Defining and Executing Practice Sessions

In order to define a practice session, users must first create and save a VE in which the practice will take place. The VE can include furniture and different robots. After that a new practice has to be created including the name of the practice, the name of the VE and the file with the instructions. Once the practice has been defined, users must complete the assigment.

Each user connected to the simulation can program a different robot and test the program. When all needed programs are ready, only one user must define a task to coordinate all programs. From this point of view a task is considered as a set of coordinated programs to perform a more sofisticated activity with the cooperation of several robots. The programs and tasks can be saved on a DB for future practice sessions.

Table 1. Control structures and instructions.

Control structure or instruction	**Description**
if...else	Decision control structure.
while	Cycling control structure.
$move(p)$	Moves robot to position p.
$move(f, g)$	Moves dof f by g units.
$speed(v)$	Defines speed v.
$flag(v)$	Assigns value v to the robot's flag.
$wcellflag(n, v)$	Writes value v to the nth flag of the environment.
$rcellflag(n, v)$	Reads the nth flag of the environment and stores the value in v.
$wait(m)$	Waits m miliseconds.

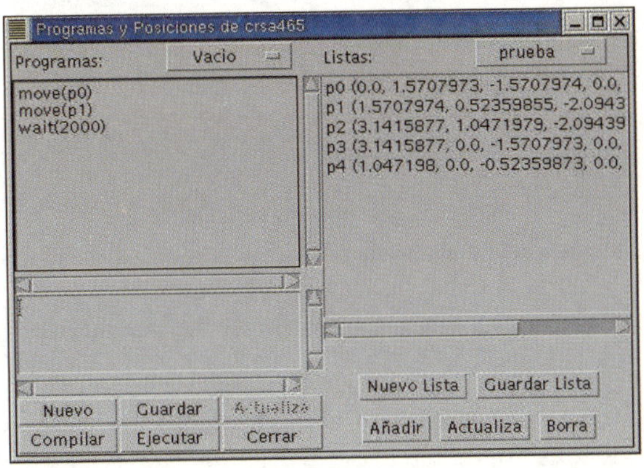

Fig. 8. GUI for programming and defining positions

For programming a robot, the GUI shown in the Fig. 8 must be used. There are two steps for programing a robot: teaching positions and writing the program. The first step, teaching positions, applies only for manipulators and is an optional step.

1. Teaching positions.
   In real manipulators, a teach pendant is used to define fixed positions that will be needed for programming a sequence of movements to that fixed positions. This feature is also implemented in the practice sessions application. The user can move each degree of freedom in a manipulator

until the desired position is reached; then the user must save the position. Each position is labeled as p_n, where n is a natural number. Once several positions are defined, the user can save a list. The list will be accesible to all practice sessions.

2. Writing a program.
   The developed programming language includes the instructions and control structures shown on Table 1. Users can create new programs and change existing ones at any time during the simulation. Each single program is saved on a file and compiled "on the fly" when the user runs the program for the first time. Each robot has an internal state flag that can be read by other robots (see instruction *flag* on Table 1). There are also a set of ten flags that corresponds to the environment that can be set and read by all robots. All these flags are useful to coordinate running programs in different robots (e.g. a robot can wait until a flag is set by other robot to continue the execution of the program). Each program is saved in a file and in the DB, and will be accesible to all practice sessions.

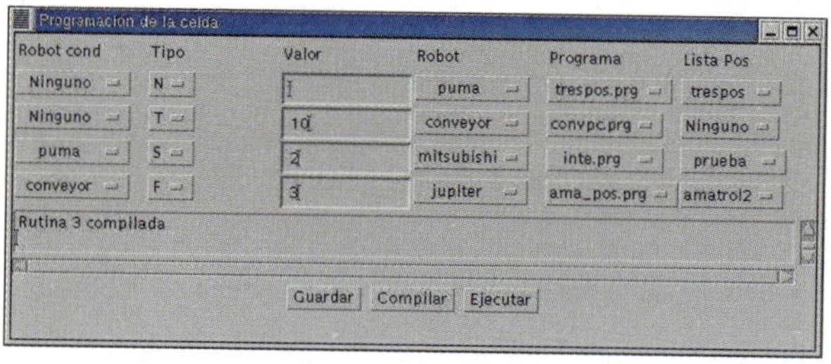

Fig. 9. GUI for defining a task with all the starting conditions

For coordinating the programs in a single task, only one user must define the starting conditions for each program using the GUI showed in Fig. 9. The interface shows six columns. The first three columns define the starting condition, and the last three define the robot, the program and the list of positions. There are four starting conditions:

- None (N): There is no condition to run the program. If all programs has no starting conditions, all programs will be executed in parallel.
- Time (T): The program will start after t seconds from the begining of the task. Time t is specified in the textbox next to the T.
- Status (S): The program will start when the robot indicated in the first column is in a given state. There are two possible states: a value of 1 means that a robot has started a program, and value of 2 means that a robot has finished a program.
- Flag (F): The program will start when the robot indicated in the first column has its flag in a certain value.

Once a user has defined a task, this can be saved for future changes. Only one task can be saved for each practice session.

Hints on Implementation

In order to develop the application previously described, several classes must be implemented. Basically these classes were divided in different components (each one is a Java package) depending on the functionality: classes for the GUI, classes for the DB connection, classes for open and saving practice session, classes for opening and saving lists of positions classes for opening, saving, compiling and running programs, and classes for opening and saving a task.

The main issue in the practice sessions application was to define a way to allow programs to be compiled and executed "on the fly", this means without having a predefined set of programs or instructions. The compiler for the programming language was developed using the Java Compiler Compiler (javacc), which is a tool that allows to define the lexical and grammatical rules using an script file and after that generates the java code for the compiler. The programming language, besides of the instructions and control structures shown on Table 1, allows to define integer variable and to perform arithmetical operations with those variables. The input of this compiler is a program of the defined programming language and the output is the java code for that program.

Inside the application, a user writes a program in the GUI and compiles it, therfore, the text of the program is saved in a file '.prg' (compiler input), the compiler is invoked and a file '.java' is written (compiler output). All programs written by users will have a *run* method in the '.java' file. After that, if the user wants to run the program in the simulation, then happens the following:

- A '.class' file is generated using the *Runtime* Java class (specifically the *exec* method). The input will be the '.java' file.
- The corresponding class is executed in the simulation with a reflexion mechanism. The '.class' file corresponding to the user's program is instanciated with the Java class called *Class*. After that the *run* method is invoked.

In this way, this mechanism is like adding new code to the simulation to perform new tasks with the robots in the VE.

Example

There is a manufacturing cell in the simulation with several manipulators (Mitsubishi EXR, Amatrol Jupiter XL and a Puma 560). The task is the following: make Puma to take a piece from the conveyor, put the piece on a lathe, and after a defined time return the piece on the conveyor again. Then move the conveyor until the piece is close enough of Jupiter robot. Jupiter

Fig. 10. Sequence of a practice session in a workcell with manipulators

robot must take a stamp from a box and put it on the piece. After that the conveyor must arrive to the Mitsubishi robot. Mitsubishi robot must take a cloth from a box and clean the piece. Finally the conveyor must return to the end. In Fig. 10 there is a sequence of snapshots taken during the execution of the task previously defined.

Down is an example of the code in the programming language defined, that is a part of the code for the conveyor in this example of practice.

```
int a //declaration of variable a
a = 0
flag(1) //The internal state flag for the conveyor is set to 1
        //this flag will be read by other robot.
rcelflag(1,a) //Read the environment flag number 1 using variable a
//The conveyor will wait until flag number 1 is set by other robot
while(a == 0)
{
  wait(500)
  rcelflag(1,a)
}
move(p1) //move to position defined by p1
move(1,1000) //move degree of freedom 1 in 1000 units
```

Part of the conveyor code

6.3 Reactive Agents

This application was developed to add agents with certain behavior. In this particular case robots become robotics agents using a reactive architecture. The problem described by [Ste99] was selected and implemented for the reactive agents application. The problem is to explore an unknown terrain using several wheeled mobile robots and to pick up rock samples that have to be returned to a space base.

At the beginning of simulation, all robots leave the space base and start the exploration. Robots do not have a map of the terrain and do not know where the rocks are. However robots have sensors to determine if in front of it there is an obstacle or a pile of rocks, and a sensor to "hear" a radio signal emitted by the space base. When a robot finds a rock, it has to return to the space base (following the radio signal) to leave the rock and then returns to explore again. There is no communication between robots and each robot works independently, but when a robot finds a pile of rocks and returns to the space base, it leaves a trail of crumbs that can be used by it or other robot to find the pile of rocks again.

Reactive Architecture for the Robot Agents

There are modes to perform the exploration: individual and cooperative. In the individual mode, robots works independently without leaving the trail of crumbs, so this means that each robot will explore always as the first time robot left the space base. On the other hand, when robots explore and leave crumbs, the next time a robot senses a trail of crumbs will follow it and go directly to a pile of rocks. Both mode use a layered reactive architecture defined as follows:

1. Avoiding obstacles layer: Robot rotates 45 degrees to left or right at random if an obstacle is detected in front of it.
2. Taking rock to the space base layer: If the robot is in the space base, then drops the rock. In the individual mode, if the robot is not in the space base, using the radio signal determines in which direction will be the space base in order to go closer to it. In the cooperative mode, before going to the next position according to the radio signal, the robot drops a pair of crumbs.
3. Collecting samples layer: When a rock or pile of rocks is detected in front of a robot, the robot takes one rock sample.
4. Collecting crumbs layer: This layer is only defined in the cooperative mode. If the robot detects a pair of crumbs, the robot takes only one crumb and moves in the direction indicated by crumbs.
5. Exploring layer: If the robot does not detect anything in front of it, then the robot moves randomly to explore the terrain.

The priority of the layer goes from the first layer to the last one, so avoiding obstacles layer is the one with the highest priority. Because of this layers definition, agents cannot determine when there are no more rocks, so they will continue the exploration until the user stops the simulation.

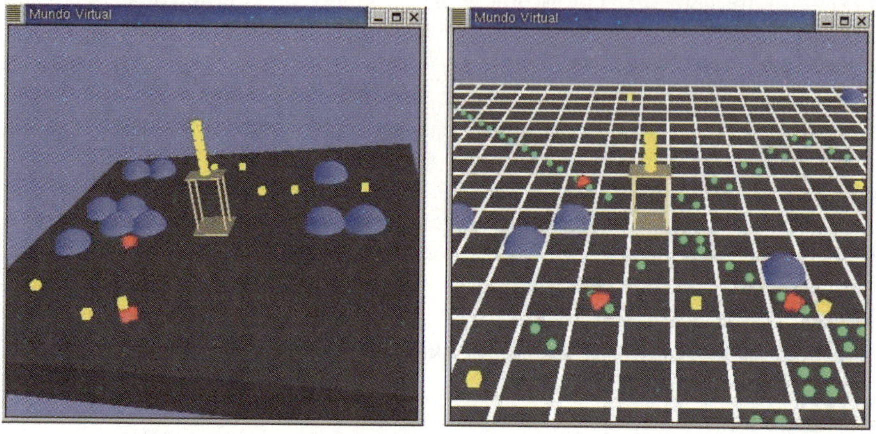

Fig. 11. Reactive Agents examples for individual and cooperative modes

In Fig. 11. there are two snapshots of the simulation, the first one is for the individual mode and the second one is for the cooperative mode. In the VE, the obstacles are represented by blue spheres, rocks by yellow cubes and crumbs by small green spheres. The space base is located in the center of the terrain and the robots are colored in red (they represent an AmigoBot).

Implementation Issues

Since this application is based on the OODVR# architecture, all objects and reactive agents that will be part of the VE are entities. The robots, the rocks and the crumbs will be dynamic entities; and the terrain, the space base and the obstacles will be static entities. In fact the rocks, crumbs and robots are considered as mobile robots in the simulation because they need to have motion capabilities in the environment.

For the integration of this application, an autonomous agent module was developed. This module models the terrain as a regular grid with cells, in which each cell contains information about the environment. Each cell has several boolean variables indicating if there is or no an obstacle, a robot, crumbs or rocks inside the cell. Robots can only sense the cell in front of it. For the radio signal emitted from the space base, in each cell there is a value indicating the proximity between the current cell and the space base, so with this information each robot can determine which is the more convenient cell to get closer to the space base. In this case, robots can sense the radio signal in the eight neighboring cells. This information on the grid can be read by the robots in order to determine which layer has to be activated and define which action the robot will perform. Once a layer is activated, it inhibits the execution of the ones with less prioriy.

The dynamic entities (one entity for each robot, rock and crumb) are teleoperated automatically by the autonomous agent module by only sending to the core module when each entity needs to move according to the behavior specified by the active layer. In this application, the simulation runs having hundreds of entities, and the system performs better under Linux, due to the fact that each entity has its own thread and Windows didn't work well with many threads.

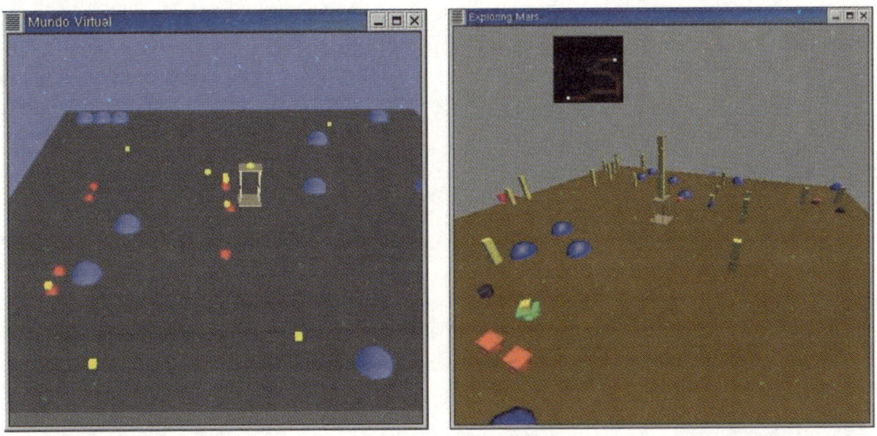

Fig. 12. Reactive Agents examples for direct communication and specialized modes

Other two different cooperative modes were implemented using direct communication between the robot agents instead of crumbs, they are called direct communication mode and specialized mode. Direct communication is performed through a blackboard in which each robot can write and read messages. In the first new mode, the only difference with respect to the crumbs case is the direct communication. In the second new mode, besides the direct communication, there are several types of specialized robots with different load capabilities and with different functions. The robots with different load capabilities can carry an specific number of rock samples more than only one. Robot with different functions are: explorer robots (black robots) and loader robots (red robots if they are not loading and green if they are). Explorer robots only find the pile of rock samples and communicate the position to the loaders robots making a map (black square) of the unknown environment. First of this last two operation modes were also implemented in Java, but because the experiments with a lot of entities ran too slow in Java, last mode were implemented in C++ using OpenGL. Fig. 12 shows these last two modes, in the direct communication mode (left side) when one robot notify to the others one position with samples, several robots go to that position forming a line.

6.4 Path Planning

The path planning application was integrated as a new module of the system. The key idea behind integrating a path planning module is to show that some external developed systems can be integrated as module of the system with only a few modifications (component represented by the package 'motionp').

The path planning module was developed as an independent application by Arechavaleta-Servin and Swain-Oropeza in [ASSO03]. That application was easily integrated to our CVE. This application solves the problem of finding a collision free path between two positions (called initial and goal positions) of the robot in a known environment. The implemented module solves the path planning problem using three different types of roadmaps. The idea of a roadmap is to have a set of roads in the environment that allows the conection between different configurations. Usually a single roadmap is defined for an environment and when a robot wants to go from an initial position to a goal position, the robot first has to find a path between the initial position to the roadmap, then travels using the roadmap and finally has to find a path between the roadmap and the goal configuration. Here is assumed that the environment is populated with polygonal obstacles.

Fig. 13. Visibility graph, probabilistic and visibility probabilistic roadmaps

Three different types of roadmaps were implemented on that path planning application: Visibility Graph roadmap, Probabilistic roadmap and Visibility Probabilistic roadmap. After the roadmap is built, the Dijkstra's algorithm is used to find the shortest path in the graph. Examples of the three types of roadmaps are shown on Fig. 13.

Implementation Issues

The original path planning application was developed in C/C++ with OpenGL for the graphic display of the environment. The application was integrated as a new module, so the result is: a user with a mobile robot in the simulation will be able to see the robot following the resultant path in the VE after running the path planning module. Besides, if there is a real robot associated to a virtual one and the VE is a representation of the real environment in which the robot is, users connected to the simulation will be able to see the execution of the path in both the VE an in the real robot telemonitored.

These are classes that were modified or added in order to integrate this module as a part of the system and to allow the display of the robot performing a path in the VE:

- Add information on the data module: Information about which kind of robots will have access to this module. Until now, only wheeled mobile robots are able to use the path planning module, but this can be easily modified for manipulators and legged robots.
- Add widgets in the GUI: For the mobile robots, there is an extra button label as 'MotionP' that allows the execution of a path planning algorithm.
- Add the corresponding classes in Java: The main program of the C/C++ application must be converted into a native method using JNI, that will communicate with a Java class that is part of the virtual entity during the path following. The class *MotionPlanning* is the one that was added to the system; this class has a native method (called *RoadMap*) and when this method is invoked from the simulation the main program of the C/C++ is executed.

The main advantage of having the implementation of the path planning module in this way is that the algorithm developed in C++ can be replaced by other one without recompiling the system in Java. It is only needed to recompile the C++ code.

Fig. 14 shows three images: the real environment with the real robot, one snapshot of the simulation showing the path planning application in OpenGL

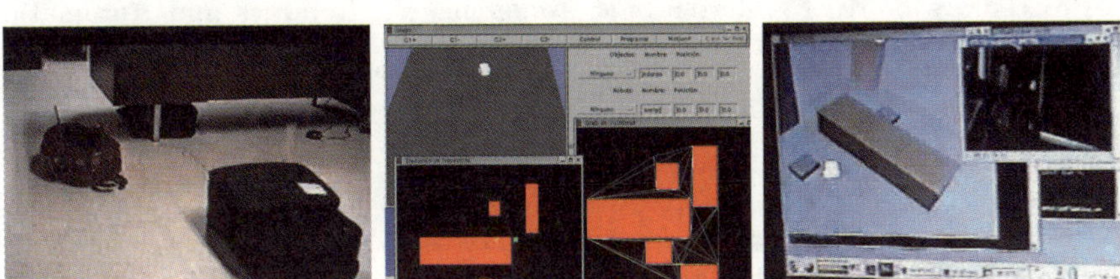

Fig. 14. Path planning for a real and virtual mobile robot

integrated in the CVE system, and the last one has the VE of the real environment and a window where the robot is monitored with its camera onboard.

7 Conclusions

We have designed and implemented a Multi-Robot-Multi-Operator Collaborative Virtual Environment useful for teaching robotics concepts and for collaborative research. In our system, we have integrated four different applications: teleoperation, programming practice sessions, reactive agents and path planning, and maybe more can be added. The main contribution is that our CVE system is the first distributed CVE for robotics applications that allows nOmR teleoperation of robots of three different types: manipulators, wheeled mobile robots and legged robots (combining SOSR, MOSR, SOMR and MOMR systems). As future work, we will perform test with students in robotics classes, to add more applications like inverse kinematics and more robots of each type defined, and to improve performance of implementation.

References

[AMMGR03] M. Alencastre-Miranda, L. Munoz-Gomez, and I. Rudomin, *Teleoperating robots in multiuser virtual environments*, Proceedings of 4th Mexican International Conference on Computer Science (September 2003), pp. 314–321.

[ASSO03] G. Arechavaleta-Servin and R. Swain-Oropeza, *Searching motion planning strategies for a mobile robot*, Proceedings of IASTED Robotics and Applications (RA'03) (June 2003), pp. 128–133.

[BF02] D. Brugali and M. E. Fayad, *Distributed computing in robotics and automation*, IEEE Transactions on Robotics and Automation Vol. 18 (August 2002), no. 4, pp. 409–420.

[BP99] K. Brown and D. Petersen (eds.), *Ready-to-run java 3d*, John Wiley & Sons Inc., New York, 1999.

[BTN00] P. G. Backes, K. S. Tso, J. Norris, G. K. Tharp, J. T. Slostad, Bonitz R. G., and K. S. Ali, *Internet-based operations for the mars polar lander mission*, Proceedings of IEEE International Conference on Robotics and Automation (ICRA'00) Vol. 2 (April 2000), pp. 2025–2032.

[BTT97] P.G. Backes, G.K. Tharp, and K.S.; Tso, *The web interface for telescience (wits)*, Proceedings of IEEE International Conference on Robotics and Automation (ICRA'97) Vol. 1 (April 1997), pp. 411–417.

[CPMTT99] T. K. Capin, I. S. Pandzic, N. Magnetat-Thalmann, and D. Thalmann (eds.), *Avatars in networked virtual environments*, Wiley, New York, 1999.

[Die01] S. Diehl (ed.), *Distributed virtual worlds. foundations and implementation techniques using vrml, java and corba*, Springer-Verlag, Berlin Heidelberg, 2001.

[Don98] D. L. Donald, *A tutorial on ergonomic and process modeling using quest and igrip*, Proceedings of the 1998 Winter Simulation Conference (December 1998), pp. 297–302.

[DQBM98] L. R. De Queiroz, M. Bergerman, R. C. Machado, S. S. Bueno, and A. Elfes, *A robotics and computer vision virtual laboratory*, Proceedings of 5th International Workshop on Advanced Motion Control (1998), pp. 694–699.

[Gol00] K. Goldberg (ed.), *The robot in the garden. telerobotics and telepistemology in the age of the internet*, MIT Press, Cambridge, Massachussets, 2000.

[Gor98] R. Gordon (ed.), *Essential jni: Java native interface*, Prentice Hall, Upper Saddle River, NJ, 1998.

[GS02] K. Goldberg and R. Seigwart (eds.), *Beyond webcams: An introduction to online robots*, MIT Press, Cambridge, 2002.

[GSL03] K. Goldberg, D. Song, and A. Levandowski, *Collaborative teleoperation using networked spatial dynamic voting*, Proceedings of the IEEE Vol. 91 (March 2003), no. 3, pp. 430–439.

[GT99] R. Gordon and S. Talley (eds.), *Essential jmf: Java media framework*, Prentice Hall, Upper Saddle River, NJ, 1999.

[HHF95] B. Hine, P. Hontalas, T. Font, L. Piguet, E. Nygren, and A. Kline, *Vevi: A virtual environment teleoperations interface for planetary exploration*, Proceedings of 25th International Conference on Environmental Systems (July 1995).

[HSH99] M. Hughes, M. Shoffner, and D. Hamner (eds.), *Java network programming: a complete guide to networking, streams, and distributed computing*, Manning, Greenwich, CT, 1999.

[KWJ99] F. Kuhl, R. Weatherly, and Dahmann J. (eds.), *Creating computer simulation systems: An introduction to the high level architecture*, Prentice Hall PTR, Upper Saddle River, NJ, 1999.

[LHHY03] Z.-J. Liu, Y.-L. Huang, P. Huang, and P. Yang, *Core data schedule in single-operator multi-agent network robot system*, Proceedings of the 2nd International Conference on Machine Learning and Cybernetics (November 2003), pp. 746–750.

[MGAMR03] L. Munoz-Gomez, M. Alencastre-Miranda, and I. Rudomin, *Defining and executing practice sessions in a robotics virtual laboratory*, Proceedings of 4th Mexican International Conference on Computer Science (September 2003), pp. 159–165.

[MGAMR04] L. Munoz-Gomez, M. Alencastre-Miranda, I. Rudomin, R. Swain-Oropeza, G. Arechavaleta, and J. Ramrez-Uresti, *Extending oodvr, a collaborative virtual robotics environment*, Proceedings of 1st Workshop on Virtual Laboratories at IX IBERO-AMERICAN WORKSHOPS ON ARTIFICIAL INTELLIGENCE(IBERAMIA'04) (November 2004), pp. 409–418.

[Mic04] O. Michel, *Webots: Professional mobile robot simulation*, International Journal of Advanced Robotic Systems Vol. 1 (March 2004), no. 1, pp. 39–42.

[MSDP02] R. Marin, P. J. Sanz, and A. P. Del Pobil, *A predictive interface based on virtual and augmented reality task specification in a web telerobotic system*, Proceedings of the IEEE International Conference on Intelligent Robots and Systems (IROS'02) Vol. 3 (September-October 2002), pp. 3005–3010.

[MSM97] O. Michel, P. Saucy, and F. Mondada, *Khepontheweb: an experimental demonstrator in telerobotics virtual reality*, Proceedings of the International Conference on Virtual Systems and MultiMedia (September 1997), pp. 90–98.

[MZP99] M. R. Macedonia, M. J. Zyda, D. R. Pratt, P. T. Barham, and S. Zeswitz, *Npsnet: A network software architecture for large scale virtual environments*, Presence Vol. 3 (Fall 1999), no. 4, pp. 265–287.

[NPV05] J. F. Norris, M. W. Powell, M. A. Vona, P. G. Backes, and J. V. Wick, *Mars exploration rover operations with the science activity planner*, Proceedings of IEEE International Conference on Robotics and Automation (ICRA'05) (April 2005), pp. 4629–4634.

[OKC99] K. Ohba, S. Kawabata, N. Y. Chong, K. Komoriya, T. Matsumaru, N. Matsuhira, K. Takase, and K. Tanie, *Remote collaboration through time delay in multiple teleoperation*, Proceedings of the IEEE/RSJ International Conference on Intelligent Robots and Systems (October 1999), pp. 1866–1871.

[PK01] E. Pitt and McNiff K. (eds.), *java.rmi the remote method invocation guide*, Pearson Education, Harlow, England, 2001.

[PLC04] G. Pisanich, L.Plice, C.Neukom, L. Flckiger, and M. Wagner, *Mission simulation facility: Simulation support for autonomy development*, Proceedings of 42nd AIAA Aerospace Science Conference (January 2004).

[Say98] C. Sayers (ed.), *Remote control robotics*, Springer-Verlag, New York, 1998.

[SK99] A. Speck and H. Klaeren, *Robosim: Java 3d robot visualization*, Proceedings of 25th Annual Conference of the IEEE Vol. 2 (November-December 1999), pp. 821–826.

[SLL01] T. Simeon, J. P. Laumond, and F. Lamiraux, *Move3d: a generic platform for path planning*, Proceedings of 4th IEEE International Symposium on Assambly and Task Planning (May 2001), pp. 25–30.

[Ste99] L. Steels, *Cooperation between distributed agents through self organization*, Proceedings of 1st European Workshop on Modelling Autonomous Agents in a Multi-Agent World (1999), pp. 175–196.

[SWND05] D. Shreiner, M. Woo, J. Neider, and T. Davis (eds.), *Opengl programming guide*, Addison Wesley, Upper Saddle River, NJ, 2005.

[SZ99] S. Singhal and M. Zyda (eds.), *Networked virtual environments*, Addison Wesley, Reading, MA, 1999.

[TCB04] J. Tan, G. J. Clapworthy, and I. R. Belousov, *The integration of a virtual environment and 3d modelling tools in a networked robot system*, International Journal of Image and Graphics (October 2004), pp. 1–20.

Modularity and Mobility of Distributed Control Software for Networked Mobile Robots

Liam Cragg, Huosheng Hu, and Norbert Voelker

University of Essex, Department of Computer Science, Colchester, CO4 3SQ, UK.
{lmcrag, hhu, norbert}@essex.ac.uk

1 Introduction

Recently, more and more intelligent robotic systems have been embedded into the Internet for service, security, and entertainment, including distributed mobile robots. These networked robots have captured the interest of many researchers worldwide [Gol05]. Apart from operation in a hazardous environment (a traditional telerobotic area), networked robots have been applied in a wide range of real-world applications, including tele-manufacturing, tele-training, tele-surgery, traffic control, space exploration, disaster rescue, health care, and as museum guides [HTCV04]).

Networked robots developed using distributed computing paradigms, normally cover large areas, communicating via wired or wireless networks. To effectively control these robots, four control software architectures can be adopted. These are: client/server, in which robots contain knowledge and resources to perform a task and can be controlled from a simple remote client; remote evaluation in which robots have resources, but need to obtain knowledge from a client to perform a task; code on demand in which robots contain resources but act as clients to access knowledge at a server to perform a task; or mobile agents, in which robots contain resources, and a task can be performed by moving a mobile agent containing knowledge and results between robots.

Mobile agents (MA) are a new distributed computing paradigm, which have not yet been extensively employed in multiple robot architectures e.g. [VCMC01] and [Pap99]. As part of wider research into the development of multiple robot architectures for hazardous environments, we have employed mobile agents to implement autonomous and remote control architectures. A number of experiments have been conducted to assess potential functional advantages, specifically through the use of mobile agent mobility. In this chapter we describe our implementations and experimental results in the use of mobile agents for multi-robot architecture development.

D. Brugali (Ed.): Software Eng. for Experimental Robotics. STAR 30, pp. 459–484, 2007.

The rest of this chapter is structured as follows: Section 2 provides some background information on mobile agents and how they can be employed in the software engineering process in robotics. Section 3 describes two types of mobile agent implementation, i.e. autonomous and telecontrol. Section 4 describes experiments conducted with these architectures. Section 5 is used to discuss the benefits and challenges which we believe MAs offer a multiple robot system developer. Finally, Section 6 provides some conclusions.

2 Mobile Agents

2.1 Mobile Agent Definition

The concept of an *agent* is widely used in computer science, AI and robotics, often with different meanings. In this chapter the term *agent* is used to mean a software component which exists within a bounded execution environment, but which can control its own execution and interaction i.e. it may be autonomous, dynamic and intelligent (able to learn). A *mobile agent* can move (or migrate) between computing nodes within a distributed computing system.

2.2 Distributed Computing Architectures

In addition to the mobile agent architecture, there are three other main types of distributed computing architecture. These are:

- *Client/Server* - as shown in Fig. 1(a), in which a server (S), has knowledge (K), and resources (R), to perform a task. A client (C) sends a message to the server requesting task execution. The server performs the task using its knowledge and resources, and returns a result to the client.
- *Remote Evaluation* - as shown in Fig. 1(b), in which a client has knowledge, while a server has the resources to perform a task. For the client to obtain the result to a task, it must send its knowledge to the server. The server uses this knowledge and its own resources to perform the task, before returning a result to the client.
- *Code on Demand* - as shown in Fig. 1(c), in which a client has resources, while a server contains knowledge to perform a task. For the client to obtain a result to a task, it must send a message to the server to request knowledge. The client uses the server's knowledge, and its own resources, to perform the task and obtain a result.

In the *mobile agent architecture* in Fig. 1(d) an agent server (AS), is host to a mobile agent (MA), which contains knowledge and results from previous tasks, but not the resources it requires to perform a new task. Another agent server contains the required resources. Instead of forwarding knowledge to the new agent server, the mobile agent moves to access the resource. When the

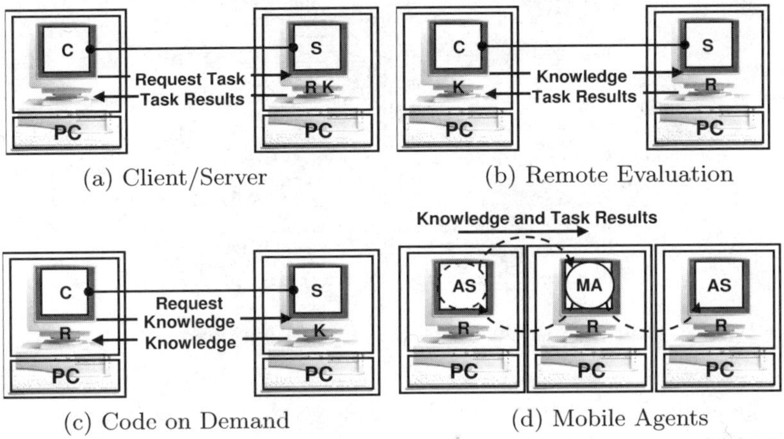

Fig. 1. Distributed Computing Architectures

mobile agent has completed its task, it can remain at the new server, or move to another agent server carrying its knowledge and task results.

2.3 Mobile Agent Environment

The execution environment within which mobile agents operate has the following components and characteristics:

- *Execution Platform* - mobile agents exist within an execution environment, provided by multiple agent servers as shown in Fig. 2. Agent servers provide *places* in which an agent can reside and interface with functionality provided by the server computer.
- *Programming Language* - most agent servers and mobile agents are written in Java which as an interpreted computer language can execute on a variety of heterogeneous computer platforms.
- *Location Identification* - multiple host computers can provide a distributed (but unified) mobile agent execution environment. Each computer can host a single agent server which can be uniquely identified using the host's IP address. An agent server may contain multiple places and agents.
- *Programming Characteristics* - basic mobile agents are lightweight computing components because many functions are available from a function library provided by the agent server, including agent creation, destruction, cloning, migration and fault tolerant storage to persistent media e.g. a hard drive.
- *Execution Characteristics* - most agent servers make extensive use of multi-threading to execute agent activity. Mobile agents run in an independent thread, allowing multiple agents to execute concurrently. A

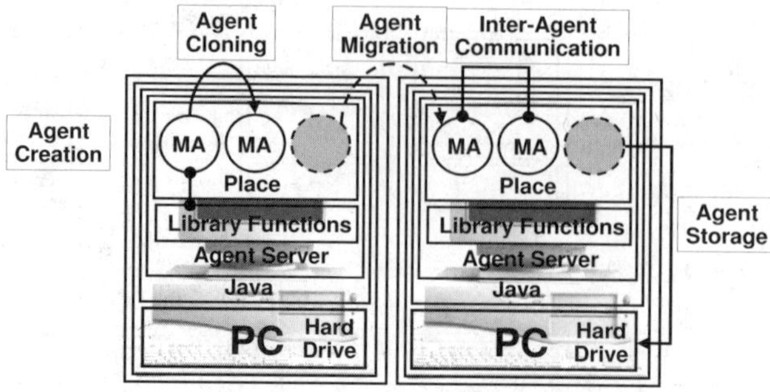

Fig. 2. Mobile Agent Execution Environment

mobile agent itself may be multi-threaded so that it can execute parallel activities.

- *Agent Communication* - Mobile agents can make use of distributed communication mechanisms: sockets (plain or SSL), RMI and CORBA. Mobile agents may also migrate to a resource location to communicate locally and reduce network load.

2.4 Mobile Agent Applications

Mobile agents have been used in a range of applications, as described in [Mil99], [GCKR02], [Lan98] and [MSS01] to play the following roles:

- *Implementation of Code or Actions at Remote Locations* - mobile agents can encapsulate protocols, data, actions, processes and temporary or permanent software and distribute it to remote network locations in order to update system functionality or perform a task.
- *Reduction of Communication and Latency across Unreliable or Limited Bandwidth Communication Mechanisms* - by migrating to a remote location and conducting a sequence of actions locally which would otherwise involve communication across an unreliable communication mechanism.
- *Provision of Personalised Virtual Presence for Users* - MAs can be used as a software embodiment of a user because of their ability to perform intelligent actions remotely and asynchronously.
- *Provision of Dynamic Network Management or Monitoring Functionality* - dynamic, intelligent, autonomous, mobile properties allow MAs to manage or monitor computer networks in which the topology of the network may change.

2.5 Mobile Agent Software Development

This section examines the characteristics of mobile agents as a software development environment for distributed robotic systems, in terms of software engineering methodology, security, and testing and debugging.

Software Engineering Methodology - due to their mobility, and ability to move knowledge and task results, mobile agents can provide the functionality found in all other distributed computing architectures combined. This means that in order to provide equivalent functionality to a mobile agent architecture, an architecture would need to be constructed from an amalgamation of all other distributed computing architectures, i.e. client/server, remote evaluation and code on demand.

A simplified view, showing comparable mobile agent, and amalgamated architectures, is shown in Fig. 3. It can be seen from Fig. 3(a), that in the amalgamated architecture functionality at distributed computing nodes is heterogeneous, and dedicated communication links are required to allow task execution. In contrast, in the mobile agent architecture shown in Fig. 3(b), functionality at distributed nodes is homogeneous, and dedicated communication links arc not required to allow task execution. Direct communication is not specifically required, mobility can be used instead.

Pfleeger [Pfl01], states that the characteristics of good design in software engineering are component independence, exception identification and handling, fault prevention, and fault tolerance; and that a good way to improve a design is to reduce its complexity. Fig. 3 shows that the mobile agent environment is less complex and more intuitive to the system designer, because there is greater component independence and homogeneity of functionality i.e. mobile agents are modular and easy to understand, functionality is encapsulated and more loosely coupled than in an equivalent amalgamation of the other distributed computing architectures, as dedicated communication links between distributed system components are not required (because mobile agents can migrate to the source of a resource once, rather than needing to communicate across a communication channel multiple times to perform a task).

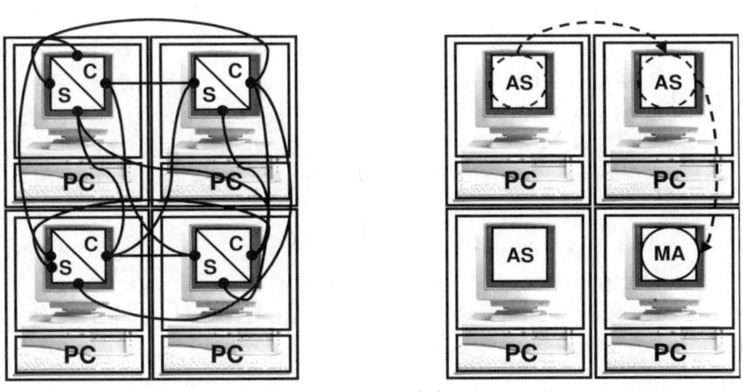

(a) Amalgamation of all Other Architectures (b) Mobile Agent Environment

Fig. 3. Distributed Computing Architecture Comparison

In addition, because they provide a unified development environment (i.e. provide the functionality of all other distributed computing architectures already built into a single unified environment), this means that a system developer developing a simple architecture in a mobile agent environment, can easily later extend their architecture to perform more complex tasks, without needing to radically change their architectural structure. A mobile agent based multiple robot architecture can therefore be more easily extended than one developed with other distributed computing architectures, or an amalgamation of such architectures.

Security - an issue often raised surrounding the use of mobile agents is security. Reliable interactions between a mobile agent and robot can be enforced and secured by making use of similar types of security measures found in existing e-commerce and Internet applications. These are often made available via the underlying mobile agent platform. Mobile agent platforms can have external and internal security mechanisms.

- External security mechanisms protect mobile agent and agent server interactions from external threats in agent-to-agent or agent-server to agent-server communication. External security mechanisms can include the use of a *Secure Socket Layer* (SSL) employing X.509 certificates and *Message Authentication Codes* (MACS) for authentication and integrity checking of encrypted transmitted data. Internal security is used to protect underlying host computers and legitimate mobile agents from malicious agent attack.
- Internal security can employ many of the security features built in to the Java programming language [Inc05]. Such security features include, codesource-based and identity-based access control policies, in which agents are only allowed access to particular agent, agent server, or host services, as agents operate in secured domains, which shield underlying agent server or host functionality.

Testing and Debugging - a final issue that might be of interest to a DHRI system developer, is how the code of a mobile agent, which interacts locally with a remote robot, might be tested and debugged. Testing and debugging can be achieved using a number of mechanisms. These include Textual and Graphical User Interfaces (TUIs and GUIs), robot simulators, and remote login software.

Mobile agent platforms often provide textual and/or graphical user interfaces for the management of stationery and mobile agents. Such user interfaces allow system developers to monitor the execution of agents at run-time, providing information on the location of executing agents, and their current behaviour.

On a remote robot, which may have no monitor or keyboard, such TUIs and GUIs can still be employed, using remote login and programs such as Exceed [Hum05]. Exceed will allow a user to display on a local PC, TUIs and GUIs executing on a remote robot. This approach can be adopted once an agent has been developed sufficiently for it to be executed on the remote robot.

To increase the speed of testing and debugging at a development stage, as with wider robotics research, testing and debugging can be conducted on local networked PC's using simulated robot software. Modern research robots available with simulation software, while not providing real world test data, will allow initial development, debugging, and testing, to be performed in an accessible and efficient way.

Having described mobile agent technology, the following section describes our implementation of mobile agent based architectures for networked robots; beginning with a description of the motivation for our research.

3 Implementation

Our research has had two main objectives (1) to investigate the development of an architecture to control networked robots in hazardous environments, specifically nuclear decommissioning characterisation [Cra05] and (2) to investigate the characteristics of mobile agents as a development environment for multiple robot architectures.

In pursuit of the first objective, we implemented three control architectures: an autonomous architecture, an implementation of the ALLIANCE architecture [Par98], a telecontrol architecture based on the work of Ali [Ali99] and an AI architecture employing reinforcement learning [SB98]. The aim was to amalgamate these three architectures into a single control architecture that had elements of autonomous and AI control for increased efficiency and telecontrol for safety.

In pursuit of the second objective, we developed a mobile agent based development environment for multiple robots, to enable the implementation of the architectures. We conducted a number of experiments to see if any benefits could be derived from the implementation of these architectures in the MA environment. We specifically focused on the role mobile agent mobility might provide, through experiments in architecture adaptability, fault tolerance and dynamic control allocation. The architectures which employed this functionality were the autonomous and telecontrol architectures, while the AI architecture was implemented to increase efficiency with respect to our first objective. Therefore we only describe the autonomous and telecontrol architectures in more detail in this chapter, for more information on the AI architecture please see [CH05] and [Cra05]

3.1 Mobile Agent Environment for Programming Networked Robots

Our mobile agent environment for programming networked robots is shown in Figure 4. It contains three networked PC's, two running WindowsXP operat-

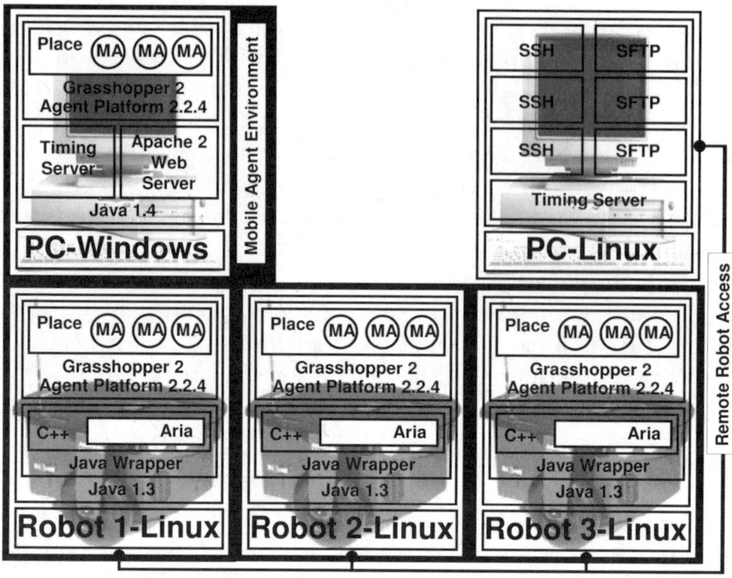

Fig. 4. Mobile Agent Development Environment for Networked Robots

ing system and one running Linux, all contain the Java programming language v1.4. The Windows PCs contained the Grasshopper 2 agent platform which was used to host mobile agents, a timing server to time experiments, and Apache Web Server v2.0 to serve Java mobile agent .class files to Grasshopper agent servers. The Linux PC contained the secure shell program (SSH), to allow remote login, and the secure file transfer program (SFTP), to allow downloading of files, to networked robots. Three ActivMedia [Rob05] Pioneer 2DX mobile robots (each running Linux, and network accessible via wireless LAN) were used to provide a networked multiple robot team.

Grasshopper is an OMG MASIF compliant Java based mobile agent server (which can be purchased from [IKV05]). It provides an execution environment and function library for Java based mobile agents. The function library provides functions for the creation, destruction, cloning, migration, and persistent storage of agents. A mobile agent developer overrides a live() function in which they incorporate their mobile agent's run-time execution code. An Apache Web Server was used to store agent Java .class files. The Grasshopper agent servers obtain .class files from this central repository.

ActivMedia's ARIA robot control software was used to control the low level behaviour of robots. It allows various levels of control to be executed on ActivMedia Pioneer 2DX robots; it is written in C++ but provided with a native language interface via a Java wrapper, which allows commands to be written in Java based mobile agents directly.

The Timing Server is a multi-threaded Java based server, which listens on known ports, and registers the time of any connection. Because it is not possible to precisely synchronise the internal clock of distributed computers, we employed a single timing server during each experiment, to generate consistent times for experiment duration calculation.

3.2 Autonomous Architecture

Our autonomous architecture is shown in Fig. 5. It is a mobile agent implementation of the ALLIANCE architecture [Par98], containing three main types of agent:

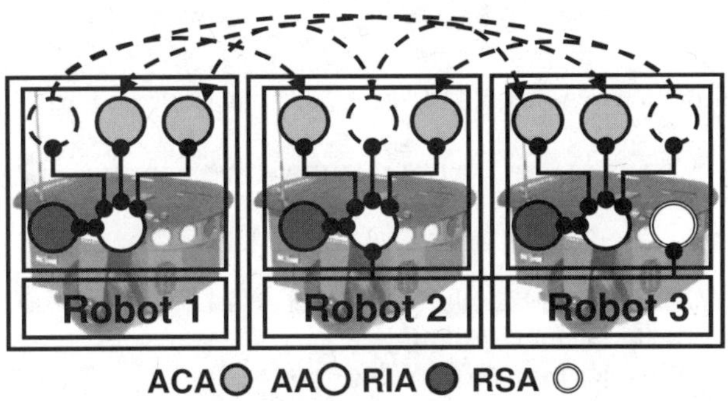

Fig. 5. Mobile Agent Implementation of the ALLIANCE Architecture

- *ALLIANCE Agent (AA)* - the AA contains the ALLIANCE action selection mechanism, and executes behaviour on the robot on which it resides. One resides on each robot in the multiple robot team. AAs engage in: calculation of motivational behaviours (the action selection mechanism), execution of behaviour sets (the behaviour execution mechanism) and management of inter-robot communication (for efficient fault tolerant behaviour) as shown in Fig. 6. Motivation values are calculated using the calculation shown in Table 1 which relies on inter and intra robot input.
- *ALLIANCE Communication Agent (ACA)* - the main purpose of the ACA is to provide an inter-robot communication mechanism between robots. The ACA carries current robot activity and ID information between robots.
- *Robot Interface Agent (RIA)* - a single RIA resides on each robot. RIAs provide an interface to robot functionality for AAs, allowing movement and sensor feedback functions. RIAs shield AAs from underlying robot platforms via a standard control interface.

For more detail on this architecture, please see [CH04a].

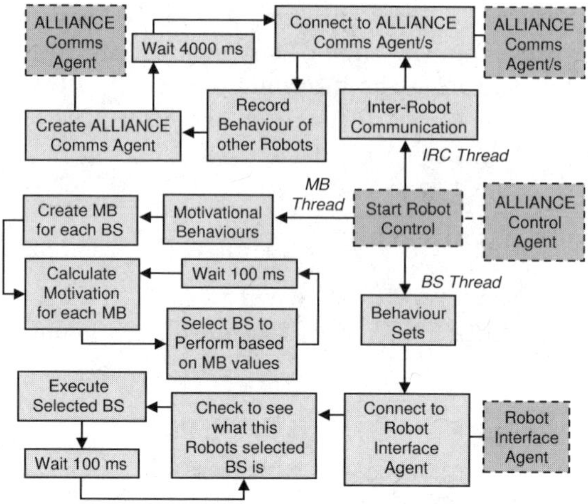

Fig. 6. Flowchart showing ALLIANCE Agent Control Thread Behaviour

Table 1. Code showing ALLIANCE Agent Motivation Calculation

```
//motivation calculation - using ALLIANCE's formal model
private int motivation(){
    int newMotivationValue = 0;
    newMotivationValue = ((motivation+impatience)
            * activitySuppression
            * sensoryFeedback
            * acquiescence
            * impatienceReset);
    return newMotivationValue;}}
```

3.3 Telecontrol Architecture

Our telecontrol architecture is shown in Fig. 7. The architecture contains four types of agent as follows:

- *Local Display Agent (LDA)* - the LDA located on a local PC (PC-L) provides a robot position GUI to show an operator the position of remote robots.
- *Remote Display Agent (RDA)* - the RDA located on a remote PC (PC-R) obtains Cartesian coordinate position information from robots and uses this to update the Robot Position GUI at the PC-L.
- *Local Control Agent (LCA)* - the LCA provides an operator at the PC-L with single or multiple control of robots via a robot control GUI.

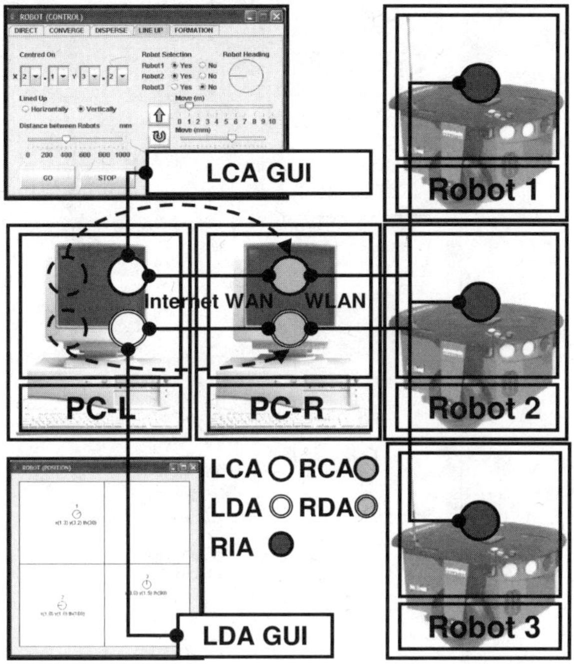

Fig. 7. Mobile Agent Architecture for Multiple Telerobot Control via the Internet

- *Remote Control Agent* - the RCA executes on robots at the remote location, robot control strategies offered by the LCA as shown in Fig. 8.

For more detail on this architecture please see [CH04b].

4 Experiments

Due to their implementation in a mobile agent environment, we were able to supplement the functionality of our autonomous and telecontrol architectures by using mobile agents to provide additional adaptability, fault tolerance, and dynamic control allocation functionality.

To test whether these functional extensions provided benefits over traditional implementations, we conducted a number of experiments, these are described in the following section. We use mobile agent mobility in the experiments in the following ways:

- *Adaptability Experiments* - using mobile agent mobility to automatically update control code at run time in multiple robots.
- *Fault Tolerance Experiments* - using mobile agent mobility to move mission level data from failing system components in the event of their failure.

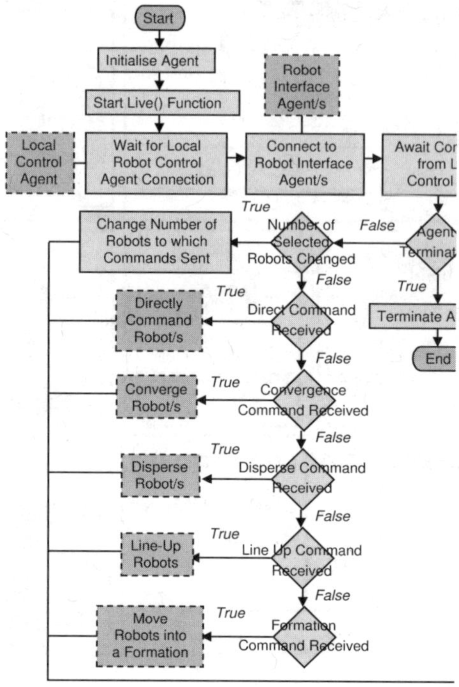

Fig. 8. Flowchart showing High Level Remote Control Agent Behaviour

- *Dynamic Control Allocation Experiments* - using mobile agent mobility to affect the control characteristics of a multiple robot system (i.e. the ability to effectively control distributed multiple robots in the presence of network delay by appropriate positioning of a control agent).

4.1 Adaptability Experiments

In adaptability experiments we employed our autonomous architecture, in the mobile agent development environment with three wirelessly networked pioneer robots. 25 trials were conducted for each set of experiments (a trial for each experiment is described in the following section). In these experiments we used two set-ups.

- *Set-up 1* - used our ALLIANCE mobile agent implementation, in which mobile agents were employed to allow existing ALLIANCE agents, to be dynamically updated by new ALLIANCE agents at run-time.
- *Set-up 2* - used a traditional implementation of ALLIANCE, in which ALLIANCE was implemented in stationary programs, and the secure shell program (SSH) used for remote robot login, and the secure file transfer program (SFTP), used for file transfer to allow updating of control programs on robots.

In each experiment set-up, we conducted a number of activities described in the following section. Table 2 and Figure 9(a), show the architecture adaptability experiment activities for Set-up 1 representing one trial, while Table 3 and Figure 9(b), show the architecture adaptability experiment activities for Set-up 2 which represent one trial.

Table 2. Architecture Adaptability Experiment Activities (Mobile Agent Implementation)

Step Number	Activity
1	ALLIANCE agents (AA1) control multiple robots via robot interface agents (RIA)
2	A grasshopper platform is started on an external windows PC, which automatically creates a new ALLIANCE agent (AA2), the time at which this occurs is recorded on the external PC
3	After initialisation, the new ALLIANCE agent automatically makes simultaneous copies of itself on the multiple robots, and removes itself from the external PC
4	The copies of the new ALLIANCE agent, communicate with their existing local ALLIANCE agent on their robot
5	Due to this communication, the existing ALLIANCE agents remove themselves from the robot which they are occupying
6	The copies of the new ALLIANCE agent communicate with their local robot interface agent to take control of their local robot, once a connection has been made, they communicate with the timing server on the external PC, on which the connection time is recorded. The time between Step (1) and Step (6) is recorded as the duration for 1 trial.

Experiments were conducted using Set-up 1 and Set-up 2, in which teams containing 1, 2, and 3 robots were employed. Architecture adaptability experiment results are shown in Figure 10.

The architecture adaptability results show that using mobile agents, control code can be updated more quickly than when using a traditional approach. Using our mobile agent implementation, a 1, 2, and 3 robot team had its control code updated in, on average 7.19, 8.58, and 9.97 seconds respectively, while using a traditional implementation, the same size robot teams took on average, 25.08, 36.96 and 52.64 seconds to update. The mobile agent implementations are much quicker, because mobile agents can execute in parallel, as they run in independent threads, while in a traditional implementation robots are updated in series.

In our traditional implementation, we tried to use automated scripts wherever possible, to reduce or remove human performance factors, and to speed up and automate the updating process. Despite this, it can be seen that mobile agents in our experiments update code much more quickly than the traditional

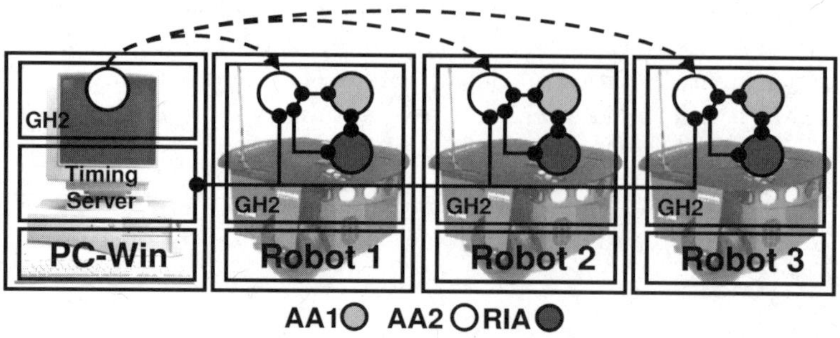

(a) Mobile Agent Implementation

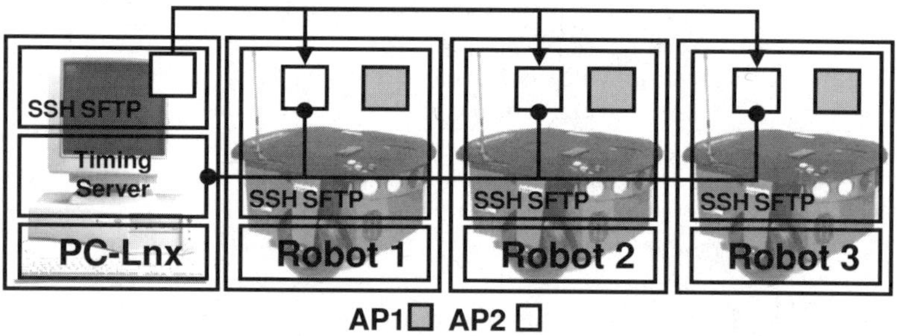

(b) Traditional Implementation

Fig. 9. Architecture Adaptability Experiment Activities

Table 3. Architecture Adaptability Experiment Activities (Traditional Implementation)

Step Number	Activity
1	ALLIANCE programs (AP1) control multiple robots directly, the time at which this occurs is recorded on an external PC
2	SFTP is then used to download a new ALLIANCE program (AP2) to one of the robots from the external PC
3	SSH is then used to stop the existing ALLIANCE program, on the robot to which the new ALLIANCE program has been downloaded
4	SSH is then used to start the new ALLIANCE program, on the robot to which the new ALLIANCE program has been downloaded, which then communicates with the timing server on the external PC, on which the connection time is recorded
5	Steps (2)-(4) are repeated until a new ALLIANCE program has been installed and started on all robots in a team. The time between Step (1) and Step (4) (for the final robot) is recorded as the duration for 1 trial.

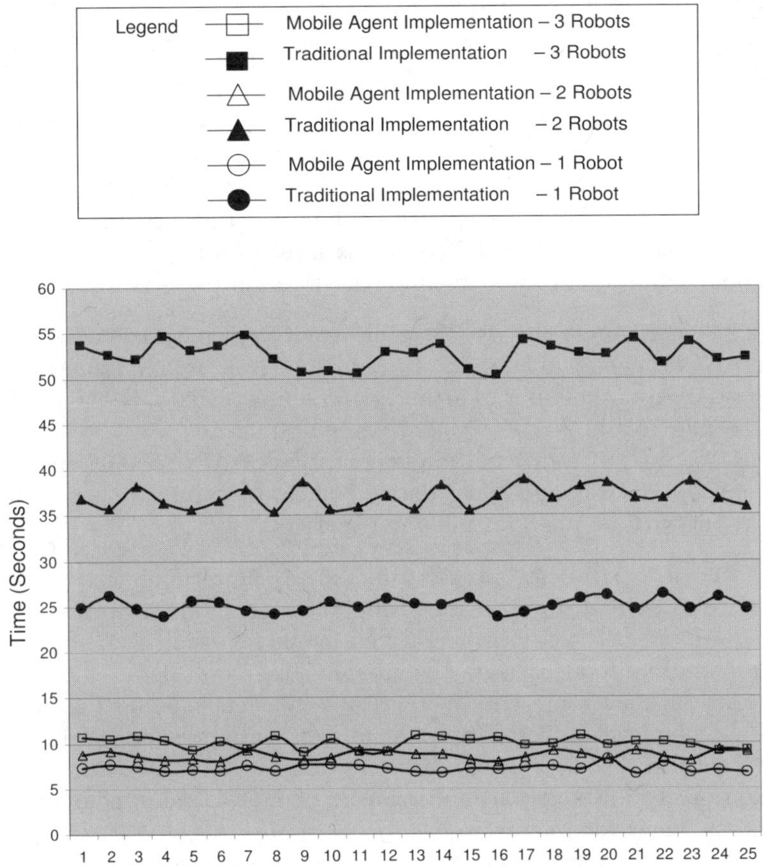

Fig. 10. Architecture Adaptability Experiment Results

implementation. The mobile agent implementation is both quicker, and fully automated, thereby reducing system down time in the event that control code needs to be updated, and reducing the burden on a supervisor.

Due to the parallel execution of mobile agents, one would expect, as robot team number increases, that the mobile agent to traditional implementation control code updating time ratio, would widen, i.e. a mobile agent implementation, relative to a traditional implementation, would be more effective at updating control code as robot number increases.

After architecture adaptability experiments were conducted, we performed mission level fault tolerance experiments.

4.2 Fault Tolerance Experiments

In fault tolerance experiments we employed our autonomous architecture, in the mobile agent development environment with three wirelessly networked pioneer robots. 25 trials were conducted for each set of experiments (a trial for each experiment is described in the following section).

These experiments were designed to investigate how a mobile agent implementation might affect the mission level fault tolerance characteristics of a multiple robot team, (i.e. the ability to retain mission level data in a system despite robot failures). In these experiments we used two set-ups.

- *Set-up 1* - used our ALLIANCE mobile agent implementation, in which an ALLIANCE agent is able to autonomously migrate from a failing robot to a secondary robot in order to maintain mission level data, before migrating on to a new robot.
- *Set-up 2* - used a traditional implementation of ALLIANCE, in which a failing robot loses mission level data when it fails, and is replaced using a new robot with no previous mission experience.

Both sets of experiments were conducted in three robot teams (i.e. one failing, one temporary, and one new robot). In each experiment set-up, we conducted a number of activities described in the following section.

Table 4 and Figure 11(a), show the mission level fault tolerance experiment activities for Set-up 1 representing one trial, while Table 5 and Figure 11(b), show the mission level fault tolerance experiment activities for Set-up 2 which represent one trial.

Mission level fault tolerance experiment results are shown in the chart in Figure 12. The fault tolerance experiment results, show that using mobile agents to retain the state and data of a failing robot takes on average 14.39 seconds, while using a traditional implementation, a new robot can be started in, on average 5.65 seconds. As in the architecture adaptability experiments, scripts were used wherever possible to automate the traditional implementation. This increased the starting speed for the new robot.

These results show that there is some time overhead in using the mobility characteristics of mobile agents to retain mission level data, relative to a traditional implementation. However, the time taken to obtain the mission level data, which is retained in the mobile agent implementation, and lost in the traditional implementation, needs to be taken into consideration.

4.3 Dynamic Control Allocation Experiments

To conduct these experiments we employed our telecontrol architecture, in our mobile agent development environment with three wirelessly networked real pioneer robots, and conducted 25 trials in each set of experiments (a trial for each experiment will be described in the following sections); robots were controlled in a 4x4m lab environment without obstacles, and moved as

Table 4. Mission Level Fault Tolerance Experiment Activities (Mobile Agent Implementation)

Step Number	Activity
1	An ALLIANCE agent (AA), controlling robot 1 via a robot interface agent (RIA) detects a terminal fault, and communicates with the timing server on an external PC, on which the connection time is recorded
2	This ALLIANCE agent then migrates to robot 2
3	On arrival at robot 2, the ALLIANCE agent waits, and does not attempt to take control of the robot
4	A replacement for robot 1, robot *1 starts. It contains a robot start-up agent (RSA), which communicates with the ALLIANCE agent waiting on robot 2 to inform it that a replacement for robot 1 has started
5	The ALLIANCE agent waiting on robot 2 migrates to the new robot *1
6	The ALLIANCE agent communicates with the local robot interface agent on robot *1 to take control. Once a connection has been made, it communicates with the timing server on the external PC, on which the connection time is recorded. The time between Step (1) and Step (6) is recorded as the duration for 1 trial.

Table 5. Mission Level Fault Tolerance Experiment Activities (Traditional Implementation)

Step Number	Activity
1	An ALLIANCE program (AP), controlling robot 1 directly, detects a terminal fault and communicates with the timing server on an external PC, on which the connection time is recorded
2	A replacement for robot 1, robot *1 starts, and SFTP is used to download a new ALLIANCE program to it
3	SSH is then used to start this ALLIANCE program on robot *1
4	The ALLIANCE program communicates with the new robot *1 directly to take control. Once a connection has been made, it communicates with the timing server on the external PC, on which the connection time is recorded. The time between Step (1) and Step (4) is recorded as the duration for 1 trial.

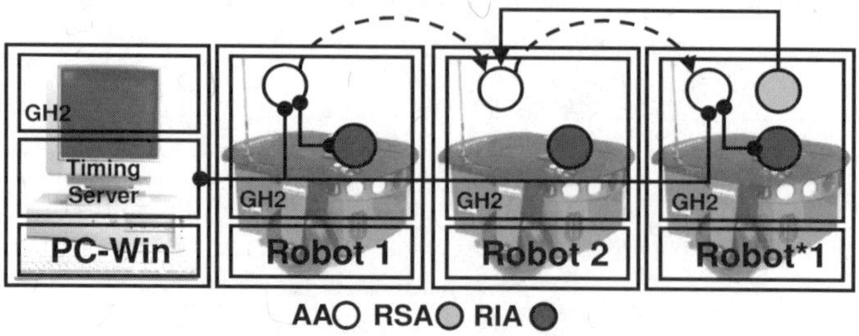

(a) Mobile Agent Implementation

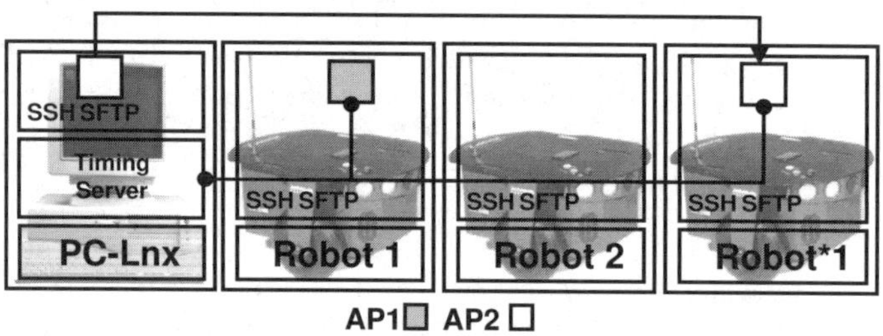

(b) Traditional Implementation

Fig. 11. Mission Level Fault Tolerance Experiment Activities

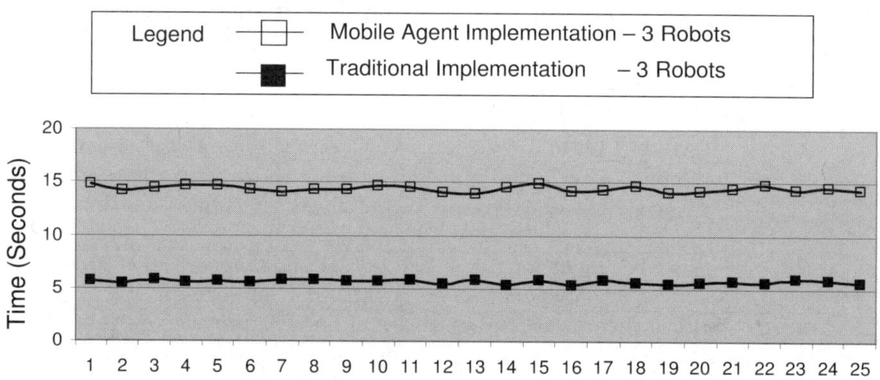

Fig. 12. Mission Level Fault Tolerance Experiment Results

shown in Figure 13 in four multiple telerobot control task types as defined and employed by Ali in [Ali99].

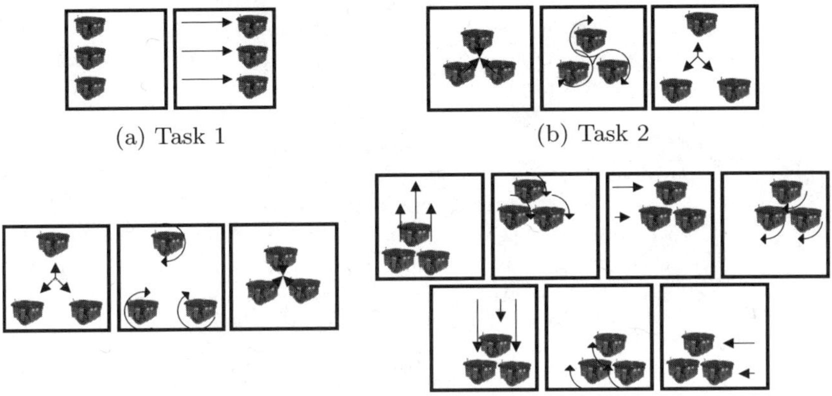

(a) Task 1 (b) Task 2

(c) Task 3 (d) Task 4

Fig. 13. Dynamic Control Allocation Experiment Tasks

These experiments were designed to investigate, the effect that using mobile agent mobility to dynamically move supervisory group control close to a robot team (at the remote PC), has on control efficiency; relative to the same control implemented close to a supervisor (on the local PC), when Internet communication between local and remote locations is subject to delay. In these experiments we used two set-ups.

- *Set-up 1* - used our implementation of multiple telerobot control via the Internet, to provide a mobile agent containing supervisory group control for a supervisor to control a team of telerobots; in which the mobile agent is located close to the telerobots at the remote PC.
- *Set-up 2* - used our implementation of multiple telerobot control via the Internet, to provide a mobile agent containing supervisory group control for a supervisor to control a team of telerobots; in which the mobile agent is located close to the supervisor at the local PC.

In each experiment set-up, we conducted a number of activities described in the following section. Table 6 and Figure 14(a), show the dynamic allocation of control experiment activities for Set-up 1 representing one trial, while Table 7 and Figure 14(b), show the dynamic allocation of control experiment activities for Set-up 2 which represent one trial.

Dynamic control allocation experiment results are shown in Fig. 15 and 16. The dynamic control allocation experiment results show that using mobile agents to encapsulate supervisory group control (and move it close to a robot

Table 6. Dynamic Control Allocation Experiment Activities (Supervisory Group Control Close to Telerobots (Remote PC))

Step Number	Activity
1	Robots are moved to their starting positions (for task 1, 2, 3, or 4, as shown in Fig. 13)
2	The remote control agent is moved close to the telerobots (the remote PC), and simulated Internet delay between local and remote PC, set as either low (no additional delay), medium (low + 3000ms) or high (low + 6000ms)
3	Parameters are set on the control panel GUI so that supervisory group control can be employed
4	A command is sent from the control panel GUI to the remote control agent, which encapsulates the supervisory group control
5	The remote control agent coordinates the low-level control of the robots in the robot group, to move them to the end positions for the specified task
6	Robots reach their end positions.
7	The time between Step (1) and Step (6) is recorded as the duration for 1 trial.

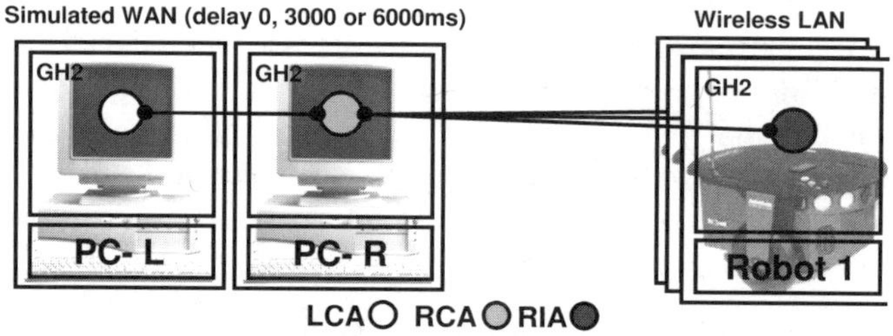

(a) Supervisory Group Control close to Telerobots (Remote PC)

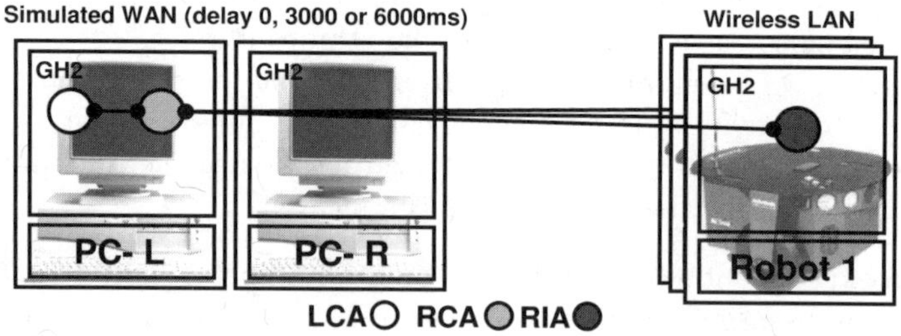

(b) Supervisory Group Control close to the Supervisor (Local PC)

Fig. 14. Dynamic Control Allocation Experiment Activities

Table 7. Dynamic Control Allocation Experiment Activities (Supervisory Group Control close to Supervisor (Local PC))

Step Number	Activity
1	Robots are moved to their starting positions (for task 1, 2, 3, or 4, as shown in Fig. 13)
2	The remote control agent is moved close to the supervisor (the local PC) and simulated Internet delay between local and remote PC set as either, low (no additional delay), medium (low + 3000ms) or high (low + 6000ms)
3	Parameters are set on the control panel GUI so that supervisory group control can be employed
4	A command is sent from the control panel GUI to the remote control agent, which encapsulates the supervisory group control
5	The remote control agent coordinates the low-level control of the robots in the robot group, to move them to the end positions for the specified task
6	Robots reach their end positions.
7	The time between Step (1) and Step (6) is recorded as the duration for 1 trial.

team in an architecture in which there is delay in the communication mechanism), allows task execution to be conducted more quickly than if more of the commands are sent across the communication mechanism from the supervisor location.

Fig. 15 and 16 show that when there is low delay, there is little difference in time taken, between the supervisor and robot team locations. As simulated Internet delay is increased to 3000ms between supervisor and robot team locations, there is a clear impact on the time taken to complete each of the four tasks. As simulated Internet delay is increased to 6000ms, this impact becomes more pronounced. The difference in performance occurs, because when supervisory group control is located close to the supervisor, more low level control commands must be sent across the simulated Internet WAN, rather than Internet LAN connection (Internet WAN being subject to greater levels of delay).

In addition, because there can be a greater discrepancy between the execution of commands or feedback received across the simulated Internet WAN connection, robot movement is more unpredictable. The main effect of locating supervisory group control close to the supervisor at the local PC is to increase task execution time. This is due to increased command execution delay, and increased command execution/feedback delay variability, which decreases applied control predictability.

Locating supervisory group control at the supervisor location on the local PC, led to a system which was inherently unsafe, and incapable of predictable control. In a real Internet connection, where delay is unpredictable, this would

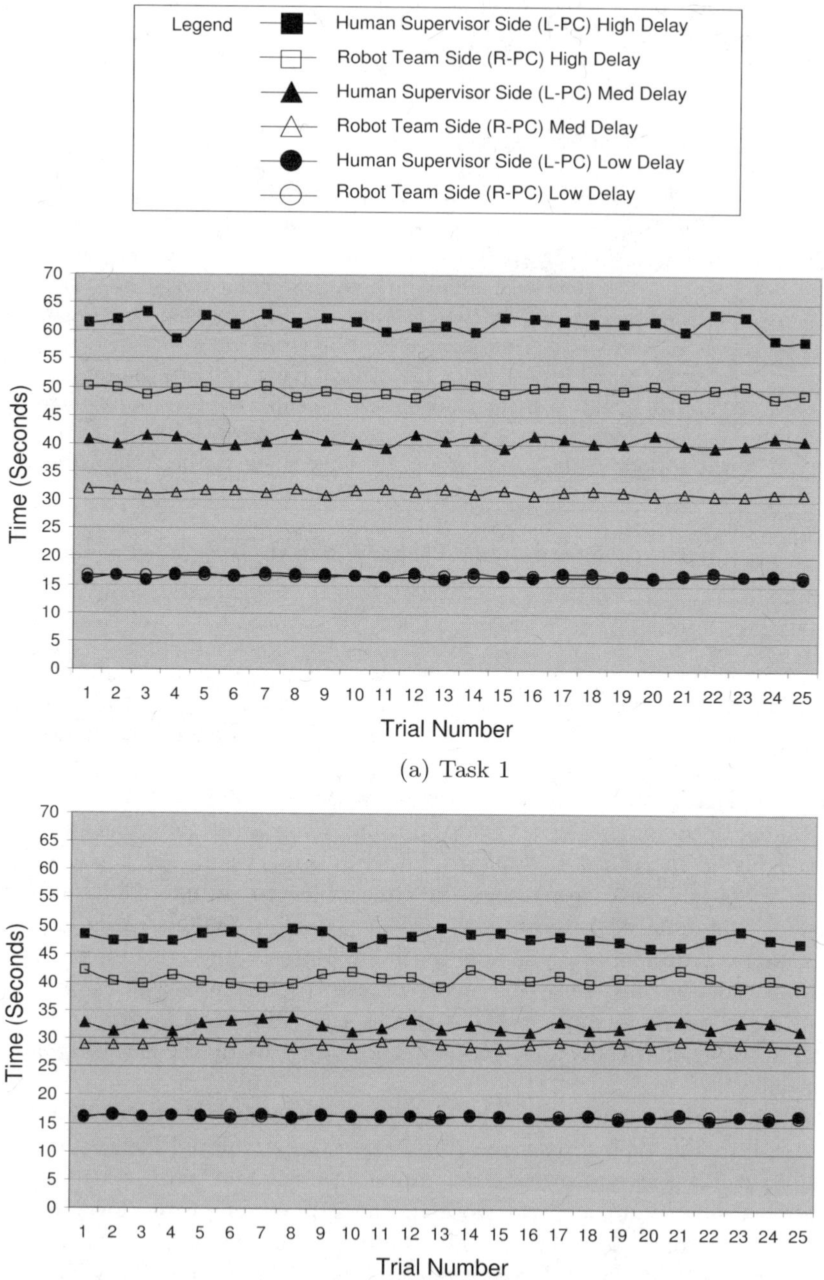

(a) Task 1

(b) Task 2

Fig. 15. Dynamic Control Allocation Experiment Results (Task 1 and 2)

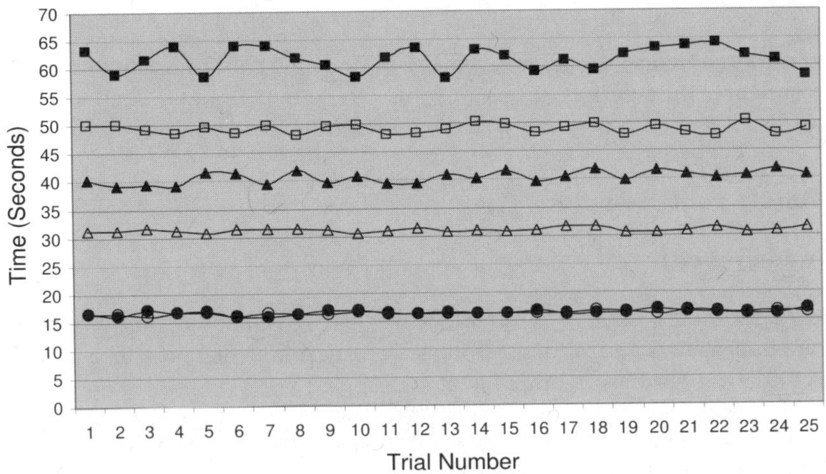

(a) Task 3

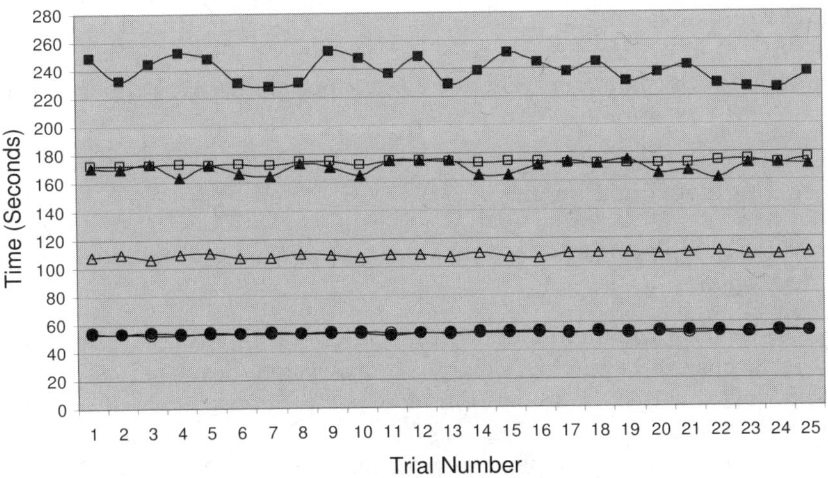

(b) Task 4

Fig. 16. Dynamic Control Allocation Experiment Results (Task 3 and 4)

make control of the robot team more difficult. By moving the control closer to the robot team, using mobile agent mobility, control problems caused by Internet delay are reduced allowing more efficient control of the robot team.

5 Discussion

Based on our experience of mobile agent based multiple robot implementations and experiments, this section discusses some of the benefits and challenges in the use of mobile agents in multi-robot control architectures.

5.1 Mobile Agent Benefits

Functional Benefits

Through experiments using our architectures we have shown that a mobile agent implementation can add the following functional benefits to a multi-robot architecture through mobile agent mobility:

- *Adaptability* - easy automatic updating of control code in a multiple robot system at development or run-time.
- *Fault Tolerance* - ability to automatically and dynamically move important mission data away from failing robot system components for safe storage or reapplication at a later time.
- *Dynamic Control Allocation* - the most effective control can be applied despite network communication delay or changes in network topology, by dynamic MA positioning.

Software Engineering Benefits

MAs provide the following software engineering benefits for distributed robot system developers:

- Because the functionality of all other distributed computing architectures is already built in to the MA environment a system developer can easily extend their architecture to perform more complex tasks without needing to radically change its structure.
- Mobile agents are modular and easy to understand, functionality is encapsulated and more loosely coupled than in an equivalent amalgamation of client/server, remote evaluation or code on demand architectures.

5.2 Mobile Agent Challenges

While mobile agents provide robotisists with many advantages, they have some limitations.

- *Resource Use Challenges* - the mobile agent environment consumes a certain level of resources in the architecture e.g. to maintain agent servers demands that robots should have reasonable processing power.

- *Application Challenges* - mobile agent mobility is useful only when large scale quantities of data need to be moved because there is a time overhead in moving a mobile agent relative to message passing in a networked architecture.
- *Programming Challenges* - system developers must learn the additional structure and functionality of a mobile agent development environment when they may already have experience of client/server, remote evaluation and code on demand applications.

6 Conclusions

In this chapter, we have examined the use of mobile agents for the control of multiple robot architectures, using autonomous and telecontrol architectures developed in a mobile agent based development environment for networked robots. A number of experiments with these architectures were conducted to examine the characteristics of implementing them in a mobile agent environment.

Through this examination and our experience of using mobile agents as a robot architecture development environment, we have found that mobile agents provide a novel software engineering methodology which compares favourably with an amalgamation of other distributed computing architectures. Employing mobile agent's mobility, the adaptability and fault tolerance of a multiple robot architecture can be improved, and control can be dynamically positioned within an architecture at the most appropriate location.

Mobile agents in general make a multiple robot architecture more dynamic and adaptable. They provide the functionality of all other distributed computing architectures combined and as such provide a system developer with this functionality already built in. Combined with a modular software engineering methodology this makes mobile agents an effective framework within which to development multiple robot architectures.

References

[Ali99] K. Ali, *Multiagent telerobotics: Matching systems to tasks*, Ph.D. thesis, Georgia Tech Mobile Robot Lab, USA, 1999.

[CH04a] L. Cragg and H. Hu, *Implementing alliance in networked robots using mobile agents*, Proceedings of 5th IFAC Symposium on Intelligent Autonomous Vehicles, Instituto Superior Tecnico (Lisbon, Portugal), 10th-12th December 2004.

[CH04b] L. Cragg and H. Hu, *Implementing multiple robot architectures using mobile agents*, Proceedings of Control 2004 (Bath, England), 6th-9th September 2004.

[CH05] L. Cragg and H. Hu, *Application of reinforcement learning to mobile robot in reaching recharging station operation*, IPROMS 2005 (On-Line Conference), 5th-14th July 2005.

[Cra05] L. Cragg, *Application of mobile agents to networked robots in hazardous environments*, Ph.D. thesis, Department of Computer Science, University of Essex, UK, September 2005.

[GCKR02] R. S. Gray, G. Cybenko, D. Kotz, and D. Rus, *Mobile agents: Motivations and state of the art systems*, Tech. report, TR2000-365, Department of Computer Science, Dartmouth College, USA, 2002.

[Gol05] K. Goldberg, *Ieee society of robotics and automation technical committee on: Networked robots*, http://www.ieor.berkeley.edu/ goldberg/tc/, 2005.

[HTCV04] H. Hu, P.W. Tsui, L. Cragg, and N. Voelker, *Agent architecture for multi-robot cooperation over the internet*, International Journal of Integrated Computer-aided Engineering, Vol. 11, No. 3, 2004.

[Hum05] Hummingbird, *Exceed homepage*, http://www.hummingbird.com/, 2005.

[IKV05] IKV, *Homepage*, http://www.ikv.de/, IKV++ Technologies AG, 2005.

[Inc05] Sun MicroSystems Inc., *Java homepage*, http://java.sun.com, 2005.

[Lan98] D. B. Lange, *Mobile objects and mobile agents: The future of distributed computing?*, In Proceedings of 12th European Conference on Object-Oriented Programmin (20-24 July 1998).

[Mil99] D. S. Milojicic, *Trend wars: Mobile agent applications*, IEEE Concurrency, Vol. 7, No. 3, Pages: 80-90 (1999).

[MSS01] P. Marques, P. Simões, L. M. Silva, F. Boavida, and J. G. Silva, *Providing applications with mobile agent technology*, Proceedings of the 4th IEEE International Conference on Open Architectures and Network Programming (Anchorage, Alaska), April 2001.

[Pap99] T. Papaioannou, *Mobile agents: Are they useful for establishing a virtual presence in space?*, Agents with adjustable Autonomy Symposia, AAAI Spring Symp, 1999.

[Par98] L. E. Parker, *Alliance: An architecture for fault tolerant multi-robot cooperation*, IEEE Transactions on Robotics and Automation, 14 (2), 1998.

[Pfl01] S. L. Pfleeger, *Software engineering: Theory and practice*, 2nd ed., Prentice Hall, 2001.

[Rob05] ActivMedia Robotics, *Homepage*, http://robots.activmedia.com, 2005.

[SB98] R. Sutton and A. G. Barto, *Reinforcement learning: An introduction*, MIT Press, 1998.

[VCMC01] W. Vieira, L. M. Camarinha-Matos, and O. Castolo, *Fitting autonomy and mobile agents*, Proceedings of the 8th IEEE International Conference on Emerging Technologies and Factory Automation, Vol. 2, Antibes-Juan les Pins, France, 15-18 October 2001.

Sidebar – Java3D for Web-Based Robot Control

Igor Belousov

Hewlett-Packard 52/1, Kosmodamianskaya Nab., 115054 Moscow, Russia
igor.belousov@hp.com

1 Introduction

Robot control via Internet is a perspective direction of scientific research having important practical value. Recent period could be considered as a breakthrough moment to extend traditional use of Internet media from information exchanges to control of physical devices, including robots.

Internet limitations on communication bandwidth and data exchange rate make robot control based on TV images inefficient because of the slow responses made by the system to the operator's actions. One way of providing suitable control conditions for the operator is by using an on-line Java3D model of the robot and environment [BCC01] - a dynamic, Java3D-based virtual environment in which graphical models of the robot and objects are used to provide real-time visualisation of the current state of the robot site. Data transmission is restricted to small parcels defining the current coordinates of the robot and the objects; this has proved effective in practice, in contrast to TV images, which would have caused substantial communication delays. Under this regime, the robot can be successfully controlled, even when communication rates are extremely low (0.1-0.5 KB/sec). Moreover, the use of a 3D virtual environment significantly simplifies the remote performance of operations as it allows changes of viewpoint, zooming, the use of semitransparent images, etc. Web control interface is presented on figure 1

The use of open technology, Java3D, allows the virtual control environment to run on any type of computer platform. Moreover, any user of the Internet can use this environment for controlling the robots from within a Web page using a standard Web browser. Since Java3D is based on the OpenGL library, it runs rapidly on computers with OpenGL acceleration boards. For the scene in figure 1, we achieved a refresh rate of 25 frames/sec on a medium PC.

Java3D was developed by Sun Microsystems (java.sun.com), current version is Java3D 1.3.2 for Java2 (JDK 5.0). The following Java3D features makes it ideal instrument for development robotics software:

D. Brugali (Ed.): Software Eng. for Experimental Robotics. STAR 30, pp. 485–490, 2007.

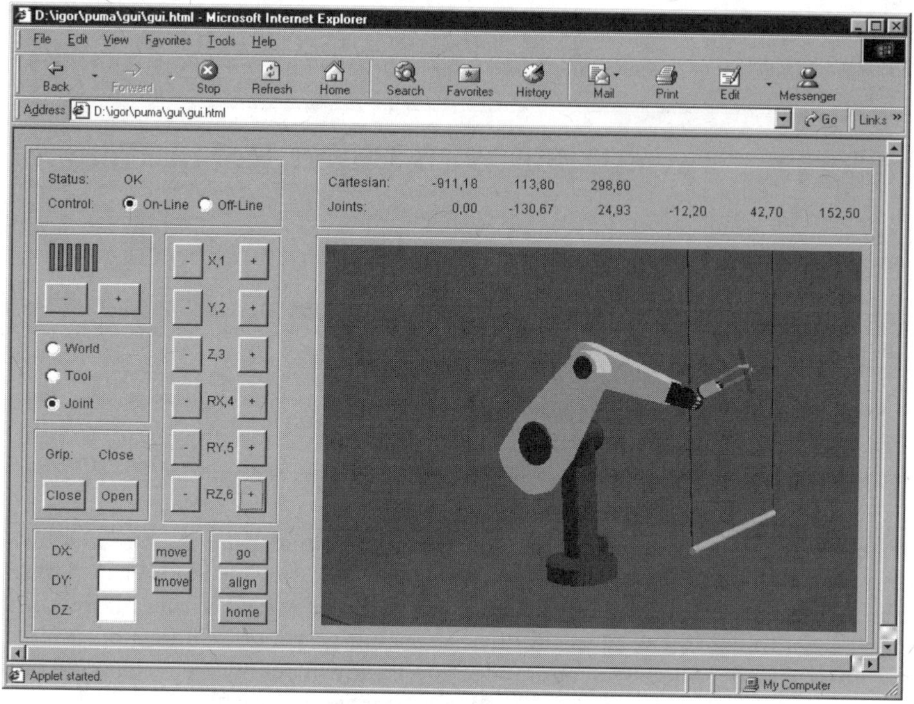

Fig. 1. Web interface for control of PUMA robot

- interactive 3D-visualisation in the network (applets)
- high performance (OpenGL at low level) " 3D sound support
- LOD
- collision detection
- support of external 3D formats - 3D Studio (3DS), AutoCAD (DXF), VRML 97 (WRL), Wavefront (OBJ), etc.
- VR ready (head-mounted stereo displays, data-gloves and position trackers)
- supported platforms: Windows, Sun/Solaris, IBM/AIX, SGI IRIX, HP/UX, Linux
- free distribution and outstanding user support
- support of mobile devices (phones etc.)

Such Java3D feature as built-in collision detection algorithm has been used to provide operational security of the system. Before sending the control command to the robot site this command was simulated at the client site at the Java3D environment. If collision is detected, the corresponding command is cancelled. During the realisation of this procedure a bug in Java3D collision detection algorithm has been found (wrong collision was detected when the objects are close to each other). It was reported to and fixed by the Sun's Java3D team.

Based on Java3D, the systems for control via the Internet of the PUMA 560 and CRS A465 manipulators and mobile robot Nomadic XR4000 [ABea00] have been developed (see figure 2). Systems were successfully tested under real Internet conditions from different locations in Russia, U.K., France and Korea [BC02]. The use of on-line Java3D-based control environment allowed to solve the most complicated task of robot control via Internet - it's interaction with fast moving objects [BSC05].

Fig. 2. Examples of the developed Java3D models of the different robots

Let's consider an example of Java3D program for generation of the model of the robot with several revolute joints. Program contains several classes. Base steps needed to create any 3D scene using Java3D are performed at the class Robot. Besides that constructor of the class Envir is invoked there. In the class Envir some objects of the robot working environment are defined, and constructor of the class Base is invoked. Class Base defines the fixed robot base and also is used to join to this base the first link of the manipulator, class Link1. Angle Link1.angle defines position of the first link relatively robot base. The particularity of the class Link1 - procedure draw(), which defines transformation of rotation corresponding to the change of the angle Link1.angle. Joining the second link Link2 is performed in the class Link1. This process continues for the rest joints and links of manipulator.

We can notice that it is needed to describe the 3D scene as a graph. Java3D graph should have a tree structure. The nodes of the graph are some classes of Java3D, for example geometric primitives or transformations. Links between the nodes are defined by the "parent-child" relations. In the above example the root of the tree is the node objRoot. Its "child" is the mouse transformation node. Then we have class Envir, etc. Hierarchy is presented in figure 3

```
public class Robot extends Applet {
 public BranchGroup createSceneGraph() {
  BranchGroup objRoot = new BranchGroup();
  TransformGroup tgMouse = HF.setMouseBehav(objRoot);
  Envir envir = new Envir();
  tgMouse.addChild(envir.getBG());
  return objRoot;
  }
 public Robot() {
  Canvas3D c = new Canvas3D(null);
  add("Center", c);
  BranchGroup scene = createSceneGraph();
  SimpleUniverse u = new SimpleUniverse(c);
  u.getViewingPlatform().setNominalViewingTransform();
  u.addBranchGraph(scene);
  }
}
```

Code structure of class *Robot*

```
public class Envir {
 private BranchGroup envirBG;
 public Envir() {
  envirBG = new BranchGroup();
  Box floor = new Box(2.0, 0.0, .0,
  Box.GENERATE_NORMALS,
    HF.createColorMaterialAppearance(Colors.green));
  envirBG.addChild(floor);
  Cylinder cyl = new Cylinder(0.05,
    0.2,Cylinder.GENERATE_NORMALS,
    HF.createColorMaterialAppearance(Colors.red));
  TransformGroup tgTranslCyl = HF.getTranslTG(0.8, 0.15, -0.1);
  envirBG.addChild(tgTranslCyl);
  tgTranslCyl.addChild(cyl);
  Base base = new Base();
  TransformGroup tgTranslBase = HF.getTranslTG(0.0, 0.2, 0.0);
  envirBG.addChild(tgTranslBase);
  tgTranslBase.addChild(base.getBG());
  }
 public BranchGroup getBG() {
  return envirBG;
  }
}
```

Code structure of class *Envir*

```
public class Base {
 private BranchGroup baseBG;
 public Base() {
  baseBG = new BranchGroup();
  Cylinder baseCyl=new Cylinder(R, H, Cylinder.GENERATE_NORMALS,
   HF.createColorMaterialAppearance(Colors.black));
  baseBG.addChild(baseCyl);
  Link1 link1 = new Link1();
  TransformGroup tgTranslLink1=HF.getTranslTG(0.0, 0.25, 0.0);
  Link1.tgRot = HF.getRotTG(2, Link1.angle);
Link1.tgRot.setCapability(TransformGroup.ALLOW_TRANSFORM_WRITE)
  baseBG.addChild(tgTranslLink1);
  tgTranslLink1.addChild(Link1.tgRot);
  Link1.tgRot.addChild(link1.getBG());
  }
 public BranchGroup getBG() {
  return baseBG;
 }
}
```

Code structure of class *Base*

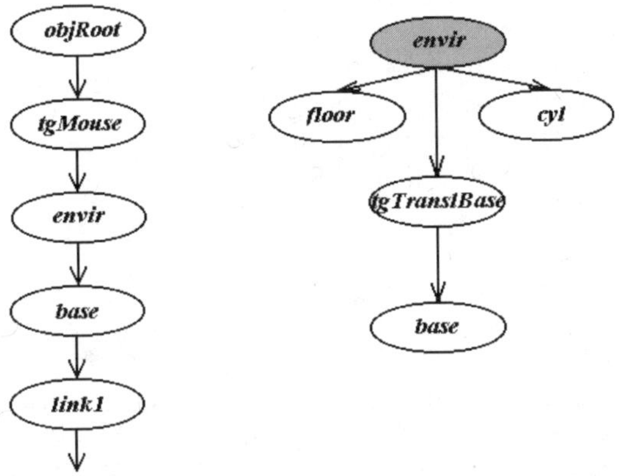

Fig. 3. Java3D scene graph.

Lets note that each node could be also presented as the tree, combined from the simple nodes. Thus, the developing of the Java3D program is generation of corresponding scene graph and its description using related functions of the Java3D language. This scene graph defines the program specifications.

```
public class Link1 {
 public static double angle = 0.0;
 private BranchGroup link1BG;
 public static Transform3D rot = new Transform3D();
 public static TransformGroup tgRot;
 public Link1() {
  link1BG = new BranchGroup();
  Cylinder baseCyl=new Cylinder(0.1, 0.2,
Cylinder.GENERATE_NORMALS,
   HF.createColorMaterialAppearance(Colors.wheat));
  link1BG.addChild(baseCyl);
  Link2 link2 = new Link2();
  TransformGroup tgTranslLink2 = HF.getTranslTG(0.0, 0.1, 0.0);
  Link2.tgRot = HF.getRotTG(3, Link2.angle);
Link2.tgRot.setCapability(TransformGroup.ALLOW_TRANSFORM_WRITE)
   link1BG.addChild(tgTranslLink2);
  tgTranslLink2.addChild(Link2.tgRot);
  Link2.tgRot.addChild(link2.getBG());
 }
 public BranchGroup getBG() {
  return link1BG;
 }
 public static void draw() {
  rot.rotY(angle);
  tgRot.setTransform(rot);
 }
}
```

Code structure of class *Link*1

References

[ABea00] R. Alami, I.R. Belousov, and et al., *Diligent: towards a human-friendly navigation system*, Proc. IEEE/RSJ Intern. Conf. on Intelligent Robots and Systems IROS'2000 (Takamatsu, Japan, October 30 - November 5), 2000.

[BC02] I. Belousov and G. Clapworthy, *Remote programming and java3d visualisation for internet robotics*, SPIE's International Technical Group Newsletter (Vol. 11, No. 1, February 2002, p. 8), 2002.

[BCC01] I. Belousov, G. Clapworthy, and R. Chellali, *Virtual reality tools for internet robotics*, Proc. IEEE Intern. Conf. on Robotics and Automation ICRA'2001 (Seoul, Korea, May, 2001, pp. 1878-1883), 2001.

[BSC05] I. Belousov, V. Sazonov, and S Chebukov, *Web-based teleoperation of the robot interacting with fast moving objects*, Proc. of the IEEE Int. Conf. on Robotics and Automation ICRA'2005 (Barcelona, Spain), 2005.

Springer Tracts in Advanced Robotics

Edited by B. Siciliano, O. Khatib, and F. Groen

Vol. 29: Secchi, C.; Stramigioli, S.; Fantuzzi, C.
Control of Interactive Robotic Interfaces – A Port-Hamiltonian Approach
225 p. 2007 [978-3-540-49712-7]

Vol. 28: Thrun, S.; Brooks, R.; Durrant-Whyte, H. (Eds.)
Robotics Research – Results of the 12th International Symposium ISRR
602 p. 2007 [978-3-540-48110-2]

Vol. 27: Montemerlo, M.; Thrun, S.
FastSLAM – A Scalable Method for the Simultaneous Localization and Mapping Problem in Robotics
120 p. 2007 [978-3-540-46399-3]

Vol. 26: Taylor, G.; Kleeman, L.
Visual Perception and Robotic Manipulation – 3D Object Recognition, Tracking and Hand-Eye Coordination
218 p. 2007 [978-3-540-33454-5]

Vol. 25: Corke, P.; Sukkarieh, S. (Eds.)
Field and Service Robotics – Results of the 5th International Conference
580 p. 2006 [978-3-540-33452-1]

Vol. 24: Yuta, S.; Asama, H.; Thrun, S.; Prassler, E.; Tsubouchi, T. (Eds.)
Field and Service Robotics – Recent Advances in Research and Applications
550 p. 2006 [978-3-540-32801-8]

Vol. 23: Andrade-Cetto, J,; Sanfeliu, A.
Environment Learning for Indoor Mobile Robots – A Stochastic State Estimation Approach to Simultaneous Localization and Map Building
130 p. 2006 [978-3-540-32795-0]

Vol. 22: Christensen, H.I. (Ed.)
European Robotics Symposium 2006
209 p. 2006 [978-3-540-32688-5]

Vol. 21: Ang Jr., H.; Khatib, O. (Eds.)
Experimental Robotics IX – The 9th International Symposium on Experimental Robotics
618 p. 2006 [978-3-540-28816-9]

Vol. 20: Xu, Y.; Ou, Y.
Control of Single Wheel Robots
188 p. 2005 [978-3-540-28184-9]

Vol. 19: Lefebvre, T.; Bruyninckx, H.; De Schutter, J.
Nonlinear Kalman Filtering for Force-Controlled Robot Tasks
280 p. 2005 [978-3-540-28023-1]

Vol. 18: Barbagli, F.; Prattichizzo, D.; Salisbury, K. (Eds.)
Multi-point Interaction with Real and Virtual Objects
281 p. 2005 [978-3-540-26036-3]

Vol. 17: Erdmann, M.; Hsu, D.; Overmars, M.; van der Stappen, F.A (Eds.)
Algorithmic Foundations of Robotics VI
472 p. 2005 [978-3-540-25728-8]

Vol. 16: Cuesta, F.; Ollero, A.
Intelligent Mobile Robot Navigation
224 p. 2005 [978-3-540-23956-7]

Vol. 15: Dario, P.; Chatila R. (Eds.)
Robotics Research – The Eleventh International Symposium
595 p. 2005 [978-3-540-23214-8]

Vol. 14: Prassler, E.; Lawitzky, G.; Stopp, A.; Grunwald, G.; Hägele, M.; Dillmann, R.; Iossifidis. I. (Eds.)
Advances in Human-Robot Interaction
414 p. 2005 [978-3-540-23211-7]

Vol. 13: Chung, W.
Nonholonomic Manipulators
115 p. 2004 [978-3-540-22108-1]

Vol. 12: Iagnemma K.; Dubowsky, S.
Mobile Robots in Rough Terrain – Estimation, Motion Planning, and Control with Application to Planetary Rovers
123 p. 2004 [978-3-540-21968-2]

Vol. 11: Kim, J.-H.; Kim, D.-H.; Kim, Y.-J.; Seow, K.-T.
Soccer Robotics
353 p. 2004 [978-3-540-21859-3]

Vol. 10: Siciliano, B.; De Luca, A.; Melchiorri, C.; Casalino, G. (Eds.)
Advances in Control of Articulated and Mobile Robots
259 p. 2004 [978-3-540-20783-2]

Vol. 9: Yamane, K.
Simulating and Generating Motions of Human Figures
176 p. 2004 [978-3-540-20317-9]

Vol. 8: Baeten, J.; De Schutter, J.
Integrated Visual Servoing and Force Control – The Task Frame Approach
198 p. 2004 [978-3-540-40475-0]

Vol. 7: Boissonnat, J.-D.; Burdick, J.; Goldberg, K.; Hutchinson, S. (Eds.)
Algorithmic Foundations of Robotics V
577 p. 2004 [978-3-540-40476-7]

Vol. 6: Jarvis, R.A.; Zelinsky, A. (Eds.)
Robotics Research – The Tenth International Symposium
580 p. 2003 [978-3-540-00550-6]

Vol. 5: Siciliano, B.; Dario, P. (Eds.)
Experimental Robotics VIII – Proceedings of the 8th International Symposium ISER02
685 p. 2003 [978-3-540-00305-2]

Vol. 4: Bicchi, A.; Christensen, H.I.; Prattichizzo, D. (Eds.)
Control Problems in Robotics
296 p. 2003 [978-3-540-00251-2]

Vol. 3: Natale, C.
Interaction Control of Robot Manipulators – Six-degrees-of-freedom Tasks
120 p. 2003 [978-3-540-00159-1]

Vol. 2: Antonelli, G.
Underwater Robots – Motion and Force Control of Vehicle-Manipulator Systems
268 p. 2006 [978-3-540-31752-4]

Vol. 1: Caccavale, F.; Villani, L. (Eds.)
Fault Diagnosis and Fault Tolerance for Mechatronic Systems – Recent Advances
191 p. 2003 [978-3-540-44159-5]

Printing: Krips bv, Meppel
Binding: Stürtz, Würzburg